NEWTON'S PRINCIPIA

NEWTON'S PRINCIPIA.

THE

MATHEMATICAL PRINCIPLES

OF

NATURAL PHILOSOPHY,

BY SIR ISAAC NEWTON;

TRANSLATED INTO ENGLISH BY ANDREW MOTTE.

TO WHICH IS ADDED

NEWTON'S SYSTEM OF THE WORLD

GYAN

Published by

GYAN

Gyan Publishing House
5, Ansari Road
Daryaganj, New Delhi-110002
Phone: 011-47034999, 9811692060
E-mail: books@gyanbooks.com

Distribution Network
gyanbooks.com
India, USA, Canada, UK, Australia

ISBN: 978-81-212-5539-4 (HB)
First Published, 1687

2nd Impression 2024

Printed at: Gyan Press, Delhi.

NEWTON'S PRINCIPIA
Author: ISAAC NEWTON, Translator: ANDREW MOTTE

SIR ISAAC NEWTON.

NEW YORK. DANIEL ADEE. 1848.

NEWTON'S PRINCIPIA.

THE

MATHEMATICAL PRINCIPLES

OF

NATURAL PHILOSOPHY,

BY SIR ISAAC NEWTON;

TRANSLATED INTO ENGLISH BY ANDREW MOTTE.

TO WHICH IS ADDED

NEWTON'S SYSTEM OF THE WORLD;

With a Portrait taken from the Bust in the Royal Observatory at Greenwich.

FIRST AMERICAN EDITION, CAREFULLY REVISED AND CORRECTED,

WITH A LIFE OF THE AUTHOR, BY N. W. CHITTENDEN, M. A., &c.

NEW-YORK.
PUBLISHED BY DANIEL ADEE, 45 LIBERTY STREET.

DEDICATION.

TO THE

TEACHERS OF THE NORMAL SCHOOL

OF THE STATE OF NEW-YORK.

GENTLEMEN :—

A stirring freshness in the air, and ruddy streaks upon the horizon of the moral world betoken the grateful dawning of a new era. The days of a drivelling instruction are departing. With us is the opening promise of a better time, wherein genuine manhood doing its noblest work shall have adequate reward. TEACHER is the highest and most responsible office man can fill. Its dignity is, and will yet be held commensurate with its duty—a duty boundless as man's intellectual capacity, and great as his moral need—a duty from the performance of which shall emanate an influence not limited to the *now* and the *here*, but which surely will, as time flows into eternity and space into infinity, roll up, a measureless curse or a measureless blessing, in inconceivable swellings along the infinite curve. It is an office that should be esteemed of even sacred import in this country. Ere long a hundred millions, extending from the Atlantic to the Pacific, from Baffin's Bay to that of Panama, shall call themselves American citizens. What a field for those two master-passions of the human soul—the love of Rule, and the love of Gain! How shall our liberties continue to be preserved from the graspings of Ambition and the corruptions of Gold? Not by Bills of Rights,

Constitutions, and Statute Books; but alone by the rightly cultivated hearts and heads of the People. They must themselves guard the Ark. It is yours to fit them for the consecrated charge. Look well to it: for you appear clothed in the majesty of great power! It is yours to fashion, and to inform; to save, and to perpetuate. You are the Educators of the People: you are the prime Conservators of the public weal. Betray your trust, and the sacred fires would go out, and the altars crumble into dust: knowledge become lost in tradition, and Christian nobleness a fable! As you, therefore, are multiplied in number, elevated in consideration, increased in means, and fulfill, well and faithfully, all the requirements of true Teachers, so shall our favoured land lift up her head among the nations of the earth, and call herself blessed.

In conclusion, Gentlemen, to you, as the conspicuous leaders in the vast and honourable labour of Educational Reform, and Popular Teaching, the First American Edition of the Principia of Newton—the greatest work of the greatest Teacher—is most respectfully dedicated.

N. W. CHITTENDEN.

INTRODUCTION TO THE AMERICAN EDITION.

THAT the PRINCIPIA of Newton should have remained so generally unknown in this country to the present day is a somewhat remarkable fact; because the name of the author, learned with the very elements of science, is revered at every hearth-stone where knowledge and virtue are of chief esteem, while, abroad, in all the high places of the land, the character which that name recalls is held up as the noblest illustration of what MAN may be, and may do, in the possession and manifestation of pre-eminent intellectual and moral worth; because the work is celebrated, not only in the history of one career and one mind, but in the history of all achievement and human reason itself; because of the spirit of inquiry, which has been aroused, and which, in pursuing its searchings, is not always satisfied with stopping short of the fountain-head of any given truth; and, finally, because of the earnest endeavour that has been and is constantly going on, in many sections of the Republic, to elevate the popular standard of education and give to scientific and other efforts a higher and a better aim.

True, the PRINCIPIA has been hitherto inaccessible to popular use. A few copies in Latin, and occasionally one in English may be found in some of our larger libraries, or in the possession of some ardent disciple of the great Master. But a dead language in the one case, and an enormous price in both, particularly in that of the English edition, have thus far opposed very sufficient obstacles to the wide circulation of the work. It is now, however, placed within the reach of all. And in performing this labour, the utmost care has been taken, by collation, revision, and otherwise, to render the First American Edition the most accurate and beautiful in our language. "Le plus beau monument que l' on puisse élever à la gloire de Newton, c'est une bonne édition de ses ouvrages:" and a monument like unto that we would here set up. The PRINCIPIA, above all, glows with the immortality of a transcendant mind. Marble and brass dissolve and pass away; but the true creations of genius endure, in time and beyond time, forever: high upon the adamant of the indestructible, they send forth afar and near, over the troublous waters of life, a pure, unwavering, quenchless light whereby the myriad myriads of barques, richly laden with reason, intelligence and various faculty, are guided through the night and the storm, by the beetling shore and the hidden rock, the breaker and the shoal, safely into havens calm and secure.

To the teacher and the taught, the scholar and the student, the devotee of Science and the worshipper of Truth, the PRINCIPIA must ever continue to be of inestimable value. If to educate means, not so much to store the memory with symbols and facts, as to bring forth the faculties of the soul and develope them to the full by healthy nurture and a hardy discipline, then, what so effective to the accomplishment of that end as the study of Geometrical Synthesis? The Calculus, in some shape or other, is, indeed, necessary to the successful prosecution of researches in the higher branches of philosophy. But has not the Analytical encroached upon the Synthetical, and Algorithmic Formulæ been employed when not requisite, either for the evolution of truth, or even its apter illustration? To each method belongs, undoubtedly, an appropriate use. Newton, himself the inventor of Fluxions, censured the handling of Geometrical subjects by Algebraical calculations; and the maturest opinions which he expressed were additionally in favour of the Geometrical Method. His preference, so strongly marked, is not to be reckoned a mere matter of taste; and his authority should bear with preponderating weight upon the decision of every instructor in adopting what may be deemed the best plan to insure the completest mental development. Geometry, the vigorous product of remote time; blended with the earliest aspirations of Science and the earliest applications of Art; as well in the measures of music as in the movement of spheres; as wholly in the structure of the atom as in that of the world; directing MOTION and shaping APPEARANCE; in a word, at the moulding of the created all, is, in comprehensive view, the outward form of that Inner Harmony of which and in which all things are. Plainly, therefore, this noble study has other and infinitely higher uses than to increase the power of abstraction. A more general and thorough cultivation of it should be strenuously insisted on. Passing from the pages of Euclid or Legendre, might not the student be led, at the suitable time, to those of the PRINCIPIA wherein Geometry may be found in varied use from the familiar to the sublime? The profoundest and the happiest results, it is believed, would attend upon this enlargement of our Educational System.

Let the PRINCIPIA, then, be gladly welcomed into every Hall where a TRUE TEACHER presides. And they who are guided to the diligent study of this incomparable work, who become strengthened by its reason, assured by its evidence, and enlightened by its truths, and who rise into loving communion with the great and pure spirit of its author, will go forth from the scenes of their pupilage, and take their places in the world as strong-minded, right-hearted men—such men as the Theory of our Government contemplates and its practical operation absolutely demands.

LIFE OF

SIR ISAAC NEWTON.

Nec fas est proprius mortali attingere Divos.—HALLEY.

FROM the thick darkness of the middle ages man's struggling spirit emerged as in new birth; breaking out of the iron control of that period; growing strong and confident in the tug and din of succeeding conflict and revolution, it bounded forwards and upwards with resistless vigour to the investigation of physical and moral truth; ascending height after height; sweeping afar over the earth, penetrating afar up into the heavens; increasing in endeavour, enlarging in endowment; every where boldly, earnestly out-stretching, till, in the AUTHOR of the PRINCIPIA, one arose, who, grasping the master-key of the universe and treading its celestial paths, opened up to the human intellect the stupendous realities of the material world, and, in the unrolling of its harmonies, gave to the human heart a new song to the goodness, wisdom, and majesty of the all-creating, all-sustaining, all-perfect God.

Sir Isaac Newton, in whom the rising intellect seemed to attain, as it were, to its culminating point, was born on the 25th of December, O. S. 1642—Christmas day—at Woolsthorpe, in the parish of Colsterworth, in Lincolnshire. His father, John Newton, died at the age of thirty-six, and only a few months after his marriage to Harriet Ayscough, daughter of James Ayscough, of Rutlandshire. Mrs. Newton, probably wrought upon by the early loss of her husband, gave premature birth to her only and posthumous child, of which, too, from its extreme diminutiveness, she appeared likely to be soon bereft. Happily, it was otherwise decreed! The tiny infant, on whose little lips the breath of life so doubtingly hovered, lived;—lived to a vigorous maturity, to a hale old age;—lived to become the boast of his country, the wonder of his time, and the "ornament of his species."

Beyond the grandfather, Robert Newton, the descent of Sir Isaac cannot with certainty be traced. Two traditions were held in the family: one, that they were of Scotch extraction; the other, that they came originally from Newton, in Lancashire, dwelling, for a time, however, at Westby, county of Lincoln, before the removal to and purchase of Woolsthorpe—about a hundred years before this memorable birth.

The widow Newton was left with the simple means of a comfortable subsistence. The Woolsthorpe estate together with a small one which she possessed at Sewstern, in Leicestershire, yielded her an income of some eighty pounds; and upon this limited sum, she had to rely chiefly for the support of herself, and the education of her child. She continued his nurture for three years, when, marrying again, she confided the tender charge to the care of her own mother.

Great genius is seldom marked by precocious development; and young Isaac, sent, at the usual age, to two day schools at Skillington and Stoke, exhibited no unusual traits of character. In his twelfth year, he was placed at the public school at Grantham, and boarded at the house of Mr. Clark, an apothecary. But even in this excellent seminary, his mental acquisitions continued for a while unpromising enough: study apparently had no charms for him; he was very inattentive, and ranked low in the school. One day, however, the boy immediately above our seemingly dull student gave him a severe kick in the stomach; Isaac, deeply affected, but with no outburst of passion, betook himself, with quiet, incessant toil, to his books; he quickly passed above the offending classmate; yet there he stopped not; the strong spirit was, for once and forever, awakened, and, yielding to its noble impulse, he speedily took up his position at the head of all.

His peculiar character began now rapidly to unfold itself. Close application grew to be habitual. Observation alternated with reflection. "A sober, silent, thinking lad," yet, the wisest and the kindliest, the indisputable leader of his fellows. Generosity, modesty, and a love of truth distinguished him then as ever afterwards. He did not often join his classmates in play; but he would contrive for them various amusements of a scientific kind. Paper kites he introduced; carefully determining their best form and proportions, and the position and number of points whereby to attach the string. He also invented paper lanterns; these served ordinarily to guide the way to school in winter mornings, but occasionally for quite another purpose; they were attached to the tails of kites in a dark night, to the dismay of the country people dreading portentous comets, and to the immeasureable delight of his companions. To him, however, young as he was, life seemed to have become an earnest thing. When not occupied with his studies, his mind would be engrossed with mechanical contrivances; now imitating, now inventing. He became singularly skilful in the use of his little saws, hatchets, hammers, and other tools. A windmill was erected near Grantham; during the operations of the workmen, he was frequently present; in a short time, he had completed a perfect working model of it, which elicited general admiration. Not content, however, with this exact imitation, he conceived the idea of employing, in the place of sails, animal power; and, adapting the construction of his mill accordingly, he enclosed in it a mouse, called the miller, and which by acting on a sort of treadwheel, gave motion to the machine. He invented, too, a mechanical carriage—having four wheels, and put in motion with a handle worked by the person sitting inside. The measurement of time early drew his attention. He first constructed a water clock, in proportions somewhat like an old-fashioned house clock. The index of the dial plate was turned by a piece of wood acted upon by dropping water. This instrument, though long used by himself, and by Mr. Clark's family, did not satisfy his inquiring mind. His thoughts rose to the sun; and, by careful and oft-repeated observations of the solar movements, he subsequently formed many dials. One of these, named *Isaac's dial*, was the accurate result of years' labour, and was frequently referred to for the hour of the day by the country people.

May we not discern in these continual efforts—the diligent research, the patient meditation, the aspiring glance, and the energy of discovery—the stirring elements of that wondrous spirit, which, clear, calm, and great, moved, in after years, through deep onward through deep of Nature's mysteries, unlocking her strongholds, dispelling darkness, educing order—everywhere silently conquering.

Newton had an early and decided taste for drawing. Pictures, taken sometimes from copies, but often from life, and drawn, coloured and framed by himself, ornamented his apartment. He was skilled also, in poetical composition, "excelled in making verses;" some of these were borne in remembrance and repeated, seventy years afterward, by Mrs. Vincent, for whom, in early youth, as Miss Storey, he formed an ardent attachment. She was the sister of a physician resident near Woolsthorpe; but Newton's intimate acquaintance with her began at Grantham, where they were both numbered among the inmates of the same house. Two or three years younger than himself, of great personal beauty, and unusual talent, her society afforded him the greatest pleasure; and their youthful friendship, it is believed, gradually rose to a higher passion; but inadequacy of fortune prevented their union. Miss Storey was afterwards twice married; Newton, never; his esteem for her continued unabated during life, accompanied by numerous acts of attention and kindness.

In 1656, Newton's mother was again left a widow, and took up her abode once more at Woolsthorpe. He was now fifteen years of age, and had made great progress in his studies; but she, desirous of his help, and from motives of economy, recalled him from school. Business occupations, however, and the management of the farm, proved utterly distasteful to him. When sent to Grantham Market on Saturdays, he would betake himself to his former lodgings in the apothecary's garret, where some of Mr. Clark's old books employed his thoughts till the aged and trustworthy servant had executed the family commissions and announced the necessity of return: or, at other times, our young philosopher would seat himself under a hedge, by the wayside, and continue his studies till the same faithful personage—proceeding alone to the town and completing the day's business—stopped as he returned. The more immediate affairs of the farm received no better attention. In fact, his passion for study grew daily more absorbing, and his dislike for every other occupation more intense. His mother, therefore, wisely resolved to give him all the advantages which an education could confer. He was sent back to Grantham school, where he remained for some months in busy preparation for his academical studies. At the recommendation of one of his uncles, who had himself studied at Trinity College, Cambridge, Newton proceeded thither, and was duly admitted, on the 5th day of June 1660, in the eighteenth year of his age.

The eager student had now entered upon a new and wider field; and we find him devoting himself to the pursuit of knowledge with amazing ardour and perseverance. Among other subjects, his attention was soon drawn to that of Judicial Astrology. He exposed the folly of this pseudo-science by erecting a figure with the aid of one or two of the problems of Euclid;—and thus began his study of the Mathematics. His researches into this science were prosecuted with unparallelled vigour and success. Regarding the propositions contained in Euclid as self-evident truths, he passed rapidly over this ancient system—a step which he afterward much regretted—and mastered, without further preparatory study, the Analytical Geometry of Descartes. Wallis's Arithmetic of Infinites, Saunderson's Logic, and the Optics of Kepler, he also studied with great care; writing upon them many comments; and, in these notes on Wallis's work was undoubtedly the germ of his fluxionary calculus. His progress was so great that he found himself more profoundly versed than his tutor in many branches of learning. Yet his acquisitions were not gotten with the rapidity of intuition; but they were thoroughly made and firmly secured. Quickness of apprehension, or intellectual nimbleness did not belong to him. He saw too far: his insight was too deep. He dwelt fully, cautiously upon the least subject; while to the consideration of the greatest, he brought a massive strength joined with a matchless clearness, that, regardless of the merely trivial or unimportant, bore with unerring sagacity upon the prominences of the subject, and, grappling with its difficulties, rarely failed to surmount them.

His early and fast friend, Dr. Barrow—in compass of invention only inferior to Newton—who had been elected Professor of Greek in the University, in 1660, was made Lucasian Professor of Mathematics in 1663, and soon afterward delivered his Optical Lectures: the manuscripts of these were revised by Newton, and several oversights corrected, and many important suggestions made by him; but they were not published till 1669.

In the year 1665, he received the degree of Bachelor of Arts; and, in 1666, he entered upon those brilliant and imposing discoveries which have conferred inappreciable benefits upon science, and immortality upon his own name.

Newton, himself, states that he was in possession of his Method of Fluxions, "in the year 1666, or before." Infinite quantities had long been a subject of profound investigation; among the ancients by Archimedes, and Pappus of Alexandria; among the moderns by Kepler, Cavaleri, Roberval, Fermat and Wallis. With consummate ability Dr. Wallis had improved upon the labours of his predecessors: with a higher power, Newton moved forwards from where Wallis stopped. Our author first invented his celebrated BINOMIAL THEOREM. And then, applying this Theorem to the rectification of curves, and to the determination of the surfaces and contents of solids, and the position of their centres of gravity, he discovered the general principle of deducing the areas of curves from the ordinate, by considering the area as a nascent quantity, increasing by continual fluxion in the proportion of the length of the ordinate, and supposing the abscissa to increase uniformly in proportion to the time. Regarding lines as generated by the motion of points, surfaces by the motion of lines, and solids by the motion of surfaces, and considering that the ordinates, abscissae, &c., of curves thus formed, vary according to a regular law depending on the equation of the curve, he deduced from this equation the velocities with which these quantities are generated, and obtained by the rules of infinite series, the ultimate value required. To the velocities with which every line or quantity is generated, he gave the name of FLUXIONS, and to the lines or quantities themselves, that of FLUENTS. A discovery that successively baffled the acutest and strongest intellects:—that, variously modified, has proved of incalculable service in aiding to develope the most abstruse and the highest truths in Mathematics and Astronomy: and that was of itself enough to render any name illustrious in the crowded Annals of Science.

At this period, the most distinguished philosophers were directing all their energies to the subject of light and the improvement of the refracting telescope. Newton, having applied himself to the grinding of "optic glasses of other figures than spherical," experienced the impracticability of executing such lenses; and conjectured that their defects, and consequently those of refracting telescopes, might arise from some other cause than the imperfect convergency of rays to a single point. He accordingly "procured a triangular glass prism to try therewith the celebrated phenomena of colours." His experiments, entered upon with zeal, and conducted with that industry, accuracy, and patient thought, for which he was so remarkable, resulted in the grand conclusion, that LIGHT WAS NOT HOMOGENEOUS, BUT CONSISTED OF RAYS, SOME OF WHICH WERE MORE REFRANGIBLE THAN OTHERS. This profound and beautiful discovery opened up a new era in the History of Optics. As bearing, however, directly upon the construction of telescopes, he saw that a lens refracting exactly like a prism would necessarily bring the different rays to different foci, at different distances from the glass, confusing and rendering the vision indistinct. Taking for granted that all bodies produced spectra of equal length, he dismissed all further consideration of the refracting instrument, and took up the principle of reflection. Rays of all colours, he found, were reflected regularly, so that the angle of reflection was equal to the angle of incidence, and hence he concluded that *optical instruments might be brought to any degree of perfection imaginable*, provided reflecting specula of the requisite figure and finish could be obtained. At this stage of his optical researches, he was forced to leave Cambridge on account of the plague which was then desolating England.

He retired to Woolsthorpe. The old manor-house, in which he was born, was situated in a beautiful little valley, on the west side of the river Witham; and here in the quiet home of his boyhood, he passed his days in serene contemplation, while the stalking pestilence was hurrying its tens of thousands into undistinguishable graves.

Towards the close of a pleasant day in the early autumn of 1666, he was seated alone beneath a tree, in his garden, absorbed in meditation. He was a slight young man; in the twenty-fourth year of his age; his countenance mild and full of thought. For a century previous, the science of Astronomy had advanced with rapid strides. The human mind had risen from the gloom and bondage of the middle ages, in unparalleled vigour, to unfold the system, to investigate the phenomena, and to establish the laws of the heavenly bodies. Copernicus, Tycho Brahe, Kepler, Galileo, and others had prepared and lighted the way for him who was to give to their labour its just value, and to their genius its true lustre. At his bidding isolated facts were to take order as parts of one harmonious whole, and sagacious conjectures grow luminous in the certain splendour of demonstrated truth. And this ablest man had come—was here. His mind, familiar with the knowledge of past effort, and its unequalled faculties developed in transcendant strength, was now moving on to the very threshold of its grandest achievement. Step by step the untrodden path was measured, till, at length, the entrance seemed disclosed, and the tireless explorer to stand amid the first opening wonders of the universe.

The nature of gravity—that mysterious power which causes all bodies to descend towards the centre of the earth—had, indeed, dawned upon him. And reason busily united link to link of that chain which was yet to be traced joining the least to the vastest, the most remote to the nearest, in one harmonious bond. From the bottoms of the deepest caverns to the summits of the highest mountains, this power suffers no sensible change: may not its action, then, extend to the moon? Undoubtedly: and further reflection convinced him that such a power might be sufficient for retaining that luminary in her orbit round the earth. But, though this power suffers no sensible variation, in the little change of distance from the earth's centre, at which we may place ourselves, yet, at the distance of the moon, may not its force undergo more or less diminution? The conjecture appeared most probable: and, in order to estimate what the degree of diminution might be, he considered that if the moon be retained in her orbit by the force of gravity, the primary planets must also be carried round the sun by the like power; and, by comparing the periods of the several planets with their distances from the sun, he found that, if they were held in their courses by any power like gravity, its strength must decrease in the duplicate proportion of the increase of distance. In forming this conclusion, he supposed the planets to move in perfect circles, concentric to the sun. Now was this the law of the moon's motion? Was such a force, emanating from the earth and directed to the moon, sufficient, when diminished as the square of the distance, to retain her in her orbit? To ascertain this master-fact, he compared the space through which heavy bodies fall, in a second of time, at a given distance from the centre of the earth, namely, at its surface, with the space through which the moon falls, as it were, to the earth, in the same time, while revolving in a circular orbit. He was absent from books; and, therefore, adopted, in computing the earth's diameter, the common estimate of sixty miles to a degree of latitude as then in use among geographers and navigators. The result of his calculations did not, of course, answer his expectations; hence, he concluded that some other cause, beyond the reach of observation—analogous, perhaps, to the vortices of Descartes—joined its action to that of the power of gravity upon the moon. Though by no means satisfied, he yet abandoned awhile further inquiry, and remained totally silent upon the subject.

These rapid marches in the career of discovery, combined with the youth of Newton, seem to evince a penetration the most lively, and an invention the most exuberant. But in him there was a conjunction of influences as extraordinary as fortunate. Study, unbroken, persevering and profound carried on its informing and disciplining work upon a genius, natively the greatest, and rendered freest in its movements, and clearest in its vision, through the untrammelling and enlightening power of religion. And, in this happy concurrence, are to be sought the elements of those amazing abilities, which, grasping, with equal facility, the minute and the stupendous, brought these successively to light, and caused science to make them her own.

In 1667, Newton was made a Junior Fellow; and, in the year following, he took his degree of Master of Arts, and was appointed to a Senior Fellowship.

On his return to Cambridge, in 1668, he resumed his optical labours. Having thought of a delicate method of polishing metal, he proceeded to the construction of his newly projected reflecting telescope; a small specimen of which he actually made with his own hands. It was six inches long; and magnified about forty times;—a power greater than a refracting instrument of six feet tube could exert with distinctness. Jupiter, with his four satellites, and the horns, or moon-like phases of Venus were plainly visible through it. THIS WAS THE FIRST REFLECTING TELESCOPE EVER EXECUTED AND DIRECTED TO THE HEAVENS. He gave an account of it, in a letter to a friend, dated February 23d, 1668-9—a letter which is also remarkable for containing the first allusion to his discoveries "concerning the nature of light." Encouraged by the success of his first experiment, he again executed with his own hands, not long afterward, a second and superior instrument of the same kind. The existence of this having come to the knowledge of the Royal Society of London, in 1671, they requested it of Newton for examination. He accordingly sent it to them. It excited great admiration; it was shown to the king; a drawing and description of it was sent to Paris; and the telescope itself was carefully preserved in the Library of the Society. Newton lived to see his invention in public use, and of eminent service in the cause of science.

In the spring of 1669, he wrote to his friend Francis Aston, Esq., then about setting out on his travels, a letter of advice and directions; it was dated May 18th, and is interesting as exhibiting some of the prominent features in Newton's character. Thus:—

"Since in your letter you give me so much liberty of spending my judgment about what may be to your advantage in travelling, I shall do it more freely than perhaps otherwise would have been decent. First, then, I will lay down some general rules, most of which, I believe, you have considered already; but if any of them be new to you, they may excuse the rest; if none at all, yet is my punishment more in writing than yours in reading.

"When you come into any fresh company. 1. Observe their humours. 2. Suit your own carriage thereto, by which insinuation you will make their converse more free and open. 3. Let your discourse be more in queries and doubtings than peremptory assertions or disputings, it being the design of travellers to learn, not to teach. Besides, it will persuade your acquaintance that you have the greater esteem of them, and so make them more ready to communicate what they know to you; whereas nothing sooner occasions disrespect and quarrels than peremptoriness. You will find little or no advantage in seeming wiser or much more ignorant than your company. 4. Seldom discommend any-thing though never so bad, or do it but moderately, lest you be unexpectedly forced to an unhandsome retraction. It is safer to commend any thing more than it deserves, than to discommend a thing so much as it deserves; for commendations meet not so often with oppositions, or, at least, are not usually so ill resented by men that think otherwise, as discommendations; and you will insinuate into men's favour by nothing sooner than seeming to approve and commend what they like; but beware of doing it by comparison. 5. If you be affronted, it is better, in a foreign country, to pass it by in silence, and with a jest, though with some dishonour, than to endeavour revenge; for, in the first case, your credit's ne'er the worse when you return into England, or come into other company that have not heard of the quarrel. But, in the second case, you may bear the marks of the quarrel while you live, if you outlive it at all. But, if you find yourself unavoidably engaged, 'tis best, I think, if you can command your passion and language, to keep them pretty evenly at some certain moderate pitch, not much heightening them to exasperate your adversary, or provoke his friends, nor letting them grow overmuch dejected to make him insult. In a word, if you can keep reason above passion, that and watchfulness will be your best defendants. To which purpose you may consider, that, though such excuses as this—He provok't me so much I could not forbear—may pass among friends, yet amongst strangers they are insignificant, and only argue a traveller's weakness.

"To these I may add some general heads for inquiries or observations, such as at present I can think on. As, 1. To observe the policies, wealth, and state affairs of nations, so far as a solitary traveller may conveniently do. 2. Their impositions upon all sorts of people, trades, or commodities, that are remarkable. 3. Their laws and customs, how far they differ from ours. 4. Their trades and arts wherein they excel or come short of us in England. 5. Such fortifications as you shall meet with, their fashion, strength, and advantages for defence, and other such military affairs as are considerable. 6. The power and respect belonging to their degrees of nobility or magistracy. 7. It will not be time misspent to make a catalogue of the names and excellencies of those men that are most wise, learned, or esteemed in any nation. 8. Observe the mechanism and manner of guiding ships. 9. Observe the products of Nature in several places, especially in mines, with the circumstances of mining and of extracting metals or minerals out of their ore, and of refining them; and if you meet with any transmutations out of their own species into another (as out of iron into copper, out of any metal into quicksilver, out of one salt into another, or into an insipid body, &c.), those, above all, will be worth your noting, being the most luciferous, and many times lucriferous experiments, too, in philosophy. 10. The prices of diet and other things. 11. And the staple commodities of places.

"These generals (such as at present I could think of), if they will serve for nothing else, yet they may assist you in drawing up a model to regulate your travels by. As for particulars, these that follow are all that I can now think of, viz.; whether at Schemnitium, in Hungary (where there are mines of gold, copper, iron, vitriol, antimony, &c.), they change iron into copper by dissolving it in a vitriolate water, which they find in cavities of rocks in the mines, and then melting the slimy solution in a strong fire, which in the cooling proves copper. The like is said to be done in other places, which I cannot now remember; perhaps, too, it may be done in Italy. For about twenty or thirty years agone there was a certain vitriol came from thence (called Roman vitriol), but of a nobler virtue than that which is now called by that name; which vitriol is not now to be gotten, because, perhaps, they make a greater gain by some such trick as turning iron into copper with it than by selling it. 2. Whether, in Hungary, Sclavonia, Bohemia, near the town Eila, or at the mountains of Bohemia near Silesia, there be rivers whose waters are impregnated with gold; perhaps, the gold being dissolved by some corrosive water like *aqua regis*, and the solution carried along with the stream, that runs through the mines. And whether the practice of laying mercury in the rivers, till it be tinged with gold, and then straining the mercury through leather, that the gold may stay behind, be a secret yet, or openly practised. 3. There is newly contrived, in Holland, a mill to grind glasses plane withal, and I think polishing them too; perhaps it will be worth the while to see it. 4. There is in Holland one —— Borry, who some years since was imprisoned by the Pope, to have extorted from him secrets (as I am told) of great worth, both as to medicine and profit, but he escaped into Holland, where they have granted him a guard. I think he usually goes clothed in green. Pray inquire what you can of him, and whether his ingenuity be any profit to the Dutch. You may inform yourself whether the Dutch have any tricks to keep their ships from being all worm-eaten in their voyages to the Indies. Whether pendulum clocks do any service in finding out the longitude, &c.

"I am very weary, and shall not stay to part with a long compliment, only I wish you a good journey, and God be with you."

It was not till the month of June, 1669, that our author made known his Method of Fluxions. He then communicated the work which he had composed upon the subject, and entitled, Analysis per Equationes numero terminorum Infinitas, to his friend Dr. Barrow. The latter, in a letter dated 20th of the same month, mentioned it to Mr. Collins, and transmitted it to him, on the 31st of July thereafter. Mr. Collins greatly approved of the work; took a copy of it; and sent the original back to Dr. Barrow. During the same and the two following years, Mr.

Collins, by his extensive correspondence, spread the knowledge of this discovery among the mathematicians in England, Scotland, France, Holland and Italy.

Dr. Barrow, having resolved to devote himself to Theology, resigned the Lucasian Professorship of Mathematics, in 1669, in favour of Newton, who accordingly received the appointment to the vacant chair.

During the years 1669, 1670, and 1671, our author, as such Professor, delivered a course of Optical Lectures. Though these contained his principal discoveries relative to the different refrangibility of light, yet the discoveries themselves did not become publicly known, it seems, till he communicated them to the Royal Society, a few weeks after being elected a member thereof, in the spring of 1671-2. He now rose rapidly in reputation, and was soon regarded as foremost among the philosophers of the age. His paper on light excited the deepest interest in the Royal Society, who manifested an anxious solicitude to secure the author from the "arrogations of others," and proposed to publish his discourse in the monthly numbers in which the Transactions were given to the world. Newton, gratefully sensible of these expressions of esteem, willingly accepted of the proposal for publication. He gave them also, at this time, the results of some further experiments in the decomposition and re-composition of light:—that the same degree of refrangibility always belonged to the same colour, and the same colour to the same degree of refrangibility: that the seven different colours of the spectrum were original, or simple, and that *whiteness*, or white light was a compound of all these seven colours.

The publication of his new doctrines on light soon called forth violent opposition as to their soundness. Hooke and Huygens—men eminent for ability and learning—were the most conspicuous of the assailants. And though Newton effectually silenced all his adversaries, yet he felt the triumph of little gain in comparison with the loss his tranquillity had sustained. He subsequently remarked in allusion to this controversy—and to one with whom he was destined to have a longer and a bitterer conflict—"I was so persecuted with discussions arising from the publication of my theory of light, that I blamed my own imprudence for parting with so substantial a blessing as my quiet to run after a shadow."

In a communication to Mr. Oldenburg, Secretary of the Royal Society, in 1672, our author stated many valuable suggestions relative to the construction of REFLECTING MICROSCOPES which he considered even more capable of improvement than telescopes. He also contemplated, about the same time, an edition of Kinckhuysen's Algebra, with notes and additions; partially arranging, as an introduction to the work, a treatise, entitled, A Method of Fluxions; but he finally abandoned the design. This treatise, however, he resolved, or rather consented, at a late period of his life, to put forth separately; and the plan would probably have been carried into execution had not his death intervened. It was translated into English, and published in 1736 by John Colson, Professor of Mathematics in Cambridge.

Newton, it is thought, made his discoveries concerning the INFLECTION and DIFFRACTION of light before 1674. The phenomena of the inflection of light had been first discovered more than ten years before by Grimaldi. And Newton began by repeating one of the experiments of the learned Jesuit—admitting a beam of the sun's light through a small pin hole into a dark chamber: the light diverged from the aperture in the form of a cone, and the shadows of all bodies placed in this light were larger than might have been expected, and surrounded with three coloured fringes, the nearest being widest, and the most remote the narrowest. Newton, advancing upon this experiment, took exact measures of the diameter of the shadow of a human hair, and of the breadth of the fringes, at different distances behind it, and discovered that these diameters and breadths were not proportional to the distances at which they were measured. He hence supposed that the rays which passed by the edge of the hair were deflected or turned aside from it, as if by a repulsive force, the nearest rays suffering the greatest, the more remote a less degree of deflection.. In explanation of the coloured fringes, he queried: whether the rays which differ in refrangibility do not differ also in flexibility, and whether they are not, by these different inflections, separated from one another, so as after separation to make the colours in the three fringes above described? Also, whether the rays, in passing by the edges and sides of bodies, are not bent several times backwards and forwards with an eel-like motion—the three fringes arising from three such bendings? His inquiries on this subject were here interrupted and never renewed.

His Theory of the COLOURS of NATURAL BODIES was communicated to the Royal Society, in February, 1675. This is justly regarded as one of the profoundest of his speculations. The fundamental principles of the Theory in brief, are:—That bodies possessing the greatest refractive powers reflect the greatest quantity of light; and that, at the confines of equally refracting media, there is no reflection. That the minutest particles of almost all natural bodies are in some degree transparent. That between the particles of bodies there are pores, or spaces, either empty or filled with media of a less density than the particles themselves. That these particles, and pores or spaces, have some definite size. Hence he deduced the Transparency, Opacity, and colours of natural bodies. Transparency arises from the particles and their pores being too small to cause reflection at their common surfaces—the light all passing through; Opacity from the opposite cause of the particles and their pores being sufficiently large to reflect the light which is "stopped or stifled" by the multitude of reflections; and colours from the particles, according to their several sizes, reflecting rays of one colour and transmitting those of another—or in other words, the colour that meets the eye is the colour reflected, while all the other rays are transmitted or absorbed.

Analogous in origin to the colours of natural bodies, he considered the COLOURS OF THIN PLATES. This subject was interesting and important, and had attracted considerable investigation. He, however, was the first to determine the law of the production of these colours, and, during the same year made known the results of his researches herein to the Royal Society. His mode of procedure in these experiments was simple and curious. He placed a double convex lens of a large known radius of curvature, upon the flat surface of a plano-convex object glass. Thus, from their point of contact at the centre, to the circumference of the lens, he obtained plates of air, or spaces varying from the extremest possible thinness, by slow degrees, to a considerable thickness. Letting the light fall, every different thickness of this plate of air gave different colours—the point of contact of the lens and glass forming the centre of numerous concentric colored rings. Now the radius of curvature of the lens being known, the thickness of the plate of air, at any given point, or where any particular colour appeared, could be exactly determined. Carefully noting, therefore, the order in which the different colours appeared, he measured, with the nicest accuracy, the different thicknesses at which the most luminous parts of the rings were produced, whether the medium were air, water, or mica—all these substances giving the same colours at different thicknesses;—the ratio of which he also ascertained. From the phenomena observed in these experiments, Newton deduced his Theory of Fits of EASY REFLECTION AND TRANSMISSION of light. It consists in supposing that every particle of light, from its first discharge from a luminous body, possesses, at equally distant intervals, dispositions to be reflected from, or transmitted through the surfaces of bodies upon which it may fall. For instance, if the rays are in a Fit of Easy Reflection, they are on reaching the surface, repelled, thrown off, or reflected from it; if, in a Fit of Easy Transmission, they are attracted, drawn in, or transmitted through it. By this Theory of Fits, our author likewise explained the colours of thick plates.

He regarded light as consisting of small material particles emitted from shining substances. He thought that these particles could be re-combined into solid matter, so that "gross bodies and light were convertible into one another;" that the particles of light and the particles of solid bodies acted mutually upon each other; those of light agitating and heating those of solid bodies, and the latter attracting and repelling the former. Newton was the first to suggest the idea of the POLARIZATION of light.

In the paper entitled *An Hypothesis Explaining Properties of Light*, December, 1675, our author first introduced his opinions respecting Ether—opinions which he afterward abandoned and again permanently resumed—" A most subtle spirit which pervades" all bodies, and is expanded through all the heavens. It is electric, and almost, if not quite immeasurably elastic and rare. " By the force and action of which spirit the particles of bodies mutually attract one another, at near distances, and cohere, if contiguous; and electric bodies operate at greater distances, as well repelling as attracting the neighbouring corpuscles; and light is emitted, reflected, refracted, inflected and heats bodies; and all sensation is excited, and the members of animal bodies move at the command of the will, namely, by the vibrations of this spirit, mutually propagated along the solid filaments of the nerves, from the outward organs of sense to the brain, and from the brain into the muscles." This " spirit" was no *anima mundi;* nothing further from the thought of Newton; but was it not, on his part, a partial recognition of, or attempt to reach an ultimate material force, or primary element, by means of which, " in the roaring loom of time," this material universe, God's visible garment, may be woven for us?

The Royal Society were greatly interested in the results of some experiments, which our author had, at the same time, communicated to them relative to the excitation of electricity in glass; and they, after several attempts and further direction from him, succeeded in re-producing the same phenomena.

One of the most curious of Newton's minor inquiries related to the connexion between the refractive powers and chemical composition of bodies. He found on comparing the refractive powers and the densities of many different substances, that the former were very nearly proportional to the latter, in the same bodies. Unctuous and sulphureous bodies were noticed as remarkable exceptions—as well as the *diamond*—their refractive powers being two or three times greater in respect of their densities than in the case of other substances, while, as among themselves, the one was generally proportional to the other. He hence inferred as to the diamond a great degree of combustibility;—a conjecture which the experiments of modern chemistry have shown to be true.

The chemical researches of our author were probably pursued with more or less diligence from the time of his witnessing some of the practical operations in that science at the Apothecary's at Grantham. DE NATURA ACIDORUM is a short chemical paper, on various topics, and published in Dr. Horsley's Edition of his works. TABULA QUANTITATUM ET GRADUUM COLORIS was inserted in the Philosophical Transactions; it contains a comparative scale of temperature from that of melting ice to that of a small kitchen coal-fire. He regarded fire as a body heated so hot as to emit light copiously; and flame as a vapour, fume, or exhalation heated so hot as to shine. To elective attraction, by the operation of which the small particles of bodies, as he conceived, act upon one another, at distances so minute as to escape observation, he ascribed all the various chemical phenomena of precipitation, combination, solution, and crystallization, and the mechanical phenomena of cohesion and capillary attraction. Newton's chemical views were illustrated and confirmed, in part, at least, in his own life-time. As to the structure of bodies, he was of opinion "that the smallest particles of matter may cohere by the strongest attractions, and compose bigger particles of weaker virtue; and many of these may cohere and compose bigger particles whose virtue is still weaker; and so on for divers successions, until the progression end in the biggest particles, on which the operations in chemistry and the colours of natural bodies depend, and which by adhering, compose bodies of sensible magnitude."

There is good reason to suppose that our author was a diligent student of the writings of Jacob Behmen; and that in conjunction with a relative, Dr. Newton, he was busily engaged, for several months in the earlier part of life, in quest of the philosopher's tincture. "Great Alchymist," however, very imperfectly describes the character of Behmen, whose researches into things material and things spiritual, things human and things divine, afford the strongest evidence of a great and original mind.

More appropriately here, perhaps, than elsewhere, may be given Newton's account of some curious experiments, made in his own person, on the action of light upon the retina. Locke, who was an intimate friend of our author, wrote to him for his opinion on a certain fact stated in Boyle's Book of Colours. Newton, in his reply, dated June 30th, 1691, narrates the following circumstances, which probably took place in the course of his optical researches. Thus :—

"The observation you mention in Mr. Boyle's Book of Colours I once tried upon myself with the hazard of my eyes. The manner was this; I looked a very little while upon the sun in the looking-glass with my right eye, and then turned my eyes into a dark corner of my chamber, and winked, to observe the impression made, and the circles of colours which encompassed it, and how they decayed by degrees, and at last vanished. This I repeated a second and a third time. At the third time, when the phantasm of light and colours about it were almost vanished, intending my fancy upon them to see their last appearance, I found, to my amazement, that they began to return, and by little and little to become as lively and vivid as when I had newly looked upon the sun. But when I ceased to intend my fancy upon them, they vanished again. After this, I found, that as often as I went into the dark, and intended my mind upon them, as when a man looks earnestly to see anything which is difficult to be seen, I could make the phantasm return without looking any more upon the sun; and the oftener I made it return, the more easily I could make it return again. And, at length, by repeating this, without looking any more upon the sun, I made such an impression on my eye, that, if I looked upon the clouds, or a book, or any bright object, I saw upon it a round bright spot of light like the sun, and, which is still stranger, though I looked upon the sun with my right eye only, and not with my left, yet my fancy began to make an impression upon my left eye, as well us upon my right. For if I shut my right eye, or looked upon a book, or the clouds, with my left eye, I could see the spectrum of the sun almost as plain as with my right eye, if I did but intend my fancy a little while upon it; for at first, if I shut my right eye, and looked with my left, the spectrum of the sun did not appear till I intended my fancy upon it; but by repeating, this appeared every time more easily. And now, in a few hours' time, I had brought my eyes to such a pass, that I could look upon no bright object with either eye, but I saw the sun before me, so that I durst neither write nor read; but to recover the use of my eyes, shut myself up in my chamber made dark, for three days together, and used all means to divert my imagination from the sun. For if I thought upon him, I presently saw his picture, though I was in the dark. But by keeping in the dark, and employing my mind about other things, I began in three or four days to have some use of my eyes again; and by forbearing to look upon bright objects, recovered them pretty well, though not so well but that, for some months after, the spectrum of the sun began to return as often as I began to meditate upon the phenomena, even though I lay in bed at midnight with my curtains drawn. But now I have been very well for many years, though I am apt to think, if I durst venture my eyes, I could still make the phantasm return by the power of my fancy. This story I tell you, to let you understand, that in the observation related by Mr. Boyle, the man's fancy probably concurred with the impression made by the sun's light to produce that phantasm of the sun which he constantly saw in bright objects. And so your question about the cause of phantasm involves another about the power of fancy, which I must confess is too hard a knot for me to untie. To place this effect in a constant motion is hard, because the sun ought then to appear perpetually. It seems rather to consist in a disposition of the sensorium to move the imagination strongly, and to be easily moved, both by the imagination and by the light, as often as bright objects are looked upon."

Though Newton had continued silent, yet his thoughts were by no means inactive upon the vast subject of the planetary motions. The idea of Universal Gravitation, first caught sight of, so to speak, in the garden at Woolsthorpe, years ago, had gradually expanded upon him. We find him, in a letter to Dr. Hooke, Secretary of the Royal Society, dated in November, 1679, proposing to verify the motion of the earth by direct experiment, namely, by the observation of the path pursued by a body falling from a considerable height. He had concluded that the path would be spiral; but Dr. Hooke maintained that it would be an eccentric ellipse in vacuo, and an ellipti-spiral in a resisting medium. Our author, aided by this correction of his error, and by the discovery that a projectile would move in an elliptical orbit when under the influence of a force varying inversely as the square of the distance, was led to discover "the theorem by which he afterwards examined the ellipsis;" and to demonstrate the celebrated proposition that a planet acted upon by an attractive force varying inversely as the squares of the distances will describe an elliptical orbit, in one of whose foci the attractive force resides.

When he was attending a meeting of the Royal Society, in June 1682, the conversation fell upon the subject of the measurement of a degree of the meridian, executed by M. Picard, a French Astronomer, in 1679. Newton took a memorandum of the result; and afterward, at the earliest opportunity, computed from it the diameter of the earth: furnished with these new data, he resumed his calculation of 1666. As he proceeded therein, he saw that his early expectations were now likely to be realized; the thick rushing, stupendous results overpowered him; he became unable to carry on the process of calculation, and intrusted its completion to one of his friends. The discoverer had, indeed, grasped the master-fact. The law of falling bodies at the earth's surface was at length identified with that which guided the moon in her orbit. And so his Great Thought, that had for sixteen years loomed up in dim, gigantic outline, amid the first dawn of a plausible hypothesis, now stood forth, radiant and not less grand, in the mid-day light of demonstrated truth.

It were difficult, nay impossible to imagine, even, the influence of a result like this upon a mind like Newton's. It was as if the keystone had been fitted to the glorious arch by which his spirit should ascend to the outskirts of infinite space—spanning the immeasurable—weighing the imponderable—computing the incalculable—mapping out the marchings of the planets, and the far-wanderings of the comets, and catching, bring back to earth some clearer notes of that higher melody which, as a sounding voice, bears perpetual witness to the design and omnipotence of a creating Deity.

Newton, extending the law thus obtained, composed a series of about twelve propositions on the motion of the primary planets about the sun. These were sent to London, and communicated to the Royal Society about the end of 1683. At or near this period, other philosophers, as Sir Christopher Wren, Dr. Halley, and Dr. Hooke, were engaged in investigating the same subject; but with no definite or satisfactory results. Dr. Halley, having seen, it is presumed, our author's propositions, went in August, 1684, to Cambridge to consult with him upon the subject. Newton assured him that he had brought the demonstration to perfection. In November, Dr. Halley received a copy of the work; and, in the following month, announced it to the Royal Society, with the author's promise to have it entered upon their Register. Newton, subsequently reminded by the Society of his promise, proceeded in the diligent preparation of the work, and, though suffering an interruption of six weeks, transmitted the manuscript of the first book to London before the end of April. The work was entitled PHILOSOPHIÆ NATURALIS PRINCIPIA MATHEMATICA, dedicated to the Royal Society, and presented thereto on the 28th of April, 1685-6. The highest encomiums were passed upon it; and the council resolved, on the 19th of May, to print it at the expense of the Society, and under the direction of Dr. Halley. The latter, a few days afterward, communicated these steps to Newton, who, in a reply, dated the 20th of June, holds the following language:—"The proof you sent me I like very well. I designed the whole to consist of three books; the second was finished last summer, being short, and only wants transcribing, and drawing the cuts fairly. Some new propositions I have since thought on, which I can as well let alone. The third wants the theory of comets. In autumn last, I spent two months in calculation to no purpose for want of a good method, which made me afterward return to the first book, and enlarge it with diverse propositions, some relating to comets, others to other things found out last winter. The third I now design to suppress. Philosophy is such an impertinently litigious lady, that a man had as good be engaged in law-suits as have to do with her. I found it so formerly, and now I can no sooner come near her again, but she gives me warning. The first two books without the third will not so well bear the title of *Philosophiæ Naturalis*

Principia Mathematicia ; and thereupon I had altered it to this, *De Motu Corporum Libri duo.* But after second thought I retain the former title. It will help the sale of the book, which I ought not to diminish now 'tis yours."

This "warning" arose from some pretensions put forth by Dr. Hooke. And though Newton gave a minute and positive refutations of such claims, yet, to reconcile all differences, he generously added to Prop. IV. Cor. 6, Book I., a Scholium, in which Wren, Hooke and Halley are acknowledged to have independently deduced the law of gravity from the second law of Kepler.

The suppression of the third book Dr. Halley could not endure to see. "I must again beg you" says he, "not to let your resentments run so high as to deprive us of your third book, wherein your applications of your mathematical doctrine to the theory of comets, and several curious experiments, which, as I guess by what you write ought to compose it, will undoubtedly render it acceptable to those who will call themselves philosophers without mathematics, which are much the greater number." To these solicitations Newton yielded. There were no "resentments," however, as we conceive, in his "design to suppress." He sought peace; for he loved and valued it above all applause. But, in spite of his efforts for tranquillity's sake, his course of discovery was all along molested by ignorance or presumptuous rivalry.

The publication of the great work now went rapidly forwards. The second book was sent to the Society, and presented on the 2d March; the third, on the 6th April; and the whole was completed and published in the month of May, 1686-7. In the second Lemma of the second book, the fundamental principle of his fluxionary calculus was, for the first time, given to the world; but its algorithm or notation did not appear till published in the second volume of Dr. Wallis's works, in 1693.

And thus was ushered into existence The PRINCIPIA—a work to which pre-eminence above all the productions of the human intellect has been awarded—a work that must be esteemed of priceless worth so long as Science has a votary, or a single worshipper be left to kneel at the altar of Truth.

The entire work bears the general title of THE MATHEMATICAL PRINCIPLES OF NATURAL PHILOSOPHY. It consists of three books: the first two, entitled, OF THE MOTION OF BODIES, are occupied with the laws and conditions of motions and forces, and are illustrated with many scholia treating of some of the most general and best established points in philosophy, such as the density and resistance of bodies, spaces void of matter, and the motion of sound and light. From these principles, there is deduced, in the third book, drawn up in as popular a style as possible and entitled, OF THE SYSTEM OF THE WORLD, the constitution of the system of the world. In regard to this book, the author says—"I had, indeed, composed the third Book in a popular method, that it might be read by many; but afterwards, considering that such as had not sufficently entered into the principles could not easily discover the strength of the consequences, nor lay aside the prejudices to which they had been many years accustomed, therefore, to prevent disputes which might be raised upon such accounts, I chose to reduce the substance of this Book into the form of Propositions (in the mathematical way), which should be read by those only who had first made themselves masters of the principles established in the preceding Books: not that I would advise any one to the previous study of every Proposition of those Books."—"It is enough if one carefully reads the Definitions, the Laws of Motion, and the three first Sections of the first Book. He may then pass on to this Book, and consult such of the remaining Propositions of the first two Books, as the references in this, and his occasions shall require." So that "The System of the World" is composed both "in a popular method," and in the form of mathematical Propositions.

The principle of Universal Gravitation, namely, *that every particle of matter is attracted by, or gravitates to, every other particle of matter, with a force inversely proportional to the squares of their distances*—is the discovery which characterizes The PRINCIPIA. This principle the author deduced from the motion of the moon, and the three laws of Kepler—laws, which Newton, in turn, by his greater law, demonstrated to be true.

From the first law of Kepler, namely, the proportionality of the areas to the times of their description, our author inferred that the force which retained the planet in its orbit was always directed to the sun; and from the second, namely, that every planet moves in an ellipse with the sun in one of its foci, he drew the more general inference that the force by which the planet moves round that focus varies inversely as the square of its distance therefrom: and he demonstrated that a planet acted upon by such a force could not move in any other curve than a conic section; showing when the moving body would describe a circular, an elliptical, a parabolic, or hyperbolic orbit. He demonstrated, too, that this force, or attracting, gravitating power resided in every, the least particle; but that, in spherical masses, it operated as if confined to their centres; so that, one sphere or body will act upon another sphere or body, with a force directly proportional to the quantity of matter, and inversely as the square of the distance between their centres; and that their velocities of mutual approach will be in the inverse ratio of their quantities of matter. Thus he grandly outlined the Universal Law. Verifying its truth by the motions of terrestrial bodies, then by those of the moon and other secondary orbs, he finally embraced, in one mighty generalization, the entire Solar System—all the movements of all its bodies—planets, satellites and comets—explaining and harmonizing the many diverse and theretofore inexplicable phenomena.

Guided by the genius of Newton, we see sphere bound to sphere, body to body, particle to particle, atom to mass, the minutest part to the stupendous whole—each to each, each to all, and all to each—in the mysterious bonds of a ceaseless, reciprocal influence. An influence whose workings are shown to be alike present in the globular dew-drop, or oblate-spheroidal earth; in the falling shower, or vast heaving ocean tides; in the flying thistle-down, or fixed, ponderous rock; in the swinging pendulum, or time-measuring sun; in the varying and unequal moon, or earth's slowly retrograding poles; in the uncertain meteor, or blazing comet wheeling swiftly away on its remote, yet determined round. An influence, in fine, that may link system to system through all the star-glowing firmament; then firmament to firmament; aye, firmament to firmament, again and again, till, converging home, it may be, to some ineffable centre, where more presently dwells He who inhabiteth immensity, and where infinitudes meet and eternities have their conflux, and where around move, in softest, swiftest measure, all the countless hosts that crowd heaven's fathomless deeps.

And yet Newton, amid the loveliness and magnitude of Omnipotence, lost not sight of the Almighty One. A secondary, however universal, was not taken for the First Cause. An impressed force, however diffused and powerful, assumed not the functions of the creating, giving Energy. Material beauties, splendours, and sublimities, however rich in glory, and endless in extent, concealed not the attributes of an intelligent Supreme. From the depths of his own soul, through reason and the WORD, he had risen, *à priori*, to God: from the heights of Omnipotence, through the design and law of the builded universe, he proved *à posteriori*, a Deity. "I had," says he, "an eye upon such principles as might work, with considering men, for the belief of a Deity," in writing the PRINCIPIA; at the conclusion whereof, he teaches that—"this most beautiful system of the sun, planets and comets, could only proceed from the counsel and dominion of an intelligent and powerful Being. And if the fixed stars are the centres of other like systems, these, being formed by the like wise counsels, must be all subject to the dominion of One; especially since the light of the fixed stars is of the same nature with the light of the sun, and from every system light passes into all other systems: and lest the systems of the fixed stars should, by their gravity, fall on each other mutually, he hath placed those systems at immense distances one from another.

"This Being governs all things, not as the soul of the world, but as Lord over all; and on account of his dominion he is wont to be called Lord God παντοκρατωρ or Universal Ruler; for God is a relative word, and has a respect to servants; and Deity is the dominion of God, not over his own body, as those imagine who fancy God to be the soul of the world, but over servants. The Supreme God is a Being eternal, infinite, absolutely perfect; but a being, however perfect, without dominion, cannot be said to be Lord God; for we say, my God, your God, the God of Israel, the God of Gods, and Lord of Lords; but we do not say, my Eternal, your Eternal, the Eternal of Israel, the Eternal of Gods; we do not say my Infinite, or my Perfect: these are titles which have no respect to servants. The word God usually signifies Lord; but every Lord is not God. It is the dominion of a spiritual Being which constitutes a God; a true, supreme, or imaginary dominion makes a true, supreme, or imaginary God. And from his true dominion it follows that the true God is a living, intelligent and powerful Being; and from his other perfections, that he is supreme or most perfect. He is eternal and infinite, omnipotent and omniscient; that is, his duration reaches from eternity to eternity; his presence from infinity to infinity; he governs all things and knows all things, that are or can be done. He is not eternity or infinity, but eternal and infinite; he is not duration and space, but he endures and is present. He endures forever and is everywhere present; and by existing always and everywhere, he constitutes duration and space. Since every particle of space is *always*, and every indivisible moment of duration is *everywhere*, certainly the Maker and Lord of things cannot be *never* and *nowhere*. Every soul that has perception is, though in different times and different organs of sense and motion, still the same indivisible person. There are given successive parts in duration, co-existent parts in space, but neither the one nor the other in the person of a man, or his thinking principle; and much less can they be found in the thinking substance of God. Every man, so far as he is a thing that has perception, is one and the same man during his whole life, in all and each of his organs of sense. God is one and the same God, always and everywhere. He is omnipresent, not *virtually* only, but also *substantially;* for virtue cannot subsist without substance. In him are all things contained and moved; yet neither affects the other; God suffers nothing from the motion of bodies; bodies find no resistance from the omnipresence of God. It is allowed by all that the Supreme God exists necessarily; and by the same necessity he exists *always* and *everywhere*. Whence also he is all similar, all eye, all ear, all brain, all arm, all power to perceive, to understand, and to act; but in a manner not at all human, in a manner not at all corporeal, in a manner utterly unknown to us. As a blind man has no idea of colours, so have we no idea of the manner by which the all-wise God perceives and understands all things. He is utterly void of all body, and bodily figure, and can therefore neither be seen, nor heard, nor touched; nor ought he to be worshipped under the representation of any corporeal thing. We have ideas of his attributes, but what the real substance of anything is we know not. In bodies we see only their figures and colours, we hear only the sounds, we touch only their outward surfaces, we smell only the smells, and taste only the savours; but their inward substances are not to be known, either by our senses, or by any reflex act of our minds: much less, then, have we any idea of the substance of God. We know him only by his most wise and excellent contrivances of things, and final causes; we admire him for his perfections; but we reverence and adore him on account of his dominion; for we adore him as his servants; and a god without dominion, providence, and final causes, is nothing else but Fate and Nature. Blind metaphysical necessity, which is certainly the same always and everywhere, could produce no variety of things. All that diversity of natural things which we find suited to different times and places could arise from nothing but the ideas and will of a Being necessarily existing."

Thus, the diligent student of science, the earnest seeker of truth, led, as through the courts of a sacred Temple, wherein, at each step, new wonders meet the eye, till, as a crowning grace, they stand before a Holy of Holies, and learn that all science and all truth are one which hath its beginning and its end in the knowledge of Him whose glory the heavens declare, and whose handiwork the firmament showeth forth.

The introduction of the pure and lofty doctrines of the PRINCIPIA was perseveringly resisted. Descartes, with his system of vortices, had sown plausibly to the imagination, and error had struck down deeply, and shot up luxuriantly, not only in the popular, but in the scientific mind. Besides the idea—in itself so simple and so grand—that the great masses of the planets were suspended in empty space, and retained in their orbits by an invisible influence residing in the sun—was to the ignorant a thing inconceivable, and to the learned a revival of the occult qualities of the ancient physics. This remark applies particularly to the continent. Leibnitz misapprehended; Huygens in part rejected; John Bernouilli opposed; and Fontenelle never received the doctrines of the PRINCIPIA. So that, the saying of Voltaire is probably true, that though Newton survived the publication of his great work more than forty years, yet, at the time of his death, he had not above twenty followers out of England.

But in England, the reception of our author's philosophy was rapid and triumphant. His own labours, while Lucasian Professor; those of his successors in that Chair—Whiston and Saunderson; those of Dr. Samuel Clarke, Dr. Laughton, Roger Cotes, and Dr. Bentley; the experimental lectures of Dr. Keill and Desaguliers; the early and powerful exertions of David Gregory at Edinburgh, and of his brother James Gregory at St. Andrew's, tended to diffuse widely in England and Scotland a knowledge of, and taste for the truths of the PRINCIPIA. Indeed, its mathematical doctrines constituted, from the first, a regular part of academical instruction; while its physical truths, given to the public in popular lectures, illustrated by experiments, had, before the lapse of twenty years, become familiar to, and adopted by the general mind. Pemberton's popular "View of Sir Isaac Newton's Philosophy" was published, in 1728; and the year afterward, an English translation of the PRINCIPIA, and System of the World, by Andrew Motte. And since that period, the labours of Le Seur and Jacquier, of Thorpe, of Jebb, of Wright and others have greatly contributed to display the most hidden treasures of the PRINCIPIA.

About the time of the publication of the Principia, James II., bent on re-establishing the Romish Faith, had, among other illegal acts, ordered by mandamus, the University of Cambridge to confer the degree of Master of Arts upon an ignorant monk. Obedience to this mandate was resolutely refused. Newton was one of the nine delegates chosen to defend the independence of the University. They appeared before the High Court;—and successfully: the king abandoned his design. The prominent part which our author took in these proceedings, and his eminence in the scientific world, induced his proposal as one of the parliamentary representatives of the University. He was elected, in 1688, and sat in the Convention Parliament till its dissolution. After the first year, however, he seems to have given little or no attention to his parliamentary duties, being seldom absent from the University till his appointment in the Mint, in 1695.

Newton began his theological researches sometime previous to 1691; in the prime of his years, and in the matured vigour of his intellectual powers. From his youth, as we have seen, he had devoted himself with an activity the most unceasing, and an energy almost superhuman to the discovery of physical truth;—giving to Philosophy a new foundation, and to Science a new temple. To pass on, then, from the consideration of the material, more directly to that of the spiritual, was a natural, nay, with so large and devout a soul, a necessary advance. The Bible was to him of inestimable worth. In the elastic freedom, which a pure and unswerving faith in Him of Nazareth gives, his mighty faculties enjoyed the only completest scope for development. His original endowment, however great, combined with a studious application, however profound, would never, without this liberation from the dominion of passion and sense, have enabled him to attain to that wondrous concentration and grasp of intellect, for which Fame has as yet assigned him no equal. Gratefully he owned, therefore, the same Author in the Book of Nature and the Book of Revelation. These were to him as drops of the same unfathomable ocean;—as outrayings of the same inner splendour;—as tones of the same ineffable voice;—as segments of the same infinite curve. With great joy he had found himself enabled to proclaim, as an interpreter, from the hieroglyphs of Creation, the existence of a God: and now, with greater joy, and in the fulness of his knowledge, and in the fulness of his strength, he laboured to make clear, from the utterances of the inspired Word, the far mightier confirmations of a Supreme Good, in all its glorious amplitude of Being and of Attribute; and to bring the infallible workings thereof plainly home to the understandings and the affections of his fellow-men; and finally to add the weight of his own testimony in favour of that Religion, whose truth is now, indeed, "girded with the iron and the rock of a ponderous and colossal demonstration."

His work, entitled, OBSERVATIONS UPON THE PROPHECIES OF HOLY WRIT, PARTICULARLY THE PROPHECIES OF DANIEL AND THE APOCALYPSE OF ST. JOHN, first published in London, in 1733 4to. consists of two parts: the one devoted to the Prophecies of Daniel, and the other to the Apocalypse of St. John. In the first part, he treats concerning the compilers of the books of the Old Testament;—of the prophetic language;—of the vision of the four beasts;—of the kingdoms represented by the feet of the image composed of iron and clay;—of the ten kingdoms represented by the ten horns of the beast;—of the eleventh horn of Daniel's fourth beast; of the power which should change times and laws;—of the kingdoms represented in Daniel by the ram and he-goat;—of the prophecy of the seventy weeks;—of the times of the birth and passion of Christ;—of the prophecy of the Scripture of Truth;—of the king who doeth according to his will, and magnified himself above every god, and honoured Mahuzzims, and regarded not the desire of women;—of the Mahuzzim, honoured by the king who doeth according to his will. In the second part, he treats of the time when the Apocalypse was written; of the scene of the vision, and the relation which the Apocalypse has to the book of the law of Moses, and to the worship of God in the temple;—of the relation which the Apocalypse has to the prophecies of Daniel, and of the subject of the prophecy itself. Newton regards the prophecies as given, not for the gratification of man's curiosity, by enabling him to foreknow; but for his conviction that the world is governed by Providence, by witnessing their fulfilment. Enough of prophecy, he thinks, has already been fulfilled to afford the diligent seeker abundant evidence of God's providence. The whole work is marked by profound erudition, sagacity and argument.

And not less learning, penetration and masterly reasoning are conspicuous in his HISTORICAL ACCOUNT OF TWO NOTABLE CORRUPTIONS OF SCRIPTURES IN A LETTER TO A FRIEND. This

Treatise, first accurately published in Dr. Horsley's edition of his works, relates to two texts: the one, 1 Epistle of St. John v. 7; the other, 1 Epistle of St. Paul to Timothy iii. 16. As this work had the effect to deprive the advocates of the doctrine of the Trinity of two leading texts, Newton has been looked upon as an Arian; but there is absolutely nothing in his writings to warrant such a conclusion.

His remaining theological works consist of the LEXICON PROPHETICUM, which was left incomplete; a Latin Dissertation on the sacred cubit of the Jews, which was translated into English, and published, in 1737, among the Miscellaneous Works of John Greaves; and FOUR LETTERS *addressed to Dr. Bentley, containing some arguments in proof of a Deity.* These Letters were dated respectively: 10th December, 1692; 17th January, 1693; 25th February, 1693; and 11th February, 1693—the fourth bearing an earlier date than the third. The best faculties and the profoundest acquirements of our author are convincingly manifest in these lucid and powerful compositions. They were published in 1756, and reviewed by Dr. Samuel Johnson.

Newton's religious writings are distinguished by their absolute freedom from prejudice. Everywhere, throughout them, there glows the genuine nobleness of soul. To his whole life, indeed, we may here fitly extend the same observation. He was most richly imbued with the very spirit of the Scriptures which he so delighted to study and to meditate upon. His was a piety, so fervent, so sincere and practical, that it rose up like a holy incense from every thought and act. His a benevolence that not only willed, but endeavoured the best for all. His a philanthropy that held in the embracings of its love every brother-man. His a toleration of the largest and the truest; condemning persecution in every, even its mildest form; and kindly encouraging each striving after excellence:—a toleration that came not of indifference—for the immoral and the impious met with their quick rebuke—but a toleration that came of the wise humbleness and the Christian charity, which see, in the nothingness of self and the almightiness of TRUTH, no praise for the ablest, and no blame for the feeblest in their strugglings upward to light and life.

In the winter of 1691-2, on returning from chapel, one morning, Newton found that a favourite little dog, called Diamond, had overturned a lighted taper on his desk, and that several papers containing the results of certain optical experiments, were nearly consumed. His only exclamation, on perceiving his loss, was, "Oh Diamond, Diamond, little knowest thou the mischief thou hast done." Dr. Brewster, in his life of our author, gives the following extract from the manuscript Diary of Mr. Abraham De La Pryme, a student in the University at the time of this occurrence.

"1692. February, 3.—What I heard to-day I must relate. There is one Mr. Newton (whom I have very oft seen), Fellow of Trinity College, that is mighty famous for his learning, being a most excellent mathematician, philosopher, divine, &c. He has been Fellow of the Royal Society these many years; and among other very learned books and tracts, he's written one upon the mathematical principles of philosophy, which has given him a mighty name, he having received, especially from Scotland, abundance of congratulatory letters for the same; but of all the books he ever wrote, there was one of colours and light, established upon thousands of experiments which he had been twenty years of making, and which had cost him many hundreds of pounds. This book which he valued so much, and which was so much talked of, had the ill luck to perish, and be utterly lost just when the learned author was almost at putting a conclusion at the same, after this manner: In a winter's morning, leaving it among his other papers on his study table while he went to chapel, the candle, which he had unfortunately left burning there, too, catched hold by some means of other papers, and they fired the aforesaid book, and utterly consumed it and several other valuable writings; and which is most wonderful did no further mischief. But when Mr. Newton came from chapel, and had seen what was done, every one thought he would have run mad, he was so troubled thereat that he was not himself for a month after. A long account of this his system of colours you may find in the Transactions of the Royal Society, which he had sent up to them long before this sad mischance happened unto him."

It will be borne in mind that all of Newton's theological writings, with the exception of the Letters to Dr. Bentley, were composed before this event which, we must conclude, from Pryme's words, produced a serious impression upon our author for about a month. But M. Biot, in his Life of Newton, relying on a memorandum contained in a small manuscript Journal of Huygens, declares this occurrence to have caused a derangement of Newton's intellect. M. Biot's opinions and deductions, however, as well as those of La Place, upon this subject, were based upon erroneous data, and have been overthrown by the clearest proof. There is not, in fact, the least evidence that Newton's reason was, for a single moment, dethroned; on the contrary, the testimony is conclusive that he was, at all times, perfectly capable of carrying on his mathematical, metaphysical and astronomical inquiries. Loss of sleep, loss of appetite, and irritated nerves will disturb somewhat the equanimity of the most serene; and an act done, or language employed, under such temporary discomposure, is not a just criterion of the general tone and strength of a man's mind. As to the accident itself, we may suppose, whatever might have been its precise nature, that it greatly distressed him, and, still further, that its shock may have originated the train of nervous derangements, which afflicted him, more or less, for two years afterward. Yet, during this very period of ill health, we find him putting forth his highest powers. In 1692, he prepared for, and transmitted to Dr. Wallis the first proposition of the Treatise on Quadratures, with examples of it in first, second and third fluxions. He investigated, in the same year, the subject of haloes; making and recording numerous and important observations relative thereto. Those profound and beautiful Letters to Dr. Bentley were written at the close of this and the beginning of the next year. In October, 1693, Locke, who was then about publishing a second edition of his work on the Human Understanding, requested Newton to reconsider his opinions on innate ideas. And in 1694, he was zealously occupied in perfecting his lunar theory; visiting Flamstead, at the Royal Observatory of Greenwich, in September, and obtaining a series of lunar observations; and commencing, in October, a correspondence with that distinguished practical Astronomer, which continued till 1698.

We now arrive at the period when Newton permanently withdrew from the seclusion of a collegiate, and entered upon a more active and public life. He was appointed Warden of the Mint, in 1695, through the influence of Charles Montague, Chancellor of the Exchequer, and afterward Earl of Halifax. The current coin of the nation had been adulterated and debased, and Montague undertook a re-coinage. Our author's mathematical and chemical knowledge proved eminently useful in accomplishing this difficult and most salutary reform. In 1699, he was promoted to the Mastership of the Mint—an office worth twelve or fifteen hundred pounds per annum, and which he held during the remainder of his life. He wrote, in this capacity, an official Report on the Coinage, which has been published; he also prepared a Table of Assays of Foreign Coins, which was printed at the end of Dr. Arbuthnot's Tables of Ancient Coins, Weights, and Measures, in 1727.

Newton retained his Professorship at Cambridge till 1703. But he had, on receiving the appointment of Master of the Mint, in 1699, made Mr. Whiston his deputy, with all the emoluments of the office; and, on finally resigning, procured his nomination to the vacant Chair.

In January 1697, John Bernouilli proposed to the most distinguished mathematicians of Europe two problems for solution. Leibnitz, admiring the beauty of one of them, requested the time for solving it to be extended to twelve months—twice the period originally named. The delay was readily granted. Newton, however, sent in, the day after he received the problems, a solution of them to the President of the Royal Society. Bernouilli obtained solutions from Newton, Leibinitz and the Marquis De L'Hopital; but Newton's though anonymous, he immediately recognised "*tanquam ungue leonem*," as the lion is known by his claw. We may mention here the famous problem of the trajectories proposed by Leibnitz, in 1716, for the purpose of "feeling the pulse of the English Analysts." Newton received the problem about five o'clock in the afternoon, as he was returning from the

Mint; and though it was extremely difficult and he himself much fatigued, yet he completed its solution, the same evening before he went to bed.

The history of these problems affords, by direct comparison, a striking illustration of Newton's vast superiority of mind. That amazing concentration and grasp of intellect, of which we have spoken, enabled him to master speedily, and, as it were, by a single effort, those things, for the achievement of which, the many would essay utterly in vain, and the very, very few attain only after long and renewed striving. And yet, with a modesty as unparalleled as his power, he attributed his successes, not to any extraordinary sagacity, but solely to industry and patient thought. He kept the subject of consideration constantly before him, and waited till the first dawning opened gradually into a full and clear light; never quitting, if possible, the mental process till the object of it were wholly gained. He never allowed this habit of meditation to appear in his intercourse with society; but in the privacy of his own chamber, or in the midst of his own family, he gave himself up to the deepest abstraction. Occupied with some interesting investigation, he would often sit down on his bedside, after he rose, and remain there, for hours, partially dressed. Meal-time would frequently come and pass unheeded; so that, unless urgently reminded, he would neglect to take the requisite quantity of nourishment. But notwithstanding his anxiety to be left undisturbed, he would, when occasion required, turn aside his thoughts, though bent upon the most intricate research, and then, when leisure served, again direct them to the very point where they ceased to act: and this he seemed to accomplish not so much by the force of his memory, as by the force of his inventive faculty, before the vigorous intensity of which, no subject, however abstruse, remained long unexplored.

He was elected a member of the Royal Academy of Sciences at Paris, in 1699, when that distinguished Body were empowered, by a new charter, to admit a small number of foreign associates. In 1700, he communicated to Dr. Halley a description of his reflecting instrument for observing the moon's distance from the fixed stars. This description was published in the Philosophical

Transactions, in 1742. The instrument was the same as that produced by Mr. Hadley, in 1731, and which, under the name of Hadley's Quadrant, has been of so great use in navigation. On the assembling of the new Parliament, in 1701, Newton was re-elected one of the members for the University of Cambridge. In 1703, he was chosen President of the Royal Society of London, to which office he was annually re-elected till the period of his decease—about twenty-five years afterward.

Our author unquestionably devoted more labour to, and, in many respects, took a greater pride in his Optical, than his other discoveries. This science he had placed on a new and indestructible basis; and he wished not only to build, but to perfect the costly and glowing structure. He had communicated, before the publication of the PRINCIPIA, his most important researches on light to the Royal Society, in detached papers which were inserted in successive numbers of the Transactions; but he did not publish a connected view of these labours till 1704, when they appeared under the title of OPTICS: OR, A TREATISE ON THE REFLEXIONS, REFRACTIONS, INFLEXIONS AND COLOURS OF LIGHT. To this, but to no subsequent edition, were added two Mathematical Treatises, entitled, TRACTATUS DUO DE SPECIEBUS ET MAGNITUDINE FIGURARUM CURVILINEARUM; the one bearing the title TRACTATUS DE QUADRATURA CURVARUM; and the other, that of ENUMERATIO LINEARUM TERTII ORDINIS. The publication of these Mathematical Treatises was made necessary in consequence of plagiarisms from the manuscripts of them loaned by the author to his friends. Dr. Samuel Clarke published a Latin translation of the Optics, in in 1706; whereupon he was presented by Newton, as a mark of his grateful approbation, with five hundred pounds, or one hundred pounds for each of his children. The work was afterward translated into French. It had a remarkably wide circulation, and appeared, in several successive editions, both in England and on the Continent. There is displayed, particularly on this Optical Treatise, the author's talent for simplifying and communicating the profoundest speculations. It is a faculty rarely united to that of the highest invention. Newton possessed both; and thus that mental perfectness which enabled him to create, to combine, and to teach, and so render himself, not the "ornament" only, but inconceivably more, the pre-eminent benefactor of his species.

The honour of knighthood was conferred on our author in 1705. Soon afterward, he was a candidate again for the Representation of the University, but was defeated by a large majority. It is thought that a more pliant man was preferred by both ministers and electors. Newton was always remarkable for simplicity of dress, and his only known departure from it was on this occasion, when he is said to have appeared in a suit of laced clothes.

The Algebraical Lectures which he had, during nine years, delivered at Cambridge, were published by Whiston, in 1707, under the title of ARITHMETICA UNIVERSALIS, SINE DE COMPOSITIONE ET RESOLUTIONE ARITHMETICA LIBER. This publication is said to have been a breach of confidence on Whiston's part. Mr. Ralphson, not long afterward, translated the work into English; and a second edition of it, with improvements by the author, was issued at London, 1712, by Dr. Machin. Subsequent editions, both in English and Latin, with commentaries, have been published.

In June, 1709, Newton intrusted the superintendence of a second edition of the PRINCIPIA to Roger Cotes, Plumian Professor of Astronomy at Cambridge. The first edition had been sold off for some time. Copies of the work had become very rare, and could only be obtained at several times their original cost. A great number of letters passed between the author and Mr. Cotes during the preparation of the edition, which finally appeared in May, 1713. It had many alterations and improvements, and was accompanied by an admirable Preface from the pen of Cotes.

Our author's early Treatise, entitled, ANALYSIS PER EQUATIONES NUMERO TERMINORUM INFINITAS, as well as a small Tract, bearing the title of METHODUS DIFFERENTIALIS, was published, with his consent, in 1711. The former of these, and the Treatise De Quadratura Curvarum, translated into English, with a large commentary, appeared in 1745. His work, entitled, ARTIS ANALYTICÆ SPECIMINA, VEL GEOMETRIA ANALYTICA, was first given to the world in the edition of Dr. Horsley, 1779.

It is a notable fact, in Newton's history, that he never voluntarily published any one of his purely mathematical writings. The cause of this unwillingness in some, and, in other instances, of his indifference, or, at least, want of solicitude to put forth his works may be confidently sought for in his repugnance to everything like contest or dispute. But, going deeper than this aversion, we find, underlying his whole character and running parallel with all his discoveries, that extraordinary humility which always preserved him in a position so relatively just to the behests of time and eternity, that the infinite value of truth, and the utter worthlessness of fame, were alike constantly present to him. Judging of his course, however, in its more temporary aspect, as bearing upon his immediate quiet, it seemed the most unfortunate. For an early publication, especially in the case of his Method of Fluxions, would have anticipated all rivalry, and secured him from the contentious claims of Leibnitz. Still each one will solve the problem of his existence in his own way, and, with a man like Newton, his own, as we conceive, could be no other than the best way. The conduct of Leibnitz in this affair is quite irreconcilable with the stature and strength of the man; giant-like, and doing nobly, in many ways, a giant's work, yet cringing himself into the dimensions and performances of a common calumniator. Opening in 1699, the discussion in question continued till the close of Leibnitz's life, in 1716. We give the summary of the case as contained in the Report of the Committee of the Royal Society, the deliberately weighed opinion of which has been adopted as an authoritative decision in all countries.

"We have consulted the letters and letter books in the custody of the Royal Society, and those found among the papers of Mr. John Collins, dated between the years 1669 and 1677, inclusive; and showed them to such as knew and avouched the hands of Mr. Barrow, Mr. Collins, Mr. Oldenburg, and Mr. Leibnitz; and compared those of Mr. Gregory with one another, and with copies of some of them taken in the hand of Mr. Collins; and have extracted from them what relates to the matter referred to us; all which extracts, herewith delivered to you, we believe to be genuine and authentic. And by these letters and papers we find:—

"I. Mr. Leibnitz was in London in the beginning of the year 1673; and went thence in or about March, to Paris, where he kept a correspondence with Mr. Collins, by means of Mr. Oldenburg, till about September, 1676, and then returned, by London and Amsterdam, to Hanover: and that Mr. Collins was very free in communicating to able mathematicians what he had received from Mr. Newton and Mr. Gregory.

"II. That when Mr. Leibnitz was the first time in London, he contended for the invention of another differential method, properly so called; and, notwithstanding he was shown by Dr. Pell that it was Newton's method, persisted in maintaining it to be his own invention, by reason that he had found it by himself without knowing what Newton had done before, and had much improved it. And we find no mention of his having any other differential method than Newton's before his letter of the 21st of June, 1677, which was a year after a copy of Mr. Newton's letter, of the 10th of December, 1672, had been sent to Paris to be communicated to him; and above four years after Mr. Collins began to communicate that letter to his correspondents; in which letter the method of fluxions was sufficiently described to any intelligent person.

"III. That by Mr. Newton's letter, of the 13th of June, 1676, it appears that he had the method of fluxions above five years before the writing of that letter. And by his Analysis per Æquationes numero Terminorum Infinitas, communicated by Dr. Barrow to Mr. Collins, in July, 1669, we find that he had invented the method before that time.

"IV. That the differential method is one and the same with the method of fluxions, excepting the name and mode of notation; Mr. Leibnitz calling those quantities differences which Mr. Newton calls moments, or fluxions; and marking them with a letter *d*—a mark not used by Mr. Newton.

"And, therefore, we take the proper question to be, not who invented this or that method, but, who was the first inventor of the method? And we believe that those who have reputed Mr. Leibnitz the first inventor knew little or nothing of his correspondence with Mr. Collins and Mr. Oldenburg long before, nor of Mr.

Newton's having that method above fifteen years before Mr. Leibnitz began to publish it in the Acta Eruditorum of Leipsic.

"For which reason we reckon Mr. Newton the first inventor; and are of opinion that Mr. Keill, in asserting the same, has been no ways injurious to Mr. Leibnitz. And we submit to the judgment of the Society, whether the extract and papers, now presented to you, together with what is extant, to the same purpose, in Dr. Wallis's third volume, may not deserve to be made public."

This Report, with the collection of letters and manuscripts, under the title of COMMERCIUM EPISTOLICUM D. JOHANNIS COLLINS ET ALIORUM DE ANALYSI PROMOTA JUSSU SOCIETATIS REGIÆ EDITUM, appeared accordingly in the early part of 1713. Its publication seemed to infuse additional bitterness into the feelings of Leibnitz, who descended to unfounded charges and empty threats. He had been privy counsellor to the Elector of Hanover, before that prince was elevated to the British throne; and in his correspondence, in 1715 and 1716, with the Abbé Conti, then at the court of George I., and with Caroline, Princess of Wales, he attacked the doctrines of the PRINCIPIA, and indirectly its author, in a manner very discreditable to himself, both as a learned and as an honourable man. His assaults, however, were triumphantly met; and, to the complete overthrow of his rival pretensions, Newton was induced to give the finishing blow. The verdict is universal and irreversible that the English preceded the German philosopher, by at least ten years, in the invention of fluxions. Newton could not have borrowed from Leibnitz; but Leibnitz might have borrowed from Newton. A new edition of the Commercium Epistolicum was published in 1722-5 (?); but neither in this, nor in the former edition, did our author take any part. The disciples, enthusiastic, capable and ready, effectually shielded, with the buckler of Truth, the character of the Master, whose own conduct throughout was replete with delicacy, dignity and justice. He kept aloof from the controversy—in which Dr. Keill stood forth as the chief representative of the Newtonian side—till the very last, when, for the satisfaction of the King, George I., rather than for his own, he consented to put forth his hand and firmly secure his rights upon a certain and impregnable basis.

A petition to have inventions for promoting the discovery of the longitude at sea, suitably rewarded, was presented to the House of Commons, in 1714. A committee, having been appointed to investigate the subject, called upon Newton and others for their opinions. That of our author was given in writing. A report, favourable to the desired measure, was then taken up, and a bill for its adoption subsequently passed.

On the ascension of George I., in 1714, Newton became an object of profound interest at court. His position under government, his surpassing fame, his spotless character, and, above all, his deep and consistent piety, attracted the reverent regard of the Princess of Wales, afterward queen-consort to George II. She was a woman of a highly cultivated mind, and derived the greatest pleasure from conversing with Newton and corresponding with Leibnitz. One day, in conversation with her, our author mentioned and explained a new system of chronology, which he had composed at Cambridge, where he had been in the habit "of refreshing himself with history and chronology, when he was weary with other studies." Subsequently, in the year 1718, she requested a copy of this interesting and ingenious work. Newton, accordingly, drew up an abstract of the system from the separate papers in which it existed, and gave it to her on condition that it should not be communicated to any other person. Sometime afterward she requested that the Abbé Conti might be allowed to have a copy of it. The author consented: and the abbé received a copy of the manuscript, under the like injunction and promise of secrecy. This manuscript bore the title of "A short Chronicle, from the First Memory of Things in Europe, to the Conquest of Persia, by Alexander the Great."

After Newton took up his residence in London, he lived in a style suited to his elevated position and rank. He kept his carriage, with an establishment of three male and three female servants. But to everything like vain show and luxury he was utterly averse. His household affairs, for the last twenty years of his life, were under the charge of his niece, Mrs. Catherine Barton, wife and widow of Colonel Barton—a woman of great beauty and accomplishment—and subsequently married to John Conduit, Esq. At home Newton was distinguished by that dignified and gentle hospitality which springs alone from true nobleness. On all proper occasions, he gave splendid entertainments, though without ostentation. In society, whether of the palace or the cottage, his manner was self-possessed and urbane; his look benign and affable; his speech candid and modest; his whole air undisturbedly serene. He had none of what are usually called the singularities of genius; suiting himself easily to every company—except that of the vicious and wicked; and speaking of himself and others, naturally, so as never even to be suspected of vanity. There was in him, if we may be allowed the expression, a WHOLENESS of nature, which did not admit of such imperfections and weakness—the circle was too perfect, the law too constant, and the disturbing forces too slight to suffer scarcely any of those eccentricities which so interrupt and mar the movements of many bright spirits, rendering their course through the world more like that of the blazing meteor than that of the light and life-imparting sun. In brief, the words GREATNESS and GOODNESS could not, humanly speaking, be more fitly employed than when applied as the pre-eminent characteristics of this pure, meek and venerable sage.

In the eightieth year of his age, Newton was seized with symptoms of stone in the bladder. His disease was pronounced incurable. He succeeded, however, by means of a strict regimen, and other precautions, in alleviating his complaint, and procuring long intervals of ease. His diet, always frugal, was now extremely temperate, consisting chiefly of broth, vegetables, and fruit, with, now and then, a little butcher meat. He gave up the use of his carriage, and employed, in its stead, when he went out, a chair. All invitations to dinner were declined; and only small parties were received, occasionally, at his own house.

In 1724 he wrote to the Lord Provost of Edinburgh, offering to contribute twenty pounds yearly toward the salary of Mr. Maclaurin, provided he accepted the assistant Professorship of Mathematics in the University of that place. Not only in the cause of ingenuity and learning, but in that of religion—in relieving the poor and assisting his relations, Newton annually expended large sums. He was generous and charitable almost to a fault. Those, he would often remark, who gave away nothing till they died, never gave at all. His wealth had become considerable by a prudent economy; but he regarded money in no other light than as one of the means wherewith he had been intrusted to do good, and he faithfully employed it accordingly.

He experienced, in spite of all his precautionary measures, a return of his complaint in the month of August, of the same year, 1724, when he passed a stone the size of pea; it came from him in two pieces, the one at the distance of two days from the other. Tolerable good health then followed for some months. In January, 1725, however, he was taken with a violent cough and inflammation of the lungs. In consequence of this attack, he was prevailed upon to remove to Kensington, where his health greatly improved. In February following, he was attacked in both feet with the gout, of the approach of which he had received, a few years before, a slight warning, and the presence of which now produced a very beneficial change in his general health. Mr. Conduit, his nephew, has recorded a curious conversation which took place, at or near this time, between himself and Sir Isaac.

"I was, on Sunday night, the 7th March, 1724-5, at Kensington, with Sir Isaac Newton, in his lodgings, just after he was out of a fit of the gout, which he had had in both of his feet, for the first time, in the eighty-third year of his age. He was better after it, and his head clearer and memory stronger than I had known them for some time. He then repeated to me, by way of discourse, very distinctly, though rather in answer to my queries, than in one continued narration, what he had often hinted to me before, viz.: that it was his conjecture (he would affirm nothing) that there was a sort of revolution in the heavenly bodies; that the vapours and light, emitted by the sun, which had their sediment, as water and other matter, had gathered themselves, by degrees, into a body, and attracted more matter from the planets, and at last made a secondary planet (viz.: one of those that go round another planet), and then, by gathering to them, and attracting more matter, became a primary planet; and then, by increasing still, became a comet, which, after certain revolutions, by coming nearer and nearer to the sun, had all its volatile parts condensed, and became a matter fit to recruit and replenish the sun (which must waste by the constant heat and light it emitted), as a faggot would this fire if put into it (we were sitting by a wood fire), and that that would probably be the effect of the comet of 1680, sooner or later; for, by the observations made upon it, it appeared, before it came near the sun, with a tail only two or three degrees long; but, by the heat it contracted, in going so near the sun, it seemed to have a tail of thirty or forty degrees when it went from it; that he could not say when this comet would drop into the sun; it might perhaps have five or six revolutions more first, but whenever it did it would so much increase the heat of the sun that this earth would be burned, and no animals in it could live. That he took the three phenomena, seen by Hipparchus, Tycho Brahe, and Kepler's disciples, to have been of this kind, for he could not otherwise account for an extraordinary light, as those were, appearing, all at once, among the the fixed stars (all which he took to be suns, enlightening other planets, as our sun does ours), as big as Mercury or Venus seems to us, and gradually diminishing, for sixteen months, and then sinking into nothing. He seemed to doubt whether there were not intelligent beings, superior to us, who superintended these revolutions of the heavenly bodies, by the direction of the Supreme Being. He appeared also to be very clearly of opinion that the inhabitants of this world were of short date, and alledged, as one reason for that opinion, that all arts, as letters, ships, printing, needle, &c., were discovered within the memory of history, which could not have happened if the world had been eternal; and that there were visible marks of ruin upon it which could not be effected by flood only. When I asked him how this earth could have been repeopled if ever it had undergone the same fate it was threatened with hereafter, by the comet of 1680, he answered, that required the power of a Creator. He said he took all the planets to be composed of the same matter with this earth, viz.: earth, water, stones, &c., but variously concocted. I asked him why he would not publish his conjectures, as conjectures, and instanced that Kepler had communicated his; and though he had not gone near so far as Kepler, yet Kepler's guesses were so just and happy that they had been proved and demonstrated by him. His answer was, "I do not deal in conjectures." But, on my talking to him about the four observations that had been made of the comet of 1680, at 574 years' distance, and asking him the particular times, he opened his *Principia*, which laid on the table, and showed me the particular periods, viz.: 1st. The Julium Sidus, in the time of Justinian, in 1106, in 1680.

"And I, observing that he said there of that comet, 'incidet in corpus solis,' and in the next paragraph adds, 'stellæ fixæ refici possunt,' told him I thought he owned there what we had been talking about, viz.: that the comet would drop into the sun, and that fixed stars were recruited and replenished by comets when they dropped into them; and, consequently, that the sun would be recruited too; and asked him why he would not own as fully what he thought of the sun as well as what he thought of the fixed stars. He said, 'that concerned us more;' and, laughing, added, that he had said enough for people to know his meaning."

In the summer of 1725, a French translation of the chronological MS., of which the Abbé Conti had been permitted, some time previous, to have a copy, was published at Paris, in violation of all good faith. The *Punic* Abbé had continued true to his promise of secrecy while he remained in England; but no sooner did he reach Paris than he placed the manuscript into the hands of M. Freret, a learned antiquarian, who translated the work, and accompanied it with an attempted refutation of the leading points of the system. In November, of the same year, Newton received a presentation copy of this publication, which bore the title of ABREGE DE CHRONOLOGIE DE M. LE CHEVALIER NEWTON, FAIT PAR LUI-MEME, ET TRADUIT SUR LE MANUSCRIPT ANGLAIS. Soon afterward a paper entitled, REMARKS ON THE OBERVATIONS MADE ON A CHRONOLOGICAL INDEX OF SIR ISAAC NEWTON, TRANSLATED INTO FRENCH BY THE OBSERVATOR, AND PUBLISHED AT PARIS, was drawn up by our author, and printed in the Philosophical Transactions for 1725. It contained a history of the whole matter, and a triumphant reply to the objections of M. Freret. This answer called into the field a fresh antagonist, Father Soueiet, whose five dissertations on this subject were chiefly remarkable for the want of knowledge and want of decorum, which they displayed. In consequence of these discussions, Newton was induced to prepare his larger work for the press, and had nearly completed it at the time of his death. It was published in 1728, under the title of THE CHRONOLOGY OF THE ANCIENT KINGDOMS AMENDED, TO WHICH IS PREFIXED A SHORT CHRONICLE FROM THE FIRST MEMORY OF THINGS IN EUROPE TO THE CONQUEST OF PERSIA BY ALEXANDER THE GREAT. It consists of six chapters: 1. On the Chronology of the Greeks; according to Whiston, our author wrote out eighteen copies of this chapter with his own hand, differing little from one another. 2. Of the Empire of Egypt; 3. Of the Assyrian Empire; 4. Of the two contemporary Empires of the Babylonians and Medes; 5. A Description of the Temple of Solomon; 6. Of the Empire of the Persians; this chapter was not found copied with the other five, but as it was discovered among his papers, and appeared to be a continuation of the same work, the Editor thought proper to add it thereto. Newton's LETTER TO A PERSON OF DISTINCTION WHO HAD DESIRED HIS OPINION OF THE LEARNED BISHOP LLOYD'S HYPOTHESIS CONCERNING THE FORM OF THE MOST ANCIENT YEAR, closes this enumeration of his Chronological Writings.

A third edition of the PRINCIPIA appeared in 1726, with many changes and additions. About four years were consumed in its preparation and publication, which were under the superintendance of Dr. Henry Pemberton, an accomplished mathematician, and the author of "A VIEW OF SIR ISAAC NEWTON'S PHILOSOPHY." 1728. This gentleman enjoyed numerous opportunities of conversing with the aged and illustrious author. "I found," says Pemberton, "he had read fewer of the modern mathematicians than one could have expected; but his own prodigious invention readily supplied him with what he might have an occasion for in the pursuit of any subject he undertook. I have often heard him censure the handling geometrical subjects by algebraic calculations; and his book of Algebra he called by the name of Universal Arithmetic, in opposition to the injudicious title of Geometry, which Descartes had given to the treatise, wherein he shows how the geometer may assist his invention by such kind of computations. He thought Huygens the most elegant of any mathematical writer of modern times, and the most just imitator of the ancients. Of their taste and form of demonstration, Sir Isaac always professed himself a great admirer. I have heard him even censure himself for not following them yet more closely than he did; and speak with regret of his mistake at the beginning of his mathematical studies, in applying himself to the works of Descartes and other algebraic writers, before he had considered the elements of Euclid with that attention which so excellent a writer deserves."

"Though his memory was much decayed," continues Dr. Pemberton, "he perfectly understood his own writings." And even this failure of memory, we would suggest, might have been more apparent than real, or, in medical terms, more the result of functional weakness than organic decay. Newton seems never to have confided largely to his memory: and as this faculty manifests the most susceptibility to cultivation; so, in the neglect of due exercise, it more readily and plainly shows a diminution of its powers.

Equanimity and temperance had, indeed, preserved Newton singularly free from all mental and bodily ailment. His hair was, to the last, quite thick, though as white as silver. He never made use of spectacles, and lost but one tooth to the day of his death. He was of middle stature, well-knit, and, in the latter part of his life, somewhat inclined to be corpulent. Mr. Conduit says, "he had a very lively and piercing eye, a comely and gracious aspect, with a fine head of hair, white as silver, without any baldness, and when his peruke was off was a venerable sight." According to Bishop Atterbury, "in the whole air of his face and make there was nothing of that penetrating sagacity which appears in his compositions. He had something rather languid in his look and manner which did not raise any great expectation in those who did not know him." Hearne remarks, "Sir Isaac was a man of no very promising aspect. He was a short, well-set man. He was full of thought, and spoke very little in company, so that his conversation was not agreeable. When he rode in his coach, one arm would be out of his coach on one side and the other on the other." These different accounts we deem easily reconcilable. In the rooms of the Royal Society, in the street, or in mixed assemblages, Newton's demeanour—always courteous, unassuming and kindly—still had in it the overawings of a profound repose and reticency, out of which the communicative spirit, and the "lively and piercing eye" would only gleam in the quiet and unrestrained freedom of his own fire-side.

"But this I immediately discovered in him," adds Pemberton, still further, "which at once both surprised and charmed me. Neither his extreme great age, nor his universal reputation had rendered him stiff in opinion, or in any degree elated. Of this I had occasion to have almost daily experience. The remarks I continually sent him by letters on his Principia, were received with the utmost goodness. These were so far from being anyways displeasing to him, that, on the contrary, it occasioned him to speak many kind things of me to my friends, and to honour me with a public testimony of his good opinion." A modesty, openness, and generosity, peculiar to the noble and comprehensive spirit of Newton. "Full of wisdom and perfect in beauty," yet not lifted up by pride nor corrupted by ambition. None, however, knew so well as himself the stupendousness of his discoveries in comparison with all that had been previously achieved; and none realized so thoroughly as himself the littleness thereof in comparison with the vast region still unexplored. A short time before his death he uttered this memorable sentiment:—"I do not know what I may appear to the world; but to myself I seem to have been only like a boy playing on the sea-shore, and diverting myself in now and then finding a smoother pebble or a prettier shell than ordinary, while the great ocean of truth lay all undiscovered before me." How few ever reach the shore even, much less find "a smoother pebble or a prettier shell!"

Newton had now resided about two years at Kensington; and

the air which he enjoyed there, and the state of absolute rest, proved of great benefit to him. Nevertheless he would occasionally go to town. And on Tuesday, the 28th of February, 1727, he proceeded to London, for the purpose of presiding at a meeting of the Royal Society. At this time his health was considered, by Mr. Conduit, better than it had been for many years. But the unusual fatigue he was obliged to suffer, in attending the meeting, and in paying and receiving visits, speedily produced a violent return of the affection in the bladder. He returned to Kensington on Saturday, the 4th of March. Dr. Mead and Dr. Cheselden attended him; they pronounced his disease to be the stone, and held out no hopes of recovery. On Wednesday, the 15th of March, he seemed a little better; and slight, though groundless, encouragement was felt that he might survive the attack. From the very first of it, his sufferings had been intense. Paroxysm followed paroxysm, in quick succession: large drops of sweat rolled down his face; but not a groan, not a complaint, not the least mark of peevishness or impatience escaped him: and during the short intervals of relief, he even smiled and conversed with his usual composure and cheerfulness. The flesh quivered, but the heart quaked not; the impenetrable gloom was settling down: the Destroyer near; the portals of the tomb opening, still, amid this utter wreck and dissolution of the mortal, the immortal remained serene, unconquerable: the radiant light broke through the gathering darkness; and Death yielded up its sting, and the grave its victory. On Saturday morning, 18th, he read the newspapers, and carried on a pretty long conversation with Dr. Mead. His senses and faculties were then strong and vigorous; but at six o'clock, the same evening, he became insensible; and in this state he continued during the whole of Sunday, and till Monday, the 20th, when he expired, between one and two o'clock in the morning, in the eighty-fifth year of his age.

And these were the last days of Isaac Newton. Thus closed the career of one of earth's greatest and best men. His mission was fulfilled. Unto the Giver, in many-fold addition, the talents were returned. While it was yet day he had worked; and for the night that quickly cometh he was not unprepared. Full of years, and full of honours, the heaven-sent was recalled; and, in the confidence of a "certain hope," peacefully he passed away into the silent depths of Eternity.

His body was placed in Westminster Abbey, with the state and ceremonial that usually attended the interment of the most distinguished. In 1731, his relatives, the inheritors of his personal estate, erected a monument to his memory in the most conspicuous part of the Abbey, which had often been refused by the dean and chapter to the greatest of England's nobility. During the same year a medal was struck at the Tower in his honour; and, in 1755, a full-length statue of him, in white marble, admirably executed, by Roubiliac, at the expense of Dr. Robert Smith, was erected in the ante-chamber of Trinity College, Cambridge. There is a painting executed in the glass of one of the windows of the same college, made pursuant to the will of Dr. Smith, who left five hundred pounds for that purpose.

Newton left a personal estate of about thirty-two thousand pounds. It was divided among his four nephews and four nieces of the half blood, the grand-children of his mother, by the Reverend Mr. Smith. The family estates of Woolsthorpe and Sustern fell to John Newton, the heir-at-law, whose great grand-father was Sir Isaac's uncle. Before his death he made an equitable distribution of his two other estates: the one in Berkshire to the sons and daughter of a brother of Mrs. Conduit; and the other, at Kensington, to Catharine, the only daughter of Mr. Conduit, and who afterward became Viscountess Lymington. Mr. Conduit succeeded to the offices of the Mint, the duties of which he had discharged during the last two years of Sir Isaac's life.

Our author's works are found in the collection of Castilion, Berlin, 1744, 4to. 8 tom.; in Bishop Horsley's Edition, London, 1779, 4to. 5 vol.; in the Biographia Brittannica, &c. Newton also published *Bern. Varenii* Geographia, &c., 1681, 8vo. There are, however, numerous manuscripts, letters, and other papers, which have never been given to the world: these are preserved, in various collections, namely, in the library of Trinity College, Cambridge; in the library of Corpus Christi College, Oxford; in the library of Lord Macclesfield; and, lastly and chiefly, in the possession of the family of the Earl of Portsmouth, through the Viscountess Lymington.

Everything appertaining to Newton has been kept and cherished with peculiar veneration. Different memorials of him are preserved in Trinity College, Cambridge; in the rooms of the Royal Society, of London; and in the Museum of the Royal Society of Edinburgh.

The manor-house, at Woolsthorpe, was visited by Dr. Stukeley, in October, 1721, who, in a letter to Dr. Mead, written in 1727, gave the following description of it:—" 'Tis built of stone, as is the way of the country hereabouts, and a reasonably good one. They led me up stairs and showed me Sir Isaac's study, where I supposed he studied, when in the country, in his younger days, or perhaps when he visited his mother from the University. I observed the shelves were of his own making, being pieces of deal boxes, which probably he sent his books and clothes down in on those occasions. There were, some years ago, two or three hundred books in it of his father-in-law, Mr. Smith, which Sir Isaac gave to Dr. Newton, of our town." The celebrated apple-tree, the fall of one of the apples of which is said to have turned the attention of Newton to the subject of gravity, was destroyed by the wind about twenty years ago; but it has been preserved in the form of a chair. The house itself has been protected with religious care. It was repaired in 1798, and a tablet of white marble put up in the room where our author was born, with the following inscription:—

"Sir Isaac Newton, son of John Newton, Lord of the Manor of Woolsthorpe, was born in this room, on the 25th of December, 1642."

Nature and Nature's Laws were hid in night,
God said, "Let Newton be," and all was light.

THE PRINCIPIA.

THE AUTHOR'S PREFACE.

SINCE the ancients (as we are told by *Pappus*), made great account of the science of mechanics in the investigation of natural things; and the moderns, laying aside substantial forms and occult qualities, have endeavoured to subject the phænomena of nature to the laws of mathematics, I have in this treatise cultivated mathematics so far as it regards philosophy. The ancients considered mechanics in a twofold respect; as rational, which proceeds accurately by demonstration; and practical. To practical mechanics all the manual arts belong, from which mechanics took its name. But as artificers do not work with perfect accuracy, it comes to pass that mechanics is so distinguished from geometry, that what is perfectly accurate is called geometrical; what is less so, is called mechanical. But the errors are not in the art, but in the artificers. He that works with less accuracy is an imperfect mechanic; and if any could work with perfect accuracy, he would be the most perfect mechanic of all; for the description of right lines and circles, upon which geometry is founded, belongs to mechanics. Geometry does not teach us to draw these lines, but requires them to be drawn; for it requires that the learner should first be taught to describe these accurately, before he enters upon geometry; then it shows how by these operations problems may be solved. To describe right lines and circles are problems, but not geometrical problems. The solution of these problems is required from mechanics; and by geometry the use of them, when so solved, is shown; and it is the glory of geometry that from those few principles, brought from without, it is able to produce so many things. Therefore geometry is founded in mechanical practice, and is nothing but that part of universal mechanics which accurately proposes and demonstrates the art of measuring. But since the manual arts are chiefly conversant in the moving of bodies, it comes to pass that geometry is commonly referred to their magnitudes, and mechanics to their motion. In this sense rational mechanics will be the science of motions resulting from any forces whatsoever, and of the forces required to produce any motions, accurately proposed and demonstrated. This part of mechanics was cultivated by the ancients in the five powers which relate to manual arts, who considered gravity (it not being a manual power), no otherwise than as it moved weights by those powers. Our design not respecting arts, but philosophy, and our subject not manual but natural powers, we consider chiefly those things which relate to gravity, levity, elastic force, the resistance of fluids, and the like forces, whether attractive or impulsive; and therefore we offer this work as the mathematical principles of philosophy; for all the difficulty of philosophy seems to consist in this—from the phænomena of motions to investigate the forces of nature, and then from these forces to demonstrate the other phænomena; and to this end the general propositions in the first and second book are directed. In the third book we give an example of this in the explication of the System of the World; for by the propositions mathematically demonstrated in the former books, we in the third derive from the celestial phænomena the forces of gravity with which bodies tend to the sun and the several planets. Then from these forces, by other propositions which are also mathematical, we deduce the motions of the planets, the comets, the moon, and the sea. I wish we could derive the rest of the phænomena of nature by the same kind of reasoning from mechanical principles; for I am induced by many reasons to suspect that they may all depend upon certain forces by which the particles of bodies, by some causes hitherto unknown, are either mutually impelled towards each other, and cohere in regular figures, or are repelled and recede from each other; which forces being unknown, philosophers have hitherto attempted the search of nature in vain; but I hope the principles here laid down will afford some light either to this or some truer method of philosophy.

In the publication of this work the most acute and universally learned Mr. Edmund Halley not only assisted me with his pains in correcting the press and taking care of the schemes, but it was to his solicitations that its becoming public is owing; for when he had obtained of me my demonstrations of the figure of the celestial orbits, he continually pressed me to communicate the same to the *Royal Society*, who afterwards, by their kind encouragement and entreaties, engaged me to think of publishing them. But after I had begun to consider the inequalities of the lunar motions, and had entered upon some other things relating to the laws and measures of gravity, and other forces; and the figures that would be described by bodies attracted according to given laws; and the motion of several bodies moving among themselves; the motion of bodies in resisting mediums; the forces, densities, and motions, of mediums; the orbits of the comets, and such like;

I deferred that publication till I had made a search into those matters, and could put forth the whole together. What relates to the lunar motions (being imperfect), I have put all together in the corollaries of Prop. 66, to avoid being obliged to propose and distinctly demonstrate the several things there contained in a method more prolix than the subject deserved, and interrupt the series of the several propositions. Some things, found out after the rest, I chose to insert in places less suitable, rather than change the number of the propositions and the citations. I heartily beg that what I have here done may be read with candour; and that the defects in a subject so difficult be not so much reprehended as kindly supplied, and investigated by new endeavours of my readers.

ISAAC NEWTON.

Cambridge, Trinity College, May 8, 1686.

In the second edition the second section of the first book was enlarged. In the seventh section of the second book the theory of the resistances of fluids was more accurately investigated, and confirmed by new experiments. In the third book the moon's theory and the præcession of the equinoxes were more fully deduced from their principles; and the theory of the comets was confirmed by more examples of the calculation of their orbits, done also with greater accuracy.

In this third edition the resistance of mediums is somewhat more largely handled than before; and new experiments of the resistance of heavy bodies falling in air are added. In the third book, the argument to prove that the moon is retained in its orbit by the force of gravity is enlarged on; and there are added new observations of Mr. Pound's of the proportion of the diameters of *Jupiter* to each other: there are, besides, added Mr. Kirk's observations of the comet in 1680; the orbit of that comet computed in an ellipsis by Dr. Halley; and the orbit of the comet in 1723, computed by Mr. Bradley.

BOOK I.

THE MATHEMATICAL PRINCIPLES OF NATURAL PHILOSOPHY.

THE

MATHEMATICAL PRINCIPLES

OF

NATURAL PHILOSOPHY.

DEFINITIONS.

DEFINITION I.

The quantity of matter is the measure of the same, arising from its density and bulk conjunctly.

THUS air of a double density, in a double space, is quadruple in quantity; in a triple space, sextuple in quantity. The same thing is to be understood of snow, and fine dust or powders, that are condensed by compression or liquefaction; and of all bodies that are by any causes whatever differently condensed. I have no regard in this place to a medium, if any such there is, that freely pervades the interstices between the parts of bodies. It is this quantity that I mean hereafter everywhere under the name of body or mass. And the same is known by the weight of each body; for it is proportional to the weight, as I have found by experiments on pendulums, very accurately made, which shall be shewn hereafter.

DEFINITION II.

The quantity of motion is the measure of the same, arising from the velocity and quantity of matter conjunctly.

The motion of the whole is the sum of the motions of all the parts; and therefore in a body double in quantity, with equal velocity, the motion is double; with twice the velocity, it is quadruple.

DEFINITION III.

The vis insita, *or innate force of matter, is a power of resisting, by which every body, as much as in it lies, endeavours to persevere in its present state, whether it be of rest, or of moving uniformly forward in a right line.*

This force is ever proportional to the body whose force it is; and differs nothing from the inactivity of the mass, but in our manner of conceiving it. A body, from the inactivity of matter, is not without difficulty put out of its state of rest or motion. Upon which account, this *vis insita*, may, by a most significant name, be called *vis inertiæ*, or force of inactivity. But a body exerts this force only, when another force, impressed upon it, endeavours to change its condition; and the exercise of this force may be considered both as resistance and impulse; it is resistance, in so far as the body, for maintaining its present state, withstands the force impressed; it is impulse, in so far as the body, by not easily giving way to the impressed force of another, endeavours to change the state of that other. Resistance is usually ascribed to bodies at rest, and impulse to those in motion; but motion and rest, as commonly conceived, are only relatively distinguished; nor are those bodies always truly at rest, which commonly are taken to be so.

DEFINITION IV.

An impressed force is an action exerted upon a body, in order to change its state, either of rest, or of moving uniformly forward in a right line.

This force consists in the action only; and remains no longer in the body, when the action is over. For a body maintains every new state it acquires, by its *vis inertiæ* only. Impressed forces are of different origins; as from percussion, from pressure, from centripetal force.

DEFINITION V.

A centripetal force is that by which bodies are drawn or impelled, or any way tend, towards a point as to a centre.

Of this sort is gravity, by which bodies tend to the centre of the earth; magnetism, by which iron tends to the loadstone; and that force, whatever it is, by which the planets are perpetually drawn aside from the rectilinear motions, which otherwise they would pursue, and made to revolve in curvilinear orbits. A stone, whirled about in a sling, endeavours to recede from the hand that turns it; and by that endeavour, distends the sling, and that with so much the greater force, as it is revolved with the greater velocity, and as soon as ever it is let go, flies away. That force which opposes itself to this endeavour, and by which the sling perpetually draws back the stone towards the hand, and retains it in its orbit, because it is directed to the hand as the centre of the orbit, I call the centripetal force. And the same thing is to be understood of all bodies, revolved in any orbits. They all endeavour to recede from the centres of their orbits; and were it not for the opposition of a contrary force which restrains them to, and detains them in their orbits, which I therefore call centripetal, would fly off in right lines, with an uniform motion. A projectile, if it was not for the force of gravity, would not deviate towards the earth, but would go off from it in a right line, and that with an uniform motion, if the resistance of the air was taken away. It is by its gravity that it is drawn aside perpetually from its rectilinear course, and made to deviate towards the earth, more or less, according to the force of its gravity, and the velocity of its motion. The less its gravity is, for the quantity of its matter, or the greater the velocity with which it is projected, the less will it deviate from a rectilinear course, and the farther it will go. If a leaden ball, projected from the top of a mountain by the force of gunpowder with a given velocity, and in a direction parallel to the horizon, is carried in a curve line to the distance of two miles before it falls to the ground; the same, if the resistance of the air were taken away, with a double or decuple velocity, would fly twice or ten times as far. And by increasing the velocity, we may at pleasure increase the distance to which it might be projected, and diminish the curvature of the line, which it might describe, till at last it should fall at the distance of 10, 30, or 90 degrees, or even might go quite round the whole earth before it falls; or lastly, so that it might never fall to the earth, but go forward into the celestial spaces, and proceed in its motion *in infinitum*. And after the same manner that a projectile, by the force of gravity, may be made to revolve in an orbit, and go round the whole earth, the moon also, either by the force of gravity, if it is endued with gravity, or by any other force, that impels it towards the earth, may be perpetually drawn aside towards the earth, out of the rectilinear way, which by its innate force it would pursue; and would be made to revolve in the orbit which it now describes; nor could the moon without some such force, be retained in its orbit. If this force was too small, it would not sufficiently turn the moon out of a rectilinear course: if it was too great, it would turn it too much, and draw down the moon from its orbit towards the earth. It is necessary, that the force be of a just quantity, and it belongs to the mathematicians to find the force, that may serve exactly to retain a body in a given orbit, with a given velocity; and *vice versa*, to determine the curvilinear way, into which a body projected from a given place, with a given velocity, may be made to deviate from its natural rectilinear way, by means of a given force.

The quantity of any centripetal force may be considered as of three kinds; absolute, accelerative, and motive.

DEFINITION VI.

The absolute quantity of a centripetal force is the measure of the same, proportional to the efficacy of the cause that propagates it from the centre, through the spaces round about.

Thus the magnetic force is greater in one load-stone and less in another, according to their sizes and strength of intensity.

DEFINITION VII.

The accelerative quantity of a centripetal force is the measure of the same, proportional to the velocity which it generates in a given time.

Thus the force of the same load-stone is greater at a less distance, and less at a greater: also the force of gravity is greater in valleys, less on tops of exceeding high mountains; and yet less (as shall hereafter be shown), at greater distances from the body of the earth; but at equal distances, it is the same everywhere; because (taking away, or allowing for, the resistance of the air), it equally accelerates all falling bodies, whether heavy or light, great or small.

DEFINITION VIII.

The motive quantity of a centripetal force, is the measure of the same, proportional to the motion which it generates in a given time.

Thus the weight is greater in a greater body, less in a less body; and, in the same body, it is greater near to the earth, and less at remoter distances. This sort of quantity is the centripetency, or propension of the whole body towards the centre, or, as I may say, its weight; and it is always known by the quantity of an equal and contrary force just sufficient to hinder the descent of the body.

These quantities of forces, we may, for brevity's sake, call by the names of motive, accelerative, and absolute forces; and, for distinction's sake, consider them, with respect to the bodies that tend to the centre; to the places of those bodies; and to the centre of force towards which they tend; that is to say, I refer the motive force to the body as an endeavour and propensity of the whole towards a centre, arising from the propensities of the several parts taken together; the accelerative force to the place of the body, as a certain power or energy diffused from the centre to all places around to move the bodies that are in them; and the absolute force to the centre, as endued with some cause, without which those motive forces would not be propagated through the spaces round about; whether that cause be some central body (such as is the load-stone, in the centre of the magnetic force, or the earth in the centre of the gravitating force), or anything else that does not yet appear. For I here design only to give a mathematical notion of those forces, without considering their physical causes and seats.

Wherefore the accelerative force will stand in the same relation to the motive, as celerity does to motion. For the quantity of motion arises from the celerity drawn into the quantity of matter; and the motive force arises from the accelerative force drawn into the same quantity of matter. For the sum of the actions of the accelerative force, upon the several particles of the body, is the motive force of the whole. Hence it is, that near the surface of the earth, where the accelerative gravity, or force productive of gravity, in all bodies is the same, the motive gravity or the weight is as the body: but if we should ascend to higher regions, where the accelerative gravity is less, the weight would be equally diminished, and would always be as the product of the body, by the accelerative gravity. So in those regions, where the accelerative gravity is diminished into one half, the weight of a body two or three times less, will be four or six times less.

I likewise call attractions and impulses, in the same sense, accelerative, and motive; and use the words attraction, impulse or propensity of any sort towards a centre, promiscuously, and indifferently, one for another; considering those forces not physically, but mathematically: wherefore, the reader is not to imagine, that by those words, I anywhere take upon me to define the kind, or the manner of any action, the causes or the physical reason thereof, or that I attribute forces, in a true and physical sense, to certain centres (which are only mathematical points); when at any time I happen to speak of centres as attracting, or as endued with attractive powers.

SCHOLIUM.

Hitherto I have laid down the definitions of such words as are less known, and explained the sense in which I would have them to be understood in the following discourse. I do not define time, space, place and motion, as being well known to all. Only I must observe, that the vulgar conceive those quantities under no other notions but from the relation they bear to sensible objects. And thence arise certain prejudices, for the removing of which, it will be convenient to distinguish them into absolute and relative, true and apparent, mathematical and common.

I. Absolute, true, and mathematical time, of itself, and from its own nature flows equably without regard to anything external, and by another name is called duration: relative, apparent, and common time, is some sensible and external (whether accurate or unequable) measure of duration by the means of motion, which is commonly used instead of true time; such as an hour, a day, a month, a year.

II. Absolute space, in its own nature, without regard to anything external, remains always similar and immovable. Relative space is some movable dimension or measure of the absolute spaces; which our senses determine by its position to bodies; and which is vulgarly taken for immovable space; such is the dimension of a subterraneous, an æreal, or celestial space, determined by its position in respect of the earth. Absolute and relative space, are the same in figure and magnitude; but they do not remain always numerically the same. For if the earth, for instance, moves, a space of our air, which relatively and in respect of the earth remains always the same, will at one time be one part of the absolute space into which the air passes; at another time it will be another part of the same, and so, absolutely understood, it will be perpetually mutable.

III. Place is a part of space which a body takes up, and is according to the space, either absolute or relative. I say, a part of space; not the situation, nor the external surface of the body. For the places of equal solids are always equal; but their superfices, by reason of their dissimilar figures, are often unequal. Positions properly have no quantity, nor are they so much the places themselves, as the properties of places. The motion of the whole is the same thing with the sum of the motions of the parts; that is, the translation of the whole, out of its place, is the same thing with the sum of the translations of the parts out of their places; and therefore the place of the whole is the same thing with the sum of the places of the parts, and for that reason, it is internal, and in the whole body.

IV. Absolute motion is the translation of a body from one absolute place into another; and relative motion, the translation from one relative place into another. Thus in a ship under sail, the relative place of a body is that part of the ship which the body possesses; or that part of its cavity which the body fills, and which therefore moves together with the ship: and relative rest is the continuance of the body in the same part of the ship, or of its cavity. But real, absolute rest, is the continuance of the body in the same part of that immovable space, in which the ship itself, its cavity, and all that it contains, is moved. Wherefore, if the earth is really at rest, the body, which relatively rests in the ship, will really and absolutely move with the same velocity which the ship has on the earth. But if the earth also moves, the true and absolute motion of the body will arise, partly from the true motion of the earth, in immovable space; partly from the relative motion of the ship on the earth; and if the body moves also relatively in the ship; its true motion will arise, partly from the true motion of the earth, in immovable space, and partly from the relative motions as well of the ship on the earth, as of the body in the ship; and from these relative motions will arise the relative motion of the body on the earth. As if that part of the earth, where the ship is, was truly moved toward the east, with a velocity of 10010 parts; while the ship itself, with a fresh gale, and full sails, is carried towards the west, with a velocity expressed by 10 of those parts; but a sailor walks in the ship towards the east, with 1 part of the said velocity; then the sailor will be moved truly in immovable space towards the east, with a velocity of 10001 parts, and relatively on the earth towards the west, with a velocity of 9 of those parts.

Absolute time, in astronomy, is distinguished from relative, by the equation or correction of the vulgar time. For the natural days are truly unequal, though they are commonly considered as equal, and used for a measure of time; astronomers correct this inequality for their more accurate deducing of the celestial motions. It may be, that there is no such thing as an equable motion, whereby time may be accurately measured. All motions may be accelerated and retarded, but the true, or equable, progress of absolute time is liable to no change. The duration or perseverance of the existence of things remains the same, whether the motions are swift or slow, or none at all: and therefore it ought to be distinguished from what are only sensible measures thereof; and out of which we collect it, by means of the astronomical equation. The necessity of which equation, for determining the times of a phænomenon, is evinced as well from the experiments of the pendulum clock, as by eclipses of the satellites of *Jupiter*.

As the order of the parts of time is immutable, so also is the order of the parts of space. Suppose those parts to be moved out of their places, and they will be moved (if the expression may be allowed) out of themselves. For times and spaces are, as it were, the places as well of themselves as of all other things. All things are placed in time as to order of succession; and in space as to order of situation. It is from their essence or nature that they are places; and that the primary places of things should be moveable, is absurd. These are therefore the absolute places; and translations out of those places, are the only absolute motions.

But because the parts of space cannot be seen, or distinguished from one another by our senses, therefore in their stead we use sensible measures of them. For from the positions and distances of things from any body considered as immovable, we define all places; and then with respect to such places, we estimate all motions, considering bodies as transferred from some of those places into others. And so, instead of absolute places and motions, we use relative ones; and that without any inconvenience in common affairs; but in philosophical disquisitions, we ought to abstract from our senses, and consider things themselves, distinct from what are only sensible measures of them. For it may be that there is no body really at rest, to which the places and motions of others may be referred.

But we may distinguish rest and motion, absolute and relative, one from the other by their properties, causes and effects. It is a property of rest, that bodies really at rest do rest in respect to one another. And therefore as it is possible, that in the remote regions of the fixed stars, or perhaps far beyond them, there may be some body absolutely at rest; but impossible to know, from the position of bodies to one another in our regions. whether any of these do keep the same position to that remote body; it follows that absolute rest cannot be determined from the position of bodies in our regions.

It is a property of motion, that the parts, which retain given positions to their wholes, do partake of the motions of those wholes. For all the parts of revolving bodies endeavour to recede from the axis of motion; and the impetus of bodies moving forward, arises from the joint impetus of all the parts. Therefore, if surrounding bodies are moved, those that are relatively at rest within them, will partake of their motion. Upon which account, the true and absolute motion of a body cannot be determined by the translation of it from those which only seem to rest; for the external bodies ought not only to appear at rest, but to be really at rest. For otherwise, all included bodies, beside their translation from near the surrounding ones, partake likewise of their true motions; and though that translation were not made they would not be really at rest, but only seem to be so. For the surrounding bodies stand in the like relation to the surrounded as the exterior part of a whole does to the interior, or as the shell does to the kernel; but, if the shell moves, the kernel will also move, as being part of the whole, without any removal from near the shell.

A property, near akin to the preceding, is this, that if a place is moved, whatever is placed therein moves along with it; and therefore a body, which is moved from a place in motion, partakes also of the motion of its place. Upon which account, all motions, from places in motion, are no other than parts of entire and absolute motions; and every entire motion is composed of the motion of the body out of its first place, and the motion of this place out of its place; and so on, until we come to some immovable place, as in the before-mentioned example of the sailor. Wherefore, entire and absolute motions can be no otherwise determined than by immovable places; and for that reason I did before refer those absolute motions to immovable places, but relative ones to movable places. Now no other places are immovable but those that, from infinity to infinity, do all retain the same given position one to another; and upon this account must ever remain unmoved; and do thereby constitute immovable space.

The causes by which true and relative motions are distinguished, one from the other, are the forces impressed upon bodies to generate motion. True motion is neither generated nor altered, but by some force impressed upon the body moved; but relative motion may be generated or altered without any force impressed upon the body. For it is sufficient only to impress some force on other bodies with which the former is compared, that by their giving way, that relation may be changed, in which the relative rest or motion of this other body did consist. Again, true motion suffers always some change from any force impressed upon the moving body; but relative motion does not necessarily undergo any change by such forces. For if the same forces are likewise impressed on those other bodies, with which the comparison is made, that the relative position may be preserved, then that condition will be preserved in which the relative motion consists. And therefore any relative motion may be changed when the true motion remains unaltered, and the relative may be preserved when the true suffers some change. Upon which accounts, true motion does by no means consist in such relations.

The effects which distinguish absolute from relative motion are, the forces of receding from the axis of circular motion. For there are no such forces in a circular motion purely relative, but in a true and absolute circular motion, they are greater or less, according to the quantity of the motion. If a vessel, hung by a long cord, is so often turned about that the cord is strongly twisted, then filled with water, and held at rest together with the water; after, by the sudden action of another force, it is whirled about the contrary way, and while the cord is untwisting itself, the vessel continues for some time in this motion; the surface of the water will at first be plain, as before the vessel began to move; but the vessel, by gradually communicating its motion to the water, will make it begin sensibly to revolve, and recede by little and little from the middle, and ascend to the sides of the vessel, forming itself into a concave figure (as I have experienced), and the swifter the motion becomes, the higher will the water rise, till at last, performing its revolutions in the same times with the vessel, it becomes relatively at rest in it. This ascent of the water shows its endeavour to recede from the axis of its motion; and the true and absolute circular motion of the water, which is here directly contrary to the relative, discovers itself, and may be measured by this endeavour. At first, when the relative motion of the water in the vessel was greatest, it produced no endeavour to recede from the axis; the water showed no tendency to the circumference, nor any ascent towards the sides of the vessel, but remained of a plain surface, and therefore its true circular motion had not yet begun. But afterwards, when the relative motion of the water had decreased, the ascent thereof towards the sides of the vessel proved its endeavour to recede from the axis; and this endeavour showed the real circular motion of the water perpetually increasing, till it had acquired its greatest quantity, when the water rested relatively in the vessel. And therefore this endeavour does not depend upon any translation of the water in respect of the ambient bodies, nor can true circular motion be defined by such translation. There is only one real circular motion of any one revolving body, corresponding to only one power of endeavouring to recede from its axis of motion, as its proper and adequate effect; but relative motions, in one and the same body, are innumerable, according to the various relations it bears to external bodies, and like other relations, are altogether destitute of any real effect, any otherwise than they may perhaps partake of that one only true motion. And therefore in their system who suppose that our heavens, revolving below the sphere of the fixed stars, carry the planets along with them; the several parts of those heavens, and the planets, which are indeed relatively at rest in their heavens, do yet really move. For they change their position one to another (which never happens to bodies truly at rest), and being carried together with their heavens, partake of their motions, and as parts of revolving wholes, endeavour to recede from the axis of their motions.

Wherefore relative quantities are not the quantities themselves, whose names they bear, but those sensible measures of them (either accurate or inaccurate), which are commonly used instead of the measured quantities themselves. And if the meaning of words is to be determined by their use, then by the names time, space, place and motion, their measures are properly to be understood; and the expression will be unusual, and purely mathematical, if the measured quantities themselves are meant. Upon which account, they do strain the sacred writings, who there interpret those words for the measured quantities. Nor do those less defile the purity of mathematical and philosophical truths, who confound real quantities themselves with their relations and vulgar measures.

It is indeed a matter of great difficulty to discover, and effectually to distinguish, the true motions of particular bodies from the apparent; because the parts of that immovable space, in which those motions are performed, do by no means come under the observation of our senses. Yet the thing is not altogether desperate; for we have some arguments to guide us, partly from the apparent motions, which are the differences of the true motions; partly from the forces, which are the causes and effects of the true motions. For instance, if two globes, kept at a given distance one from the other by means of a cord that connects them, were revolved about their common centre of gravity, we might, from the tension of the cord, discover the endeavour of the globes to recede from the axis of their motion, and from thence we might compute the quantity of their circular motions. And then if any equal forces should be impressed at once on the alternate faces of the globes to augment or diminish their circular motions, from the increase or decrease of the tension of the cord, we might infer the increment or decrement of their motions; and thence would be found on what faces those forces ought to be impressed, that the motions of the globes might be most augmented; that is, we might discover their hindermost faces, or those which, in the circular motion, do follow. But the faces which follow being known, and consequently the opposite ones that precede, we should likewise know the determination of their motions. And thus we might find both the quantity and the determination of this circular motion, even in an immense vacuum, where there was nothing external or sensible with which the globes could be compared. But now, if in that space some remote bodies were placed that kept always a given position one to another, as the fixed stars do in our regions, we could not indeed determine from the relative translation of the globes among those bodies, whether the motion did belong to the globes or to the bodies. But if we observed the cord, and found that its tension was that very tension which the motions of the globes required, we might conclude the motion to be in the globes, and the bodies to be at rest; and then, lastly, from the translation of the globes among the bodies, we should find the determination of their motions. But how we are to collect the true motions from their causes, effects, and apparent differences; and, *vice versa*, how from the motions, either true or apparent, we may come to the knowledge of their causes and effects, shall be explained more at large in the following tract. For to this end it was that I composed it.

AXIOMS, OR LAWS OF MOTION.

LAW I.

Every body perseveres in its state of rest, or of uniform motion in a right line, unless it is compelled to change that state by forces impressed thereon.

PROJECTILES persevere in their motions, so far as they are not retarded by the resistance of the air, or impelled downwards by the force of gravity. A top, whose parts by their cohesion are perpetually drawn aside from rectilinear motions, does not cease its rotation, otherwise than as it is retarded by the air. The greater bodies of the planets and comets, meeting with less resistance in more free spaces, preserve their motions both progressive and circular for a much longer time.

LAW II.

The alteration of motion is ever proportional to the motive force impressed; and is made in the direction of the right line in which that force is impressed.

If any force generates a motion, a double force will generate double the motion, a triple force triple the motion, whether that force be impressed altogether and at once, or gradually and successively. And this motion (being always directed the same way with the generating force), if the body moved before, is added to or subducted from the former motion, according as they directly conspire with or are directly contrary to each other; or obliquely joined, when they are oblique, so as to produce a new motion compounded from the determination of both.

LAW III.

To every action there is always opposed an equal reaction: or the mutual actions of two bodies upon each other are always equal, and directed to contrary parts.

Whatever draws or presses another is as much drawn or pressed by that other. If you press a stone with your finger, the finger is also pressed by the stone. If a horse draws a stone tied to a rope, the horse (if I may so say) will be equally drawn back towards the stone: for the distended rope, by the same endeavour to relax or unbend itself, will draw the horse as much towards the stone, as it does the stone towards the horse, and will obstruct the progress of the one as much as it advances that of the other.

If a body impinge upon another, and by its force change the motion of the other, that body also (because of the equality of the mutual pressure) will undergo an equal change, in its own motion, towards the contrary part. The changes made by these actions are equal, not in the velocities but in the motions of bodies; that is to say, if the bodies are not hindered by any other impediments. For, because the motions are equally changed, the changes of the velocities made towards contrary parts are reciprocally proportional to the bodies. This law takes place also in attractions, as will be proved in the next scholium.

COROLLARY I.

A body by two forces conjoined will describe the diagonal of a parallelogram, in the same time that it would describe the sides, by those forces apart.

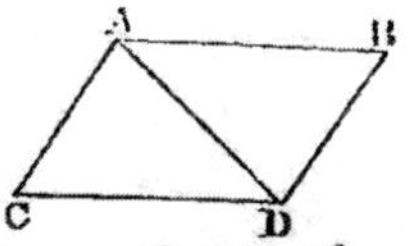

If a body in a given time, by the force M impressed apart in the place A, should with an uniform motion be carried from A to B; and by the force N impressed apart in the same place, should be carried from A to C; complete the parallelogram ABCD, and, by both forces acting together, it will in the same time be carried in the diagonal from A to D. For since the force N acts in the direction of the line AC, parallel to BD, this force (by the second law) will not at all alter the velocity generated by the other force M, by which the body is carried towards the line BD. The body therefore will arrive at the line BD in the same time, whether the force N be impressed or not; and therefore at the end of that time it will be found somewhere in the line BD. By the same argument, at the end of the same time it will be found somewhere in the line CD. Therefore it will be found in the point D, where both lines meet. But it will move in a right line from A to D, by Law I.

COROLLARY II.

And hence is explained the composition of any one direct force AD, out of any two oblique forces AC and CD; and, on the contrary, the resolution of any one direct force AD into two oblique forces AC and CD: which composition and resolution are abundantly confirmed from mechanics.

As if the unequal radii OM and ON drawn from the centre O of any wheel, should sustain the weights A and P by the cords MA and NP; and the forces of those weights to move the wheel were required. Through the centre O draw the right line KOL, meeting the cords perpendicularly in K and L; and from the centre O, with OL the greater of the distances

OK and OL, describe a circle, meeting the cord MA in D: and drawing OD, make AC parallel and DC perpendicular thereto. Now, it being indifferent whether the points K, L, D, of the cords be fixed to the plane of the wheel or not, the weights will have the same effect whether they are suspended from the points K and L, or from D and L. Let the whole force of the weight A be represented by the line AD, and let it be resolved into the forces AC and CD; of which the force AC, drawing the radius OD directly from the centre, will have no effect to move the wheel: but the other force DC, drawing the radius DO perpendicularly, will have the same effect as if it drew perpendicularly the radius OL equal to OD; that is, it will have the same effect as the weight P, if that weight is to the weight A as the force DC is to the force DA; that is (because of the similar triangles ADC, DOK), as OK to OD or OL. Therefore the weights A and P, which are reciprocally as the radii OK and OL that lie in the same right line, will be equipollent, and so remain in equilibrio; which is the well known property of the balance, the lever, and the wheel. If either weight is greater than in this ratio, its force to move the wheel will be so much greater.

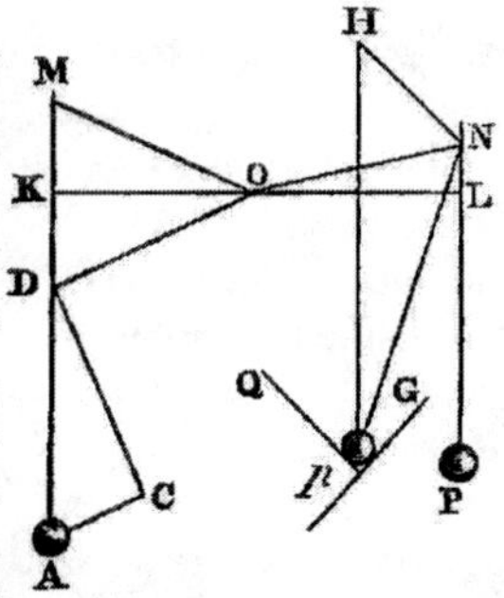

If the weight *p*, equal to the weight P, is partly suspended by the cord N*p*, partly sustained by the oblique plane *p*G; draw *p*H, NH, the former perpendicular to the horizon, the latter to the plane *p*G; and if the force of the weight *p* tending downwards is represented by the line *p*H, it may be resolved into the forces *p*N, HN. If there was any plane *p*Q, perpendicular to the cord *p*N, cutting the other plane *p*G in a line parallel to the horizon, and the weight *p* was supported only by those planes *p*Q, *p*G, it would press those planes perpendicularly with the forces *p*N, HN; to wit, the plane *p*Q with the force *p*N, and the plane *p*G with the force HN. And therefore if the plane *p*Q was taken away, so that the weight might stretch the cord, because the cord, now sustaining the weight, supplies the place of the plane that was removed, it will be strained by the same force *p*N which pressed upon the plane before. Therefore, the tension of this oblique cord *p*N will be to that of the other perpendicular cord PN as *p*N to *p*H. And therefore if the weight *p* is to the weight A in a ratio compounded of the reciprocal ratio of the least distances of the cords PN, AM, from the centre of the wheel, and of the direct ratio of *p*H to *p*N, the weights will have the same effect towards moving the wheel, and will therefore sustain each other; as any one may find by experiment.

But the weight *p* pressing upon those two oblique planes, may be considered as a wedge between the two internal surfaces of a body split by it; and hence the forces of the wedge and the mallet may be determined; for because the force with which the weight p presses the plane pQ is to the force with which the same, whether by its own gravity, or by the blow of a mallet, is impelled in the direction of the line pH towards both the planes, as pN to pH; and to the force with which it presses the other plane pG, as pN to NH. And thus the force of the screw may be deduced from a like resolution of forces; it being no other than a wedge impelled with the force of a lever. Therefore the use of this Corollary spreads far and wide, and by that diffusive extent the truth thereof is farther confirmed. For on what has been said depends the whole doctrine of mechanics variously demonstrated by different authors. For from hence are easily deduced the forces of machines, which are compounded of wheels, pullies, levers, cords, and weights, ascending directly or obliquely, and other mechanical powers; as also the force of the tendons to move the bones of animals.

COROLLARY III.

The quantity of motion, which is collected by taking the sum of the motions directed towards the same parts, and the difference of those that are directed to contrary parts, suffers no change from the action of bodies among themselves.

For action and its opposite re-action are equal, by Law III, and therefore, by Law II, they produce in the motions equal changes towards opposite parts. Therefore if the motions are directed towards the same parts, whatever is added to the motion of the preceding body will be subducted from the motion of that which follows; so that the sum will be the same as before. If the bodies meet, with contrary motions, there will be an equal deduction from the motions of both; and therefore the difference of the motions directed towards opposite parts will remain the same.

Thus if a spherical body A with two parts of velocity is triple of a spherical body B which follows in the same right line with ten parts of velocity, the motion of A will be to that of B as 6 to 10. Suppose, then, their motions to be of 6 parts and of 10 parts, and the sum will be 16 parts. Therefore, upon the meeting of the bodies, if A acquire 3, 4, or 5 parts of motion, B will lose as many; and therefore after reflexion A will proceed with 9, 10, or 11 parts, and B with 7, 6, or 5 parts; the sum remaining always of 16 parts as before. If the body A acquire 9, 10, 11, or 12 parts of motion, and therefore after meeting proceed with 15, 16, 17, or 18 parts, the body B, losing so many parts as A has got, will either proceed with 1 part, having lost 9, or stop and remain at rest, as having lost its whole progressive motion of 10 parts; or it will go back with 1 part, having not only lost its whole motion, but (if I may so say) one part more; or it will go back with 2 parts, because a progressive motion of 12 parts is taken off. And so the sums of the conspiring motions 15+1, or 16+0, and the differences of the contrary motions 17—1 and

18—2, will always be equal to 16 parts, as they were before the meeting and reflexion of the bodies. But, the motions being known with which the bodies proceed after reflexion, the velocity of either will be also known, by taking the velocity after to the velocity before reflexion, as the motion after is to the motion before. As in the last case, where the motion of the body A was of 6 parts before reflexion and of 18 parts after, and the velocity was of 2 parts before reflexion, the velocity thereof after reflexion will be found to be of 6 parts; by saying, as the 6 parts of motion before to 18 parts after, so are 2 parts of velocity before reflexion to 6 parts after.

But if the bodies are either not spherical, or, moving in different right lines, impinge obliquely one upon the other, and their motions after reflexion are required, in those cases we are first to determine the position of the plane that touches the concurring bodies in the point of concourse; then the motion of each body (by Corol. II) is to be resolved into two, one perpendicular to that plane, and the other parallel to it. This done, because the bodies act upon each other in the direction of a line perpendicular to this plane, the parallel motions are to be retained the same after reflexion as before; and to the perpendicular motions we are to assign equal changes towards the contrary parts; in such manner that the sum of the conspiring and the difference of the contrary motions may remain the same as before. From such kind of reflexions also sometimes arise the circular motions of bodies about their own centres. But these are cases which I do not consider in what follows; and it would be too tedious to demonstrate every particular that relates to this subject.

COROLLARY IV.

The common centre of gravity of two or more bodies does not alter its state of motion or rest by the actions of the bodies among themselves; and therefore the common centre of gravity of all bodies acting upon each other (excluding outward actions and impediments) is either at rest, or moves uniformly in a right line.

For if two points proceed with an uniform motion in right lines, and their distance be divided in a given ratio, the dividing point will be either at rest, or proceed uniformly in a right line. This is demonstrated hereafter in Lem. XXIII and its Corol., when the points are moved in the same plane; and by a like way of arguing, it may be demonstrated when the points are not moved in the same plane. Therefore if any number of bodies move uniformly in right lines, the common centre of gravity of any two of them is either at rest, or proceeds uniformly in a right line; because the line which connects the centres of those two bodies so moving is divided at that common centre in a given ratio. In like manner the common centre of those two and that of a third body will be either at rest or moving uniformly in a right line; because at that centre the distance between the common centre of the two bodies, and the centre of this last, is divided in a given ratio. In like manner the common centre of these three, and of a fourth body, is either at rest, or moves uniformly in a right line; because the distance between the common centre of the three bodies, and the centre of the fourth is there also divided in a given ratio, and so on *in infinitum*. Therefore, in a system of bodies where there is neither any mutual action among themselves, nor any foreign force impressed upon them from without, and which consequently move uniformly in right lines, the common centre of gravity of them all is either at rest or moves uniformly forward in a right line.

Moreover, in a system of two bodies mutually acting upon each other, since the distances between their centres and the common centre of gravity of both are reciprocally as the bodies, the relative motions of those bodies, whether of approaching to or of receding from that centre, will be equal among themselves. Therefore since the changes which happen to motions are equal and directed to contrary parts, the common centre of those bodies, by their mutual action between themselves, is neither promoted nor retarded, nor suffers any change as to its state of motion or rest. But in a system of several bodies, because the common centre of gravity of any two acting mutually upon each other suffers no change in its state by that action; and much less the common centre of gravity of the others with which that action does not intervene; but the distance between those two centres is divided by the common centre of gravity of all the bodies into parts reciprocally proportional to the total sums of those bodies whose centres they are; and therefore while those two centres retain their state of motion or rest, the common centre of all does also retain its state: it is manifest that the common centre of all never suffers any change in the state of its motion or rest from the actions of any two bodies between themselves. But in such a system all the actions of the bodies among themselves either happen between two bodies, or are composed of actions interchanged between some two bodies; and therefore they do never produce any alteration in the common centre of all as to its state of motion or rest. Wherefore since that centre, when the bodies do not act mutually one upon another, either is at rest or moves uniformly forward in some right line, it will, notwithstanding the mutual actions of the bodies among themselves, always persevere in its state, either of rest, or of proceeding uniformly in a right line, unless it is forced out of this state by the action of some power impressed from without upon the whole system. And therefore the same law takes place in a system consisting of many bodies as in one single body, with regard to their persevering in their state of motion or of rest. For the progressive motion, whether of one single body, or of a whole system of bodies, is always to be estimated from the motion of the centre of gravity.

COROLLARY V.

The motions of bodies included in a given space are the same among themselves, whether that space is at rest, or moves uniformly forwards in a right line without any circular motion.

For the differences of the motions tending towards the same parts, and the sums of those that tend towards contrary parts, are, at first (by supposition), in both cases the same; and it is from those sums and differences that the collisions and impulses do arise with which the bodies mutually impinge one upon another. Wherefore (by Law II), the effects of those collisions will be equal in both cases; and therefore the mutual motions of the bodies among themselves in the one case will remain equal to the mutual motions of the bodies among themselves in the other. A clear proof of which we have from the experiment of a ship; where all motions happen after the same manner, whether the ship is at rest, or is carried uniformly forwards in a right line.

COROLLARY VI.

If bodies, any how moved among themselves, are urged in the direction of parallel lines by equal accelerative forces, they will all continue to move among themselves, after the same manner as if they had been urged by no such forces.

For these forces acting equally (with respect to the quantities of the bodies to be moved), and in the direction of parallel lines, will (by Law II) move all the bodies equally (as to velocity), and therefore will never produce any change in the positions or motions of the bodies among themselves.

SCHOLIUM.

Hitherto I have laid down such principles as have been received by mathematicians, and are confirmed by abundance of experiments. By the first two Laws and the first two Corollaries, Galileo discovered that the descent of bodies observed the duplicate ratio of the time, and that the motion of projectiles was in the curve of a parabola; experience agreeing with both, unless so far as these motions are a little retarded by the resistance of the air. When a body is falling, the uniform force of its gravity acting equally, impresses, in equal particles of time, equal forces upon that body, and therefore generates equal velocities; and in the whole time impresses a whole force, and generates a whole velocity proportional to the time. And the spaces described in proportional times are as the velocities and the times conjunctly; that is, in a duplicate ratio of the times. And when a body is thrown upwards, its uniform gravity impresses forces and takes off velocities proportional to the times; and the times of ascending to the greatest heights are as the velocities to be taken off, and those heights are as the velocities and the times conjunctly, or in the duplicate ratio of the velocities. And if a body be projected in any direction, the motion arising from its projection is compounded with the motion arising from its gravity. As if the body A by its motion of projection alone could describe in a given time the right line AB, and with its motion of falling alone could describe in the same time the altitude AC; complete the paralellogram ABDC, and the body by that compounded motion will at the end of the time be found in the place D; and the curve line AED, which that body describes, will be a parabola, to which the right line AB will be a tangent in A; and whose ordinate BD will be as the square of the line AB.

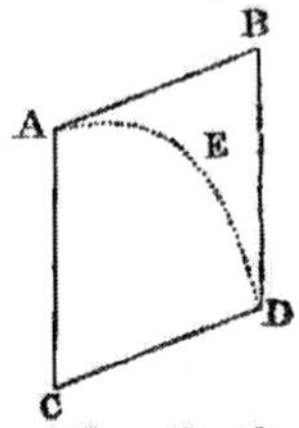

On the same Laws and Corollaries depend those things which have been demonstrated concerning the times of the vibration of pendulums, and are confirmed by the daily experiments of pendulum clocks. By the same, together with the third Law, Sir Christ. Wren, Dr. Wallis, and Mr. Huygens, the greatest geometers of our times, did severally determine the rules of the congress and reflexion of hard bodies, and much about the same time communicated their discoveries to the Royal Society, exactly agreeing among themselves as to those rules. Dr. Wallis, indeed, was something more early in the publication; then followed Sir Christopher Wren, and, lastly, Mr. Huygens. But Sir Christopher Wren confirmed the truth of the thing before the Royal Society by the experiment of pendulums, which Mr. Mariotte soon after thought fit to explain in a treatise entirely upon that subject. But to bring this experiment to an accurate agreement with the theory, we are to have a due regard as well to the resistance of the air as to the elastic force of the concurring bodies. Let the spherical bodies A, B be suspended by the parallel and equal strings AC, BD, from the centres C, D. About these centres, with those intervals, describe the semicircles EAF, GBH, bisected by the radii CA, DB. Bring the body A to any point R of the arc EAF, and (withdrawing the body B) let it go from thence, and after one oscillation suppose it to return to the point V: then RV will be the retardation arising from the resistance of the air. Of this RV let ST be a fourth part, situated in the middle, to wit, so as RS and TV may be equal, and RS may be to ST as 3 to 2: then will ST represent very nearly the retardation during the descent from S to A. Restore the body B to its place: and, supposing the body A to be let fall from the point S, the velocity thereof in the place of reflexion A, without sensible error, will be the same as if it had descended *in vacuo* from the point T. Upon which account this velocity may be represented by the chord of the arc TA. For it is a proposition well known to geometers, that the velocity of a pendulous body in the lowest point is as the chord of the arc which it has described in its descent. After

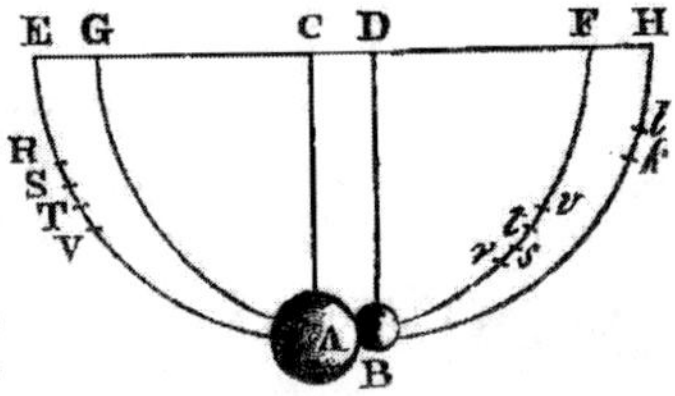

reflexion, suppose the body A comes to the place *s*, and the body B to the place *k*. Withdraw the body B, and find the place *v*, from which if the body A, being let go, should after one oscillation return to the place *r*, *st* may be a fourth part of *rv*, so placed in the middle thereof as to leave *rs* equal to *tv*, and let the chord of the arc *t*A represent the velocity which the body A had in the place A immediately after reflexion. For *t* will be the true and correct place to which the body A should have ascended, if the resistance of the air had been taken off. In the same way we are to correct the place *k* to which the body B ascends, by finding the place *l* to which it should have ascended *in vacuo*. And thus everything may be subjected to experiment, in the same manner as if we were really placed *in vacuo*. These things being done, we are to take the product (if I may so say) of the body A, by the chord of the arc TA (which represents its velocity), that we may have its motion in the place A immediately before reflexion; and then by the chord of the arc *t*A, that we may have its motion in the place A immediately after reflexion. And so we are to take the product of the body B by the chord of the arc B*l*, that we may have the motion of the same immediately after reflexion. And in like manner, when two bodies are let go together from different places, we are to find the motion of each, as well before as after reflexion; and then we may compare the motions between themselves, and collect the effects of the reflexion. Thus trying the thing with pendulums of ten feet, in unequal as well as equal bodies, and making the bodies to concur after a descent through large spaces, as of 8, 12, or 16 feet, I found always, without an error of 3 inches, that when the bodies concurred together directly, equal changes towards the contrary parts were produced in their motions, and, of consequence, that the action and reaction were always equal. As if the body A impinged upon the body B at rest with 9 parts of motion, and losing 7, proceeded after reflexion with 2, the body B was carried backwards with those 7 parts. If the bodies concurred with contrary motions, A with twelve parts of motion, and B with six, then if A receded with 2, B receded with 8; to wit, with a deduction of 14 parts of motion on each side. For from the motion of A subducting twelve parts, nothing will remain; but subducting 2 parts more, a motion will be generated of 2 parts towards the contrary way; and so, from the motion of the body B of 6 parts, subducting 14 parts, a motion is generated of 8 parts towards the contrary way. But if the bodies were made both to move towards the same way, A, the swifter, with 14 parts of motion, B, the slower, with 5, and after reflexion A went on with 5, B likewise went on with 14 parts; 9 parts being transferred from A to B. And so in other cases. By the congress and collision of bodies, the quantity of motion, collected from the sum of the motions directed towards the same way, or from the difference of those that were directed towards contrary ways, was never changed. For the error of an inch or two in measures may be easily ascribed to the difficulty of executing everything with accuracy. It was not easy to let go the two pendulums so exactly together that the bodies should impinge one upon the other in the lowermost place AB; nor to mark the places *s*, and *k*, to which the bodies ascended after congress. Nay, and some errors, too, might have happened from the unequal density of the parts of the pendulous bodies themselves, and from the irregularity of the texture proceeding from other causes.

But to prevent an objection that may perhaps be alledged against the rule, for the proof of which this experiment was made, as if this rule did suppose that the bodies were either absolutely hard, or at least perfectly elastic (whereas no such bodies are to be found in nature), I must add, that the experiments we have been describing, by no means depending upon that quality of hardness, do succeed as well in soft as in hard bodies. For if the rule is to be tried in bodies not perfectly hard, we are only to diminish the reflexion in such a certain proportion as the quantity of the elastic force requires. By the theory of Wren and Huygens, bodies absolutely hard return one from another with the same velocity with which they meet. But this may be affirmed with more certainty of bodies perfectly elastic. In bodies imperfectly elastic the velocity of the return is to be diminished together with the elastic force; because that force (except when the parts of bodies are bruised by their congress, or suffer some such extension as happens under the strokes of a hammer) is (as far as I can perceive) certain and determined, and makes the bodies to return one from the other with a relative velocity, which is in a given ratio to that relative velocity with which they met. This I tried in balls of wool, made up tightly, and strongly compressed. For, first, by letting go the pendulous bodies, and measuring their reflexion, I determined the quantity of their elastic force; and then, according to this force, estimated the reflexions that ought to happen in other cases of congress. And with this computation other experiments made afterwards did accordingly agree; the balls always receding one from the other with a relative velocity, which was to the relative velocity with which they met as about 5 to 9. Balls of steel returned with almost the same velocity: those of cork with a velocity something less; but in balls of glass the proportion was as about 15 to 16. And thus the third Law, so far as it regards percussions and reflexions, is proved by a theory exactly agreeing with experience.

In attractions, I briefly demonstrate the thing after this manner. Suppose an obstacle is interposed to hinder the congress of any two bodies A, B, mutually attracting one the other: then if either body, as A, is more attracted towards the other body B, than that other body B is towards the first body A, the obstacle will be more strongly urged by the pressure of the body A than by the pressure of the body B, and therefore will not remain in equilibrio: but the stronger pressure will prevail, and will make the system of the two bodies, together with the obstacle, to move directly towards the parts on which B lies; and in free spaces, to go forward *in infinitum* with a motion perpetually accelerated; which is absurd and contrary to the first Law. For, by the first Law, the system ought to persevere in its state of rest, or of moving uniformly forward in a right line; and therefore the bodies must equally press the obstacle, and be equally attracted one by the other. I made the experiment on the loadstone and iron. If these, placed apart in proper vessels, are made to float by one another in standing water, neither of them will propel the other; but, by being equally attracted, they will sustain each other's pressure, and rest at last in an equilibrium.

So the gravitation betwixt the earth and its parts is mutual. Let the earth FI be cut by any plane EG into two parts EGF and EGI, and their weights one towards the other will be mutually equal. For if by another plane HK, parallel to the former EG, the greater part EGI is cut into two parts EGKH and HKI, whereof HKI is equal to the part EFG, first cut off, it is evident that the middle part EGKH, will have no propension by its proper weight towards either side, but will hang as it were, and rest in an equilibrium betwixt both. But the one extreme part HKI will with its whole weight bear upon and press the middle part towards the other extreme part EGF; and therefore the force with which EGI, the sum of the parts HKI and EGKH, tends towards the third part EGF, is equal to the weight of the part HKI, that is, to the weight of the third part EGF. And therefore the weights of the two parts EGI and EGF, one towards the other, are equal, as I was to prove. And indeed if those weights were not equal, the whole earth floating in the non-resisting æther would give way to the greater weight, and, retiring from it, would be carried off *in infinitum*.

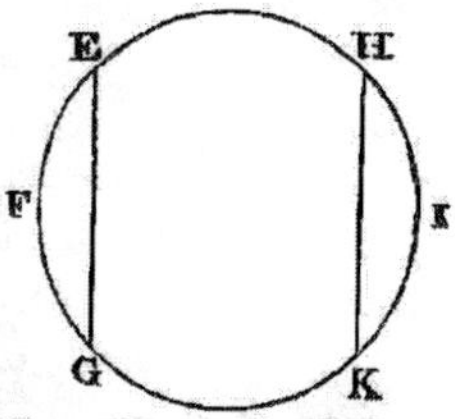

And as those bodies are equipollent in the congress and reflexion, whose velocities are reciprocally as their innate forces, so in the use of mechanic instruments those agents are equipollent, and mutually sustain each the contrary pressure of the other, whose velocities, estimated according to the determination of the forces, are reciprocally as the forces.

So those weights are of equal force to move the arms of a balance; which during the play of the balance are reciprocally as their velocities upwards and downwards; that is, if the ascent or descent is direct, those weights are of equal force, which are reciprocally as the distances of the points at which they are suspended from the axis of the balance; but if they are turned aside by the interposition of oblique planes, or other obstacles, and made to ascend or descend obliquely, those bodies will be equipollent, which are reciprocally as the heights of their ascent and descent taken according to the perpendicular; and that on account of the determination of gravity downwards.

And in like manner in the pully, or in a combination of pullies, the force of a hand drawing the rope directly, which is to the weight, whether ascending directly or obliquely, as the velocity of the perpendicular ascent of the weight to the velocity of the hand that draws the rope, will sustain the weight.

In clocks and such like instruments, made up from a combination of wheels, the contrary forces that promote and impede the motion of the wheels, if they are reciprocally as the velocities of the parts of the wheel on which they are impressed, will mutually sustain the one the other.

The force of the screw to press a body is to the force of the hand that turns the handles by which it is moved as the circular velocity of the handle in that part where it is impelled by the hand is to the progressive velocity of the screw towards the pressed body.

The forces by which the wedge presses or drives the two parts of the wood it cleaves are to the force of the mallet upon the wedge as the progress of the wedge in the direction of the force impressed upon it by the mallet is to the velocity with which the parts of the wood yield to the wedge, in the direction of lines perpendicular to the sides of the wedge. And the like account is to be given of all machines.

The power and use of machines consist only in this, that by diminishing the velocity we may augment the force, and the contrary: from whence, in all sorts of proper machines, we have the solution of this problem; *To move a given weight with a given power*, or with a given force to overcome any other given resistance. For if machines are so contrived that the velocities of the agent and resistant are reciprocally as their forces, the agent will just sustain the resistant, but with a greater disparity of velocity will overcome it. So that if the disparity of velocities is so great as to overcome all that resistance which commonly arises either from the attrition of contiguous bodies as they slide by one another, or from the cohesion of continuous bodies that are to be separated, or from the weights of bodies to be raised, the excess of the force remaining, after all those resistances are overcome, will produce an acceleration of motion proportional thereto, as well in the parts of the machine as in the resisting body. But to treat of mechanics is not my present business. I was only willing to show by those examples the great extent and certainty of the third Law of motion. For if we estimate the action of the agent from its force and velocity conjunctly, and likewise the reaction of the impediment conjunctly from the velocities of its several parts, and from the forces of resistance arising from the attrition, cohesion, weight, and acceleration of those parts, the action and reaction in the use of all sorts of machines will be found always equal to one another. And so far as the action is propagated by the intervening instruments, and at last impressed upon the resisting body, the ultimate determination of the action will be always contrary to the determination of the reaction.

BOOK I.

OF THE MOTION OF BODIES.

SECTION I.

Of the method of first and last ratios of quantities, by the help whereof we demonstrate the propositions that follow.

LEMMA I.

Quantities, and the ratios of quantities, which in any finite time converge continually to equality, and before the end of that time approach nearer the one to the other than by any given difference, become ultimately equal.

If you deny it, suppose them to be ultimately unequal, and let D be their ultimate difference. Therefore they cannot approach nearer to equality than by that given difference D; which is against the supposition.

LEMMA II.

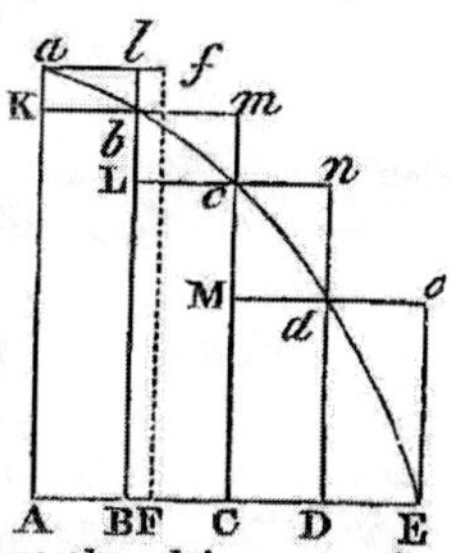

If in any figure AacE, *terminated by the right lines* Aa, AE, *and the curve* acE, *there be inscribed any number of parallelograms* Ab, Bc, Cd, *&c., comprehended under equal bases* AB, BC, CD, *&c., and the sides*, Bb, Cc, Dd, *&c., parallel to one side* Aa *of the figure; and the parallelograms* aKbl, bLcm, cMdn, *&c., are completed. Then if the breadth of those parallelograms be supposed to be diminished, and their number to be augmented* in infinitum; *I say, that the ultimate ratios which the inscribed figure* AKbLcMdD, *the circumscribed figure* AalbmcndoE, *and curvilinear figure* AabcdE, *will have to one another, are ratios of equality.*

For the difference of the inscribed and circumscribed figures is the sum of the parallelograms K*l*, L*m*, M*n*, D*o*, that is (from the equality of all their bases), the rectangle under one of their bases K*b* and the sum of their altitudes A*a*, that is, the rectangle AB*la*. But this rectangle, because its breadth AB is supposed diminished *in infinitum*, becomes less than any given space. And therefore (by Lem. I) the figures inscribed and circumscribed become ultimately equal one to the other; and much more will the intermediate curvilinear figure be ultimately equal to either. Q.E.D.

LEMMA III.

The same ultimate ratios are also ratios of equality, when the breadths, AB, BC, DC, *&c., of the parallelograms are unequal, and are all diminished* in infinitum.

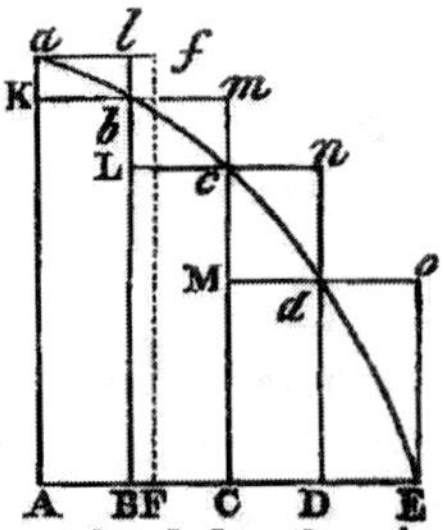

For suppose AF equal to the greatest breadth, and complete the parallelogram FA*af*. This parallelogram will be greater than the difference of the inscribed and circumscribed figures; but, because its breadth AF is diminished *in infinitum*, it will become less than any given rectangle. Q.E.D.

Cor. 1. Hence the ultimate sum of those evanescent parallelograms will in all parts coincide with the curvilinear figure.

Cor. 2. Much more will the rectilinear figure comprehended under the chords of the evanescent arcs *ab*, *bc*, *cd*, &c., ultimately coincide with the curvilinear figure.

Cor. 3. And also the circumscribed rectilinear figure comprehended under the tangents of the same arcs.

Cor. 4 And therefore these ultimate figures (as to their perimeters *ac*E), are not rectilinear, but curvilinear limits of rectilinear figures.

LEMMA IV.

If in two figures AacE, PprT, *you inscribe (as before) two ranks of parallelograms, an equal number in each rank, and, when their breadths are diminished* in infinitum, *the ultimate ratios of the parallelograms in one figure to those in the other, each to each respectively, are the same; I say, that those two figures* AacE, PprT, *are to one another in that same ratio.*

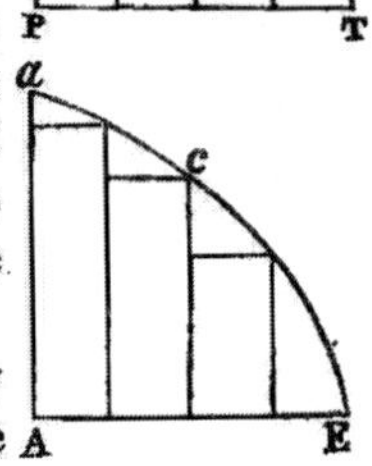

For as the parallelograms in the one are severally to the parallelograms in the other, so (by composition) is the sum of all in the one to the sum of all in the other; and so is the one figure to the other; because (by Lem. III) the former figure to the former sum, and the latter figure to the latter sum, are both in the ratio of equality. Q.E.D.

Cor. Hence if two quantities of any kind are any how divided into an equal number of parts, and those parts, when their number is augmented, and their magnitude diminished *in infinitum*, have a given ratio one to the other, the first to the first, the second to the second, and so on in order, the whole quantities will be one to the other in that same given ratio. For if, in the figures of this Lemma, the parallelograms are taken one to the other in the ratio of the parts, the sum of the parts will always be as the sum of the parallelograms; and therefore supposing the number of the parallelograms and parts to be augmented, and their magnitudes diminished *in infinitum*, those sums will be in the ultimate ratio of the parallelogram in the one figure to the correspondent parallelogram in the other; that is (by the supposition), in the ultimate ratio of any part of the one quantity to the correspondent part of the other.

LEMMA V.

In similar figures, all sorts of homologous sides, whether curvilinear or rectilinear, are proportional; and the areas are in the duplicate ratio of the homologous sides.

LEMMA VI.

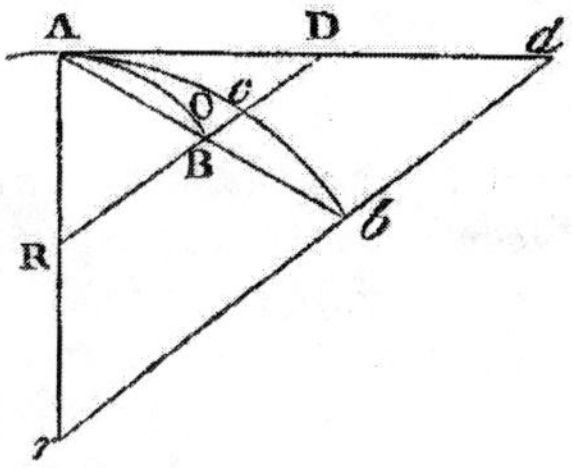

If any arc ACB, given in position is subtended by its chord AB, and in any point A, in the middle of the continued curvature, is touched by a right line AD, produced both ways; then if the points A and B approach one another and meet, I say, the angle BAD, contained between the chord and the tangent, will be diminished in infinitum, *and ultimately will vanish.*

For if that angle does not vanish, the arc ACB will contain with the tangent AD an angle equal to a rectilinear angle; and therefore the curvature at the point A will not be continued, which is against the supposition.

LEMMA VII.

The same things being supposed, I say that the ultimate ratio of the arc, chord, and tangent, any one to any other, is the ratio of equality.

For while the point B approaches towards the point A, consider always AB and AD as produced to the remote points *b* and *d*, and parallel to the secant BD draw *bd*: and let the arc A*cb* be always similar to the arc ACB. Then, supposing the points A and B to coincide, the angle *d*A*b* will vanish, by the preceding Lemma; and therefore the right lines A*b*, A*d* (which are always finite), and the intermediate arc A*cb*, will coincide, and become equal among themselves. Wherefore, the right lines AB, AD, and the intermediate arc ACB (which are always proportional to the former), will vanish, and ultimately acquire the ratio of equality. Q.E.D.

COR. 1. Whence if through B we draw BF parallel to the tangent, always cutting any right line AF passing through A in F, this line BF will be ultimately in the ratio of equality with the evanescent arc ACB; because, completing the parallelogram AFBD, it is always in a ratio of equality with AD.

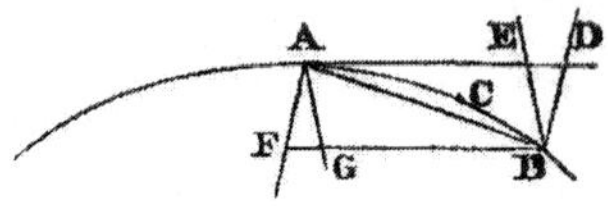

COR. 2. And if through B and A more right lines are drawn, as BE, BD, AF, AG, cutting the tangent AD and its parallel BF; the ultimate ratio of all the abscissas AD, AE, BF, BG, and of the chord and arc AB, any one to any other, will be the ratio of equality.

COR. 3. And therefore in all our reasoning about ultimate ratios, we may freely use any one of those lines for any other.

LEMMA VIII.

If the right lines AR, BR, *with the arc* ACB, *the chord* AB, *and the tangent* AD, *constitute three triangles* RAB, RACB, RAD, *and the points* A *and* B *approach and meet: I say, that the ultimate form of these evanescent triangles is that of similitude, and their ultimate ratio that of equality.*

For while the point B approaches towards the point A, consider always AB, AD, AR, as produced to the remote points *b*, *d*, and *r*, and *rbd* as drawn parallel to RD, and let the arc A*cb* be always similar to the arc ACB. Then supposing the points A and B to coincide, the angle *b*A*d* will vanish; and therefore the three triangles *r*A*b*, *r*A*cb*, *r*A*d* (which are always finite), will coincide, and on that account become both similar and equal. And therefore the triangles RAB, RACB, RAD which are always similar and proportional to these, will ultimately become both similar and equal among themselves. Q.E.D.

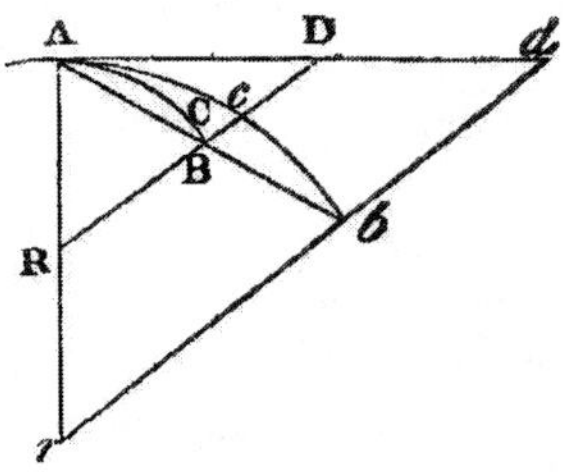

COR. And hence in all reasonings about ultimate ratios, we may indifferently use any one of those triangles for any other.

LEMMA IX.

If a right line AE, *and a curve line* ABC, *both given by position, cut each other in a given angle,* A; *and to that right line, in another given angle,* BD, CE *are ordinately applied, meeting the curve in* B, C; *and the points* B *and* C *together approach towards and meet in the point* A: *I say, that the areas of the triangles* ABD, ACE, *will ultimately be one to the other in the duplicate ratio of the sides.*

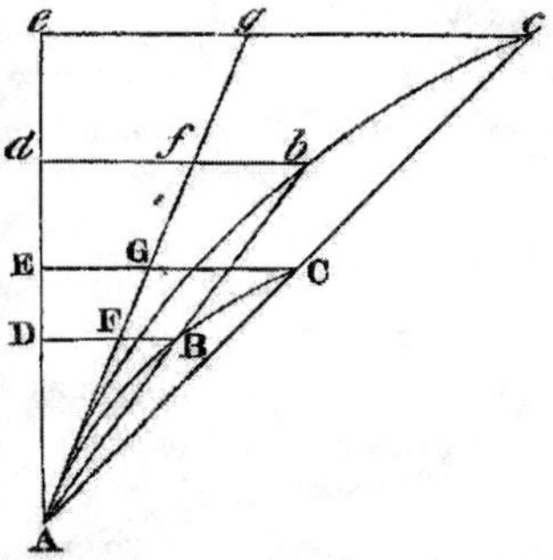

For while the points B, C, approach towards the point A, suppose always AD to be produced to the remote points *d* and *e*, so as A*d*, A*e* may be proportional to AD, AE; and the ordinates *db*, *ec*, to be drawn parallel to the ordinates DB and EC, and meeting AB and AC produced in *b* and *c*. Let the curve A*bc* be similar to the curve ABC, and draw the right line A*g* so as to touch both curves in A, and cut the ordinates DB, EC, *db* *ec*, in F, G, *f*, *g*. Then, supposing the length A*e* to remain the same, let the points B and C meet in the point A; and the angle *c*A*g* vanishing, the curvilinear areas A*bd*, A*ce* will coincide with the rectilinear areas A*fd*, A*ge*; and therefore (by Lem. V) will be one to the other in the duplicate ratio of the sides A*d*, A*e*. But the areas ABD, ACE are always proportional to these areas; and so the sides AD, AE are to these sides. And therefore the areas ABD, ACE are ultimately one to the other in the duplicate ratio of the sides AD, AE. Q.E.D.

LEMMA X.

The spaces which a body describes by any finite force urging it, whether that force is determined and immutable, or is continually augmented or continually diminished, are in the very beginning of the motion one to the other in the duplicate ratio of the times.

Let the times be represented by the lines AD, AE, and the velocities generated in those times by the ordinates DB, EC. The spaces described with these velocities will be as the areas ABD, ACE, described by those ordinates, that is, at the very beginning of the motion (by Lem. IX), in the duplicate ratio of the times AD, AE. Q.E.D.

Cor. 1. And hence one may easily infer, that the errors of bodies describing similar parts of similar figures in proportional times, are nearly as the squares of the times in which they are generated; if so be these errors are generated by any equal forces similarly applied to the bodies, and measured by the distances of the bodies from those places of the similar figures, at which, without the action of those forces, the bodies would have arrived in those proportional times.

Cor. 2. But the errors that are generated by proportional forces, similarly applied to the bodies at similar parts of the similar figures, are as the forces and the squares of the times conjunctly.

Cor. 3. The same thing is to be understood of any spaces whatsoever described by bodies urged with different forces; all which, in the very beginning of the motion, are as the forces and the squares of the times conjunctly.

COR. 4. And therefore the forces are as the spaces described in the very beginning of the motion directly, and the squares of the times inversely.

COR. 5. And the squares of the times are as the spaces described directly, and the forces inversely.

SCHOLIUM.

If in comparing indetermined quantities of different sorts one with another, any one is said to be as any other directly or inversely, the meaning is, that the former is augmented or diminished in the same ratio with the latter, or with its reciprocal. And if any one is said to be as any other two or more directly or inversely, the meaning is, that the first is augmented or diminished in the ratio compounded of the ratios in which the others, or the reciprocals of the others, are augmented or diminished. As if A is said to be as B directly, and C directly, and D inversely, the meaning is, that A is augmented or diminished in the same ratio with $B \times C \times \frac{1}{D}$, that is to say, that A and $\frac{BC}{D}$ are one to the other in a given ratio.

LEMMA XI.

The evanescent subtense of the angle of contact, in all curves which at the point of contact have a finite curvature, is ultimately in the duplicate ratio of the subtense of the conterminate arc.

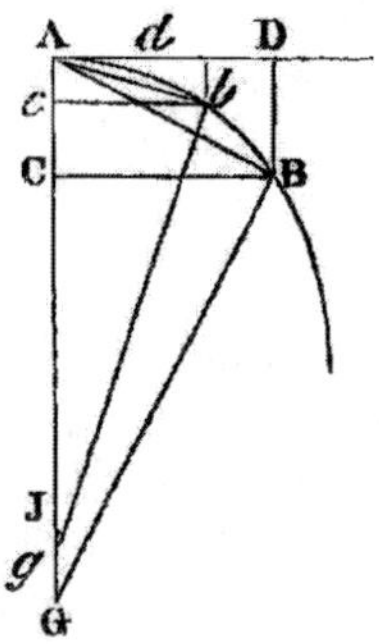

CASE 1. Let AB be that arc, AD its tangent, BD the subtense of the angle of contact perpendicular on the tangent, AB the subtense of the arc. Draw BG perpendicular to the subtense AB, and AG to the tangent AD, meeting in G; then let the points D, B, and G, approach to the points *d*, *b*, and *g*, and suppose J to be the ultimate intersection of the lines BG, AG, when the points D, B, have come to A. It is evident that the distance GJ may be less than any assignable. But (from the nature of the circles passing through the points A, B, G, A, *b*, *g*,) $AB^2 = AG \times BD$, and $Ab^2 = Ag \times bd$; and therefore the ratio of AB^2 to Ab^2 is compounded of the ratios of AG to A*g*, and of B*d* to *bd*. But because GJ may be assumed of less length than any assignable, the ratio of AG to A*g* may be such as to differ from the ratio of equality by less than any assignable difference; and therefore the ratio of AB^2 to Ab^2 may be such as to differ from the ratio of BD to *bd* by less than any assignable difference. Therefore, by Lem. I, the ultimate ratio of AB^2 to Ab^2 is the same with the ultimate ratio of BD to *bd*. Q.E.D.

CASE 2. Now let BD be inclined to AD in any given angle, and the ultimate ratio of BD to *bd* will always be the same as before, and therefore the same with the ratio of AB^2 to Ab^2. Q.E.D.

Case 3. And if we suppose the angle D not to be given, but that the right line BD converges to a given point, or is determined by any other condition whatever; nevertheless the angles D, *d*, being determined by the same law, will always draw nearer to equality, and approach nearer to each other than by any assigned difference, and therefore, by Lem. I, will at last be equal; and therefore the lines BD, *bd* are in the same ratio to each other as before. Q.E.D.

Cor. 1. Therefore since the tangents AD, A*d*, the arcs AB, A*b*, and their sines, BC, *bc*, become ultimately equal to the chords AB, A*b*, their squares will ultimately become as the subtenses BD, *bd*.

Cor. 2. Their squares are also ultimately as the versed sines of the arcs, bisecting the chords, and converging to a given point. For those versed sines are as the subtenses BD, *bd*.

Cor. 3. And therefore the versed sine is in the duplicate ratio of the time in which a body will describe the arc with a given velocity.

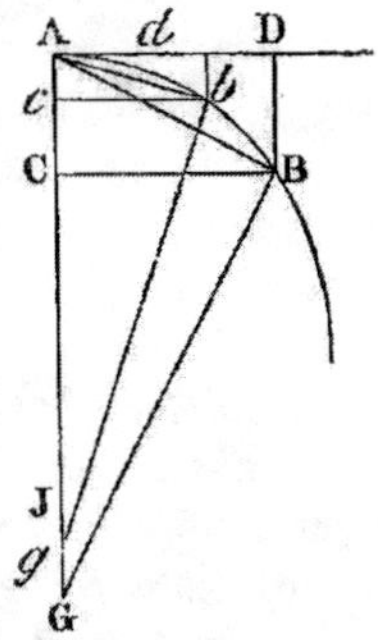

Cor. 4. The rectilinear triangles ADB, A*db* are ultimately in the triplicate ratio of the sides AD, A*d*, and in a sesquiplicate ratio of the sides DB, *db*; as being in the ratio compounded of the sides AD to DB, and of A*d* to *db*. So also the triangles ABC, A*bc* are ultimately in the triplicate ratio of the sides BC, *bc*. What I call the sesquiplicate ratio is the subduplicate of the triplicate, as being compounded of the simple and subduplicate ratio.

Cor. 5. And because DB, *db* are ultimately parallel and in the duplicate ratio of the lines AD, A*d*, the ultimate curvilinear areas ADB, A*db* will be (by the nature of the parabola) two thirds of the rectilinear triangles ADB, A*db*: and the segments AB, A*b* will be one third of the same triangles. And thence those areas and those segments will be in the triplicate ratio as well of the tangents AD, A*d*, as of the chords and arcs AB, AB.

SCHOLIUM.

But we have all along supposed the angle of contact to be neither infinitely greater nor infinitely less than the angles of contact made by circles and their tangents; that is, that the curvature at the point A is neither infinitely small nor infinitely great, or that the interval AJ is of a finite magnitude. For DB may be taken as AD^3: in which case no circle can be drawn through the point A, between the tangent AD and the curve AB, and therefore the angle of contact will be infinitely less than those of circles. And by a like reasoning, if DB be made successfully as AD^4, AD^5, AD^6, AD^7, &c., we shall have a series of angles of contact, proceeding *in infinitum*, wherein every succeeding term is infinitely less than the preceding. And if DB be made successively as AD^2, $AD^{\frac{3}{2}}$, $AD^{\frac{4}{3}}$, $AD^{\frac{5}{4}}$, $AD^{\frac{6}{5}}$, $AD^{\frac{7}{6}}$, &c., we shall have another infinite series of angles of contact, the first of which is of the same sort with those of circles, the second infinitely greater, and every succeeding one infinitely greater than the preceding. But between any two of these angles another series of intermediate angles of contact may be interposed, proceeding both ways *in infinitum*, wherein every succeeding angle shall be infinitely greater or infinitely less than the preceding. As if between the terms AD^2 and AD^3 there were interposed the series $AD^{\frac{13}{6}}$, $AD^{\frac{11}{5}}$, $AD^{\frac{9}{4}}$, $AD^{\frac{7}{3}}$, $AD^{\frac{5}{2}}$, $AD^{\frac{8}{3}}$, $AD^{\frac{11}{4}}$, $AD^{\frac{14}{5}}$, $AD^{\frac{17}{6}}$, &c. And again, between any two angles of this series, a new series of intermediate angles may be interposed, differing from one another by infinite intervals. Nor is nature confined to any bounds.

Those things which have been demonstrated of curve lines, and the superfices which they comprehend, may be easily applied to the curve superfices and contents of solids. These Lemmas are premised to avoid the tediousness of deducing perplexed demonstrations *ad absurdum*, according to the method of the ancient geometers. For demonstrations are more contracted by the method of indivisibles: but because the hypothesis of indivisibles seems somewhat harsh, and therefore that method is reckoned less geometrical, I chose rather to reduce the demonstrations of the following propositions to the first and last sums and ratios of nascent and evanescent quantities, that is, to the limits of those sums and ratios; and so to premise, as short as I could, the demonstrations of those limits. For hereby the same thing is performed as by the method of indivisibles; and now those principles being demonstrated, we may use them with more safety. Therefore if hereafter I should happen to consider quantities as made up of particles, or should use little curve lines for right ones, I would not be understood to mean indivisibles, but evanescent divisible quantities; not the sums and ratios of determinate parts, but always the limits of sums and ratios; and that the force of such demonstrations always depends on the method laid down in the foregoing Lemmas.

Perhaps it may be objected, that there is no ultimate proportion, of evanescent quantities; because the proportion, before the quantities have vanished, is not the ultimate, and when they are vanished, is none. But by the same argument, it may be alledged, that a body arriving at a certain place, and there stopping, has no ultimate velocity: because the velocity, before the body comes to the place, is not its ultimate velocity; when it has arrived, is none. But the answer is easy; for by the ultimate velocity is meant that with which the body is moved, neither before it arrives at its last place and the motion ceases, nor after, but at the very instant it arrives; that is, that velocity with which the body arrives at its last place, and with which the motion ceases. And in like manner, by the ultimate ratio of evanescent quantities is to be understood the ratio of the quantities not before they vanish, nor afterwards, but with which they vanish. In like manner the first ratio of nascent quantities is that with which they begin to be. And the first or last sum is that with which they begin and cease to be (or to be augmented or diminished). There is a limit which the velocity at the end of the motion may attain, but not exceed. This is the ultimate velocity. And there is the like limit in all quantities and proportions that begin and cease to be. And since such limits are certain and definite, to determine the same is a problem strictly geometrical. But whatever is geometrical we may be allowed to use in determining and demonstrating any other thing that is likewise geometrical.

It may also be objected, that if the ultimate ratios of evanescent quantities are given, their ultimate magnitudes will be also given: and so all quantities will consist of indivisibles, which is contrary to what *Euclid* has demonstrated concerning incommensurables, in the 10th Book of his Elements. But this objection is founded on a false supposition. For those ultimate ratios with which quantities vanish are not truly the ratios of ultimate quantities, but limits towards which the ratios of quantities decreasing without limit do always converge; and to which they approach nearer than by any given difference, but never go beyond, nor in effect attain to, till the quantities are diminished *in infinitum*. This thing will appear more evident in quantities infinitely great. If two quantities, whose difference is given, be augmented *in infinitum*, the ultimate ratio of these quantities will be given, to wit, the ratio of equality; but it does not from thence follow, that the ultimate or greatest quantities themselves, whose ratio that is, will be given. Therefore if in what follows, for the sake of being more easily understood, I should happen to mention quantities as least, or evanescent, or ultimate, you are not to suppose that quantities of any determinate magnitude are meant, but such as are conceived to be always diminished without end.

SECTION II.

Of the Invention of Centripetal Forces.

PROPOSITION I. THEOREM I.

The areas, which revolving bodies describe by radii drawn to an immovable centre of force do lie in the same immovable planes, and are proportional to the times in which they are described.

For suppose the time to be divided into equal parts, and in the first part of that time let the body by its innate force describe the right line AB. In the second part of that time, the same would (by Law I.), if not hindered, proceed directly to *c*, along the line B*c* equal to AB; so that by the radii AS, BS, *c*S, drawn to the centre, the equal areas ASB, BS*c*, would be de-

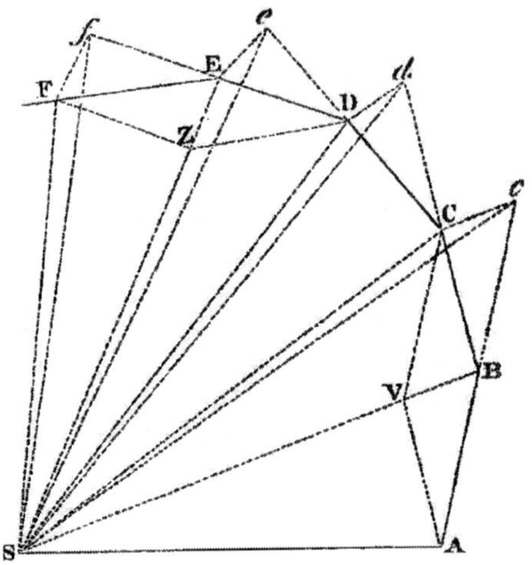

scribed. But when the body is arrived at B, suppose that a centripetal force acts at once with a great impulse, and, turning aside the body from the right line Bc, compels it afterwards to continue its motion along the right line BC. Draw cC parallel to BS meeting BC in C; and at the end of the second part of the time, the body (by Cor. I. of the Laws) will be found in C, in the same plane with the triangle ASB. Join SC, and, because SB and Cc are parallel, the triangle SBC will be equal to the triangle SBc, and therefore also to the triangle SAB. By the like argument, if the centripetal force acts successively in C, D, E, &c., and makes the body, in each single particle of time, to describe the right lines CD, DE, EF, &c., they will all lie in the same plane; and the triangle SCD will be equal to the triangle SBC, and SDE to SCD, and SEF to SDE. And therefore, in equal times, equal areas are described in one immovable plane: and, by composition, any sums SADS, SAFS, of those areas, are one to the other as the times in which they are described. Now let the number of those triangles be augmented, and their breadth diminished *in infinitum;* and (by Cor. 4, Lem. III.) their ultimate perimeter ADF will be a curve line: and therefore the centripetal force, by which the body is perpetually drawn back from the tangent of this curve, will act continually; and any described areas SADS, SAFS, which are always proportional to the times of description, will, in this case also, be proportional to those times. Q.E.D.

Cor. 1. The velocity of a body attracted towards an immovable centre, in spaces void of resistance, is reciprocally as the perpendicular let fall from that centre on the right line that touches the orbit. For the velocities in those places A, B, C, D, E, are as the bases AB, BC, CD, DE, EF, of equal triangles; and these bases are reciprocally as the perpendiculars let fall upon them.

Cor. 2. If the chords AB, BC of two arcs, successively described in equal times by the same body, in spaces void of resistance, are completed into a parallelogram ABCV, and the diagonal BV of this parallelogram, in the position which it ultimately acquires when those arcs are diminished *in infinitum*, is produced both ways, it will pass through the centre of force.

Cor. 3. If the chords AB, BC, and DE, EF, of arcs described in equal times, in spaces void of resistance, are completed into the parallelograms ABCV, DEFZ; the forces in B and E are one to the other in the ultimate ratio of the diagonals BV, EZ, when those arcs are diminished *in infinitum.* For the motions BC and EF of the body (by Cor. 1 of the Laws) are compounded of the motions B*c*, BV, and E*f*, EZ: but BV and EZ, which are equal to C*c* and F*f*, in the demonstration of this Proposition, were generated by the impulses of the centripetal force in B and E, and are therefore proportional to those impulses.

COR. 4. The forces by which bodies, in spaces void of resistance, are drawn back from rectilinear motions, and turned into curvilinear orbits, are one to another as the versed sines of arcs described in equal times; which versed sines tend to the centre of force, and bisect the chords when those arcs are diminished to infinity. For such versed sines are the halves of the diagonals mentioned in Cor. 3.

COR. 5. And therefore those forces are to the force of gravity as the said versed sines to the versed sines perpendicular to the horizon of those parabolic arcs which projectiles describe in the same time.

COR. 6. And the same things do all hold good (by Cor. 5 of the Laws), when the planes in which the bodies are moved, together with the centres of force which are placed in those planes, are not at rest, but move uniformly forward in right lines.

PROPOSITION II. THEOREM II.

Every body that moves in any curve line described in a plane, and by a radius, drawn to a point either immovable, or moving forward with an uniform rectilinear motion, describes about that point areas proportional to the times, is urged by a centripetal force directed to that point.

CASE. 1. For every body that moves in a curve line, is (by Law 1) turned aside from its rectilinear course by the action of some force that impels it. And that force by which the body is turned off from its rectilinear course, and is made to describe, in equal times, the equal least triangles SAB, SBC, SCD, &c., about the immovable point S (by Prop. XL. Book 1, Elem. and Law II), acts in the place B, according to the direction of a line par-

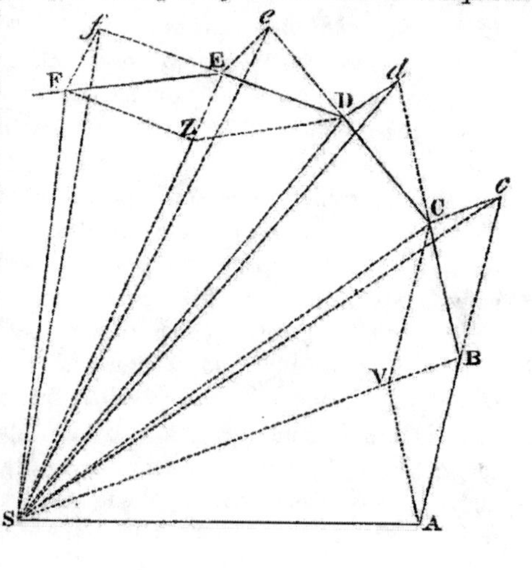

allel to *c*C, that is, in the direction of the line BS; and in the place C, according to the direction of a line parallel to *d*D, that is, in the direction of the line CS, &c.; and therefore acts always in the direction of lines tending to the immovable point S. Q.E.D.

Case. 2. And (by Cor. 5 of the Laws) it is indifferent whether the superfices in which a body describes a curvilinear figure be quiescent, or moves together with the body, the figure described, and its point S, uniformly forward in right lines.

Cor. 1. In non-resisting spaces or mediums, if the areas are not proportional to the times, the forces are not directed to the point in which the radii meet; but deviate therefrom *in consequentia*, or towards the parts to which the motion is directed, if the description of the areas is accelerated; but *in antecedentia*, if retarded.

Cor. 2. And even in resisting mediums, if the description of the areas is accelerated, the directions of the forces deviate from the point in which the radii meet, towards the parts to which the motion tends.

SCHOLIUM.

A body may be urged by a centripetal force compounded of several forces; in which case the meaning of the Proposition is, that the force which results out of all tends to the point S. But if any force acts perpetually in the direction of lines perpendicular to the described surface, this force will make the body to deviate from the plane of its motion: but will neither augment nor diminish the quantity of the described surface, and is therefore to be neglected in the composition of forces.

PROPOSITION III. THEOREM III.

Every body, that by a radius drawn to the centre of another body, howsoever moved, describes areas about that centre proportional to the times, is urged by a force compounded out of the centripetal force tending to that other body, and of all the accelerative force by which that other body is impelled.

Let L represent the one, and T the other body; and (by Cor. 6 of the Laws) if both bodies are urged in the direction of parallel lines, by a new force equal and contrary to that by which the second body T is urged, the first body L will go on to describe about the other body T the same areas as before: but the force by which that other body T was urged will be now destroyed by an equal and contrary force; and therefore (by Law I.) that other body T, now left to itself, will either rest, or move uniformly forward in a right line: and the first body L impelled by the difference of the forces, that is, by the force remaining, will go on to describe about the other body T areas proportional to the times. And therefore (by Theor. II.) the difference of the forces is directed to the other body T as its centre. Q.E.D.

COR. 1. Hence if the one body L, by a radius drawn to the other body T, describes areas proportional to the times; and from the whole force, by which the first body L is urged (whether that force is simple, or, according to Cor. 2 of the Laws, compounded out of several forces), we subduct (by the same Cor.) that whole accelerative force by which the other body is urged; the whole remaining force by which the first body is urged will tend to the other body T, as its centre.

COR. 2. And, if these areas are proportional to the times nearly, the remaining force will tend to the other body T nearly.

COR. 3. And *vice versa*, if the remaining force tends nearly to the other body T, those areas will be nearly proportional to the times.

COR. 4. If the body L, by a radius drawn to the other body T, describes areas, which, compared with the times, are very unequal; and that other body T be either at rest, or moves uniformly forward in a right line: the action of the centripetal force tending to that other body T is either none at all, or it is mixed and compounded with very powerful actions of other forces: and the whole force compounded of them all, if they are many, is directed to another (immovable or moveable) centre. The same thing obtains, when the other body is moved by any motion whatsoever; provided that centripetal force is taken, which remains after subducting that whole force acting upon that other body T.

SCHOLIUM.

Because the equable description of areas indicates that a centre is respected by that force with which the body is most affected, and by which it is drawn back from its rectilinear motion, and retained in its orbit; why may we not be allowed, in the following discourse, to use the equable description of areas as an indication of a centre, about which all circular motion is performed in free spaces?

PROPOSITION IV. THEOREM IV.

The centripetal forces of bodies, which by equable motions describe different circles, tend to the centres of the same circles; and are one to the other as the squares of the arcs described in equal times applied to the radii of the circles.

These forces tend to the centres of the circles (by Prop. II., and Cor. 2, Prop. I.), and are one to another as the versed sines of the least arcs described in equal times (by Cor. 4, Prop. I.); that is, as the squares of the same arcs applied to the diameters of the circles (by Lem. VII.); and therefore since those arcs are as arcs described in any equal times, and the diameters are as the radii, the forces will be as the squares of any arcs described in the same time applied to the radii of the circles. Q.E.D.

COR. 1. Therefore, since those arcs are as the velocities of the bodies, the centripetal forces are in a ratio compounded of the duplicate ratio of the velocities directly, and of the simple ratio of the radii inversely.

Cor. 2. And since the periodic times are in a ratio compounded of the ratio of the radii directly, and the ratio of the velocities inversely, the centripetal forces, are in a ratio compounded of the ratio of the radii directly, and the duplicate ratio of the periodic times inversely.

Cor. 3. Whence if the periodic times are equal, and the velocities therefore as the radii, the centripetal forces will be also as the radii; and the contrary.

Cor. 4. If the periodic times and the velocities are both in the subduplicate ratio of the radii, the centripetal forces will be equal among themselves; and the contrary.

Cor. 5. If the periodic times are as the radii, and therefore the velocities equal, the centripetal forces will be reciprocally as the radii; and the contrary.

Cor. 6. If the periodic times are in the sesquiplicate ratio of the radii, and therefore the velocities reciprocally in the subduplicate ratio of the radii, the centripetal forces will be in the duplicate ratio of the radii inversely; and the contrary.

Cor. 7. And universally, if the periodic time is as any power R^n of the radius R, and therefore the velocity reciprocally as the power R^{n-1} of the radius, the centripetal force will be reciprocally as the power R^{2n-1} of the radius; and the contrary.

Cor. 8. The same things all hold concerning the times, the velocities, and forces by which bodies describe the similar parts of any similar figures that have their centres in a similar position with those figures; as appears by applying the demonstration of the preceding cases to those. And the application is easy, by only substituting the equable description of areas in the place of equable motion, and using the distances of the bodies from the centres instead of the radii.

Cor. 9. From the same demonstration it likewise follows, that the arc which a body, uniformly revolving in a circle by means of a given centripetal force, describes in any time, is a mean proportional between the diameter of the circle, and the space which the same body falling by the same given force would descend through in the same given time.

SCHOLIUM.

The case of the 6th Corollary obtains in the celestial bodies (as Sir Christopher Wren, Dr. Hooke, and Dr. Halley have severally observed); and therefore in what follows, I intend to treat more at large of those things which relate to centripetal force decreasing in a duplicate ratio of the distances from the centres.

Moreover, by means of the preceding Proposition and its Corollaries, we may discover the proportion of a centripetal force to any other known force, such as that of gravity. For if a body by means of its gravity revolves in a circle concentric to the earth, this gravity is the centripetal force of that body. But from the descent of heavy bodies, the time of one entire revolution, as well as the arc described in any given time, is given (by Cor. 9 of this Prop.). And by such propositions, Mr. Huygens, in his excellent book *De Horologio Oscillatorio*, has compared the force of gravity with the centrifugal forces of revolving bodies.

The preceding Proposition may be likewise demonstrated after this manner. In any circle suppose a polygon to be inscribed of any number of sides. And if a body, moved with a given velocity along the sides of the polygon, is reflected from the circle at the several angular points, the force, with which at every reflection it strikes the circle, will be as its velocity: and therefore the sum of the forces, in a given time, will be as that velocity and the number of reflections conjunctly; that is (if the species of the polygon be given), as the length described in that given time, and increased or diminished in the ratio of the same length to the radius of the circle; that is, as the square of that length applied to the radius: and therefore the polygon, by having its sides diminished *in infinitum*, coincides with the circle, as the square of the arc described in a given time applied to the radius. This is the centrifugal force, with which the body impels the circle; and to which the contrary force, wherewith the circle continually repels the body towards the centre, is equal.

PROPOSITION V. PROBLEM I.

There being given, in any places, the velocity with which a body describes a given figure, by means of forces directed to some common centre: to find that centre.

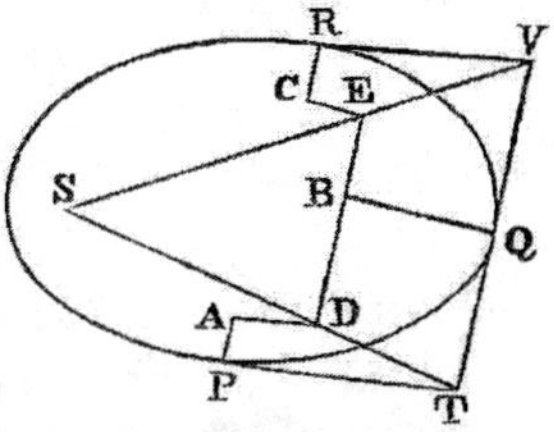

Let the three right lines PT, TQV, VR touch the figure described in as many points, P, Q, R, and meet in T and V. On the tangents erect the perpendiculars PA, QB, RC, reciprocally proportional to the velocities of the body in the points P, Q, R, from which the perpendiculars were raised; that is, so that PA may be to QB as the velocity in Q to the velocity in P, and QB to RC as the velocity in R to the velocity in Q. Through the ends A, B, C, of the perpendiculars draw AD, DBE, EC, at right angles, meeting in D and E: and the right lines TD, VE produced, will meet in S, the centre required.

For the perpendiculars let fall from the centre S on the tangents PT, QT, are reciprocally as the velocities of the bodies in the points P and Q (by Cor. 1, Prop. I.), and therefore, by construction, as the perpendiculars AP, BQ directly; that is, as the perpendiculars let fall from the point D on the tangents. Whence it is easy to infer that the points S, D, T, are in one right line. And by the like argument the points S, E, V are also in one right line; and therefore the centre S is in the point where the right lines TD, VE meet. Q.E.D.

PROPOSITION VI. THEOREM V.

In a space void of resistance, if a body revolves in any orbit about an immovable centre, and in the least time describes any arc just then nascent; and the versed sine of that arc is supposed to be drawn bisecting the chord, and produced passing through the centre of force: the centripetal force in the middle of the arc will be as the versed sine directly and the square of the time inversely.

For the versed sine in a given time is as the force (by Cor. 4, Prop. 1); and augmenting the time in any ratio, because the arc will be augmented in the same ratio, the versed sine will be augmented in the duplicate of that ratio (by Cor. 2 and 3, Lem. XI.), and therefore is as the force and the square of the time. Subduct on both sides the duplicate ratio of the time, and the force will be as the versed sine directly, and the square of the time inversely. Q.E.D.

And the same thing may also be easily demonstrated by Corol. 4, Lem. X.

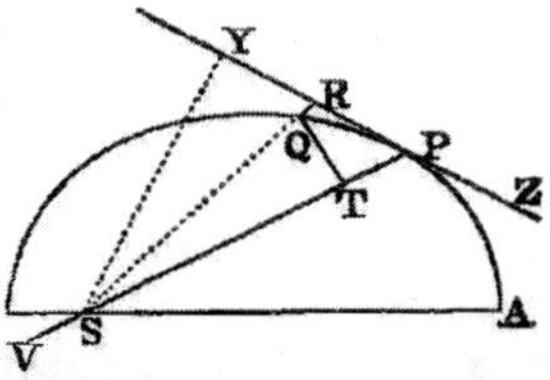

Cor. 1. If a body P revolving about the centre S describes a curve line APQ, which a right line ZPR touches in any point P; and from any other point Q of the curve, QR is drawn parallel to the distance SP, meeting the tangent in R; and QT is drawn perpendicular to the distance SP; the centripetal force will be reciprocally as the solid $\frac{SP^2 \times QT^2}{QR}$, if the solid be taken of that magnitude which it ultimately acquires when the points P and Q coincide. For QR is equal to the versed sine of double the arc QP, whose middle is P: and double the triangle SQP, or SP $\times$ QT is proportional to the time in which that double arc is described; and therefore may be used for the exponent of the time.

Cor. 2. By a like reasoning, the centripetal force is reciprocally as the solid $\frac{SY^2 \times QP^2}{QR}$; if SY is a perpendicular from the centre of force on PR the tangent of the orbit. For the rectangles SY $\times$ QP and SP $\times$ QT are equal.

COR. 3. If the orbit is either a circle, or touches or cuts a circle concentrically, that is, contains with a circle the least angle of contact or section, having the same curvature and the same radius of curvature at the point P; and if PV be a chord of this circle, drawn from the body through the centre of force; the centripetal force will be reciprocally as the solid $SY^2 \times PV$. For PV is $\frac{QP^2}{QR}$.

COR. 4. The same things being supposed, the centripetal force is as the square of the velocity directly, and that chord inversely. For the velocity is reciprocally as the perpendicular SY, by Cor. 1. Prop. I.

COR. 5. Hence if any curvilinear figure APQ is given, and therein a point S is also given, to which a centripetal force is perpetually directed, that law of centripetal force may be found, by which the body P will be continually drawn back from a rectilinear course, and, being detained in the perimeter of that figure, will describe the same by a perpetual revolution. That is, we are to find, by computation, either the solid $\frac{SP^2 \times QT^2}{QR}$ or the solid $SY^2 \times PV$, reciprocally proportional to this force. Examples of this we shall give in the following Problems.

PROPOSITION VII. PROBLEM II.

If a body revolves in the circumference of a circle; it is proposed to find the law of centripetal force directed to any given point.

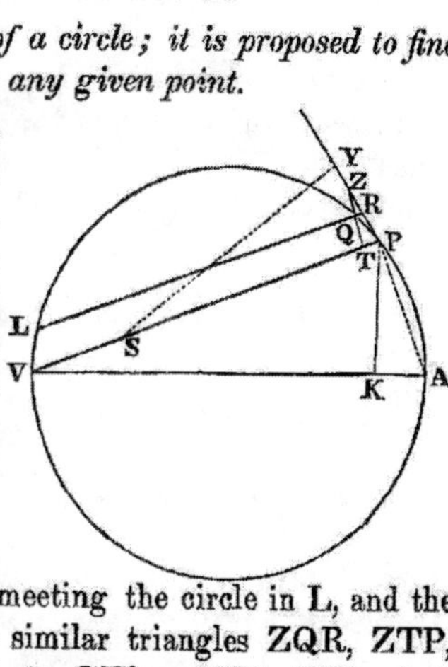

Let VQPA be the circumference of the circle; S the given point to which as to a centre the force tends; P the body moving in the circumference; Q the next place into which it is to move; and PRZ the tangent of the circle at the preceding place. Through the point S draw the chord PV, and the diameter VA of the circle: join AP, and draw QT perpendicular to SP, which produced, may meet the tangent PR in Z; and lastly, through the point Q, draw LR parallel to SP, meeting the circle in L, and the tangent PZ in R. And, because of the similar triangles ZQR, ZTP, VPA, we shall have RP^2, that is, QRL to QT^2 as AV^2 to PV^2. And therefore $\frac{QRL \times PV^2}{AV^2}$ is equal to QT^2. Multiply those equals by $\frac{SP^2}{QR}$, and the points P and Q coinciding, for RL write PV; then we shall have $\frac{SP^2 \times PV^3}{AV^2} = \frac{SP^2 \times QT^2}{QR}$. And therefore (by Cor. 1 and 5, Prop. VI.)

the centripetal force is reciprocally as $\frac{SP^2 \times PV^3}{AV^2}$; that is (because AV^2 is given), reciprocally as the square of the distance or altitude SP, and the cube of the chord PV conjunctly. Q.E.I.

The same otherwise.

On the tangent PR produced let fall the perpendicular SY; and (because of the similar triangles SYP, VPA), we shall have AV to PV as SP to SY, and therefore $\frac{SP \times PV}{AV} = SY$, and $\frac{SP^2 \times PV^3}{AV^2} = SY^2 \times PV$. And therefore (by Corol. 3 and 5, Prop. VI), the centripetal force is reciprocally as $\frac{SP^2 \times PV^3}{AV^2}$; that is (because AV is given), reciprocally as $SP^2 \times PV^3$. Q.E.I.

Cor. 1. Hence if the given point S, to which the centripetal force always tends, is placed in the circumference of the circle, as at V, the centripetal force will be reciprocally as the quadrato-cube (or fifth power) of the altitude SP.

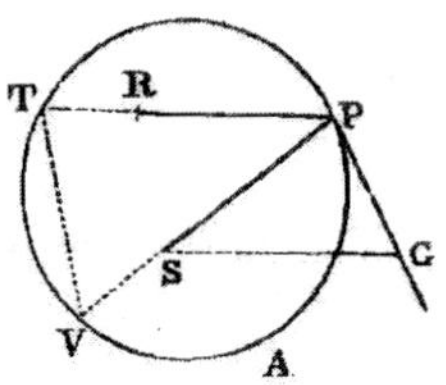

Cor. 2. The force by which the body P in the circle APTV revolves about the centre of force S is to the force by which the same body P may revolve in the same circle, and in the same periodic time, about any other centre of force R, as $RP^2 \times SP$ to the cube of the right line SG, which from the first centre of force S is drawn parallel to the distance PR of the body from the second centre of force R, meeting the tangent PG of the orbit in G. For by the construction of this Proposition, the former force is to the latter as $RP^2 \times PT^3$ to $SP^2 \times PV^3$; that is, as $SP \times RP^2$ to $\frac{SP^3 \times PV^3}{PT^3}$; or (because of the similar triangles PSG, TPV) to SG^3.

Cor. 3. The force by which the body P in any orbit revolves about the centre of force S, is to the force by which the same body may revolve in the same orbit, and the same periodic time, about any other centre of force R, as the solid $SP \times RP^2$, contained under the distance of the body from the first centre of force S, and the square of its distance from the second centre of force R, to the cube of the right line SG, drawn from the first centre of the force S, parallel to the distance RP of the body from the second centre of force R, meeting the tangent PG of the orbit in G. For the force in this orbit at any point P is the same as in a circle of the same curvature.

PROPOSITION VIII. PROBLEM III.

If a body moves in the semi-circumference PQA; *it is proposed to find the law of the centripetal force tending to a point* S, *so remote, that all the lines* PS, RS *drawn thereto, may be taken for parallels.*

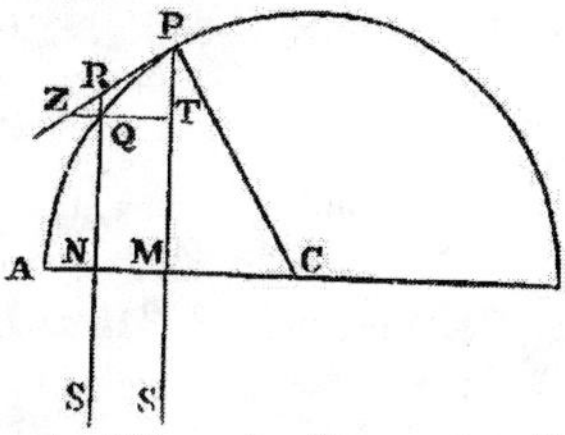

From C, the centre of the semi-circle, let the semi-diameter CA be drawn, cutting the parallels at right angles in M and N, and join CP. Because of the similar triangles CPM, PZT, and RZQ, we shall have CP^2 to PM^2 as PR^2 to QT^2; and, from the nature of the circle, PR^2 is equal to the rectangle $QR \times \overline{RN + QN}$, or, the points P, Q coinciding, to the rectangle $QR \times 2PM$. Therefore CP^2 is to PM^2 as $QR \times 2PM$ to QT^2; and $\frac{QT^2}{QR} = \frac{2PM^3}{CP^2}$, and $\frac{QT^2 \times SP^2}{QR} = \frac{2PM^3 \times SP^2}{CP^2}$. And therefore (by Corol. 1 and 5, Prop. VI.), the centripetal force is reciprocally as $\frac{2PM^3 \times SP^2}{CP^2}$; that is (neglecting the given ratio $\frac{2SP^2}{CP^2}$), reciprocally as PM^3. Q.E.I.

And the same thing is likewise easily inferred from the preceding Proposition.

SCHOLIUM.

And by a like reasoning, a body will be moved in an ellipsis, or even in an hyperbola, or parabola, by a centripetal force which is reciprocally as the cube of the ordinate directed to an infinitely remote centre of force.

PROPOSITION IX. PROBLEM IV.

If a body revolves in a spiral PQS, *cutting all the radii* SP, SQ, *&c., in a given angle; it is proposed to find the law of the centripetal force tending to the centre of that spiral.*

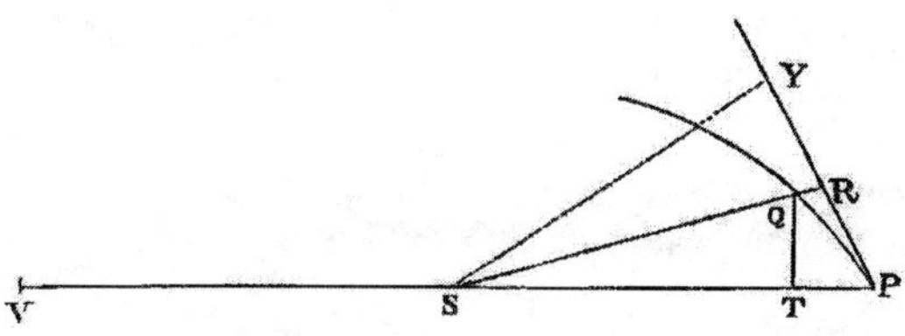

Suppose the indefinitely small angle PSQ to be given; because, then, all the angles are given, the figure SPRQT will be given in specie. Therefore the ratio $\frac{QT}{QR}$ is also given, and $\frac{QT^2}{QR}$ is as QT, that is (because the figure is given in specie), as SP. But if the angle PSQ is any way changed, the right line QR, subtending the angle of contact QPR (by Lemma XI) will be changed in the duplicate ratio of PR or QT. Therefore the ratio $\frac{QT^2}{QR}$ remains the same as before, that is, as SP. And $\frac{QT^2 \times SP^2}{QR}$ is as SP^3, and therefore (by Corol. 1 and 5, Prop. VI) the centripetal force is reciprocally as the cube of the distance SP. Q.E.I.

The same otherwise.

The perpendicular SY let fall upon the tangent, and the chord PV of the circle concentrically cutting the spiral, are in given ratios to the height SP; and therefore SP^3 is as $SY^2 \times PV$, that is (by Corol. 3 and 5, Prop. VI) reciprocally as the centripetal force.

LEMMA XII.

All parallelograms circumscribed about any conjugate diameters of a given ellipsis or hyperbola are equal among themselves.

This is demonstrated by the writers on the conic sections.

PROPOSITION X. PROBLEM V.

If a body revolves in an ellipsis; it is proposed to find the law of the centripetal force tending to the centre of the ellipsis.

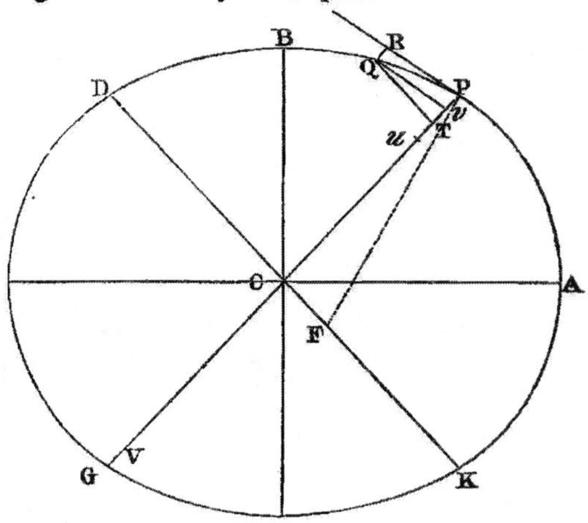

Suppose CA, CB to be semi-axes of the ellipsis; GP, DK, conjugate diameters; PF, QT perpendiculars to those diameters; Qv an ordinate to the diameter GP; and if the parallelogram QvPR be completed, then (by the properties of the conic sections) the rectangle PvG will be to Qv^2 as PC^2 to CD^2; and (because of the similar triangles QvT, PCF), Qv^2 to QT^2 as PC^2 to PF^2; and, by composition, the ratio of PvG to QT^2 is compounded of the ratio of PC^2 to CD^2, and of the ratio of PC^2 to PF^2, that is, vG to $\frac{QT^2}{Pv}$ as PC^2 to $\frac{CD^2 \times PF^2}{PC^2}$. Put QR for P$v$, and (by Lem. XII) BC × CA for CD × PF; also (the points P and Q coinciding) 2PC for vG; and multiplying the extremes and means together, we shall have $\frac{QT^2 \times PC^2}{QR}$ equal to $\frac{2BC^2 \times CA^2}{PC}$. Therefore (by Cor. 5, Prop. VI), the centripetal force is reciprocally as $\frac{2BC^2 \times CA^2}{PC}$; that is (because $2BC^2 \times CA^2$ is given), reciprocally as $\frac{1}{PC}$; that is, directly as the distance PC. QEI.

The same otherwise.

In the right line PG on the other side of the point T, take the point u so that Tu may be equal to Tv; then take uV, such as shall be to vG as DC^2 to PC^2. And because Qv^2 is to PvG as DC^2 to PC^2 (by the conic sections), we shall have $Qv^2 = Pv \times uV$. Add the rectangle uPv to both sides, and the square of the chord of the arc PQ will be equal to the rectangle VPv; and therefore a circle which touches the conic section in P, and passes through the point Q, will pass also through the point V. Now let the points P and Q meet, and the ratio of uV to vG, which is the same with the ratio of DC^2 to PC^2, will become the ratio of PV to PG, or PV to 2PC; and therefore PV will be equal to $\frac{2DC^2}{PC}$. And therefore the force by which the body P revolves in the ellipsis will be reciprocally as $\frac{2DC^2}{PC} \times PF^2$ (by Cor. 3, Prop. VI); that is (because $2DC^2 \times PF^2$ is given) directly as PC. Q.E.I.

COR. 1. And therefore the force is as the distance of the body from the centre of the ellipsis; and, *vice versa*, if the force is as the distance, the body will move in an ellipsis whose centre coincides with the centre of force, or perhaps in a circle into which the ellipsis may degenerate.

COR. 2. And the periodic times of the revolutions made in all ellipses whatsoever about the same centre will be equal. For those times in similar ellipses will be equal (by Corol. 3 and 8, Prop. IV); but in ellipses that have their greater axis common, they are one to another as the whole areas of the ellipses directly, and the parts of the areas described in the same time inversely; that is, as the lesser axes directly, and the velocities of the bodies in their principal vertices inversely; that is, as those lesser axes directly, and the ordinates to the same point of the common axes inversely; and therefore (because of the equality of the direct and inverse ratios) in the ratio of equality.

SCHOLIUM.

If the ellipsis, by having its centre removed to an infinite distance, degenerates into a parabola, the body will move in this parabola; and the force, now tending to a centre infinitely remote, will become equable. Which is *Galileo's* theorem. And if the parabolic section of the cone (by changing the inclination of the cutting plane to the cone) degenerates into an hyperbola, the body will move in the perimeter of this hyperbola, having its centripetal force changed into a centrifugal force. And in like manner as in the circle, or in the ellipsis, if the forces are directed to the centre of the figure placed in the abscissa, those forces by increasing or diminishing the ordinates in any given ratio, or even by changing the angle of the inclination of the ordinates to the abscissa, are always augmented or diminished in the ratio of the distances from the centre; provided the periodic times remain equal; so also in all figures whatsoever, if the ordinates are augmented or diminished in any given ratio, or their inclination is any way changed, the periodic time remaining the same, the forces directed to any centre placed in the abscissa are in the several ordinates augmented or diminished in the ratio of the distances from the centre.

SECTION III.

Of the motion of bodies in eccentric conic sections.

PROPOSITION XI. PROBLEM VI.

If a body revolves in an ellipsis; it is required to find the law of the centripetal force tending to the focus of the ellipsis.

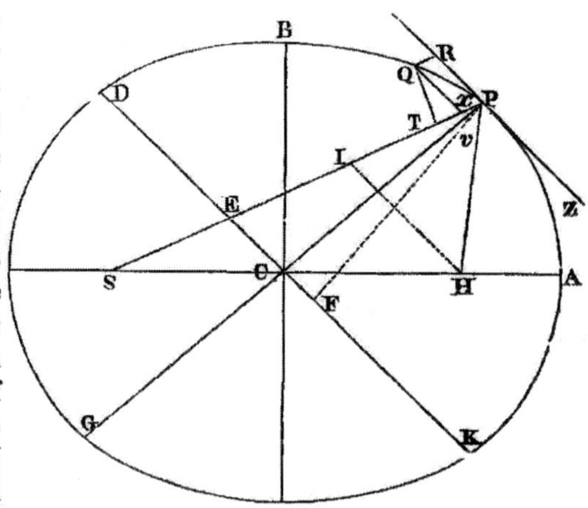

Let S be the focus of the ellipsis. Draw SP cutting the diameter DK of the ellipsis in E, and the ordinate Qv in x; and complete the parallelogram QxPR. It is evident that EP is equal to the greater semi-axis AC: for drawing HI from the other focus H of the ellipsis parallel to EC, because CS, CH are equal, ES, EI will be also equal; so that EP is the half sum of PS, PI, that is (because of the parallels HI, PR, and the equal angles IPR, HPZ), of PS, PH, which taken together are equal to the whole axis 2AC. Draw QT perpendicular to SP, and putting L for the princi al latus rectum of the ellipsis (or for

$\frac{2BC^2}{AC}$), we shall have L × QR to L × Pv as QR to Pv, that is, as PE or AC to PC; and L × Pv to GvP as L to Gv; and GvP to Qv^2 as PC^2 to CD^2; and by (Corol. 2, Lem. VII) the points Q and P coinciding, Qv^2 is to Qx^2 in the ratio of equality; and Qx^2 or Qv^2 is to QT^2 as EP^2 to PF^2, that is, as CA^2 to PF^2, or (by Lem. XII) as CD^2 to CB^2. And compounding all those ratios together, we shall have L × QR to QT^2 as AC × L × PC^2 × CD^2, or 2CB^2 × PC^2 × CD^2 to PC × Gv × CD^2 × CB^2, or as 2PC to Gv. But the points Q and P coinciding, 2PC and Gv are equal. And therefore the quantities L × QR and QT^2, proportional to these, will be also equal. Let those equals be drawn into $\frac{SP^2}{QR}$, and L × SP^2 will become equal to $\frac{SP^2 \times QT^2}{QR}$. And therefore (by Corol. 1 and 5, Prop. VI) the centripetal force is reciprocally as L × SP^2, that is, reciprocally in the duplicate ratio of the distance SP. Q.E.I.

The same otherwise.

Since the force tending to the centre of the ellipsis, by which the body P may revolve in that ellipsis, is (by Corol. 1, Prop. X.) as the distance CP of the body from the centre C of the ellipsis; let CE be drawn parallel to the tangent PR of the ellipsis; and the force by which the same body P may revolve about any other point S of the ellipsis, if CE and PS intersect in E, will be as $\frac{PE^3}{SP^2}$ (by Cor. 3, Prop. VII.); that is, if the point S is the focus of the ellipsis, and therefore PE be given as SP^2 reciprocally. Q.E.I.

With the same brevity with which we reduced the fifth Problem to the parabola, and hyperbola, we might do the like here: but because of the dignity of the Problem and its use in what follows, I shall confirm the other cases by particular demonstrations.

PROPOSITION XII. PROBLEM VII.

Suppose a body to move in an hyperbola; it is required to find the law of the centripetal force tending to the focus of that figure.

Let CA, CB be the semi-axes of the hyperbola; PG, KD other conjugate diameters; PF a perpendicular to the diameter KD; and Qv an ordinate to the diameter GP. Draw SP cutting the diameter DK in E, and the ordinate Qv in x, and complete the parallelogram QRPx. It is evident that EP is equal to the semi-transverse axis AC; for drawing HI, from the other focus H of the hyperbola, parallel to EC, because CS, CH are equal, ES, EI will be also equal; so that EP is the half difference

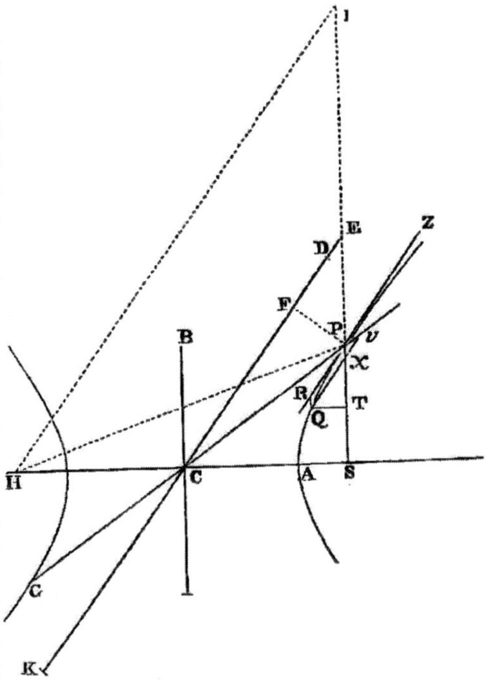

of PS, PI; that is (because of the parallels IH, PR, and the equal angles IPR, HPZ), of PS, PH, the difference of which is equal to the whole axis 2AC. Draw QT perpendicular to SP; and putting L for the principal latus rectum of the hyperbola (that is, for $\frac{2BC^2}{AC}$), we shall have L × QR to L × Pv as QR to Pv, or Px to Pv, that is (because of the similar triangles Pxv, PEC), as PE to PC, or AC to PC. And L × Pv will be to Gv × Pv as L to Gv; and (by the properties of the conic sections) the rectangle GvP is to Qv^2 as PC^2 to CD^2; and by (Cor. 2, Lem. VII.), Qv^2 to Qx^2, the points Q and P coinciding, becomes a ratio of equality; and Qx^2 or Qv^2 is to QT^2 as EP^2 to PF^2, that is, as CA^2 to PF^2, or (by Lem. XII.) as CD^2 to CB^2: and, compounding all those ratios together, we shall have L × QR to QT^2 as AC × L × PC^2 × CD^2, or $2CB^2$ × PC^2 × CD^2 to PC × Gv × CD^2 × CB^2, or as 2PC to Gv. But the points P and Q coinciding, 2PC and Gv are equal. And therefore the quantities L × QR and QT^2, proportional to them, will be also equal. Let those equals be drawn into $\frac{SP^2}{QR}$, and we shall have L × SP^2 equal to $\frac{SP^2 \times QT^2}{QR}$. And therefore (by Cor. 1 and 5, Prop. VI.) the centripetal force is reciprocally as L × SP^2, that is, reciprocally in the duplicate ratio of the distance SP. Q.E.I.

The same otherwise.

Find out the force tending from the centre C of the hyperbola. This will be proportional to the distance CP. But from thence (by Cor. 3, Prop. VII.) the force tending to the focus S will be as $\frac{PE^3}{SP^2}$, that is, because PE is given reciprocally as SP^2. Q.E.I.

And the same way may it be demonstrated, that the body having its centripetal changed into a centrifugal force, will move in the conjugate hyperbola.

LEMMA XIII.

The latus rectum of a parabola belonging to any vertex is quadruple the distance of that vertex from the focus of the figure.

This is demonstrated by the writers on the conic sections.

LEMMA XIV.

The perpendicular, let fall from the focus of a parabola on its tangent, is a mean proportional between the distances of the focus from the point of contact, and from the principal vertex of the figure.

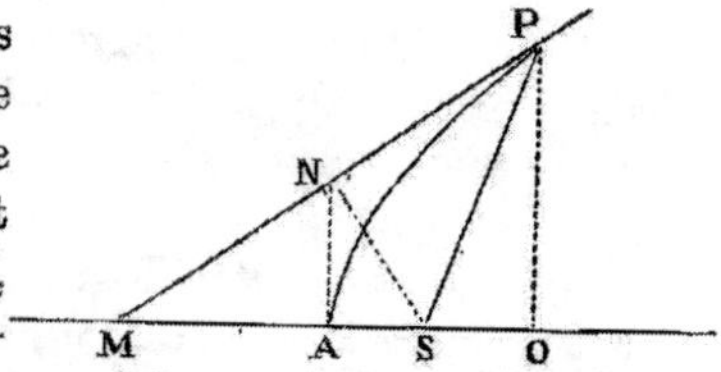

For, let AP be the parabola, S its focus, A its principal vertex, P the point of contact, PO an ordinate to the principal diameter, PM the tangent meeting the principal diameter in M, and SN the perpendicular from the focus on the tangent: join AN, and because of the equal lines MS and SP, MN and NP, MA and AO, the right lines AN, OP, will be parallel; and thence the triangle SAN will be right-angled at A, and similar to the equal triangles SNM, SNP; therefore PS is to SN as SN to SA. Q.E.D.

COR. 1. PS^2 is to SN^2 as PS to SA.

COR. 2. And because SA is given, SN^2 will be as PS.

COR. 3. And the concourse of any tangent PM, with the right line SN, drawn from the focus perpendicular on the tangent, falls in the right line AN that touches the parabola in the principal vertex.

PROPOSITION XIII. PROBLEM VIII.

If a body moves in the perimeter of a parabola; it is required to find the law of the centripetal force tending to the focus of that figure.

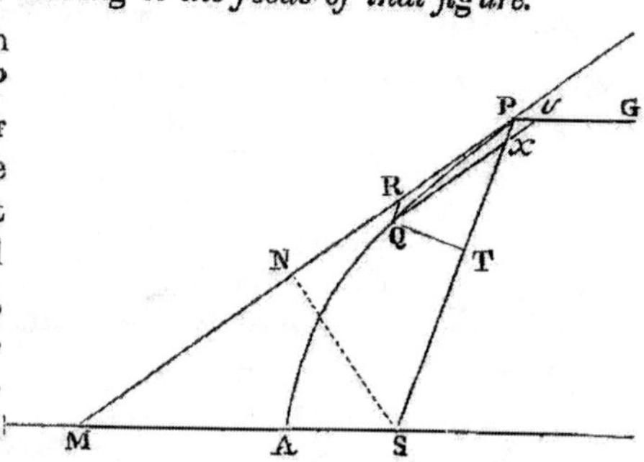

Retaining the construction of the preceding Lemma, let P be the body in the perimeter of the parabola; and from the place Q, into which it is next to succeed, draw QR parallel and QT perpendicular to SP, as also Qv parallel to the tangent, and meeting the diameter PG in v, and the distance

SP in x. Now, because of the similar triangles Pxv, SPM, and of the equal sides SP, SM of the one, the sides Px or QR and Pv of the other will be also equal. But (by the conic sections) the square of the ordinate Qv is equal to the rectangle under the latus rectum and the segment Pv of the diameter; that is (by Lem. XIII.), to the rectangle 4PS × Pv, or 4PS × QR; and the points P and Q coinciding, the ratio of Qv to Qx (by Cor. 2, Lem. VII.,) becomes a ratio of equality. And therefore Qx^2, in this case, becomes equal to the rectangle 4PS × QR. But (because of the similar triangles QxT, SPN), Qx^2 is to QT^2 as PS^2 to SN^2, that is (by Cor. 1, Lem. XIV.), as PS to SA; that is, as 4PS × QR to 4SA × QR, and therefore (by Prop. IX. Lib. V., Elem.) QT^2 and 4SA × QR are equal. Multiply these equals by $\frac{SP^2}{QR}$, and $\frac{SP^2 \times QT^2}{QR}$ will become equal to SP^2 × 4SA: and therefore (by Cor. 1 and 5, Prop. VI.), the centripetal force is reciprocally as SP^2 × 4SA; that is, because 4SA is given, reciprocally in the duplicate ratio of the distance SP. Q.E.I.

Cor. 1. From the three last Propositions it follows, that if any body P goes from the place P with any velocity in the direction of any right line PR, and at the same time is urged by the action of a centripetal force that is reciprocally proportional to the square of the distance of the places from the centre, the body will move in one of the conic sections, having its focus in the centre of force; and the contrary. For the focus, the point of contact, and the position of the tangent, being given, a conic section may be described, which at that point shall have a given curvature. But the curvature is given from the centripetal force and velocity of the body being given; and two orbits, mutually touching one the other, cannot be described by the same centripetal fórce and the same velocity.

Cor. 2. If the velocity with which the body goes from its place P is such, that in any infinitely small moment of time the lineola PR may be thereby described; and the centripetal force such as in the same time to move the same body through the space QR; the body will move in one of the conic sections, whose principal latus rectum is the quantity $\frac{QT^2}{QR}$ in its ultimate state, when the lineolæ PR, QR are diminished *in infinitum*. In these Corollaries I consider the circle as an ellipsis; and I except the case where the body descends to the centre in a right line.

PROPOSITION XIV. THEOREM VI.

If several bodies revolve about one common centre, and the centripetal force is reciprocally in the duplicate ratio of the distance of places from the centre; I say, that the principal latera recta of their orbits are in the duplicate ratio of the areas, which the bodies by radii drawn to the centre describe in the same time.

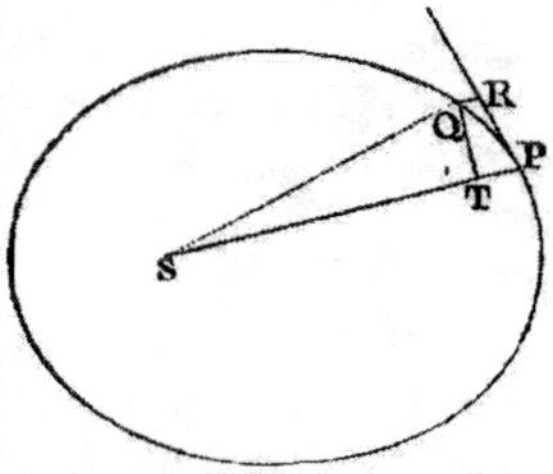

For (by Cor. 2, Prop. XIII) the latus rectum L is equal to the quantity $\frac{QT^2}{QR}$ in its ultimate state when the points P and Q coincide. But the lineola QR in a given time is as the generating centripetal force; that is (by supposition), reciprocally as SP^2. And therefore $\frac{QT^2}{QR}$ is as $QT^2 \times SP^2$; that is, the latus rectum L is in the duplicate ratio of the area $QT \times SP$. Q.E.D.

COR. Hence the whole area of the ellipsis, and the rectangle under the axes, which is proportional to it, is in the ratio compounded of the subduplicate ratio of the latus rectum, and the ratio of the periodic time. For the whole area is as the area $QT \times SP$, described in a given time, multiplied by the periodic time.

PROPOSITION XV. THEOREM VII.

The same things being supposed, I say, that the periodic times in ellipses are in the sesquiplicate ratio of their greater axes.

For the lesser axis is a mean proportional between the greater axis and the latus rectum; and, therefore, the rectangle under the axes is in the ratio compounded of the subduplicate ratio of the latus rectum and the sesquiplicate ratio of the greater axis. But this rectangle (by Cor. 3, Prop. XIV) is in a ratio compounded of the subduplicate ratio of the latus rectum, and the ratio of the periodic time. Subduct from both sides the subduplicate ratio of the latus rectum, and there will remain the sesquiplicate ratio of the greater axis, equal to the ratio of the periodic time. Q.E.D.

COR. Therefore the periodic times in ellipses are the same as in circles whose diameters are equal to the greater axes of the ellipses.

PROPOSITION XVI. THEOREM VIII.

The same things being supposed, and right lines being drawn to the bodies that shall touch the orbits, and perpendiculars being let fall on those tangents from the common focus; I say, that the velocities of the bodies are in a ratio compounded of the ratio of the perpendiculars inversely, and the subduplicate ratio of the principal latera recta directly.

From the focus S draw SY perpendicular to the tangent PR, and the velocity of the body P will be reciprocally in the subduplicate ratio of the quantity $\frac{SY^2}{L}$. For that velocity is as the infinitely small arc PQ de-

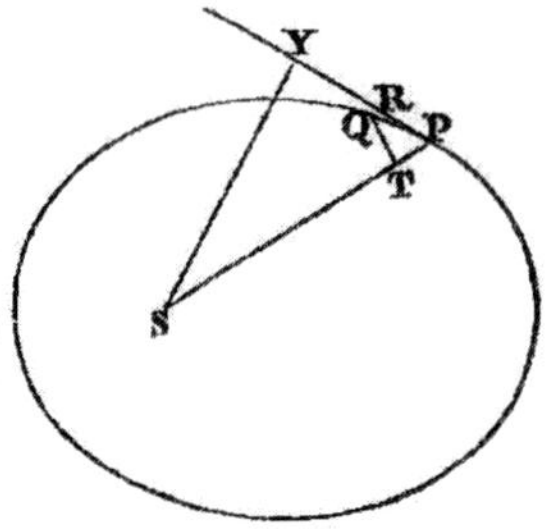

scribed in a given moment of time, that is (by Lem. VII), as the tangent PR; that is (because of the proportionals PR to QT, and SP to SY), as $\frac{SP \times QT}{SY}$; or as SY reciprocally, and SP × QT directly; but SP × QT is as the area described in the given time, that is (by Prop. XIV), in the subduplicate ratio of the latus rectum. Q.E.D.

COR. 1. The principal latera recta are in a ratio compounded of the duplicate ratio of the perpendiculars and the duplicate ratio of the velocities.

COR. 2. The velocities of bodies, in their greatest and least distances from the common focus, are in the ratio compounded of the ratio of the distances inversely, and the subduplicate ratio of the principal latera recta directly. For those perpendiculars are now the distances.

COR. 3. And therefore the velocity in a conic section, at its greatest or least distance from the focus, is to the velocity in a circle, at the same distance from the centre, in the subduplicate ratio of the principal latus rectum to the double of that distance.

COR. 4. The velocities of the bodies revolving in ellipses, at their mean distances from the common focus, are the same as those of bodies revolving in circles, at the same distances; that is (by Cor. 6, Prop. IV), reciprocally in the subduplicate ratio of the distances. For the perpendiculars are now the lesser semi-axes, and these are as mean proportionals between the distances and the latera recta. Let this ratio inversely be compounded with the subduplicate ratio of the latera recta directly, and we shall have the subduplicate ratio of the distance inversely.

COR. 5. In the same figure, or even in different figures, whose principal latera recta are equal, the velocity of a body is reciprocally as the perpendicular let fall from the focus on the tangent.

COR. 6. In a parabola, the velocity is reciprocally in the subduplicate ratio of the distance of the body from the focus of the figure; it is more variable in the ellipsis, and less in the hyperbola, than according to this ratio. For (by Cor. 2, Lem. XIV) the perpendicular let fall from the focus on the tangent of a parabola is in the subduplicate ratio of the distance. In the hyperbola the perpendicular is less variable; in the ellipsis more.

COR. 7. In a parabola, the velocity of a body at any distance from the focus is to the velocity of a body revolving in a circle, at the same distance from the centre, in the subduplicate ratio of the number 2 to 1; in the ellipsis it is less, and in the hyperbola greater, than according to this ratio. For (by Cor. 2 of this Prop.) the velocity at the vertex of a parabola is in this ratio, and (by Cor. 6 of this Prop. and Prop. IV) the same proportion holds in all distances. And hence, also, in a parabola, the velocity is everywhere equal to the velocity of a body revolving in a circle at half the distance; in the ellipsis it is less, and in the hyperbola greater.

COR. 8. The velocity of a body revolving in any conic section is to the velocity of a body revolving in a circle, at the distance of half the principal latus rectum of the section, as that distance to the perpendicular let fall from the focus on the tangent of the section. This appears from Cor. 5.

COR. 9. Wherefore since (by Cor. 6, Prop. IV), the velocity of a body revolving in this circle is to the velocity of another body revolving in any other circle reciprocally in the subduplicate ratio of the distances; therefore, *ex æquo*, the velocity of a body revolving in a conic section will be to the velocity of a body revolving in a circle at the same distance as a mean proportional between that common distance, and half the principal latus rectum of the section, to the perpendicular let fall from the common focus upon the tangent of the section.

PROPOSITION XVII. PROBLEM IX.

Supposing the centripetal force to be reciprocally proportional to the squares of the distances of places from the centre, and that the absolute quantity of that force is known; it is required to determine the line which a body will describe that is let go from a given place with a given velocity in the direction of a given right line.

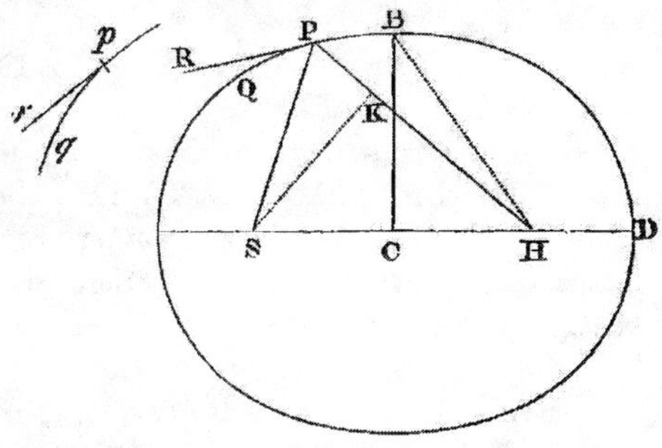

Let the centripetal force tending to the point S be such as will make the body p revolve in any given orbit pq; and suppose the velocity of this body in the place p is known. Then from the place P suppose the body P to be let go with a given velocity in the direction of the line PR; but by virtue of a centripetal force to be immediately turned aside from that right line into the conic section PQ. This, the right line PR will therefore touch in P. Suppose likewise that the right line pr touches the orbit pq in p; and if from S you suppose perpendiculars let fall on those tangents, the principal latus rectum of the conic section (by Cor. 1, Prop. XVI) will be to the principal latus rectum of that orbit in a ratio compounded of the duplicate ratio of the perpendiculars, and the duplicate ratio of the velocities; and is therefore given. Let this latus rectum be L; the focus S of the conic section is also given. Let the angle RPH be the complement of the angle RPS to two right; and the line PH, in which the other focus H is placed, is given by position. Let fall SK perpendicular on PH, and erect the conjugate semi-axis BC; this done, we shall have $SP^2 - 2KPH + PH^2 = SH^2 = 4CH^2 = 4BH^2 - 4BC^2 = \overline{SP + PH}^2 - L \times \overline{SP + PH} = SP^2 + 2SPH + PH^2 - L \times \overline{SP + PH}$. Add on both sides $2KPH - SP^2 - PH^2 + L \times \overline{SP + PH}$, and we shall have $L \times \overline{SP + PH} = 2SPH + 2KPH$, or SP + PH to PH, as 2SP + 2KP to L. Whence PH is given both in length and position. That is, if the velocity of the body in P is such that the latus rectum L is less than 2SP + 2KP, PH will lie on the same side of the tangent PR with the line SP; and therefore the figure will be an ellipsis, which from the given foci S, H, and the principal axis SP + PH, is given also. But if the velocity of the body is so great, that the latus rectum L becomes equal to 2SP + 2KP, the length PH will be infinite; and therefore, the figure will be a parabola, which has its axis SH parallel to the line PK, and is thence given. But if the body goes from its place P with a yet greater velocity, the length PH is to be taken on the other side the tangent; and so the tangent passing between the foci, the figure will be an hyperbola having its principal axis equal to the difference of the lines SP and PH, and thence is given. For if the body, in these cases, revolves in a conic section so found, it is demonstrated in Prop. XI, XII, and XIII, that the centripetal force will be reciprocally as the square of the distance of the body from the centre of force S; and therefore we have rightly determined the line PQ, which a body let go from a given place P with a given velocity, and in the direction of the right line PR given by position, would describe with such a force. Q.E.F.

Cor. 1. Hence in every conic section, from the principal vertex D, the latus rectum L, and the focus S given, the other focus H is given, by taking DH to DS as the latus rectum to the difference between the latus rectum and 4DS. For the proportion, SP + PH to PH as 2SP + 2KP to L, becomes, in the case of this Corollary, DS + DH to DH as 4DS to L, and by division DS to DH as 4DS — L to L.

Cor. 2. Whence if the velocity of a body in the principal vertex D is given, the orbit may be readily found; to wit, by taking its latus rectum to twice the distance DS, in the duplicate ratio of this given velocity to the velocity of a body revolving in a circle at the distance DS (by Cor. 3, Prop. XVI.), and then taking DH to DS as the latus rectum to the difference between the latus rectum and 4DS.

Cor. 3. Hence also if a body move in any conic section, and is forced out of its orbit by any impulse, you may discover the orbit in which it will afterwards pursue its course. For by compounding the proper motion of the body with that motion, which the impulse alone would generate, you will have the motion with which the body will go off from a given place of impulse in the direction of a right line given in position.

COR. 4. And if that body is continually disturbed by the action of some foreign force, we may nearly know its course, by collecting the changes which that force introduces in some points, and estimating the continual changes it will undergo in the intermediate places, from the analogy that appears in the progress of the series.

SCHOLIUM.

If a body P, by means of a centripetal force tending to any given point R, move in the perimeter of any given conic section whose centre is C; and the law of the centripetal force is required: draw CG parallel to the radius RP, and meeting the tangent PG of the orbit in G; and the force required (by Cor. 1, and

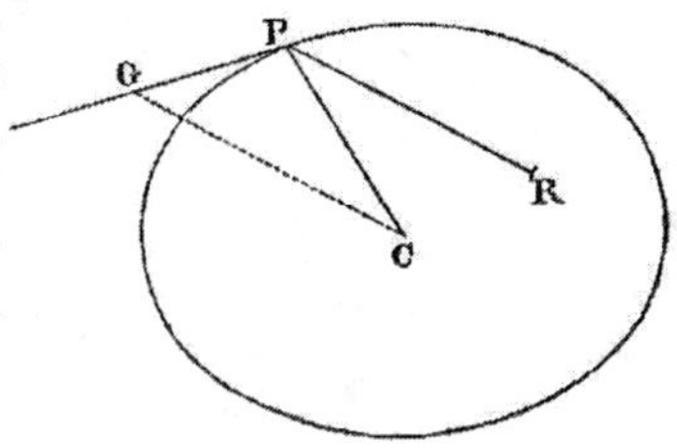

Schol. Prop. X., and Cor. 3, Prop. VII.) will be as $\frac{CG^3}{RP^2}$.

SECTION IV.

Of the finding of elliptic, parabolic, and hyperbolic orbits, from the focus given.

LEMMA XV.

If from the two foci S, H, *of any ellipsis or hyberbola, we draw to any third point* V *the right lines* SV, HV, *whereof one* HV *is equal to the principal axis of the figure, that is, to the axis in which the foci are situated, the other*, SV, *is bisected in* T *by the perpendicular* TR *let fall upon it ; that perpendicular* TR *will somewhere touch the conic section : and,* vice versa, *if it does touch it,* HV *will be equal to the principal axis of the figure.*

For, let the perpendicular TR cut the right line HV, produced, if need be, in R; and join SR. Because TS, TV are equal, therefore the right lines SR, VR, as well as the angles TRS, TRV, will be also equal. Whence the point R will be in the conic section, and the perpendicular TR will touch the same; and the contrary. Q.E.D.

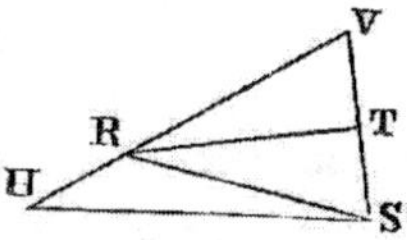

PROPOSITION XVIII. PROBLEM X.

From a focus and the principal axes given, to describe elliptic and hyperbolic trajectories, which shall pass through given points, and touch right lines given by position.

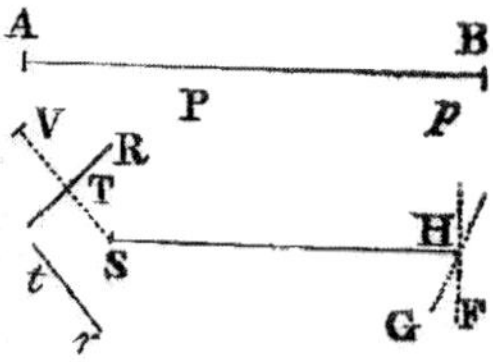

Let S be the common focus of the figures; AB the length of the principal axis of any trajectory; P a point through which the trajectory should pass; and TR a right line which it should touch. About the centre P, with the interval AB — SP, if the orbit is an ellipsis, or AB + SP, if the orbit is an hyperbola, describe the circle HG. On the tangent TR let fall the perpendicular ST, and produce the same to V, so that TV may be equal to ST; and about V as a centre with the interval AB describe the circle FH. In this manner, whether two points P, *p*, are given, or two tangents TR, *tr*, or a point P and a tangent TR, we are to describe two circles. Let H be their common intersection, and from the foci S, H, with the given axis describe the trajectory: I say, the thing is done. For (because PH + SP in the ellipsis, and PH — SP in the hyperbola, is equal to the axis) the described trajectory will pass through the point P, and (by the preceding Lemma) will touch the right line TR. And by the same argument it will either pass through the two points P, *p*, or touch the two right lines TR, *tr*. Q.E.F.

PROPOSITION XIX. PROBLEM XI.

About a given focus, to describe a parabolic trajectory, which shall pass through given points, and touch right lines given by position.

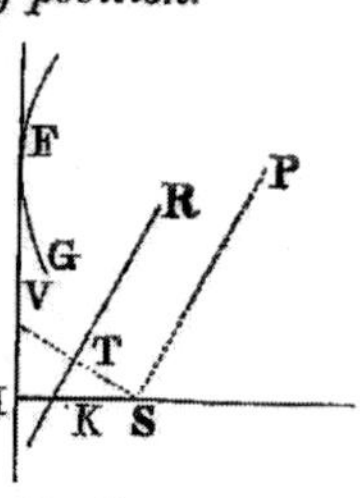

Let S be the focus, P a point, and TR a tangent of the trajectory to be described. About P as a centre, with the interval PS, describe the circle FG. From the focus let fall ST perpendicular on the tangent, and produce the same to V, so as TV may be equal to ST. After the same manner another circle *fg* is to be described, if another point *p* is given; or another point *v* is to be found, if another tangent *tr* is given; then draw the right line IF, which shall touch the two circles FG, *fg*, if two points P, *p* are given; or pass through the two points V, *v*, if two tangents TR, *tr*, are given: or touch the circle FG, and pass through the point V, if the point P and the tangent TR are given. On FI let fall the perpendicular SI, and bisect the same in K; and with the axis SK and principal vertex K describe a parabola: I say the thing is done. For this parabola (because SK is equal to IK, and SP to FP) will pass through the point P; and (by Cor. 3, Lem. XIV) because ST is equal to TV, and STR a right angle, it will touch the right line TR. Q.E.F.

PROPOSITION XX. PROBLEM XII.

About a given focus to describe any trajectory given in specie which shall pass through given points, and touch right lines given by position.

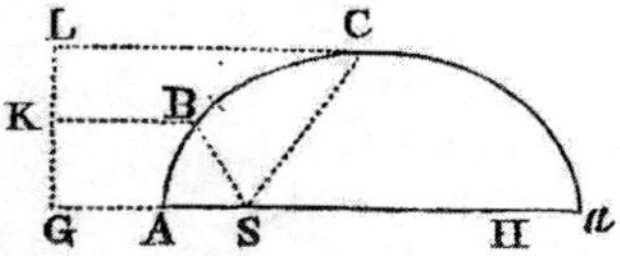

CASE 1. About the focus S it is reuired to describe a trajectory ABC, passing through two points B, C. Because the trajectory is given in specie, the ratio of the principal axis to the distance of the foci will be given. In that ratio take KB to BS, and LC to CS. About the centres B, C, with the intervals BK, CL, describe two circles; and on the right line KL, that touches the same in K and L, let fall the perpendicular SG; which cut in A and a, so that GA may be to AS, and Ga to aS, as KB to BS; and with the axis Aa, and vertices A, a, describe a trajectory: I say the thing is done. For let H be the other focus of the described figure, and seeing GA is to AS as Ga to aS, then by division we shall have Ga—GA, or Aa to aS—AS, or SH in the same ratio, and therefore in the ratio which the principal axis of the figure to be described has to the distance of its foci; and therefore the described figure is of the same species with the figure which was to be described. And since KB to BS, and LC to CS, are in the same ratio, this figure will pass through the points B, C, as is manifest from the conic sections.

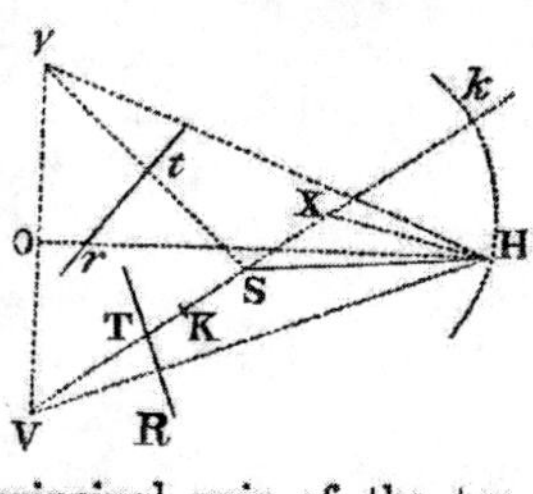

CASE 2. About the focus S it is required to describe a trajectory which shall somewhere touch two right lines TR, tr. From the focus on those tangents let fall the perpendiculars ST, St, which produce to V, v, so that TV, tv may be equal to TS, tS. Bisect Vv in O, and erect the indefinite perpendicular OH, and cut the right line VS infinitely produced in K and k, so that VK be to KS, and Vk to kS, as the principal axis of the trajectory to be described is to the distance of its foci. On the diameter Kk describe a circle cutting OH in H; and with the foci S, H, and principal axis equal to VH, describe a trajectory: I say, the thing is done. For bisecting Kk in X, and joining HX, HS, HV, Hv, because VK is to KS as Vk to kS; and by composition, as VK + Vk to KS + kS; and by division, as Vk — VK to kS — KS, that is, as 2VX to 2KX, and 2KX to 2SX, and therefore as VX to HX and HX to SX, the triangles VXH, HXS will be similar; therefore VH will be to SH as VX to XH; and therefore as VK to KS. Wherefore VH, the principal axis of the described trajectory, has the same ratio to SH, the distance of the foci, as

the principal axis of the trajectory which was to be described has to the distance of its foci; and is therefore of the same species. And seeing VH, *v*H are equal to the principal axis, and VS, *v*S are perpendicularly bisected by the right lines TR, *tr*, it is evident (by Lem. XV) that those right lines touch the described trajectory. Q.E.F.

Case. 3. About the focus S it is required to describe a trajectory, which shall touch a right line TR in a given Point R. On the right line TR let fall the perpendicular ST, which produce to V, so that TV may be equal to ST; join VR, and cut the right line VS indefinitely produced in K and *k*, so that VK may be to SK, and V*k* to S*k*, as the principal axis of the ellipsis to be described to the distance of its foci; and on the diameter K*k* describing a circle, cut the right line VR produced in H; then with the foci S, H, and principal axis equal to VH, describe a trajectory: I say, the thing is done. For VH is to SH as VK to SK,

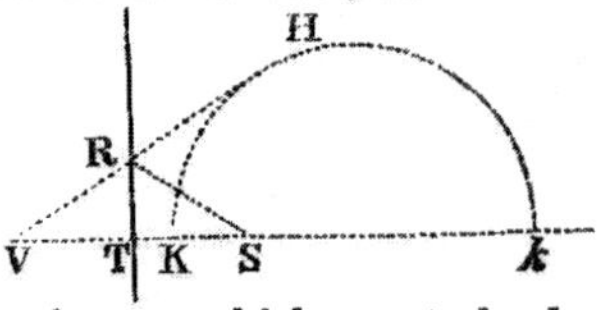

and therefore as the principal axis of the trajectory which was to be described to the distance of its foci (as appears from what we have demonstrated in Case 2); and therefore the described trajectory is of the same species with that which was to be described; but that the right line TR, by which the angle VRS is bisected, touches the trajectory in the point R, is certain from the properties of the conic sections. Q.E.F.

Case 4. About the focus S it is required to describe a trajectory APB that shall touch a right line TR, and pass through any given point P without the tangent, and shall be similar to the figure *apb*, described with the principal axis *ab*, and foci *s*, *h*. On the tangent TR let fall the perpendicular ST, which produce to V, so that TV may be equal to ST; and making the angles *hsq*, *shq*, equal to the angles VSP, SVP, about *q* as a centre, and with an interval which shall be to *ab* as SP to VS, describe a circle cutting the figure *apb* in *p*: join *sp*, and draw SH such that it may be to *sh* as SP is to *sp*, and may make the angle PSH equal to the angle *psh*, and the angle VSH equal to the angle *psq*. Then with the foci S, H, and principal axis AB, equal to the distance VH, describe a conic section: I say, the thing is done; for if *sv* is drawn so that it shall be to

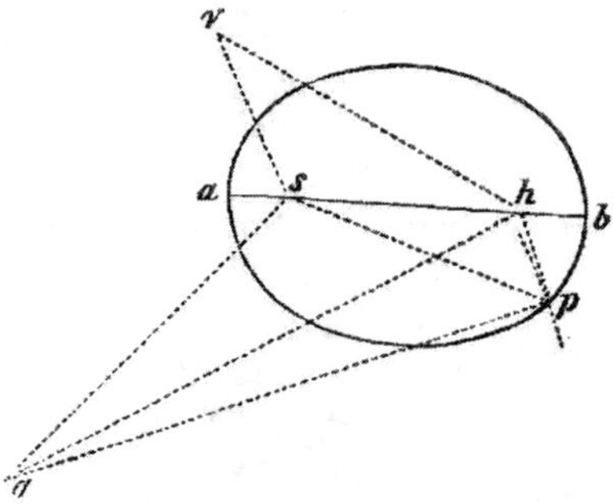

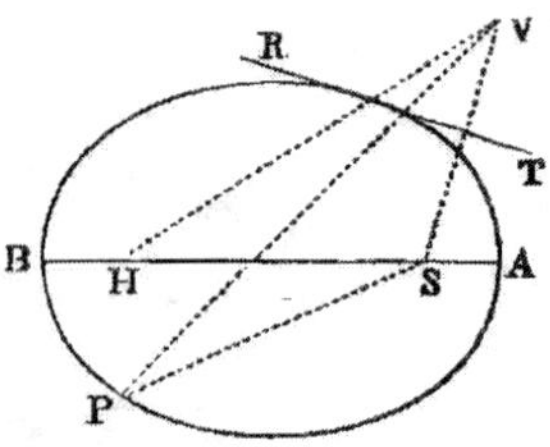

sp as *sh* is to *sq*, and shall make the angle *vsp* equal to the angle *hsq*, and the angle *vsh* equal to the angle *psq*, the triangles *svh*, *spq*, will be similar, and therefore *vh* will be to *pq* as *sh* is to *sq*; that is (because of the similar triangles VSP, *hsq*), as VS is to SP, or as *ab* to *pq*. Wherefore *vh* and *ab* are equal. But, because of the similar triangles VSH, *vsh*, VH is to SH as *vh* to *sh*; that is, the axis of the conic section now described is to the distance of its foci as the axis *ab* to the distance of the foci *sh*; and therefore the figure now described is similar to the figure *aph*. But, because the triangle PSH is similar to the triangle *psh*, this figure passes through the point P; and because VH is equal to its axis, and VS is perpendicularly bisected by the right line TR, the said figure touches the right line TR. Q.E.F.

LEMMA XVI.

From three given points to draw to a fourth point that is not given three right lines whose differences shall be either given, or none at all.

CASE 1. Let the given points be A, B, C, and Z the fourth point which we are to find; because of the given difference of the lines AZ, BZ, the locus of the point Z will be an hyperbola whose foci are A and B, and whose principal axis is the given difference. Let that axis be MN. Taking PM to MA as MN is to AB, erect PR perpendicular to AB, and let fall ZR perpendicular to PR; then from the nature of the hyperbola, ZR will be to AZ as MN is to AB. And by the like argument, the locus of the point Z will

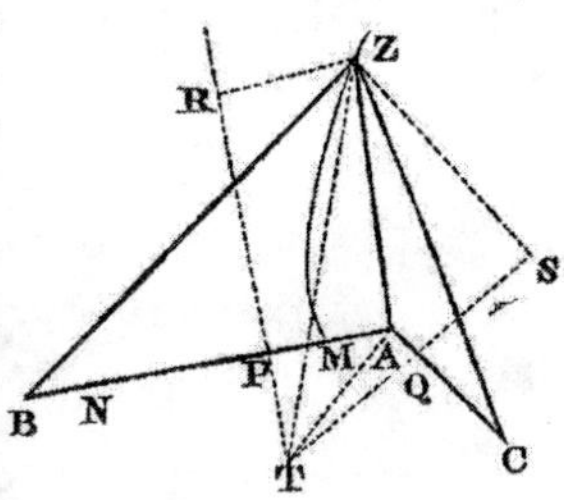

be another hyperbola, whose foci are A, C, and whose principal axis is the difference between AZ and CZ; and QS a perpendicular on AC may be drawn, to which (QS) if from any point Z of this hyperbola a perpendicular ZS is let fall (this ZS), shall be to AZ as the difference between AZ and CZ is to AC. Wherefore the ratios of ZR and ZS to AZ are given, and consequently the ratio of ZR to ZS one to the other; and therefore if the right lines RP, SQ, meet in T, and TZ and TA are drawn, the figure TRZS will be given in specie, and the right line TZ, in which the point Z is somewhere placed, will be given in position. There will be given also the right line TA, and the angle ATZ; and because the ratios of AZ and TZ to ZS are given, their ratio to each other is given also; and thence will be given likewise the triangle ATZ, whose vertex is the point Z. Q.E.I.

CASE 2. If two of the three lines, for example AZ and BZ, are equal, draw the right line TZ so as to bisect the right line AB; then find the triangle ATZ as above. Q.E.I.

Case 3. If all the three are equal, the point Z will be placed in the centre of a circle that passes through the points A, B, C. Q.E.I.

This problematic Lemma is likewise solved in Apollonius's Book of Tactions restored by Vieta.

PROPOSITION XXI. PROBLEM XIII.

About a given focus to describe a trajectory that shall pass through given points and touch right lines given by position.

Let the focus S, the point P, and the tangent TR be given, and suppose that the other focus H is to be found. On the tangent let fall the perpendicular ST, which produce to Y, so that TY may be equal to ST, and YH will be equal to the principal axis. Join SP, HP, and SP will be the difference between HP and the principal axis. After this manner, if more tangents TR are given, or more points P, we shall always determine as many lines YH, or PH, drawn from the said points Y or P, to the focus H, which either shall be equal to the axes, or differ from the axes by given lengths SP; and therefore which shall either be equal among themselves, or shall have given differences; from whence (by the preceding Lemma), that other focus H is given. But having the foci and the length of the axis (which is either YH, or, if the trajectory be an ellipsis, PH + SP; or PH — SP, if it be an hyperbola), the trajectory is given. Q.E.I.

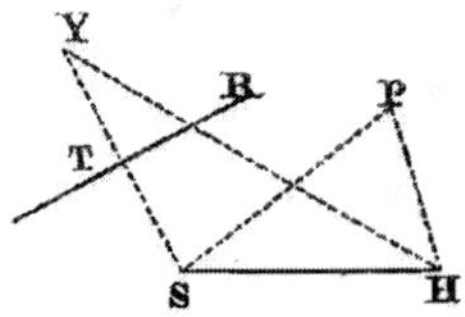

SCHOLIUM.

When the trajectory is an hyperbola, I do not comprehend its conjugate hyperbola under the name of this trajectory. For a body going on with a continued motion can never pass out of one hyperbola into its conjugate hyperbola.

The case when three points are given is more readily solved thus. Let B, C, D, be the given points. Join BC, CD, and produce them to E, F, so as EB may be to EC as SB to SC; and FC to FD as SC to SD. On EF drawn and produced let fall the perpendiculars SG, BH, and in GS produced indefinitely take GA to AS, and Ga to aS, as HB is to BS; then A will be the vertex, and Aa the principal axis of the trajectory; which, according as GA is greater than, equal to, or less than

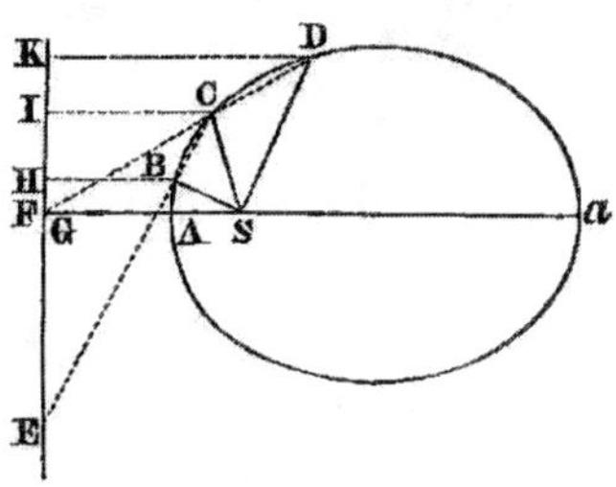

AS, will be either an ellipsis, a parabola, or an hyperbola; the point a in the first case falling on the same side of the line GF as the point A; in the second, going off to an infinite distance; in the third, falling on the other side of the line GF. For if on GF the perpendiculars CI, DK are let fall, IC will be to HB as EC to EB; that is, as SC to SB; and by permutation, IC to SC as HB to SB, or as GA to SA. And, by the like argument, we may prove that KD is to SD in the same ratio. Wherefore the points B, C, D lie in a conic section described about the focus S, in such manner that all the right lines drawn from the focus S to the several points of the section, and the perpendiculars let fall from the same points on the right line GF, are in that given ratio.

That excellent geometer M. De la Hire has solved this Problem much after the same way, in his Conics, Prop. XXV., Lib. VIII.

SECTION V.

How the orbits are to be found when neither focus is given.

LEMMA XVII.

If from any point P *of a given conic section, to the four produced sides* AB, CD, AC, DB, *of any trapezium* ABDC *inscribed in that section, as many right lines* PQ, PR, PS, PT *are drawn in given angles, each line to each side; the rectangle* PQ × PR *of those on the opposite sides* AB, CD, *will be to the rectangle* PS × PT *of those on the other two opposite sides* AC, BD, *in a given ratio.*

CASE 1. Let us suppose, first, that the lines drawn to one pair of opposite sides are parallel to either of the other sides; as PQ and PR to the side AC, and PS and PT to the side AB. And farther, that one pair of the opposite sides, as AC and BD, are parallel betwixt themselves; then the right line which bisects those parallel sides will be one of the diameters of the conic section, and will likewise bisect RQ. Let O be the point in which RQ is bisected, and PO will be an ordinate to that diameter. Produce PO to K, so that OK may be equal to PO, and OK will be an ordinate on the other side of that diameter. Since, therefore, the points A, B, P, and K are placed in the conic section, and PK cuts AB in a given angle, the rectangle PQK (by Prop. XVII., XIX., XXI. and XXIII., Book III., of Apollonius's Conics) will be to the rectangle AQB in a given ratio. But QK and PR are equal, as being the differences of the equal lines OK, OP, and OQ, OR; whence the rectangles PQK and PQ × PR are equal; and therefore the rectangle PQ × PR is to the rectangle AQB, that is, to the rectangle PS × PT in a given ratio. Q.E.D.

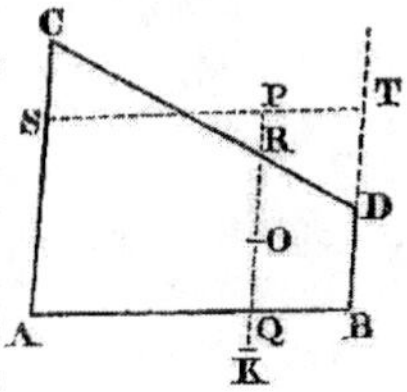

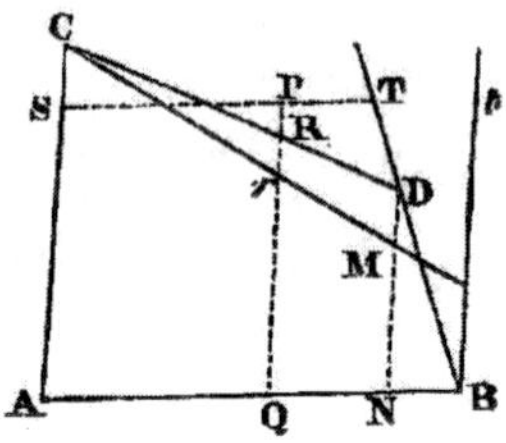

CASE 2. Let us next suppose that the opposite sides AC and BD of the trapezium are not parallel. Draw Bd parallel to AC, and meeting as well the right line ST in t, as the conic section in d. Join Cd cutting PQ in r, and draw DM parallel to PQ, cutting Cd in M, and AB in N. Then (because of the similar triangles BTt, DBN), Bt or PQ is to Tt as DN to NB. And so Rr is to AQ or PS as DM to AN. Wherefore, by multiplying the antecedents by the antecedents, and the consequents by the consequents, as the rectangle PQ × Rr is to the rectangle PS × Tt, so will the rectangle NDM be to the rectangle ANB; and (by Case 1) so is the rectangle PQ × Pr to the rectangle PS × Pt; and by division, so is the rectangle PQ × PR to the rectangle PS × PT. Q.E.D.

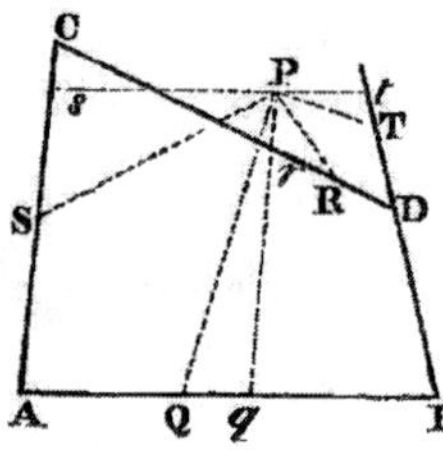

CASE 3. Let us suppose, lastly, the four lines PQ, PR, PS, PT, not to be parallel to the sides AC, AB, but any way inclined to them. In their place draw Pq, Pr, parallel to AC; and Ps, Pt parallel to AB; and because the angles of the triangles PQq, PRr, PSs, PTt are given, the ratios of PQ to Pq, PR to Pr, PS to Ps, PT to Pt will be also given; and therefore the compounded ratios PQ × PR to Pq × Pr, and PS × PT to Ps × Pt are given. But from what we have demonstrated before, the ratio of Pq × Pr to Ps × Pt is given; and therefore also the ratio of PQ × PR to PS × PT. Q.E.D.

LEMMA XVIII.

The same things supposed, if the rectangle PQ × PR *of the lines drawn to the two opposite sides of the trapezium is to the rectangle* PS × PT *of those drawn to the other two sides in a given ratio, the point* P, *from whence those lines are drawn, will be placed in a conic section described about the trapezium.*

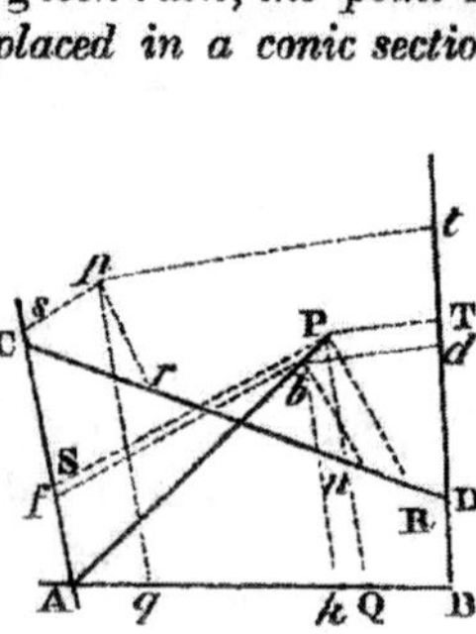

Conceive a conic section to be described passing through the points A, B, C, D, and any one of the infinite number of points P, as for example p; I say, the point P will be always placed in this section. If you deny the thing, join AP cutting this conic section somewhere else, if possible, than in P, as in b. Therefore if from those points p and b, in the given angles to the sides of the trapezium, we draw the right lines pq, pr, ps, pt, and bk, bn, bf, bd, we shall have, as bk × bn to bf × bd, so (by Lem. XVII) $pq \times pr$ to $ps \times pt$; and so (by supposition) PQ × PR to PS × PT. And because of the similar trapezia $bkAf$, PQAS, as bk to bf, so PQ to PS. Wherefore by dividing the terms of the preceding proportion by the correspondent terms of this, we shall have bn to bd as PR to PT. And therefore the equiangular trapezia Dnbd, DRPT, are similar, and consequently their diagonals Db, DP do coincide. Wherefore b falls in the intersection of the right lines AP, DP, and consequently coincides with the point P. And therefore the point P, wherever it is taken, falls to be in the assigned conic section. Q.E.D.

COR. Hence if three right lines PQ, PR, PS, are drawn from a common point P, to as many other right lines given in position, AB, CD, AC, each to each, in as many angles respectively given, and the rectangle PQ × PR under any two of the lines drawn be to the square of the third PS in a given ratio; the point P, from which the right lines are drawn, will be placed in a conic section that touches the lines AB, CD in A and C; and the contrary. For the position of the three right lines AB, CD, AC remaining the same, let the line BD approach to and coincide with the line AC; then let the line PT come likewise to coincide with the line PS; and the rectangle PS × PT will become PS^2, and the right lines AB, CD, which before did cut the curve in the points A and B, C and D, can no longer cut, but only touch, the curve in those coinciding points.

SCHOLIUM.

In this Lemma, the name of conic section is to be understood in a large sense, comprehending as well the rectilinear section through the vertex of the cone, as the circular one parallel to the base. For if the point p happens to be in a right line, by which the points A and D, or C and B are joined, the conic section will be changed into two right lines, one of which is that right line upon which the point p falls, and the other is a right line that joins the other two of the four points. If the two opposite angles of the trapezium taken together are equal to two right angles, and if the four lines PQ, PR, PS, PT, are drawn to the sides thereof at right angles, or any other equal angles, and the rectangle PQ × PR under two of the lines drawn PQ and PR, is equal to the rectangle

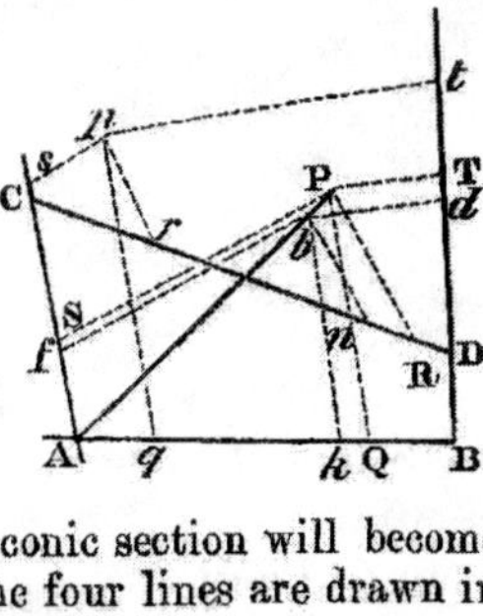

PS × PT under the other two PS and PT, the conic section will become a circle. And the same thing will happen if the four lines are drawn in any angles, and the rectangle PQ × PR, under one pair of the lines drawn, is to the rectangle PS × PT under the other pair as the rectangle under the sines of the angles S, T, in which the two last lines PS, PT are drawn to the rectangle under the sines of the angles Q, R, in which the first two

PQ, PR are drawn. In all other cases the locus of the point P will be one of the three figures which pass commonly by the name of the conic sections. But in room of the trapezium ABCD, we may substitute a quadrilateral figure whose two opposite sides cross one another like diagonals. And one or two of the four points A, B, C, D may be supposed to be removed to an infinite distance, by which means the sides of the figure which converge to those points, will become parallel; and in this case the conic section will pass through the other points, and will go the same way as the parallels *in infinitum.*

LEMMA XIX.

To find a point P *from which if four right lines* PQ, PR, PS, PT *are drawn to as many other right lines* AB, CD, AC, BD, *given by position, each to each, at given angles, the rectangle* PQ × PR, *under any two of the lines drawn, shall be to the rectangle* PS × PT, *under the other two, in a given ratio.*

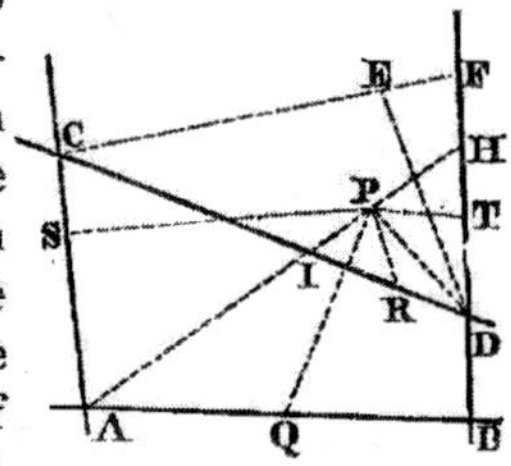

Suppose the lines AB, CD, to which the two right lines PQ, PR, containing one of the rectangles, are drawn to meet two other lines, given by position, in the points A, B, C, D. From one of those, as A, draw any right line AH, in which you would find the point P. Let this cut the opposite lines BD, CD, in H and I; and, because all the angles of the figure are given, the ratio of PQ to PA, and PA to PS, and therefore of PQ to PS, will be also given. Subducting this ratio from the given ratio of PQ × PR to PS × PT, the ratio of PR to PT will be given; and adding the given ratios of PI to PR, and PT to PH, the ratio of PI to PH, and therefore the point P will be given. Q.E.I.

Cor. 1. Hence also a tangent may be drawn to any point D of the locus of all the points P. For the chord PD, where the points P and D meet, that is, where AH is drawn through the point D, becomes a tangent. In which case the ultimate ratio of the evanescent lines IP and PH will be found as above. Therefore draw CF parallel to AD, meeting BD in F, and cut it in E in the same ultimate ratio, then DE will be the tangent; because CF and the evanescent IH are parallel, and similarly cut in E and P.

Cor. 2. Hence also the locus of all the points P may be determined. Through any of the points A, B, C, D, as A, draw AE touching the locus, and through any other point B parallel to the tangent, draw BF meeting the locus in F; and find the point F by this Lemma. Bisect BF in G, and, drawing the indefinite line AG, this will be the position of the diameter to which BG and FG are ordinates. Let this AG meet the locus in H, and AH will be its diameter or latus transversum, to which the latus rectum will be as BG^2 to AG × GH. If AG nowhere meets the locus, the line AH being infinite, the locus will be a parabola; and its latus rectum corresponding to the diameter AG will be $\frac{BG^2}{AG}$. But if it does meet it anywhere, the locus will be an hyperbola, when the points A and H are placed on the same side the point G; and an ellipsis, if the point G falls between the points A and H; unless, perhaps, the angle AGB is a right angle, and at the same time BG^2 equal to the rectangle AGH, in which case the locus will be a circle.

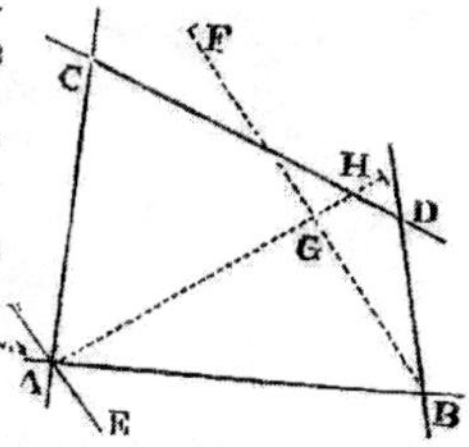

And so we have given in this Corollary a solution of that famous Problem of the ancients concerning four lines, begun by Euclid, and carried on by Apollonius; and this not an analytical calculus, but a geometrical composition, such as the ancients required.

LEMMA XX.

If the two opposite angular points A *and* P *of any parallelogram* ASPQ *touch any conic section in the points* A *and* P; *and the sides* AQ, AS *of one of those angles, indefinitely produced, meet the same conic section in* B *and* C; *and from the points of concourse* B *and* C *to any fifth point* D *of the conic section, two right lines* BD, CD *are drawn meeting the two other sides* PS, PQ *of the parallelogram, indefinitely produced in* T *and* R; *the parts* PR *and* PT, *cut off from the sides, will always be one to the other in a given ratio. And* vice versa, *if those parts cut off are one to the other in a given ratio, the locus of the point* D *will be a conic section passing through the four points* A, B, C, P.

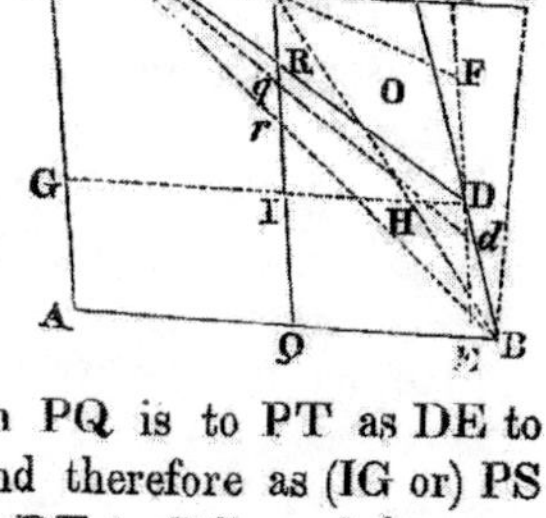

CASE 1. Join BP, CP, and from the point D draw the two right lines DG, DE, of which the first DG shall be parallel to AB, and meet PB, PQ, CA in H, I, G; and the other DE shall be parallel to AC, and meet PC, PS, AB, in F, K, E; and (by Lem. XVII) the rectangle DE × DF will be to the rectangle DG × DH in a given ratio. But PQ is to DE (or IQ) as PB to HB, and consequently as PT to DH; and by permutation PQ is to PT as DE to DH. Likewise PR is to DF as RC to DC, and therefore as (IG or) PS to DG; and by permutation PR is to PS as DF to DG; and, by compounding those ratios, the rectangle PQ × PR will be to the rectangle PS × PT as the rectangle DE × DF is to the rectangle DG × DH, and consequently in a given ratio. But PQ and PS are given, and therefore the ratio of PR to PT is given. Q.E.D.

Case 2. But if PR and PT are supposed to be in a given ratio one to the other, then by going back again, by a like reasoning, it will follow that the rectangle DE × DF is to the rectangle DG × DH in a given ratio; and so the point D (by Lem. XVIII) will lie in a conic section passing through the points A, B, C, P, as its locus. Q.E.D.

Cor. 1. Hence if we draw BC cutting PQ in *r* and in PT take P*t* to P*r* in the same ratio which PT has to PR; then B*t* will touch the conic section in the point B. For suppose the point D to coalesce with the point B, so that the chord BD vanishing, BT shall become a tangent, and CD and BT will coincide with CB and B*t*.

Cor. 2. And, vice versa, if B*t* is a tangent, and the lines BD, CD meet in any point D of a conic section, PR will be to PT as P*r* to P*t*. And, on the contrary, if PR is to PT as P*r* to P*t*, then BD and CD will meet in some point D of a conic section.

Cor. 3. One conic section cannot cut another conic section in more than four points. For, if it is possible, let two conic sections pass through the five points A, B, C, P, O; and let the right line BD cut them in the points D, *d*, and the right line C*d* cut the right line PQ in *q*. Therefore PR is to PT as P*q* to PT: whence PR and P*q* are equal one to the other, against the supposition.

LEMMA XXI.

If two moveable and indefinite right lines BM, CM *drawn through given points* B, C, *as poles, do by their point of concourse* M *describe a third right line* MN *given by position; and other two indefinite right lines* BD, CD *are drawn, making with the former two at those given points* B, C, *given angles,* MBD, MCD : *I say, that those two right lines* BD, CD *will by their point of concourse* D *describe a conic section passing through the points* B, C. *And, vice versa, if the right lines* BD, CD *do by their point of concourse* D *describe a conic section passing through the given points* B, C, A, *and the angle* DBM *is always equal to the given angle* ABC, *as well as the angle* DCM *always equal to the given angle* ACB, *the point* M *will lie in a right line given by position, as its locus.*

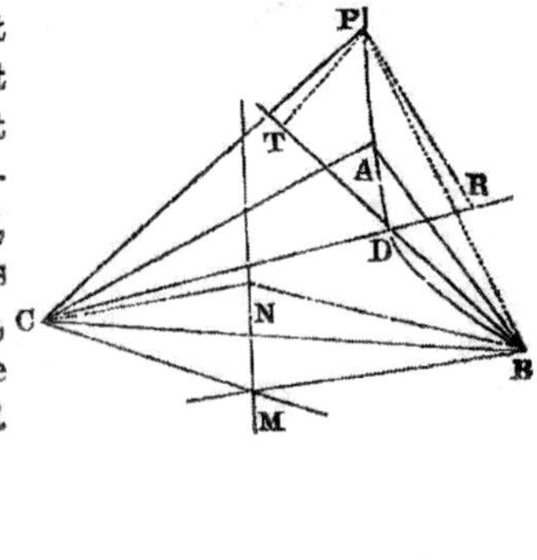

For in the right line MN let a point N be given, and when the moveable point M falls on the immoveable point N, let the moveable point D fall on an immovable point P. Join CN, BN, CP, BP, and from the point P draw the right lines PT, PR meeting BD, CD in T and R, and making the angle BPT equal to the given angle BNM, and the angle CPR equal to the given angle CNM. Wherefore since (by supposition) the angles MBD, NBP are equal, as also the angles MCD, NCP, take away the angles NBD and NCD that are common, and there will remain the angles NBM and PBT, NCM and PCR equal; and therefore the triangles NBM, PBT are similar, as also the triangles NCM, PCR. Wherefore PT is to NM as PB to NB; and PR to NM as PC to NC. But the points, B, C, N, P are immovable: wherefore PT and PR have a given ratio to NM, and consequently a given ratio between themselves; and therefore, (by Lemma XX) the point D wherein the moveable right lines BT and CR perpetually concur, will be placed in a conic section passing through the points B, C, P. Q.E.D.

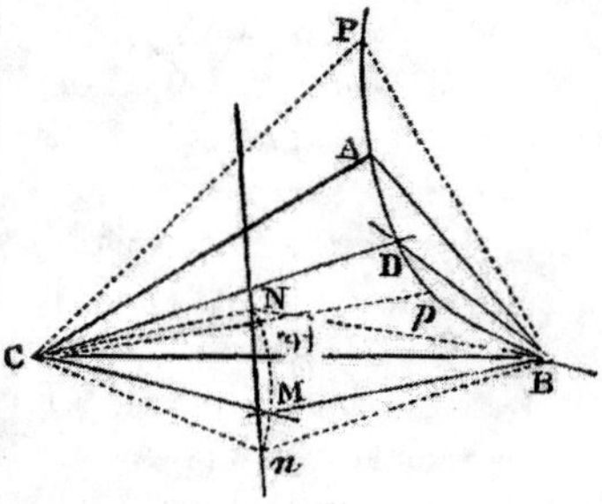

And, *vice versa*, if the moveable point D lies in a conic section passing through the given points B, C, A; and the angle DBM is always equal to the given angle ABC, and the angle DCM always equal to the given angle ACB, and when the point D falls successively on any two immovable points *p*, P, of the conic section, the moveable point M falls successively on two immovable points *n*, N. Through these points *n*, N, draw the right line *n*N: this line *n*N will be the perpetual locus of that moveable point M. For, if possible, let the point M be placed in any curve line. Therefore the point D will be placed in a conic section passing through the five points B, C, A, *p*, P, when the point M is perpetually placed in a curve line. But from what was demonstrated before, the point D will be also placed in a conic section passing through the same five points B, C, A, *p*, P, when the point M is perpetually placed in a right line. Wherefore the two conic sections will both pass through the same five points, against Corol. 3, Lem. XX. It is therefore absurd to suppose that the point M is placed in a curve line. Q.E.D.

PROPOSITION XXII. PROBLEM XIV.

To describe a trajectory that shall pass through five given points.

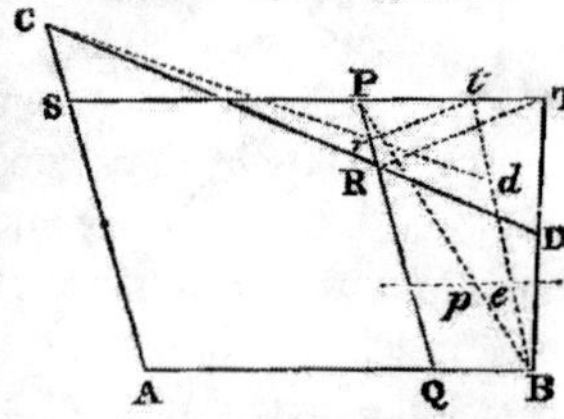

Let the five given points be A, B, C, P, D. From any one of them, as A, to any other two as B, C, which may be called the poles, draw the right lines AB, AC, and parallel to those the lines TPS, PRQ through the fourth point P. Then from the two poles B, C, draw through the fifth point D two indefinite lines BDT, CRD, meeting with the last drawn lines TPS, PRQ (the former with the former, and the latter with the latter) in T and R. Then drawing the right line *tr* parallel to TR, cutting off from the right lines PT, PR, any segments P*t*, P*r*, proportional to PT, PR; and if through their extremities, *t*, *r*, and the poles B, C, the right lines B*t*, C*r* are drawn, meeting in *d*, that point *d* will be placed in the trajectory required. For (by Lem. XX) that point *d* is placed in a conic section passing through the four points A, B, C, P; and the lines R*r*, T*t* vanishing, the point *d* comes to coincide with the point D. Wherefore the conic section passes through the five points A, B, C, P, D. Q.E.D.

The same otherwise.

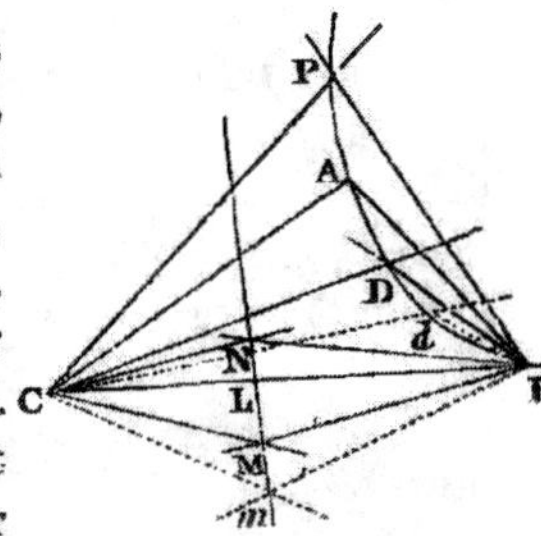

Of the given points join any three, as A, B, C; and about two of them B, C, as poles, making the angles ABC, ACB of a given magnitude to revolve, apply the legs BA, CA, first to the point D, then to the point P, and mark the points M, N, in which the other legs BL, CL intersect each other in both cases. Draw the indefinite right line MN, and let those moveable angles revolve about their poles B, C, in such manner that the intersection, which is now supposed to be *m*, of the legs BL, CL, or BM, CM, may always fall in that indefinite right line MN; and the intersection, which is now supposed to be *d*, of the legs BA, CA, or BD, CD, will describe the trajectory required, PAD*d*B. For (by Lem. XXI) the point *d* will be placed in a conic section passing through the points B, C; and when the point *m* comes to coincide with the points L, M, N, the point *d* will (by construction) come to coincide with the points A, D, P. Wherefore a conic section will be described that shall pass through the five points A, B, C, P, D. Q.E.F.

Cor. 1. Hence a right line may be readily drawn which shall be a tangent to the trajectory in any given point B. Let the point *d* come to coincide with the point B, and the right line B*d* will become the tangent required.

Cor. 2. Hence also may be found the centres, diameters, and latera recta of the trajectories, as in Cor. 2, Lem. XIX.

SCHOLIUM.

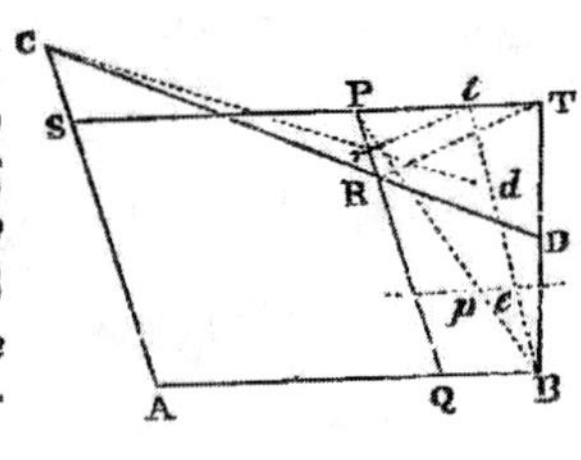

The former of these constructions will become something more simple by joining BP, and in that line, produced, if need be, taking B*p* to BP as PR is to PT; and through *p* draw the indefinite right line *pe* parallel to SPT, and in that line *pe* taking always *pe* equal to P*r*; and draw the right lines B*e*, C*r*

to meet in *d*. For since P*r* to P*t*, PR to PT, *p*B to PB, *pe* to P*t*, are all in the same ratio, *pe* and P*r* will be always equal. After this manner the points of the trajectory are most readily found, unless you would rather describe the curve mechanically, as in the second construction.

PROPOSITION XXIII. PROBLEM XV.

To describe a trajectory that shall pass through four given points, and touch a right line given by position.

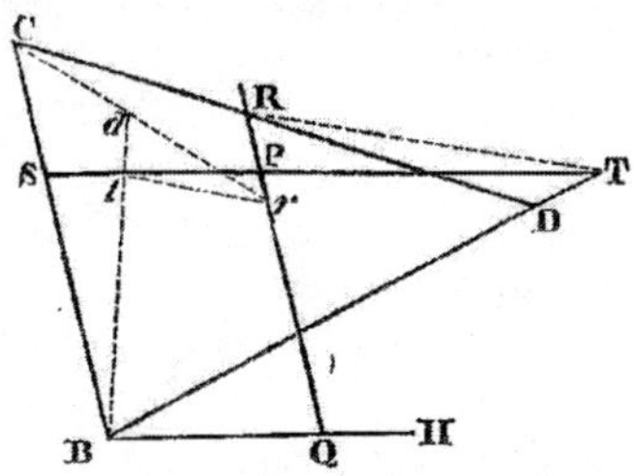

CASE 1. Suppose that HB is the given tangent, B the point of contact, and C, D, P, the three other given points. Join BC, and draw PS parallel to BH, and PQ parallel to BC; complete the parallelogram BSPQ. Draw BD cutting SP in T, and CD cutting PQ in R. Lastly, draw any line *tr* parallel to TR, cutting off from PQ, PS, the segments P*r*, P*t* proportional to PR, PT respectively; and draw C*r*, B*t* their point of concourse *d* will (by Lem. XX) always fall on the trajectory to be described.

The same otherwise.

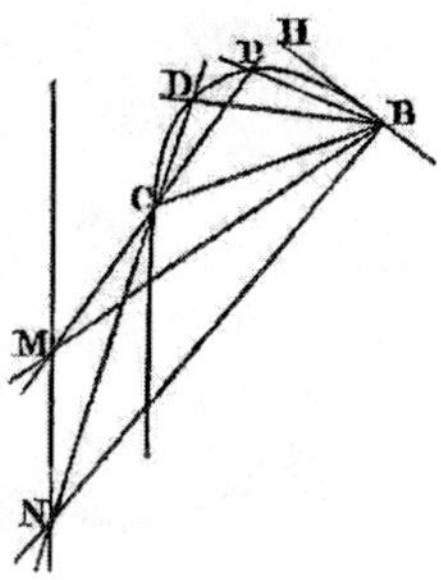

Let the angle CBH of a given magnitude revolve about the pole B, as also the rectilinear radius DC, both ways produced, about the pole C. Mark the points M, N, on which the leg BC of the angle cuts that radius when BH, the other leg thereof, meets the same radius in the points P and D. Then drawing the indefinite line MN, let that radius CP or CD and the leg BC of the angle perpetually meet in this line; and the point of concourse of the other leg BH with the radius will delineate the trajectory required.

For if in the constructions of the preceding Problem the point A comes to a coincidence with the point B, the lines CA and CB will coincide, and the line AB, in its last situation, will become the tangent BH; and therefore the constructions there set down will become the same with the constructions here described. Wherefore the concourse of the leg BH with the radius will describe a conic section passing through the points C, D, P, and touching the line BH in the point B. Q.E.F.

CASE 2. Suppose the four points B, C, D, P, given, being situated without the tangent HI. Join each two by the lines BD, CP meeting in G, and cutting the tangent in H and I. Cut the tangent in A in such manner

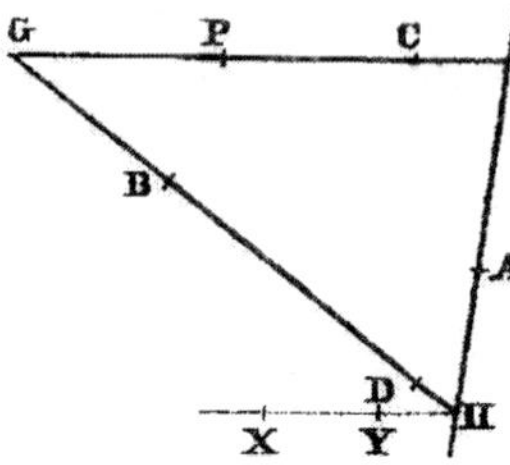

that HA may be to IA as the rectangle under a mean proportional between CG and GP, and a mean proportional between BH and HD is to a rectangle under a mean proportional between GD and GB, and a mean proportional between PI and IC, and A will be the point of contact. For if HX, a parallel to the right line PI, cuts the trajectory in any points X and Y, the point A (by the properties of the conic sections) will come to be so placed, that HA^2 will become to AI^2 in a ratio that is compounded out of the ratio of the rectangle XHY to the rectangle BHD, or of the rectangle CGP to the rectangle DGB; and the ratio of the rectangle BHD to the rectangle PIC. But after the point of contact A is found, the trajectory will be described as in the first Case. Q.E.F. But the point A may be taken either between or without the points H and I, upon which account a twofold trajectory may be described.

PROPOSITION XXIV. PROBLEM XVI.

To describe a trajectory that shall pass through three given points, and touch two right lines given by position.

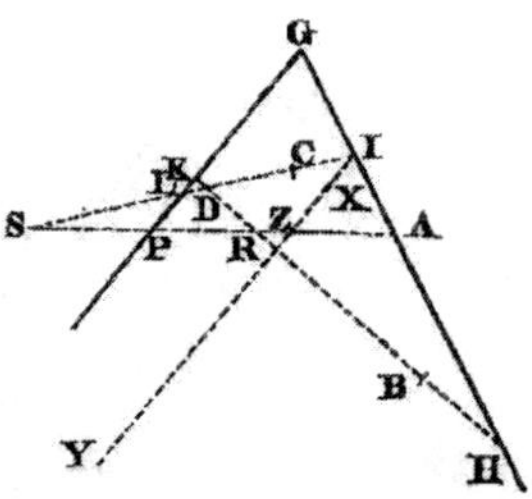

Suppose HI, KL to be the given tangents, and B, C, D, the given points. Through any two of those points, as B, D, draw the indefinite right line BD meeting the tangents in the points H, K. Then likewise through any other two of these points, as C, D, draw the indefinite right line CD meeting the tangents in the points I, L. Cut the lines drawn in R and S, so that HR may be to KR as the mean proportional between BH and HD is to the mean proportional between BK and KD; and IS to LS as the mean proportional between CI and ID is to the mean proportional between CL and LD. But you may cut, at pleasure, either within or between the points K and H, I and L, or without them; then draw RS cutting the tangents in A and P, and A and P will be the points of contact. For if A and P are supposed to be the points of contact, situated anywhere else in the tangents, and through any of the points H, I, K, L, as I, situated in either tangent HI, a right line IY is drawn parallel to the other tangent KL, and meeting the curve in X and Y, and in that right line there be taken IZ equal to a mean proportional between IX and IY, the rectangle XIY or IZ^2, will (by the properties of the conic sections) be to LP^2 as the rectangle CID is to the rectangle CLD, that is (by the construction), as SI is to SL^2, and therefore

IZ is to LP as SI to SL. Wherefore the points S, P, Z, are in one right line. Moreover, since the tangents meet in G, the rectangle XIY or IZ^2 will (by the properties of the conic sections) be to IA^2 as GP^2 is to GA^2, and consequently IZ will be to IA as GP to GA. Wherefore the points P, Z, A, lie in one right line, and therefore the points S, P, and A are in one right line. And the same argument will prove that the points R, P, and A are in one right line. Wherefore the points of contact A and P lie in the right line RS. But after these points are found, the trajectory may be described, as in the first Case of the preceding Problem. Q.E.F.

In this Proposition, and Case 2 of the foregoing, the constructions are the same, whether the right line XY cut the trajectory in X and Y, or not; neither do they depend upon that section. But the constructions being demonstrated where that right line does cut the trajectory, the constructions where it does not are also known; and therefore, for brevity's sake, I omit any farther demonstration of them.

LEMMA XXII.

To transform figures into other figures of the same kind.

Suppose that any figure HGI is to be transformed. Draw, at pleasure, two parallel lines AO, BL, cutting any third line AB, given by position, in A and B, and from any point G of the figure, draw out any right line GD, parallel to OA, till it meet the right line AB. Then from any given point O in the line OA, draw to the point D the right line OD, meeting BL in *d*; and from the point of concourse raise the right line *dg* containing any given angle with the right line BL, and having such ratio to O*d* as DG has to OD; and *g* will be the point in the new figure *hgi*, corresponding to the point G. And in like manner the several points of the first figure will give as many correspondent points of the new figure. If we therefore conceive the point G to be carried along by a continual motion through all the points of the first figure, the point *g* will be likewise carried along by a continual motion through all the points of the new figure, and describe the same. For distinction's sake, let us call DG the first ordinate, *dg* the new ordinate, AD the first abscissa, *ad* the new abscissa; O the pole, OD the abscinding radius, OA the first ordinate radius, and O*a* (by which the parallelogram OAB*a* is completed) the new ordinate radius.

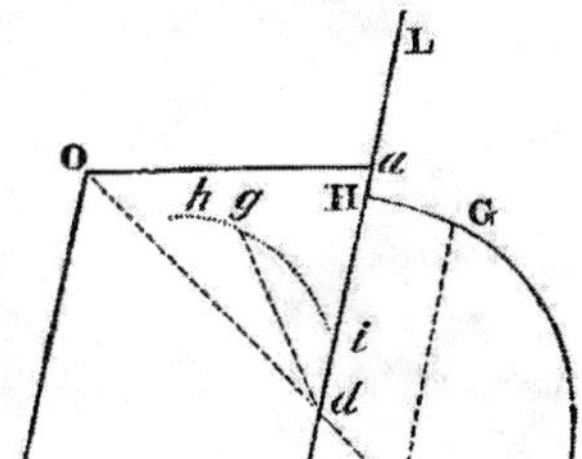

I say, then, that if the point G is placed in a right line given by position, the point *g* will be also placed in a right line given by position. If the point G is placed in a conic section, the point *g* will be likewise placed in a conic section. And here I understand the circle as one of the conic sections. But farther, if the point G is placed in a line of the third analytical order, the point g will also be placed in a line of the third order, and so on in curve lines of higher orders. The two lines in which the points G, g, are placed, will be always of the same analytical order. For as ad is to OA, so are Od to OD, dg to DG, and AB to AD; and therefore AD is equal to $\frac{\text{OA} \times \text{AB}}{ad}$, and DG equal to $\frac{\text{OA} \times dg}{ad}$. Now if the point G is placed in a right line, and therefore, in any equation by which the relation between the abscissa AD and the ordinate GD is expressed, those indetermined lines AD and DG rise no higher than to one dimension, by writing this equation $\frac{\text{OA} \times \text{AB}}{ad}$ in place of AD, and $\frac{\text{OA} \times dg}{ad}$ in place of DG, a new equation will be produced, in which the new abscissa ad and new ordinate dg rise only to one dimension; and which therefore must denote a right line. But if AD and DG (or either of them) had risen to two dimensions in the first equation, ad and dg would likewise have risen to two dimensions in the second equation. And so on in three or more dimensions. The indetermined lines, ad, dg in the second equation, and AD, DG, in the first, will always rise to the same number of dimensions; and therefore the lines in which the points G, g, are placed are of the same analytical order.

I say farther, that if any right line touches the curve line in the first figure, the same right line transferred the same way with the curve into the new figure will touch that curve line in the new figure, and *vice versa*. For if any two points of the curve in the first figure are supposed to approach one the other till they come to coincide, the same points transferred will approach one the other till they come to coincide in the new figure; and therefore the right lines with which those points are joined will become together tangents of the curves in both figures. I might have given demonstrations of these assertions in a more geometrical form; but I study to be brief.

Wherefore if one rectilinear figure is to be transformed into another, we need only transfer the intersections of the right lines of which the first figure consists, and through the transferred intersections to draw right lines in the new figure. But if a curvilinear figure is to be transformed, we must transfer the points, the tangents, and other right lines, by means of which the curve line is defined. This Lemma is of use in the solution of the more difficult Problems; for thereby we may transform the proposed figures, if they are intricate, into others that are more simple. Thus any right lines converging to a point are transformed into parallels, by taking for the first ordinate radius any right line that passes through the point of concourse of the converging lines, and that because their point of concourse is by this means made to go off *in infinitum;* and parallel lines are such as tend to a point infinitely remote. And after the problem is solved in the new figure, if by the inverse operations we transform the new into the first figure, we shall have the solution required.

This Lemma is also of use in the solution of solid problems. For as often as two conic sections occur, by the intersection of which a problem may be solved, any one of them may be transformed, if it is an hyperbola or a parabola, into an ellipsis, and then this ellipsis may be easily changed into a circle. So also a right line and a conic section, in the construction of plane problems, may be transformed into a right line and a circle.

PROPOSITION XXV. PROBLEM XVII.

To describe a trajectory that shall pass through two given points, and touch three right lines given by position.

Through the concourse of any two of the tangents one with the other, and the concourse of the third tangent with the right line which passes through the two given points, draw an indefinite right line; and, taking this line for the first ordinate radius, transform the figure by the preceding Lemma into a new figure. In this figure those two tangents will become parallel to each other, and the third tangent will be parallel to the right line that passes through the two given points. Suppose *hi, kl* to be those two parallel tangents, *ik* the third tangent, and *hl* a right line parallel thereto, passing through those points *a, b*, through which the conic section ought to pass in this new figure; and completing the parallelogram *hikl*, let the right lines *hi, ik, kl* be so cut in *c, d, e*, that *hc* may be to the square root of the rectangle *ahb, ic*, to *id*, and *ke* to *kd*, as the sum of the right lines *hi* and *kl* is to the sum of the three lines, the first whereof is the right line *ik*, and the other two are the square roots of the rectangles *ahb* and *alb;* and *c, d, e*, will be the points of contact.

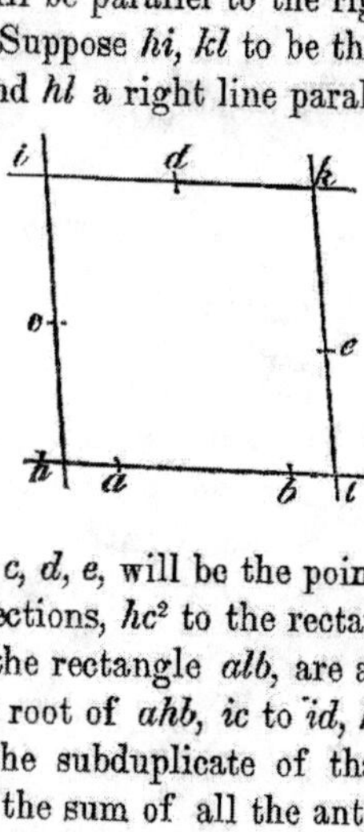

For by the properties of the conic sections, hc^2 to the rectangle *ahb*, and ic^2 to id^2, and ke^2 to kd^2, and el^2 to the rectangle *alb*, are all in the same ratio; and therefore *hc* to the square root of *ahb, ic* to *id, ke* to *kd*, and *el* to the square root of *alb*, are in the subduplicate of that ratio; and by composition, in the given ratio of the sum of all the antecedents $hi + kl$, to the sum of all the consequents $\sqrt{ahb} + ik + \sqrt{alb}$. Wherefore from that given ratio we have the points of contact *c, d, e*, in the new figure. By the inverted operations of the last Lemma, let those points be transferred into the first figure, and the trajectory will be there described by Prob. XIV. Q.E.F. But according as the points *a, b*, fall between the points *h, l*, or without them, the points *c, d, e*, must be taken either between the points, h, i, k, l, or without them. If one of the points a, b, falls between the points h, i, and the other without the points h, l, the Problem is impossible.

PROPOSITION XXVI. PROBLEM XVIII.

To describe a trajectory that shall pass through a given point, and touch four right lines given by position.

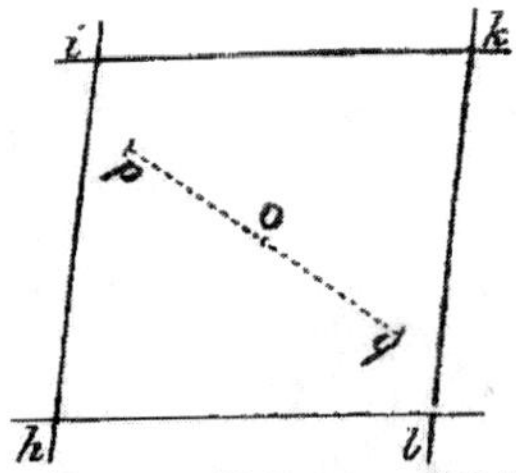

From the common intersections, of any two of the tangents to the common intersection of the other two, draw an indefinite right line; and taking this line for the first ordinate radius, transform the figure (by Lem. XXII) into a new figure, and the two pairs of tangents, each of which before concurred in the first ordinate radius, will now become parallel. Let hi and kl, ik and hl, be those pairs of parallels completing the parallelogram $hikl$. And let p be the point in this new figure corresponding to the given point in the first figure. Through O the centre of the figure draw pq: and Oq being equal to Op, q will be the other point through which the conic section must pass in this new figure. Let this point be transferred, by the inverse operation of Lem. XXII into the first figure, and there we shall have the two points through which the trajectory is to be described. But through those points that trajectory may be described by Prop. XVII.

LEMMA XXIII.

If two right lines, as AC, BD *given by position, and terminating in given points* A, B, *are in a given ratio one to the other, and the right line* CD, *by which the indeterminted points* C, D *are joined is cut in* K *in a given ratio; I say, that the point* K *will be placed in a right line given by position.*

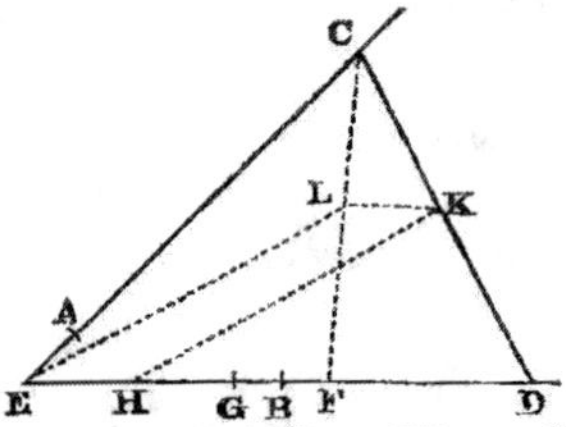

For let the right lines AC, BD meet in E, and in BE take BG to AE as BD is to AC, and let FD be always equal to the given line EG; and, by construction, EC will be to GD, that is, to EF, as AC to BD, and therefore in a given ratio; and therefore the triangle EFC will be given in kind. Let CF be cut in L so as CL may be to CF in the ratio of CK to CD; and because that is a given ratio, the triangle EFL will be given in kind, and therefore the point L will be placed in the right line EL given by position. Join LK, and the triangles CLK, CFD will be similar; and because FD is a given line, and LK is to FD in a given ratio, LK will be also given.

To this let EH be taken equal, and ELKH will be always a parallelogram. And therefore the point K is always placed in the side HK (given by position) of that parallelogram. Q.E.D.

COR. Because the figure EFLC is given in kind, the three right lines EF, EL, and EC, that is, GD, HK, and EC, will have given ratios to each other.

LEMMA XXIV.

If three right lines, two whereof are parallel, and given by position, touch any conic section; I say, that the semi-diameter of the section which is parallel to those two is a mean proportional between the segments of those two that are intercepted between the points of contact and the third tangent.

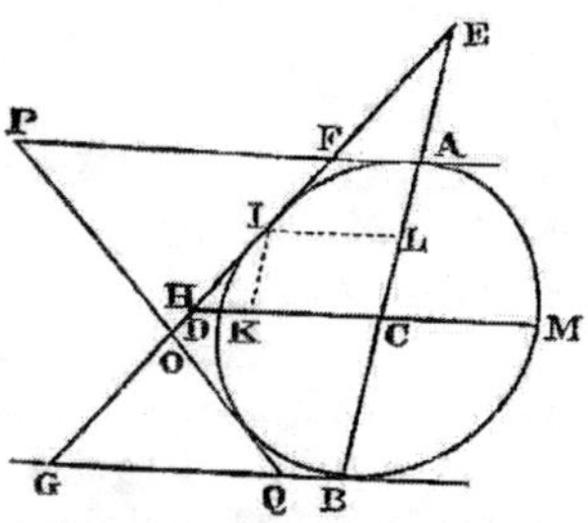

Let AF, GB be the two parallels touching the conic section ADB in A and B; EF the third right line touching the conic section in I, and meeting the two former tangents in F and G, and let CD be the semi-diameter of the figure parallel to those tangents; I say, that AF, CD, BG are continually proportional.

For if the conjugate diameters AB, DM meet the tangent FG in E and H, and cut one the other in C, and the parallelogram IKCL be completed; from the nature of the conic sections, EC will be to CA as CA to CL; and so by division, EC — CA to CA — CL, or EA to AL; and by composition, EA to EA + AL or EL, as EC to EC+CA or EB; and therefore (because of the similitude of the triangles EAF, ELI, ECH, EBG) AF is to LI as CH to BG. Likewise, from the nature of the conic sections, LI (or CK) is to CD as CD to CH; and therefore (*ex æquo perturbatè*) AF is to CD as CD to BG. Q.E.D.

COR. 1. Hence if two tangents FG, PQ meet two parallel tangents AF, BG in F and G, P and Q, and cut one the other in O; AF (*ex æquo perturbatè*) will be to BQ as AP to BG, and by division, as FP to GQ, and therefore as FO to OG.

COR. 2. Whence also the two right lines PG, FQ drawn through the points P and G, F and Q, will meet in the right line ACB passing through the centre of the figure and the points of contact A, B.

LEMMA XXV.

If four sides of a parallelogram indefinitely produced touch any conic section, and are cut by a fifth tangent; I say, that, taking those segments of any two conterminous sides that terminate in opposite angles of the parallelogram, either segment is to the side from which it is cut off as that part of the other conterminous side which is intercepted between the point of contact and the third side is to the other segment.

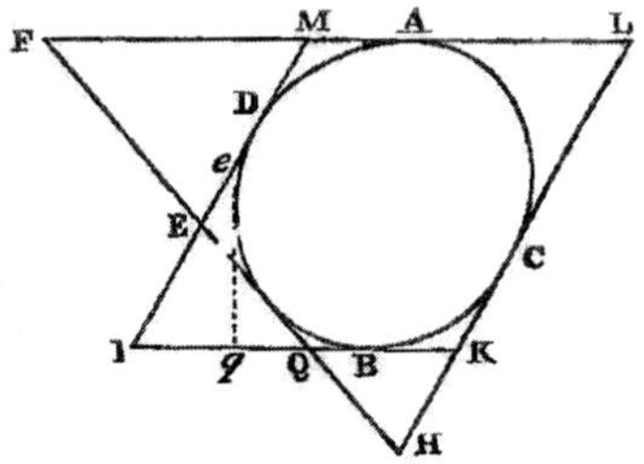

Let the four sides ML, IK, KL, MI, of the parallelogram MLIK touch the conic section in A, B, C, D; and let the fifth tangent FQ cut those sides in F, Q, H, and E; and taking the segments ME, KQ of the sides MI, KI, or the segments KH, MF of the sides KL, ML, I say, that ME is to MI as BK to KQ; and KH to KL as AM to MF. For, by Cor. 1 of the preceding Lemma, ME is to EI as (AM or) BK to BQ; and, by composition, ME is to MI as BK to KQ. Q.E.D. Also KH is to HL as (BK or) AM to AF; and by division, KH to KL as AM to MF. Q.E.D.

Cor. 1. Hence if a parallelogram IKLM described about a given conic section is given, the rectangle KQ × ME, as also the rectangle KH × MF equal thereto, will be given. For, by reason of the similar triangles KQH, MFE, those rectangles are equal.

Cor. 2. And if a sixth tangent *eq* is drawn meeting the tangents KI, MI in *q* and *e*, the rectangle KQ × ME will be equal to the rectangle K*q* × M*e*, and KQ will be to M*e* as K*q* to ME, and by division as Q*q* to E*e*.

Cor. 3. Hence, also, if E*q*, *e*Q, are joined and bisected, and a right line is drawn through the points of bisection, this right line will pass through the centre of the conic section. For since Q*q* is to E*e* as KQ to M*e*, the same right line will pass through the middle of all the lines E*q*, *e*Q, MK (by Lem. XXIII), and the middle point of the right line MK is the centre of the section.

PROPOSITION XXVII. PROBLEM XIX.

To describe a trajectory that may touch five right lines given by position.

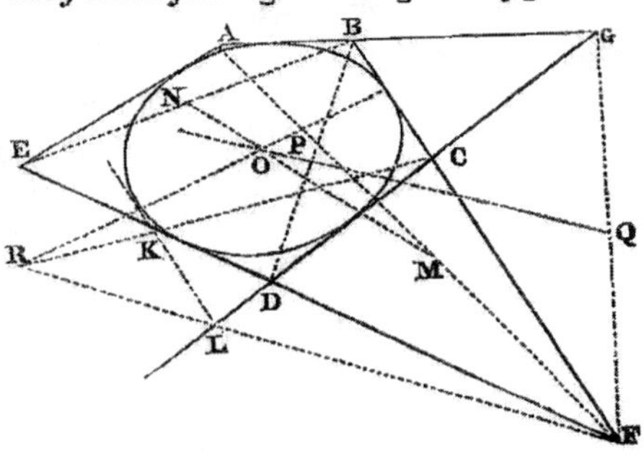

Supposing ABG, BCF, GCD, FDE, EA to be the tangents given by position. Bisect in M and N, AF, BE, the diagonals of the quadrilateral figure ABFE contained under any four of them; and (by Cor. 3, Lem. XXV) the right line MN drawn through the points of bisection will pass through the centre of the trajectory. Again, bisect in P and Q the diagonals (if I may so call them) BD, GF of the quadrilateral figure BGDF contained under any other four tangents, and the right line PQ drawn through the points of bisection will pass through the centre of the trajectory; and therefore the centre will be given in the concourse of the bisecting lines. Suppose it to be O. Parallel to any tangent BC draw KL at such distance that the centre O may be placed in the middle between the parallels; this KL will touch the trajectory to be described. Let this cut any other two tangents GCD, FDE, in L and K. Through the points C and K, F and L, where the tangents not parallel, CL, FK meet the parallel tangents CF, KL, draw CK, FL meeting in R; and the right line OR drawn and produced, will cut the parallel tangents CF, KL, in the points of contact. This appears from Cor. 2, Lem. XXIV. And by the same method the other points of contact may be found, and then the trajectory may be described by Prob. XIV. Q.E.F.

SCHOLIUM.

Under the preceding Propositions are comprehended those Problems wherein either the centres or asymptotes of the trajectories are given. For when points and tangents and the centre are given, as many other points and as many other tangents are given at an equal distance on the other side of the centre. And an asymptote is to be considered as a tangent, and its infinitely remote extremity (if we may say so) is a point of contact. Conceive the point of contact of any tangent removed *in infinitum*, and the tangent will degenerate into an asymptote, and the constructions of the preceding Problems will be changed into the constructions of those Problems wherein the asymptote is given.

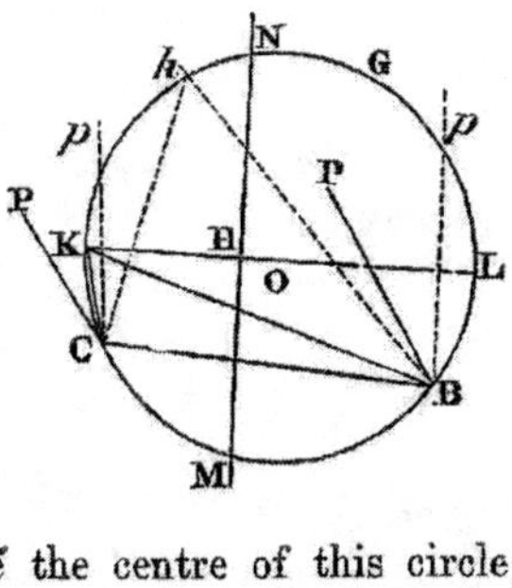

After the trajectory is described, we may find its axes and foci in this manner. In the construction and figure of Lem. XXI, let those legs BP, CP, of the moveable angles PBN, PCN, by the concourse of which the trajectory was described, be made parallel one to the other; and retaining that position, let them revolve about their poles B, C, in that figure. In the mean while let the other legs CN, BN, of those angles, by their concourse K or *k*, describe the circle BKGC. Let O be the centre of this circle; and from this centre upon the ruler MN, wherein those legs CN, BN did concur while the trajectory was described, let fall the perpendicular OH meeting the circle in K and L. And when those other legs CK, BK meet in the point K that is nearest to the ruler, the first legs CP, BP will be parallel to the greater axis, and perpendicular on the lesser; and the contrary will happen if those legs meet in the remotest point L. Whence if the centre of the trajectory is given, the axes will be given; and those being given, the foci will be readily found.

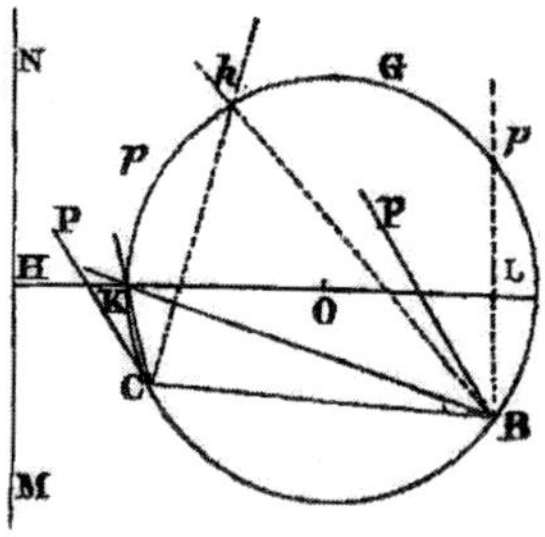

But the squares of the axes are one to the other as KH to LH, and thence it is easy to describe a trajectory given in kind through four given points. For if two of the given points are made the poles C, B, the third will give the moveable angles PCK, PBK; but those being given, the circle BGKC may be described. Then, because the trajectory is given in kind, the ratio of OH to OK, and and therefore OH itself, will be given. About the centre O, with the interval OH, describe another circle, and the right line that touches this circle, and passes through the concourse of the legs CK, BK, when the first legs CP, BP meet in the fourth given point, will be the ruler MN, by means of which the trajectory may be described. Whence also on the other hand a trapezium given in kind (excepting a few cases that are impossible) may be inscribed in a given conic section.

There are also other Lemmas, by the help of which trajectories given in kind may be described through given points, and touching given lines. Of such a sort is this, that if a right line is drawn through any point given by position, that may cut a given conic section in two points, and the distance of the intersections is bisected, the point of bisection will touch another conic section of the same kind with the former, and having its axes parallel to the axes of the former. But I hasten to things of greater use.

LEMMA XXVI.

To place the three angles of a triangle, given both in kind and magnitude, in respect of as many right lines given by position, provided they are not all parallel among themselves, in such manner that the several angles may touch the several lines.

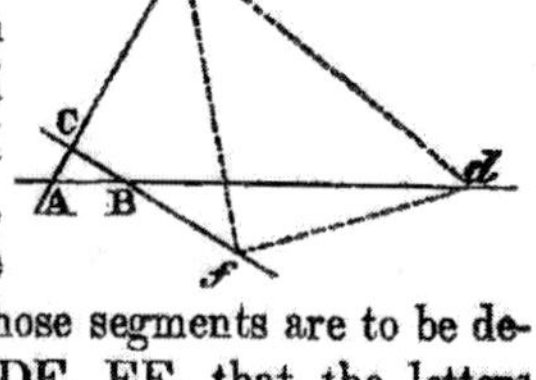

Three indefinite right lines AB, AC, BC, are given by position, and it is required so to place the triangle DEF that its angle D may touch the line AB, its angle E the line AC, and its angle F the line BC. Upon DE, DF, and EF, describe three segments of circles DRE, DGF, EMF, capable of angles equal to the angles BAC, ABC, ACB respectively. But those segments are to be described towards such sides of the lines DE, DF, EF, that the letters

DRED may turn round about in the same order with the letters BACB; the letters DGFD in the same order with the letters ABCA; and the letters EMFE in the same order with the letters ACBA; then, completing those segments into entire circles let the two former circles cut one the other in G, and suppose P and Q to be their centres. Then joining GP, PQ, take G*a* to AB as GP is to PQ; and about the centre G, with the interval G*a*, describe a circle that may cut the first circle DGE in *a*. Join *a*D cutting the second circle DFG in *b*, as well as *a*E cutting the third circle EMF in *c*. Complete the figure ABC*def* similar and equal to the figure *abc*DEF: I say, the thing is done.

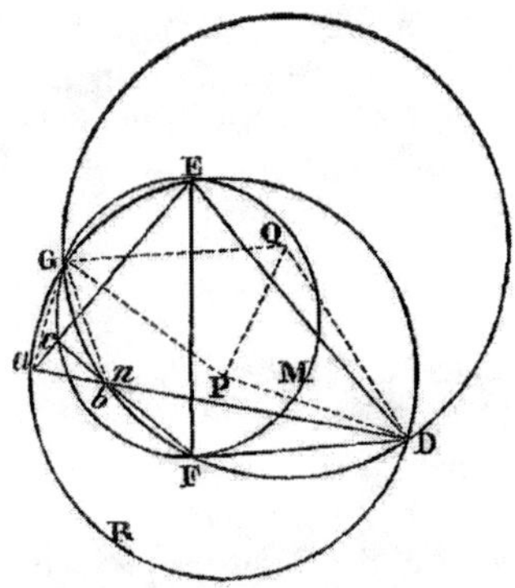

For drawing F*c* meeting *a*D in *n*, and joining *a*G, *b*G, QG, QD, PD, by construction the angle E*a*D is equal to the angle CAB, and the angle *ac*F equal to the angle ACB; and therefore the triangle *anc* equiangular to the triangle ABC. Wherefore the angle *anc* or F*n*D is equal to the angle ABC, and consequently to the angle F*b*D; and therefore the point *n* falls on the point *b*. Moreover the angle GPQ, which is half the angle GPD at the centre, is equal to the angle G*a*D at the circumference; and the angle GQP, which is half the angle GQD at the centre, is equal to the complement to two right angles of the angle G*b*D at the circumference, and therefore equal to the angle G*ba*. Upon which account the triangles GPQ, G*ab*, are similar, and G*a* is to *ab* as GP to PQ; that is (by construction), as G*a* to AB. Wherefore *ab* and AB are equal; and consequently the triangles *abc*, ABC, which we have now proved to be similar, are also equal. And therefore since the angles D, E, F, of the triangle DEF do respectively touch the sides *ab*, *ac*, *bc* of the triangle *abc*, the figure ABC*def* may be completed similar and equal to the figure *abc*DEF, and by completing it the Problem will be solved. Q.E.F.

Cor. Hence a right line may be drawn whose parts given in length may be intercepted between three right lines given by position. Suppose the triangle DEF, by the access of its point D to the side EF, and by having the sides DE, DF placed *in directum* to be changed into a right line whose given part DE is to be interposed between the right lines AB, AC given by position; and its given part DF is to be interposed between the right lines AB, BC, given by position; then, by applying the preceding construction to this case, the Problem will be solved.

PROPOSITION XXVIII. PROBLEM XX.

To describe a trajectory given both in kind and magnitude, given parts of which shall be interposed between three right lines given by position.

Suppose a trajectory is to be described that may be similar and equal to the curve line DEF, and may be cut by three right lines AB, AC, BC, given by position, into parts DE and EF, similar and equal to the given parts of this curve line.

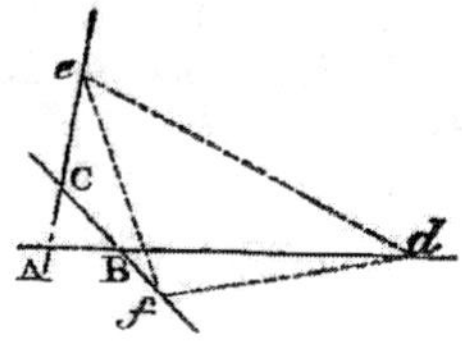

Draw the right lines DE, EF, DF: and place the angles D, E, F, of this triangle DEF, so as to touch those right lines given by position (by Lem. XXVI). Then about the triangle describe the trajectory, similar and equal to the curve DEF. Q.E.F.

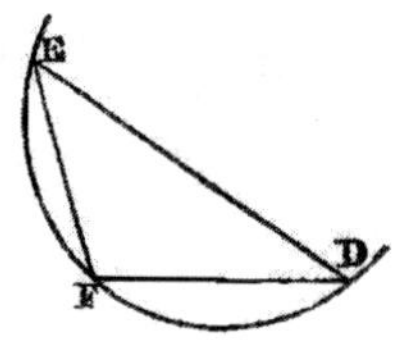

LEMMA XXVII.

To describe a trapezium given in kind, the angles whereof may be so placed, in respect of four right lines given by position, that are neither all parallel among themselves, nor converge to one common point, that the several angles may touch the several lines.

Let the four right lines ABC, AD, BD, CE, be given by position; the first cutting the second in A, the third in B, and the fourth in C; and suppose a trapezium *fghi* is to be described that may be similar to the trapezium FGHI, and whose angle *f*, equal to the given angle F, may touch the right line ABC; and the other angles *g*, *h*, *i*, equal to the other given angles, G, H, I, may touch the other lines AD, BD, CE, respectively. Join FH, and upon FG, FH, FI describe as many segments of circles FSG, FTH, FVI, the first of which FSG may be capable of an angle equal to the angle BAD; the second FTH capable of an angle equal to the angle CBD; and the third FVI of an angle equal to the angle ACE. But the segments are to be described towards those sides of the lines FG, FH, FI, that the circular order of the letters FSGF may be the same as of the letters BADB, and that the letters FTHF may turn about in the same order as the letters CBDC, and the letters FVIF in the same order as the letters ACEA. Complete the segments into entire circles, and let P be the centre of the first circle FSG, Q the centre of the second FTH. Join and produce both ways the line PQ, and in it take QR in the same ratio to PQ as BC has to AB. But QR is to be taken towards that side of the point Q, that the order of the letters P, Q, R.

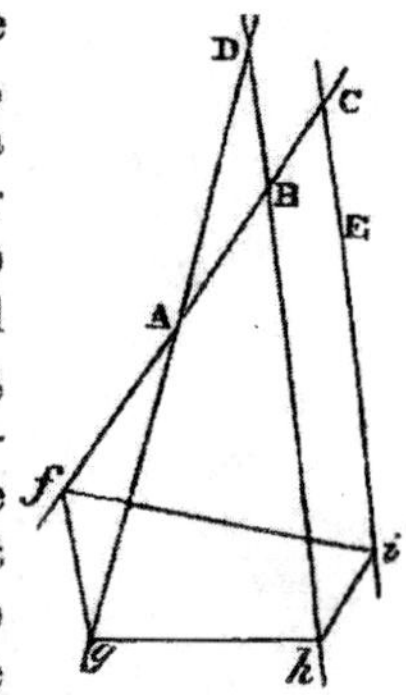

may be the same as of the letters A, B, C; and about the centre R with the interval RF describe a fourth circle FN*c* cutting the third circle FVI in *c*. Join F*c* cutting the first circle in *a*, and the second in *b*. Draw *a*G, *b*H, *c*I, and let the figure ABC*fghi* be made similar to the figure *abc*FGHI; and the trapezium *fghi* will be that which was required to be described.

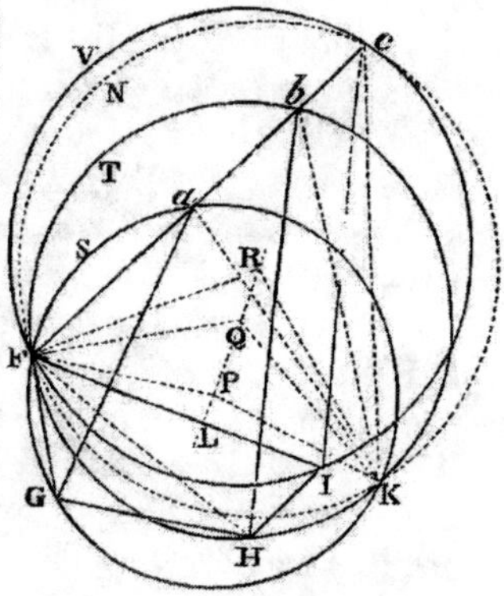

For let the two first circles FSG, FTH cut one the other in K; join PK, QK, RK, *a*K, *b*K, *c*K, and produce QP to L. The angles F*a*K, F*b*K, F*c*K at the circumferences are the halves of the angles FPK, FQK, FRK, at the centres, and therefore equal to LPK, LQK, LRK, the halves of those angles. Wherefore the figure PQRK is equiangular and similar to the figure *abc*K, and consequently *ab* is to *bc* as PQ to QR, that is, as AB to BC. But by construction, the angles *f*A*g*, *f*B*h*, *f*C*i*, are equal to the angles F*a*G, F*b*H, F*c*I. And therefore the figure ABC*fghi* may be completed similar to the figure *abc*FGHI. Which done a trapezium *fghi* will be constructed similar to the trapezium FGHI, and which by its angles *f*, *g*, *h*, *i* will touch the right lines ABC, AD, BD, CE. Q.E.F.

COR. Hence a right line may be drawn whose parts intercepted in a given order, between four right lines given by position, shall have a given proportion among themselves. Let the angles FGH, GHI, be so far increased that the right lines FG, GH, HI, may lie *in directum*; and by constructing the Problem in this case, a right line *fghi* will be drawn, whose parts *fg*, *gh*, *hi*, intercepted between the four right lines given by position, AB and AD, AD and BD, BD and CE, will be one to another as the lines FG, GH, HI, and will observe the same order among themselves. But the same thing may be more readily done in this manner.

Produce AB to K and BD to L, so as BK may be to AB as HI to GH; and DL to BD as GI to FG; and join KL meeting the right line CE in *i*. Produce *i*L to M, so as LM may be to *i*L as GH to HI; then draw MQ parallel to LB, and meeting the right line AD in *g*, and join *gi* cutting AB, BD in *f*, *h*; I say, the thing is done.

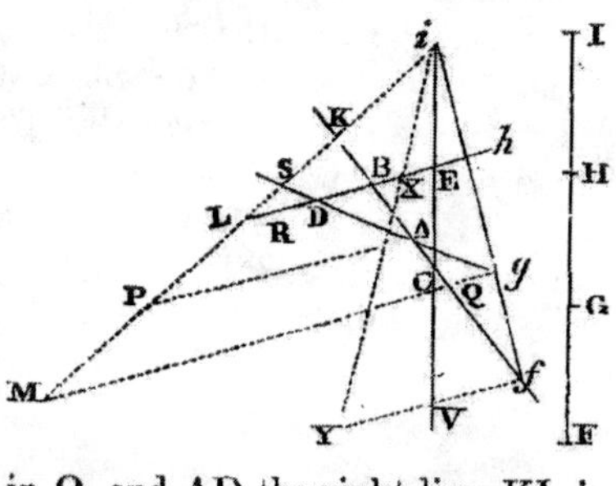

For let M*g* cut the right line AB in Q, and AD the right line KL in

S, and draw AP parallel to BD, and meeting *i*L in P, and *g*M to L*h* (*gi* to *hi*, M*i* to L*i*, GI to HI, AK to BK) and AP to BL, will be in the same ratio. Cut DL in R, so as DL to RL may be in that same ratio; and because *g*S to *g*M, AS to AP, and DS to DL are proportional; therefore (*ex æquo*) as *g*S to L*h*, so will AS be to BL, and DS to RL; and mixtly, BL — RL to L*h* — BL, as AS — DS to *g*S — AS. That is, BR is to B*h* as AD is to A*g*, and therefore as BD to *g*Q. And alternately BR is to BD as B*h* to *g*Q, or as *fh* to *fg*. But by construction the line BL was cut in D and R in the same ratio as the line FI in G and H; and therefore BR is to BD as FH to FG. Wherefore *fh* is to *fg* as FH to FG. Since, therefore, *gi* to *hi* likewise is as M*i* to L*i*, that is, as GI to HI, it is manifest that the lines FI, *fi*, are similarly cut in G and H, *g* and *h*. Q.E.F.

In the construction of this Corollary, after the line LK is drawn cutting CE in *i*, we may produce *i*E to V, so as EV may be to E*i* as FH to HI, and then draw V*f* parallel to BD. It will come to the same, if about the centre *i* with an interval IH, we describe a circle cutting BD in X, and produce *i*X to Y so as *i*Y may be equal to IF, and then draw Y*f* parallel to BD.

Sir Christopher Wren and Dr. Wallis have long ago given other solutions of this Problem.

PROPOSITION XXIX. PROBLEM XXI.

To describe a trajectory given in kind, that may be cut by four right lines given by position, into parts given in order, kind, and proportion.

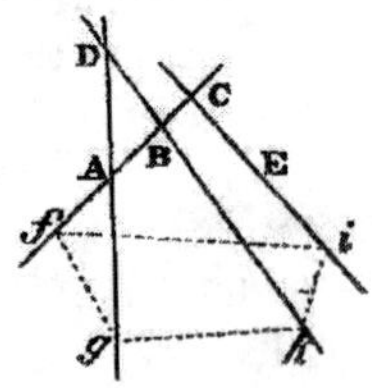

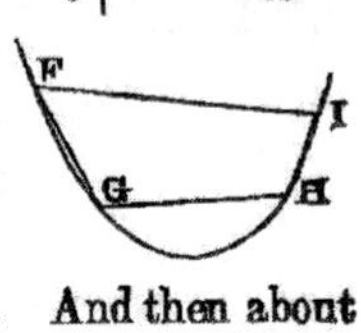

Suppose a trajectory is to be described that may be similar to the curve line FGHI, and whose parts, similar and proportional to the parts FG, GH, HI of the other, may be intercepted between the right lines AB and AD, AD, and BD, BD and CE given by position, viz., the first between the first pair of those lines, the second between the second, and the third between the third. Draw the right lines FG, GH, HI, FI; and (by Lem. XXVII) describe a trapezium *fghi* that may be similar to the trapezium FGHI, and whose angles *f*, *g*, *h*, *i*, may touch the right lines given by position AB, AD, BD, CE, severally according to their order. And then about this trapezium describe a trajectory, that trajectory will be similar to the curve line FGHI.

SCHOLIUM.

This problem may be likewise constructed in the following manner. Joining FG, GH, HI, FI, produce GF to V, and join FH, IG, and make

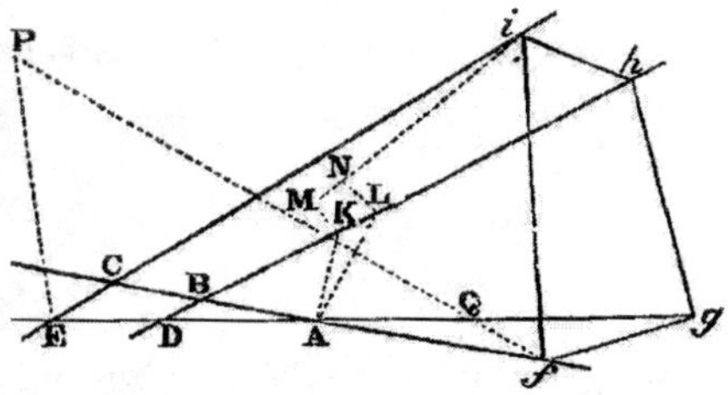

the angles CAK, DAL equal to the angles FGH, VFH. Let AK, AL meet the right line BD in K and L, and thence draw KM, LN, of which let KM make the angle AKM equal to the angle GHI, and be itself to AK as HI is to GH; and let LN make the angle ALN equal to the angle FHI, and be itself to AL as HI to FH. But AK, KM, AL, LN are to be drawn towards those sides of the lines AD, AK, AL, that the letters CAKMC, ALKA, DALND may be carried round in the same order as the letters FGHIF; and draw MN meeting the right line CE in *i*. Make the angle *i*EP equal to the angle IGF, and let PE be to E*i* as FG to GI; and through P draw PQ*f* that may with the right line ADE contain an angle PQE equal to the angle FIG, and may meet the right line AB in *f*, and join *fi*. But PE and PQ are to be drawn towards those sides of the lines CE, PE, that the circular order of the letters PE*i*P and PEQP may be the same as of the letters FGHIF; and if upon the line *fi*, in the same order of letters, and similar to the trapezium FGHI, a trapezium *fghi* is constructed, and a trajectory given in kind is circumscribed about it, the Problem will be solved.

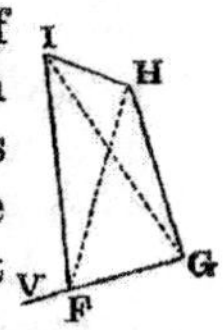

So far concerning the finding of the orbits. It remains that we determine the motions of bodies in the orbits so found.

SECTION VI.

How the motions are to be found in given orbits.

PROPOSITION XXX. PROBLEM XXII.

To find at any assigned time the place of a body moving in a given parabolic trajectory.

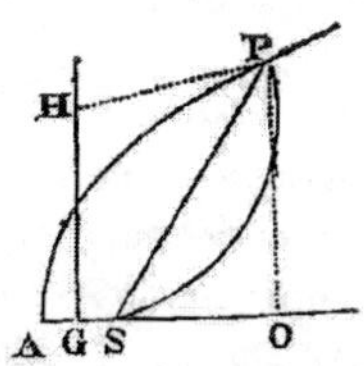

Let S be the focus, and A the principal vertex of the parabola; and suppose $4AS \times M$ equal to the parabolic area to be cut off APS, which either was described by the radius SP, since the body's departure from the vertex, or is to be described thereby before its arrival there. Now the quantity of that area to be cut off is known from the time which is proportional to it. Bisect AS in G, and erect the perpendicular GH equal to 3M, and a circle described about the centre H, with the interval HS, will cut the parabola in the place P required. For letting fall PO perpendicular on the axis, and drawing PH, there will be $AG^2 + GH^2$ ($= HP^2 = \overline{AO - AG}|^2 + \overline{PO - GH}|^2$) $= AO^2 + PO^2 - 2GAO - 2GH + PO +$

$AG^2 + GH^2$. Whence $2GH \times PO$ $(= AO^2 + PO^2 - 2GAO) = AO^2 + \frac{3}{4}PO^2$. For AO^2 write $AO \times \frac{PO^2}{4AS}$; then dividing all the terms by 3PO, and multiplying them by 2AS, we shall have $\frac{4}{3}GH \times AS$ $(= \frac{1}{6}AO \times PO + \frac{1}{2}AS \times PO = \frac{AO + 3AS}{6} \times PO = \frac{4AO - 3SO}{6} \times PO =$ to the area $\overline{APO - SPO})| =$ to the area APS. But GH was 3M, and therefore $\frac{4}{3}GH \times AS$ is $4AS \times M$. Wherefore the area cut off APS is equal to the area that was to be cut off $4AS \times M$. Q.E.D.

COR. 1. Hence GH is to AS as the time in which the body described the arc AP to the time in which the body described the arc between the vertex A and the perpendicular erected from the focus S upon the axis.

COR. 2. And supposing a circle ASP perpetually to pass through the moving body P, the velocity of the point H is to the velocity which the body had in the vertex A as 3 to 8; and therefore in the same ratio is the line GH to the right line which the body, in the time of its moving from A to P, would describe with that velocity which it had in the vertex A.

COR. 3. Hence, also, on the other hand, the time may be found in which the body has described any assigned arc AP. Join AP, and on its middle point erect a perpendicular meeting the right line GH in H.

LEMMA XXVIII.

There is no oval figure whose area, cut off by right lines at pleasure, can be universally found by means of equations of any number of finite terms and dimensions.

Suppose that within the oval any point is given, about which as a pole a right line is perpetually revolving with an uniform motion, while in that right line a moveable point going out from the pole moves always forward with a velocity proportional to the square of that right line within the oval. By this motion that point will describe a spiral with infinite circumgyrations. Now if a portion of the area of the oval cut off by that right line could be found by a finite equation, the distance of the point from the pole, which is proportional to this area, might be found by the same equation, and therefore all the points of the spiral might be found by a finite equation also; and therefore the intersection of a right line given in position with the spiral might also be found by a finite equation. But every right line infinitely produced cuts a spiral in an infinite number of points; and the equation by which any one intersection of two lines is found at the same time exhibits all their intersections by as many roots, and therefore rises to as many dimensions as there are intersections. Because two circles mutually cut one another in two points, one of those intersections is not to be found but by an equation of two dimensions, by which the other intersection may be also found. Because there may be four intersections of two conic sections, any one of them is not to be found universally, but by an equation of four dimensions, by which they may be all found together. For if those intersections are severally sought, because the law and condition of all is the same, the calculus will be the same in every case, and therefore the conclusion always the same, which must therefore comprehend all those intersections at once within itself, and exhibit them all indifferently. Hence it is that the intersections of the conic sections with the curves of the third order, because they may amount to six, come out together by equations of six dimensions; and the intersections of two curves of the third order, because they may amount to nine, come out together by equations of nine dimensions. If this did not necessarily happen, we might reduce all solid to plane Problems, and those higher than solid to solid Problems. But here I speak of curves irreducible in power. For if the equation by which the curve is defined may be reduced to a lower power, the curve will not be one single curve, but composed of two, or more, whose intersections may be severally found by different calculusses. After the same manner the two intersections of right lines with the conic sections come out always by equations of two dimensions; the three intersections of right lines with the irreducible curves of the third order by equations of three dimensions; the four intersections of right lines with the irreducible curves of the fourth order, by equations of four dimensions; and so on *in infinitum*. Wherefore the innumerable intersections of a right line with a spiral, since this is but one simple curve, and not reducible to more curves, require equations infinite in number of dimensions and roots, by which they may be all exhibited together. For the law and calculus of all is the same. For if a perpendicular is let fall from the pole upon that intersecting right line, and that perpendicular together with the intersecting line revolves about the pole, the intersections of the spiral will mutually pass the one into the other; and that which was first or nearest, after one revolution, will be the second; after two, the third; and so on: nor will the equation in the mean time be changed but as the magnitudes of those quantities are changed, by which the position of the intersecting line is determined. Wherefore since those quantities after every revolution return to their first magnitudes, the equation will return to its first form; and consequently one and the same equation will exhibit all the intersections, and will therefore have an infinite number of roots, by which they may be all exhibited. And therefore the intersection of a right line with a spiral cannot be universally found by any finite equation; and of consequence there is no oval figure whose area, cut off by right lines at pleasure, can be universally exhibited by any such equation.

By the same argument, if the interval of the pole and point by which the spiral is described is taken proportional to that part of the perimeter of the oval which is cut off, it may be proved that the length of the perimeter cannot be universally exhibited by any finite equation. But here I speak of ovals that are not touched by conjugate figures running out *in infinitum.*

Cor. Hence the area of an ellipsis, described by a radius drawn from the focus to the moving body, is not to be found from the time given by a finite equation; and therefore cannot be determined by the description of curves geometrically rational. Those curves I call geometrically rational, all the points whereof may be determined by lengths that are definable by equations; that is, by the complicated ratios of lengths. Other curves (such as spirals, quadratrixes, and cycloids) I call geometrically irrational. For the lengths which are or are not as number to number (according to the tenth Book of Elements) are arithmetically rational or irrational. And therefore I cut off an area of an ellipsis proportional to the time in which it is described by a curve geometrically irrational, in the following manner.

PROPOSITION XXXI. PROBLEM XXIII.

To find the place of a body moving in a given elliptic trajectory at any assigned time.

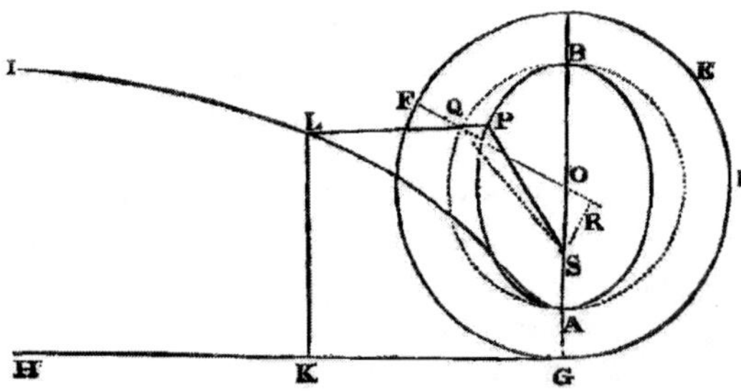

Suppose A to be the principal vertex, S the focus, and O the centre of the ellipsis APB; and let P be the place of the body to be found. Produce OA to G so as OG may be to OA as OA to OS. Erect the perpendicular GH; and about the centre O, with the interval OG, describe the circle GEF; and on the ruler GH, as a base, suppose the wheel GEF to move forwards, revolving about its axis, and in the mean time by its point A describing the cycloid ALI. Which done, take GK to the perimeter GEFG of the wheel, in the ratio of the time in which the body proceeding from A described the arc AP, to the time of a whole revolution in the ellipsis. Erect the perpendicular KL meeting the cycloid in L; then LP drawn parallel to KG will meet the ellipsis in P, the required place of the body.

For about the centre O with the interval OA describe the semi-circle AQB, and let LP, produced, if need be, meet the arc AQ in Q, and join

SQ, OQ. Let OQ meet the arc EFG in F, and upon OQ let fall the perpendicular SR. The area APS is as the area AQS, that is, as the difference between the sector OQA and the triangle OQS, or as the difference of the rectangles $\frac{1}{2}$OQ × AQ, and $\frac{1}{2}$OQ × SR, that is, because $\frac{1}{2}$OQ is given, as the difference between the arc AQ and the right line SR; and therefore (because of the equality of the given ratios SR to the sine of the arc AQ, OS to OA, OA to OG, AQ to GF; and by division, AQ—SR to GF — sine of the arc AQ) as GK, the difference between the arc GF and the sine of the arc AQ. Q.E.D.

SCHOLIUM.

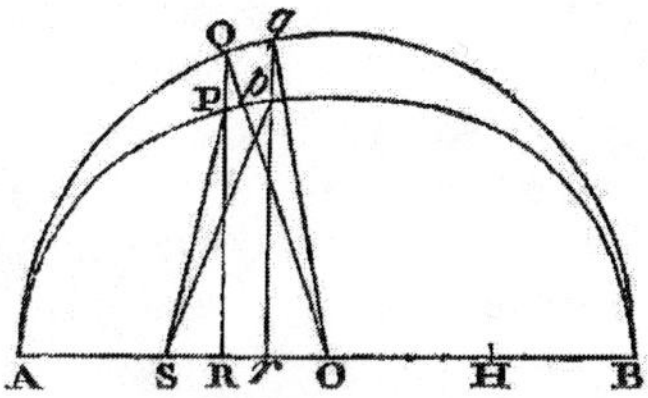

But since the description of this curve is difficult, a solution by approximation will be preferable. First, then, let there be found a certain angle B which may be to an angle of 57,29578 degrees, which an arc equal to the radius subtends, as SH, the distance of the foci, to AB, the diameter of the ellipsis. Secondly, a certain length L, which may be to the radius in the same ratio inversely. And these being found, the Problem may be solved by the following analysis. By any construction (or even by conjecture), suppose we know P the place of the body near its true place *p*. Then letting fall on the axis of the ellipsis the ordinate PR from the proportion of the diameters of the ellipsis, the ordinate RQ of the circumscribed circle AQB will be given; which ordinate is the sine of the angle AOQ, supposing AO to be the radius, and also cuts the ellipsis in P. It will be sufficient if that angle is found by a rude calculus in numbers near the truth. Suppose we also know the angle proportional to the time, that is, which is to four right angles as the time in which the body described the arc A*p*, to the time of one revolution in the ellipsis. Let this angle be N. Then take an angle D, which may be to the angle B as the sine of the angle AOQ to the radius; and an angle E which may be to the angle N — AOQ + D as the length L to the same length L diminished by the cosine of the angle AOQ, when that angle is less than a right angle, or increased thereby when greater. In the next place, take an angle F that may be to the angle B as the sine of the angle AOQ + E to the radius, and an angle G, that may be to the angle N — AOQ — E + F as the length L to the same length L diminished by the cosine of the angle AOQ + E, when that angle is less than a right angle, or increased thereby when greater. For the third time take an angle H, that may be to the angle B as the sine of the angle AOQ + E + G to the radius; and an angle I to the angle N — AOQ — E — G + H, as the length L is to the same length L diminished by the cosine of the angle AOQ + E + G, when that angle is less than a right angle, or increased thereby when greater. And so we may proceed *in infinitum.* Lastly, take the angle AOq equal to the angle AOQ + E + G + I +, &c. and from its cosine Or and the ordinate pr, which is to its sine qr as the lesser axis of the ellipsis to the greater, we shall have p the correct place of the body. When the angle N — AOQ + D happens to be negative, the sign + of the angle E must be every where changed into —, and the sign — into +. And the same thing is to be understood of the signs of the angles G and I, when the angles N — AOQ — E + F, and N — AOQ — E — G + H come out negative. But the infinite series AOQ + E + G + I +, &c. converges so very fast, that it will be scarcely ever needful to proceed beyond the second term E. And the calculus is founded upon this Theorem, that the area APS is as the difference between the arc AQ and the right line let fall from the focus S perpendicularly upon the radius OQ.

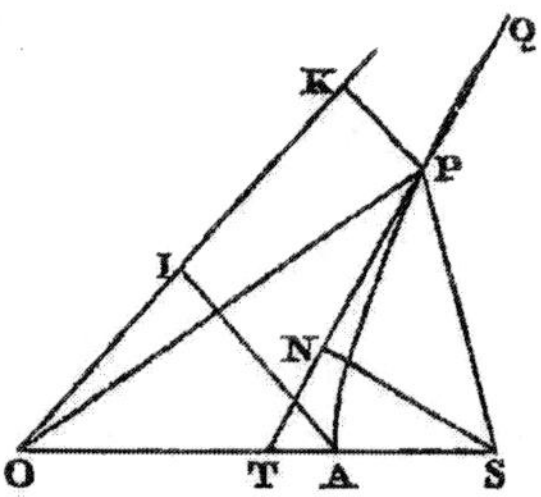

And by a calculus not unlike, the Problem is solved in the hyperbola. Let its centre be O, its vertex A, its focus S, and asymptote OK; and suppose the quantity of the area to be cut off is known, as being proportional to the time. Let that be A, and by conjecture suppose we know the position of a right line SP, that cuts off an area APS near the truth. Join OP, and from A and P to the asymptote draw AI, PK parallel to the other asymptote; and by the table of logarithms the area AIKP will be given, and equal thereto the area OPA, which subducted from the triangle OPS, will leave the area cut off APS. And by applying 2APS — 2A, or 2A — 2APS, the double difference of the area A that was to be cut off, and the area APS that is cut off, to the line SN that is let fall from the focus S, perpendicular upon the tangent TP, we shall have the length of the chord PQ. Which chord PQ is to be inscribed between A and P, if the area APS that is cut off be greater than the area A that was to be cut off, but towards the contrary side of the point P, if otherwise: and the point Q will be the place of the body more accurately. And by repeating the computation the place may be found perpetually to greater and greater accuracy.

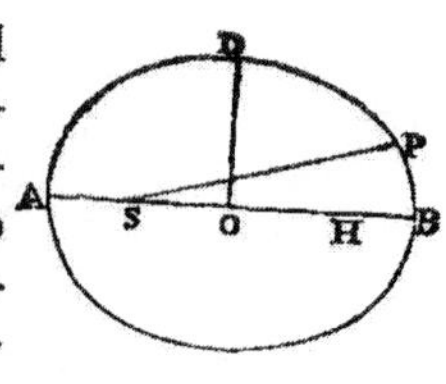

And by such computations we have a general analytical resolution of the Problem. But the particular calculus that follows is better fitted for astronomical purposes. Supposing AO, OB, OD, to be the semi-axis of the ellipsis, and L its latus rectum, and D the difference betwixt the lesser semi-axis OD, and $\frac{1}{2}$L the half of the latus rectum: let an angle Y be found, whose sine may be to the radius as the rectangle under that difference D, and AO + OD the half sum of the axes to the square of the greater axis AB. Find also an angle Z, whose sine may be to the radius as the double rectangle under the distance of the foci SH and that difference D to triple the square of half the greater semi-axis AO. Those angles being once found, the place of the body may be thus determined. Take the angle T proportional to the time in which the arc BP was described, or equal to what is called the mean motion; and an angle V the first equation of the mean motion to the angle Y, the greatest first equation, as the sine of double the angle T is to the radius; and an angle X, the second equation, to the angle Z, the second greatest equation, as the cube of the sine of the angle T is to the cube of the radius. Then take the angle BHP the mean motion equated equal to T + X + V, the sum of the angles T, V, X, if the angle T is less than a right angle; or equal to T + X — V, the difference of the same, if that angle T is greater than one and less than two right angles; and if HP meets the ellipsis in P, draw SP, and it will cut off the area BSP nearly proportional to the time.

This practice seems to be expeditious enough, because the angles V and X, taken in second minutes, if you please, being very small, it will be sufficient to find two or three of their first figures. But it is likewise sufficiently accurate to answer to the theory of the planet's motions. For even in the orbit of Mars, where the greatest equation of the centre amounts to ten degrees, the error will scarcely exceed one second. But when the angle of the mean motion equated BHP is found, the angle of the true motion BSP, and the distance SP, are readily had by the known methods.

And so far concerning the motion of bodies in curve lines. But it may also come to pass that a moving body shall ascend or descend in a right line; and I shall now go on to explain what belongs to such kind of motions.

SECTION VII.

Concerning the rectilinear ascent and descent of bodies,

PROPOSITION XXXII. PROBLEM XXIV.

Supposing that the centripetal force is reciprocally proportional to the square of the distance of the places from the centre; it is required to define the spaces which a body, falling directly, describes in given times.

CASE 1. If the body does not fall perpendicularly, it will (by Cor. 1.

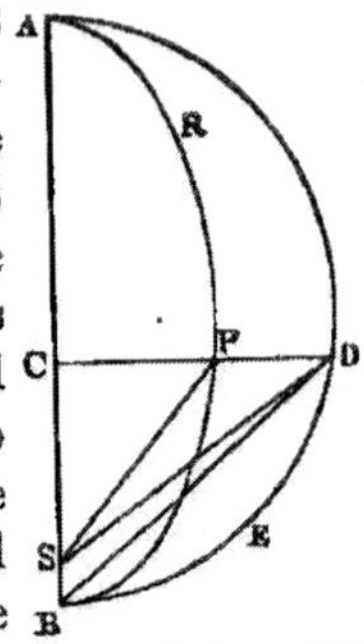

Prop. XIII) describe some conic section whose focus is placed in the centre of force. Suppose that conic section to be ARPB and its focus S. And, first, if the figure be an ellipsis, upon the greater axis thereof AB describe the semi-circle ADB, and let the right line DPC pass through the falling body, making right angles with the axis; and drawing DS, PS, the area ASD will be proportional to the area ASP, and therefore also to the time. The axis AB still remaining the same, let the breadth of the ellipsis be perpetually diminished, and the area ASD will always remain proportional to the time. Suppose that breadth to be diminished *in infinitum;* and the orbit APB in that case coinciding with the axis AB, and the focus S with the extreme point of the axis B, the body will descend in the right line AC, and the area ABD will become proportional to the time. Wherefore the space AC will be given which the body describes in a given time by its perpendicular fall from the place A, if the area ABD is taken proportional to the time, and from the point D the right line DC is let fall perpendicularly on the right line AB. Q.E.I.

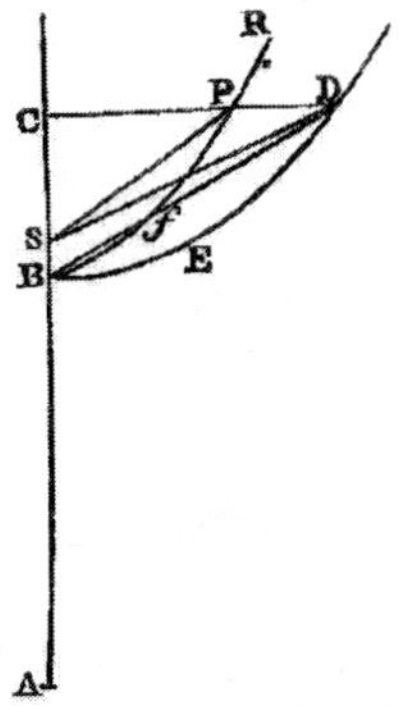

Case 2. If the figure RPB is an hyperbola, on the same principal diameter AB describe the rectangular hyperbola BED; and because the areas CSP, CB*f*P, SP*f*B, are severally to the several areas CSD, CBED, SDEB, in the given ratio of the heights CP, CD, and the area SP*f*B is proportional to the time in which the body P will move through the arc P*f*B, the area SDEB will be also proportional to that time. Let the latus rectum of the hyperbola RPB be diminished *in infinitum,* the latus transversum remaining the same; and the arc PB will come to coincide with the right line CB, and the focus S, with the vertex B, and the right line SD with the right line BD. And therefore the area BDEB will be proportional to the time in which the body C, by its perpendicular descent, describes the line CB. Q.E.I.

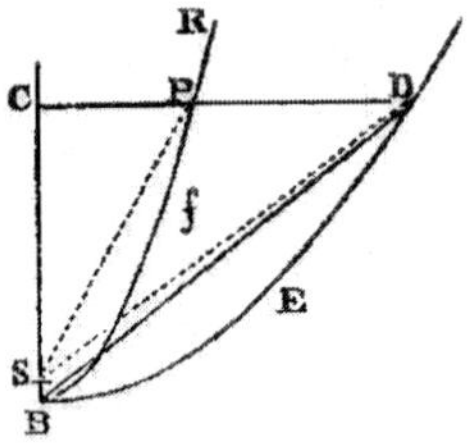

Case 3. And by the like argument, if the figure RPB is a parabola, and to the same principal vertex B another parabola BED is described, that may always remain given while the former parabola in whose perimeter the body P moves, by having its latus rectum diminished and reduced to nothing, comes to coincide with the line CB, the parabolic segment BDEB will be proportional to the time in which that body P or C will descend to the centre S or B. Q.E.I.

PROPOSITION XXXIII. THEOREM IX.

The things above found being supposed, I say, that the velocity of a falling body in any place C *is to the velocity of a body, describing a circle about the centre* B *at the distance* BC, *in the subduplicate ratio of* AC, *the distance of the body from the remoter vertex* A *of the circle or rectangular hyperbola, to* $\frac{1}{2}$AB, *the principal semi-diameter of the figure.*

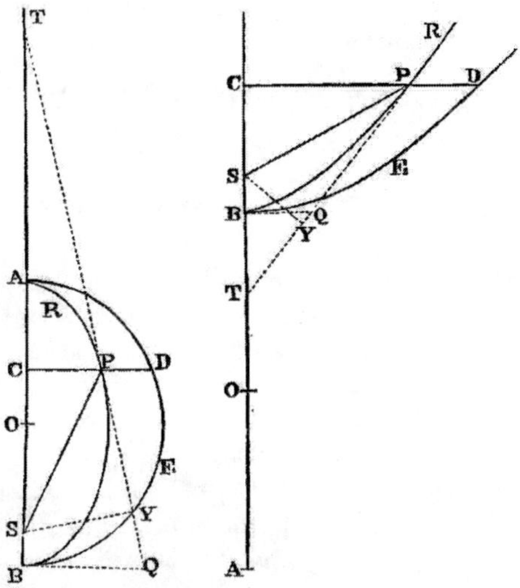

Let AB, the common diameter of both figures RPB, DEB, be bisected in O; and draw the right line PT that may touch the figure RPB in P, and likewise cut that common diameter AB (produced, if need be) in T; and let SY be perpendicular to this line, and BQ to this diameter, and suppose the latus rectum of the figure RPB to be L. From Cor. 9, Prop. XVI, it is manifest that the velocity of a body, moving in the line RPB about the centre S, in any place P, is to the velocity of a body describing a circle about the same centre, at the distance SP, in the subduplicate ratio of the rectangle $\frac{1}{2}L \times SP$ to SY^2. For by the properties of the conic sections ACB is to CP^2 as 2AO to L, and therefore $\frac{2CP^2 \times AO}{ACB}$ is equal to L. Therefore those velocities are to each other in the subduplicate ratio of $\frac{CP^2 \times AO \times SP}{ACB}$ to SY^2. Moreover, by the properties of the conic sections, CO is to BO as BO to TO, and (by composition or division) as CB to BT. Whence (by division or composition) BO — or + CO will be to BO as CT to BT, that is, AC will be to AO as CP to BQ; and therefore $\frac{CP^2 \times AO \times SP}{ACB}$ is equal to $\frac{BQ^2 \times AC \times SP}{AO \times BC}$. Now suppose CP, the breadth of the figure RPB, to be diminished *in infinitum*, so as the point P may come to coincide with the point C, and the point S with the point B, and the line SP with the line BC, and the line SY with the line BQ; and the velocity of the body now descending perpendicularly in the line CB will be to the velocity of a body describing a circle about the centre B, at the distance BC, in the subduplicate ratio of $\frac{BQ^2 \times AC \times SP}{AO \times BC}$ to SY^2, that is (neglecting the ratios of equality of SP to BC, and BQ^2 to SY^2), in the subduplicate ratio of AC to AO, or ½AB. Q.E.D.

Cor. 1. When the points B and S come to coincide, TC will become to TS as AC to AO.

Cor. 2. A body revolving in any circle at a given distance from the centre, by its motion converted upwards, will ascend to double its distance from the centre.

PROPOSITION XXXIV. THEOREM X.

If the figure BED *is a parabola, I say, that the velocity of a falling body in any place* C *is equal to the velocity by which a body may uniformly describe a circle about the centre* B *at half the interval* BC.

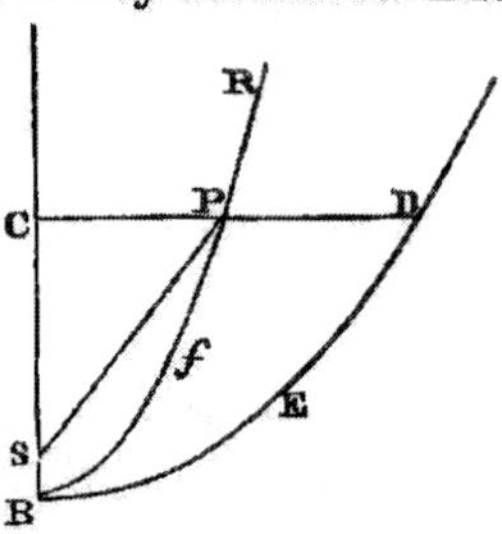

For (by Cor. 7, Prop. XVI) the velocity of a body describing a parabola RPB about the centre S, in any place P, is equal to the velocity of a body uniformly describing a circle about the same centre S at half the interval SP. Let the breadth CP of the parabola be diminished *in infinitum*, so as the parabolic arc P*f*B may come to coincide with the right line CB, the centre S with the vertex B, and the interval SP with the interval BC, and the proposition will be manifest. Q.E.D.

PROPOSITION XXXV. THEOREM XI.

The same things supposed, I say, that the area of the figure DES, *described by the indefinite radius* SD, *is equal to the area which a body with a radius equal to half the latus rectum of the figure* DES, *by uniformly revolving about the centre* S, *may describe in the same time.*

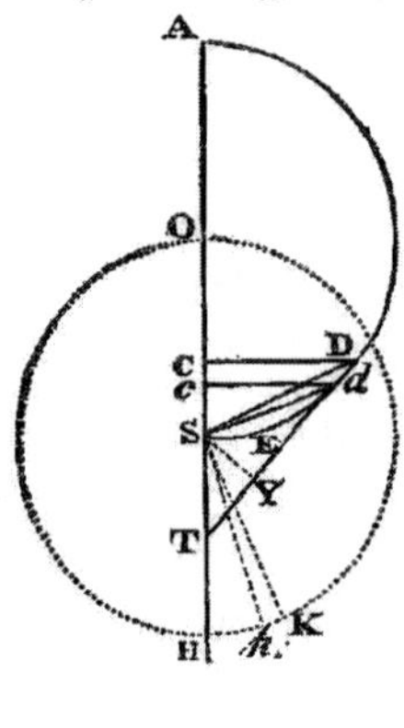

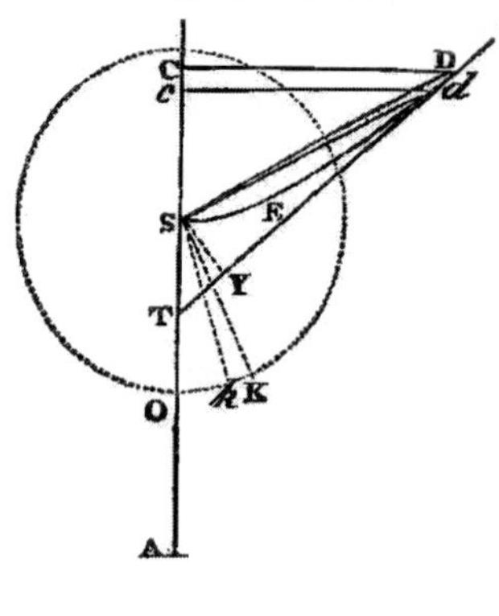

For suppose a body C in the smallest moment of time describes in falling the infinitely little line C*c*, while another body K, uniformly revolving about the centre S in the circle OK*k*, describes the arc K*k*. Erect the perpendiculars CD, *cd*, meeting the figure DES in D, *d*. Join SD, S*d*, SK, S*k*, and draw D*d* meeting the axis AS in T, and thereon let fall the perpendicular SY.

CASE 1. If the figure DES is a circle, or a rectangular hyperbola, bisect its transverse diameter AS in O, and SO will be half the latus rectum. And because TC is to TD as C*c* to D*d*, and TD to TS as CD to SY; *ex æquo* TC will be to TS as CD × C*c* to SY × D*d*. But (by Cor. 1, Prop. XXXIII) TC is to TS as AC to AO; to wit, if in the coalescence of the points D, *d*, the ultimate ratios of the lines are taken. Wherefore AC is to AO or SK as CD × C*c* to SY × D*d*. Farther, the velocity of the descending body in C is to the velocity of a body describing a circle about the centre S, at the interval SC, in the subduplicate ratio of AC to AO or SK (by Prop. XXXIII); and this velocity is to the velocity of a body describing the circle OK*k* in the subduplicate ratio of SK to SC (by Cor. 6, Prop IV); and, *ex æquo*, the first velocity to the last, that is, the little line C*c* to the arc K*k*, in the subduplicate ratio of AC to SC, that is, in the ratio of AC to CD. Wherefore CD × C*c* is equal to AC × K*k*, and consequently AC to SK as AC × K*k* to SY × D*d*, and thence SK × K*k* equal to SY × D*d*, and $\frac{1}{2}$SK × K*k* equal to $\frac{1}{2}$SY × D*d*, that is, the area KS*k* equal to the area SD*d*. Therefore in every moment of time two equal particles, KS*k* and SD*d*, of areas are generated, which, if their magnitude is diminished, and their number increased *in infinitum*, obtain the ratio of equality, and consequently (by Cor. Lem. IV), the whole areas together generated are always equal. Q.E.D.

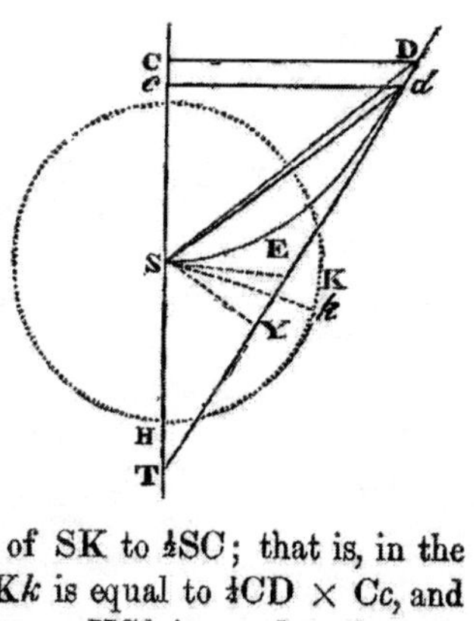

CASE 2. But if the figure DES is a parabola, we shall find, as above, CD × C*c* to SY × D*d* as TC to TS, that is, as 2 to 1; and that therefore $\frac{1}{4}$CD × C*c* is equal to $\frac{1}{2}$ SY × D*d*. But the velocity of the falling body in C is equal to the velocity with which a circle may be uniformly described at the interval $\frac{1}{2}$SC (by Prop. XXXIV). And this velocity to the velocity with which a circle may be described with the radius SK, that is, the little line C*c* to the arc K*k*, is (by Cor. 6, Prop. IV) in the subduplicate ratio of SK to $\frac{1}{2}$SC; that is, in the ratio of SK to $\frac{1}{2}$CD. Wherefore $\frac{1}{2}$SK × K*k* is equal to $\frac{1}{4}$CD × C*c*, and therefore equal to $\frac{1}{2}$SY × D*d*; that is, the area KS*k* is equal to the area SD*d*, as above. Q.E.D.

PROPOSITION XXXVI. PROBLEM XXV.

To determine the times of the descent of a body falling from a given place A.

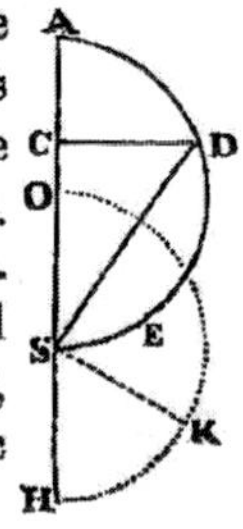

Upon the diameter AS, the distance of the body from the centre at the beginning, describe the semi-circle ADS, as likewise the semi-circle OKH equal thereto, about the centre S. From any place C of the body erect the ordinate CD. Join SD, and make the sector OSK equal to the area ASD. It is evident (by Prop. XXXV) that the body in falling will describe the space AC in the same time in which another body, uniformly revolving about the centre S, may describe the arc OK. Q.E.F.

PROPOSITION XXXVII. PROBLEM XXVI.

To define the times of the ascent or descent of a body projected upwards or downwards from a given place.

Suppose the body to go off from the given place G, in the direction of the line GS, with any velocity. In the duplicate ratio of this velocity to the uniform velocity in a circle, with which the body may revolve about

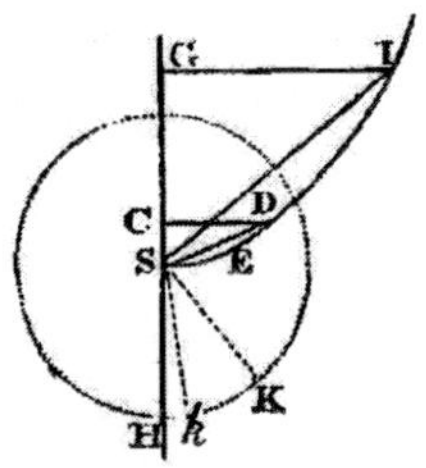

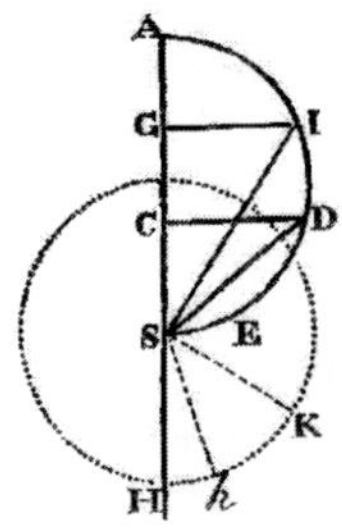

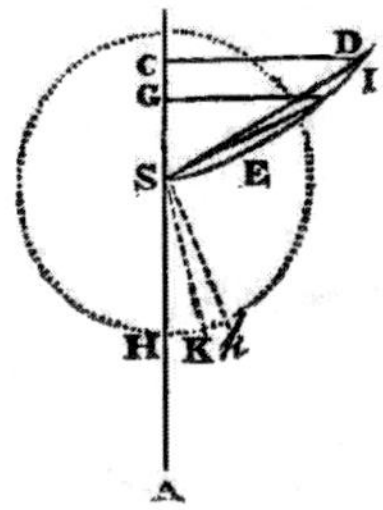

the centre S at the given interval SG, take GA to ½AS. If that ratio is the same as of the number 2 to 1, the point A is infinitely remote; in which case a parabola is to be described with any latus rectum to the vertex S, and axis SG; as appears by Prop. XXXIV. But if that ratio is less or greater than the ratio of 2 to 1, in the former case a circle, in the latter a rectangular hyperbola, is to be described on the diameter SA; as appears by Prop. XXXIII. Then about the centre S, with an interval equal to half the latus rectum, describe the circle H*k*K; and at the place G of the ascending or descending body, and at any other place C, erect the perpendiculars GI, CD, meeting the conic section or circle in I and D. Then joining SI, SD, let the sectors HSK, HS*k* be made equal to the segments SEIS, SEDS, and (by Prop. XXXV) the body G will describe the space GC in the same time in which the body K may describe the arc Kk. Q.E.F.

PROPOSITION XXXVIII. THEOREM XII.

Supposing that the centripetal force is proportional to the altitude or distance of places from the centre, I say, that the times and velocities of falling bodies, and the spaces which they describe, are respectively proportional to the arcs, and the right and versed sines of the arcs.

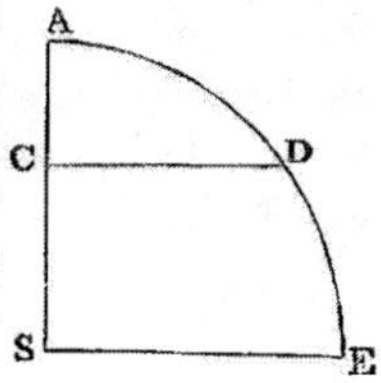

Suppose the body to fall from any place A in the right line AS; and about the centre of force S, with the interval AS, describe the quadrant of a circle AE; and let CD be the right sine of any arc AD; and the body A will in the time AD in falling describe the space AC, and in the place C will acquire the velocity CD.

This is demonstrated the same way from Prop. X, as Prop. XXXII was demonstrated from Prop. XI.

COR. 1. Hence the times are equal in which one body falling from the place A arrives at the centre S, and another body revolving describes the quadrantal arc ADE.

COR. 2. Wherefore all the times are equal in which bodies falling from whatsoever places arrive at the centre. For all the periodic times of revolving bodies are equal (by Cor. 3, Prop. IV).

PROPOSITION XXXIX. PROBLEM XXVII.

Supposing a centripetal force of any kind, and granting the quadratures of curvilinear figures; it is required to find the velocity of a body, ascending or descending in a right line, in the several places through which it passes; as also the time in which it will arrive at any place: and vice versa.

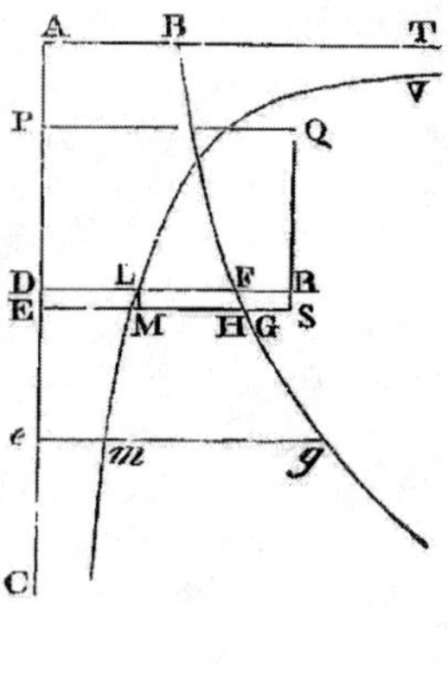

Suppose the body E to fall from any place A in the right line ADEC; and from its place E imagine a perpendicular EG always erected proportional to the centripetal force in that place tending to the centre C; and let BFG be a curve line, the locus of the point G. And in the beginning of the motion suppose EG to coincide with the perpendicular AB; and the velocity of the body in any place E will be as a right line whose square is equal to the curvilinear area ABGE. Q.E.I.

In EG take EM reciprocally proportional to

a right line whose square is equal to the area ABGE, and let VLM be a curve line wherein the point M is always placed, and to which the right line AB produced is an asymptote; and the time in which the body in falling describes the line AE, will be as the curvilinear area ABTVME. Q.E.I.

For in the right line AE let there be taken the very small line DE of a given length, and let DLF be the place of the line EMG, when the body was in D; and if the centripetal force be such, that a right line, whose square is equal to the area ABGE, is as the velocity of the descending body, the area itself will be as the square of that velocity; that is, if for the velocities in D and E we write V and V + I, the area ABFD will be as VV, and the area ABGE as VV + 2VI + II; and by division, the area DFGE as 2VI + II, and therefore $\frac{\text{DFGE}}{\text{DE}}$ will be as $\frac{2\text{VI} + \text{II}}{\text{DE}}$; that is, if we take the first ratios of those quantities when just nascent, the length DF is as the quantity $\frac{2\text{VI}}{\text{DE}}$, and therefore also as half that quantity $\frac{\text{I} \times \text{V}}{\text{DE}}$. But the time in which the body in falling describes the very small line DE, is as that line directly and the velocity V inversely; and the force will be as the increment I of the velocity directly and the time inversely; and therefore if we take the first ratios when those quantities are just nascent, as $\frac{\text{I} \times \text{V}}{\text{DE}}$, that is, as the length DF. Therefore a force proportional to DF or EG will cause the body to descend with a velocity that is as the right line whose square is equal to the area ABGE. Q.E.D.

Moreover, since the time in which a very small line DE of a given length may be described is as the velocity inversely, and therefore also inversely as a right line whose square is equal to the area ABFD; and since the line DL, and by consequence the nascent area DLME, will be as the same right line inversely, the time will be as the area DLME, and the sum of all the times will be as the sum of all the areas; that is (by Cor. Lem. IV), the whole time in which the line AE is described will be as the whole area ATVME. Q.E.D.

Cor. 1. Let P be the place from whence a body ought to fall, so as that, when urged by any known uniform centripetal force (such as gravity is vulgarly supposed to be), it may acquire in the place D a velocity equal to the velocity which another body, falling by any force whatever, hath acquired in that place D. In the perpendicular DF let there be taken DR, which may be to DF as that uniform force to the other force in the place D. Complete the rectangle PDRQ, and cut off the area ABFD equal to that rectangle. Then A will be the place

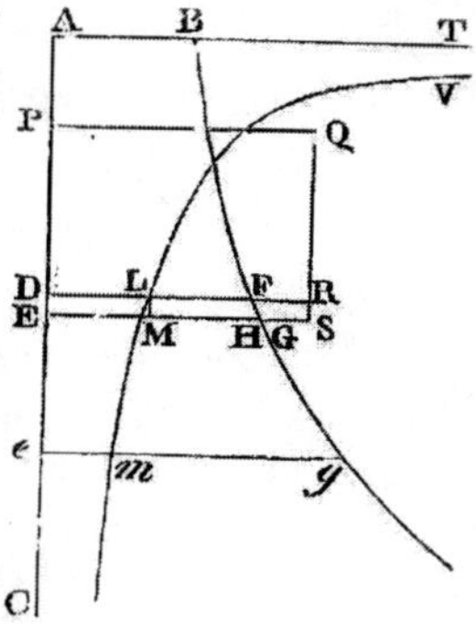

from whence the other body fell. For completing the rectangle DRSE, since the area ABFD is to the area DFGE as VV to 2VI, and therefore as ½V to I, that is, as half the whole velocity to the increment of the velocity of the body falling by the unequable force; and in like manner the area PQRD to the area DRSE as half the whole velocity to the increment of the velocity of the body falling by the uniform force; and since those increments (by reason of the equality of the nascent times) are as the generating forces, that is, as the ordinates DF, DR, and consequently as the nascent areas DFGE, DRSE; therefore, *ex æquo*, the whole areas ABFD, PQRD will be to one another as the halves of the whole velocities; and therefore, because the velocities are equal, they become equal also.

COR. 2. Whence if any body be projected either upwards or downwards with a given velocity from any place D, and there be given the law of centripetal force acting on it, its velocity will be found in any other place, as *e*, by erecting the ordinate *eg*, and taking that velocity to the velocity in the place D as a right line whose square is equal to the rectangle PQRD, either increased by the curvilinear area DF*ge*, if the place *e* is below the place D, or diminished by the same area DF*ge*, if it be higher, is to the right line whose square is equal to the rectangle PQRD alone.

COR. 3. The time is also known by erecting the ordinate *em* reciprocally proportional to the square root of PQRD + or — DF*ge*, and taking the time in which the body has described the line D*e* to the time in which another body has fallen with an uniform force from P, and in falling arrived at D in the proportion of the curvilinear area DL*me* to the rectangle 2PD × DL. For the time in which a body falling with an uniform force hath described the line PD, is to the time in which the same body has described the line PE in the subduplicate ratio of PD to PE; that is (the very small line DE being just nascent), in the ratio of PD to PD + ½DE, or 2PD to 2PD + DE, and, by division, to the time in which the body hath described the small line DE, as 2PD to DE, and therefore as the rectangle 2PD × DL to the area DLME; and the time in which both the bodies described the very small line DE is to the time in which the body moving unequably hath described the line D*e* as the area DLME to the area DL*me*; and, *ex æquo*, the first mentioned of these times is to the last as the rectangle 2PD × DL to the area DL*me*.

SECTION VIII.

Of the invention of orbits wherein bodies will revolve, being acted upon by any sort of centripetal force.

PROPOSITION XL. THEOREM XIII.

If a body, acted upon by any centripetal force, is any how moved, and another body ascends or descends in a right line, and their velocities be equal in any one case of equal altitudes, their velocities will be also equal at all equal altitudes.

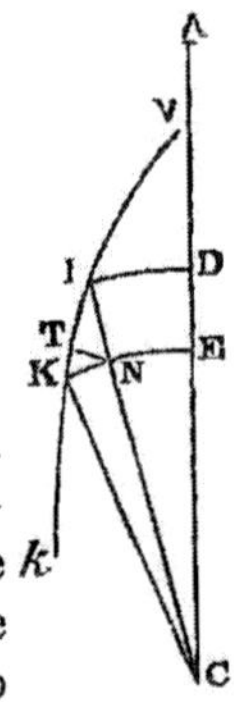

Let a body descend from A through D and E, to the centre C; and let another body move from V in the curve line VIK*k*. From the centre C, with any distances, describe the concentric circles DI, EK, meeting the right line AC in D and E, and the curve VIK in I and K. Draw IC meeting KE in N, and on IK let fall the perpendicular NT; and let the interval DE or IN between the circumferences of the circles be very small; and imagine the bodies in D and I to have equal velocities. Then because the distances CD and CI are equal, the centripetal forces in D and I will be also equal. Let those forces be expressed by the equal lineolæ DE and IN; and let the force IN (by Cor. 2 of the Laws of Motion) be resolved into two others, NT and IT. Then the force NT acting in the direction of the line NT perpendicular to the path ITK of the body will not at all affect or change the velocity of the body in that path, but only draw it aside from a rectilinear course, and make it deflect perpetually from the tangent of the orbit, and proceed in the curvilinear path ITK*k*. That whole force, therefore, will be spent in producing this effect; but the other force IT, acting in the direction of the course of the body, will be all employed in accelerating it, and in the least given time will produce an acceleration proportional to itself. Therefore the accelerations of the bodies in D and I, produced in equal times, are as the lines DE, IT (if we take the first ratios of the nascent lines DE, IN, IK, IT, NT); and in unequal times as those lines and the times conjunctly. But the times in which DE and IK are described, are, by reason of the equal velocities (in D and I) as the spaces described DE and IK, and therefore the accelerations in the course of the bodies through the lines DE and IK are as DE and IT, and DE and IK conjunctly; that is, as the square of DE to the rectangle IT into IK. But the rectangle IT × IK is equal to the square of IN, that is, equal to the square of DE; and therefore the accelerations generated in the passage of the bodies from D and I to E and K are equal. Therefore the velocities of the bodies in E and K are also equal: and by the same reasoning they will always be found equal in any subsequent equal distances. Q.E.D.

By the same reasoning, bodies of equal velocities and equal distances from the centre will be equally retarded in their ascent to equal distances. Q.E.D.

COR. 1. Therefore if a body either oscillates by hanging to a string, or by any polished and perfectly smooth impediment is forced to move in a curve line; and another body ascends or descends in a right line, and their velocities be equal at any one equal altitude, their velocities will be also equal at all other equal altitudes. For by the string of the pendulous body, or by the impediment of a vessel perfectly smooth, the same thing will be effected as by the transverse force NT. The body is neither accelerated nor retarded by it, but only is obliged to leave its rectilinear course.

COR. 2. Suppose the quantity P to be the greatest distance from the centre to which a body can ascend, whether it be oscillating, or revolving in a trajectory, and so the same projected upwards from any point of a trajectory with the velocity it has in that point. Let the quantity A be the distance of the body from the centre in any other point of the orbit; and let the centripetal force be always as the power A^{n-1}, of the quantity A, the index of which power $n-1$ is any number n diminished by unity. Then the velocity in every altitude A will be as $\sqrt{P^n - A^n}$, and therefore will be given. For by Prop. XXXIX, the velocity of a body ascending and descending in a right line is in that very ratio.

PROPOSITION XLI. PROBLEM XXVIII.

Supposing a centripetal force of any kind, and granting the quadratures of curvilinear figures, it is required to find as well the trajectories in which bodies will move, as the times of their motions in the trajectories found.

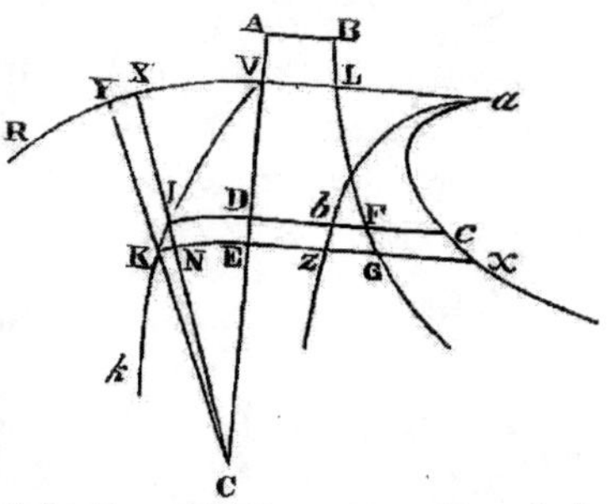

Let any centripetal force tend to the centre C, and let it be required to find the trajectory VIKk. Let there be given the circle VR, described from the centre C with any interval CV; and from the same centre describe any other circles ID, KE cutting the trajectory in I and K, and the right line CV in D and E. Then draw the right line CNIX cutting the circles KE, VR in N and X, and the right line CKY meeting the circle VR in Y. Let the points I and K be indefinitely near; and let the body go on from V through I and K to k; and let the point A be the place from whence another body is to fall, so as in the place D to acquire a velocity equal to the velocity of the first body in I. And things remaining as in Prop. XXXIX, the lineola IK, described in the least given time, will be as the velocity, and therefore as the right line whose square is equal to the area ABFD, and the triangle ICK proportional to the time will be given, and therefore KN will be reciprocally as the altitude IC; that is (if there be given any quantity Q, and the altitude IC be called A), as $\frac{Q}{A}$. This quantity $\frac{Q}{A}$ call Z, and suppose the magnitude of Q to be such that in some case $\sqrt{ABFD}$ may be to Z as IK to KN, and then in all cases $\sqrt{ABFD}$ will be to Z as IK to KN, and ABFD to ZZ as IK^2 to KN^2, and by division ABFD — ZZ to ZZ as IN^2 to KN^2, and therefore $\sqrt{ABFD - ZZ}$ to Z, or $\frac{Q}{A}$ as IN to KN; and therefore A × KN will be equal to $\frac{Q \times IN}{\sqrt{ABFD - ZZ}}$. Therefore since YX × XC is to A × KN as CX^2, to AA, the rectangle XY × XC will be equal to $\frac{Q \times IN \times CX^2}{AA\sqrt{ABFD - ZZ}}$. Therefore in the perpendicular DF let there be taken continually D*b*, D*c* equal to $\frac{Q}{2\sqrt{ABFD - ZZ}}$, $\frac{Q \times CX^2}{2AA\sqrt{ABFD - ZZ}}$ respectively, and let the curve lines *ab*, *ac*, the foci of the points *b* and *c*, be described: and from the point V let the perpendicular V*a* be erected to the line AC, cutting off the curvilinear areas VD*ba*, VD*ca*, and let the ordinates E*z*, E*x*, be erected also. Then because the rectangle D*b* × IN or D*bz*E is equal to half the rectangle A × KN, or to the triangle ICK; and the rectangle D*c* × IN or D*cx*E is equal to half the rectangle YX × XC, or to the triangle XCY; that is, because the nascent particles D*bz*E, ICK of the areas VD*ba*, VIC are always equal; and the nascent particles D*cx*E, XCY of the areas VD*ca*, VCX are always equal; therefore the generated area VD*ba* will be equal to the generated area VIC, and therefore proportional to the time; and the generated area VD*ca* is equal to the generated sector VCX. If, therefore, any time be given during which the body has been moving from V, there will be also given the area proportional to it VD*ba*; and thence will be given the altitude of the body CD or CI; and the area VD*ca*, and the sector VCX equal thereto, together with its angle VCI. But the angle VCI, and the altitude CI being given, there is also given the place I, in which the body will be found at the end of that time. Q.E.I.

Cor. 1. Hence the greatest and least altitudes of the bodies, that is, the apsides of the trajectories, may be found very readily. For the apsides are those points in which a right line IC drawn through the centre falls perpendicularly upon the trajectory VIK; which comes to pass when the right lines IK and NK become equal; that is, when the area ABFD is equal to ZZ.

COR. 2. So also the angle KIN, in which the trajectory at any place cuts the line IC, may be readily found by the given altitude IC of the body: to wit, by making the sine of that angle to radius as KN to IK; that is, as Z to the square root of the area ABFD.

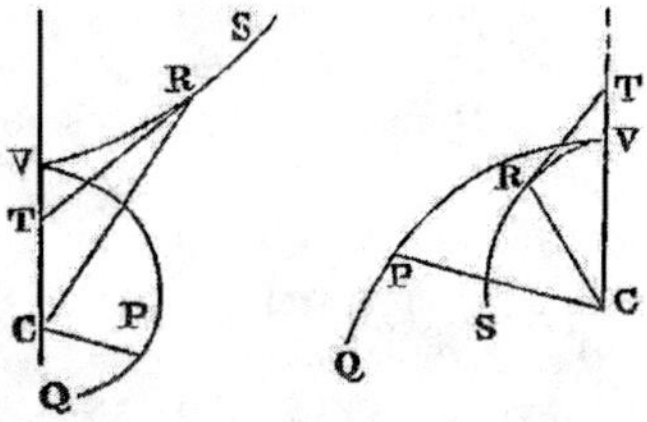

COR. 3. If to the centre C, and the principal vertex V, there be described a conic section VRS; and from any point thereof, as R, there be drawn the tangent RT meeting the axis CV indefinitely produced in the point T; and then joining CR there be drawn the right line CP, equal to the abscissa CT, making an angle VCP proportional to the sector VCR; and if a centripetal force, reciprocally proportional to the cubes of the distances of the places from the centre, tends to the centre C; and from the place V there sets out a body with a just velocity in the direction of a line perpendicular to the right line CV; that body will proceed in a trajectory VPQ, which the point P will always touch; and therefore if the conic section VRS be an hyberbola, the body will descend to the centre; but if it be an ellipsis, it will ascend perpetually, and go farther and farther off *in infinitum*. And, on the contrary, if a body endued with any velocity goes off from the place V, and according as it begins either to descend obliquely to the centre, or ascends obliquely from it, the figure VRS be either an hyperbola or an ellipsis, the trajectory may be found by increasing or diminishing the angle VCP in a given ratio. And the centripetal force becoming centrifugal, the body will ascend obliquely in the trajectory VPQ, which is found by taking the angle VCP proportional to the elliptic sector VRC, and the length CP equal to the length CT, as before. All these things follow from the foregoing Proposition, by the quadrature of a certain curve, the invention of which, as being easy enough, for brevity's sake I omit.

PROPOSITION XLII. PROBLEM XXIX.

The law of centripetal force being given, it is required to find the motion of a body setting out from a given place, with a given velocity, in the direction of a given right line.

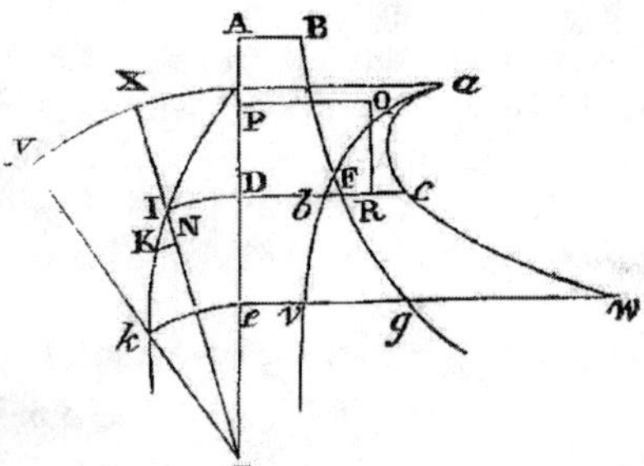

Suppose the same things as in the three preceding propositions; and let the body go off from the place I in the direction of the little line, IK, with the same velocity as another body, by falling with an uniform centripetal force from the place P, may acquire in D; and let this uniform force be to the force with which the body is at first urged in I, as DR to DF. Let the body go on towards k; and about the centre C, with the interval Ck, describe the circle ke, meeting the right line PD in e, and let there be erected the lines eg, ev, ew, ordinately applied to the curves BFg, abv, acw. From the given rectangle PDRQ and the given law of centripetal force, by which the first body is acted on, the curve line BFg is also given, by the construction of Prop. XXVII, and its Cor. 1. Then from the given angle CIK is given the proportion of the nascent lines IK, KN; and thence, by the construction of Prob. XXVIII, there is given the quantity Q, with the curve lines abv, acw; and therefore, at the end of any time Dbve, there is given both the altitude of the body Ce or Ck, and the area Dcwe, with the sector equal to it XCy, the angle ICk, and the place k, in which the body will then be found. Q.E.I.

We suppose in these Propositions the centripetal force to vary in its recess from the centre according to some law, which any one may imagine at pleasure; but at equal distances from the centre to be everywhere the same.

I have hitherto considered the motions of bodies in immovable orbits. It remains now to add something concerning their motions in orbits which revolve round the centres of force.

SECTION IX.

Of the motion of bodies in moveable orbits; and of the motion of the apsides.

PROPOSITION XLIII. PROBLEM XXX.

It is required to make a body move in a trajectory that revolves about the centre of force in the same manner as another body in the same trajectory at rest.

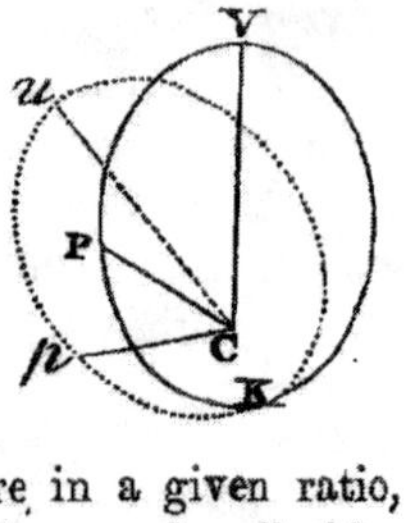

In the orbit VPK, given by position, let the body P revolve, proceeding from V towards K. From the centre C let there be continually drawn Cp, equal to CP, making the angle VCp proportional to the angle VCP; and the area which the line Cp describes will be to the area VCP, which the line CP describes at the same time, as the velocity of the describing line Cp to the velocity of the describing line CP; that is, as the angle VCp to the angle VCP, therefore in a given ratio, and therefore proportional to the time. Since, then, the area described by the line Cp in an immovable plane is proportional to the time, it is manifest that a body, being acted upon by a just quantity of centripetal force, may revolve with the point p in the curve line which the same point p, by the method just now explained, may be made to describe an immovable plane. Make the angle VCu equal to the angle PCp, and the line Cu equal to CV, and the figure uCp equal to the figure VCP, and the body being always in the point p, will move in the perimeter of the revolving figure uCp, and will describe its (revolving) arc up in the same time that the other body P describes the similar and equal arc VP in the quiescent figure VPK. Find, then, by Cor. 5, Prop. VI., the centripetal force by which the body may be made to revolve in the curve line which the point p describes in an immovable plane, and the Problem will be solved. Q.E.F.

PROPOSITION XLIV. THEOREM XIV.

The difference of the forces, by which two bodies may be made to move equally, one in a quiescent, the other in the same orbit revolving, is in a triplicate ratio of their common altitudes inversely.

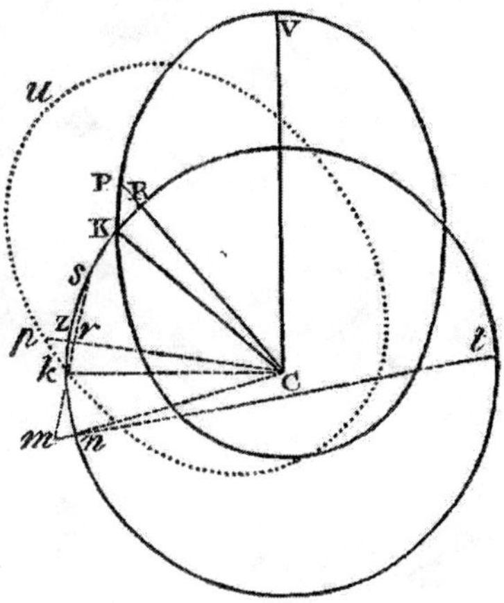

Let the parts of the quiescent orbit VP, PK be similar and equal to the parts of the revolving orbit up, pk; and let the distance of the points P and K be supposed of the utmost smallness. Let fall a perpendicular kr from the point k to the right line pC, and produce it to m, so that mr may be to kr as the angle VCp to the angle VCP. Because the altitudes of the bodies PC and pC, KC and kC, are always equal, it is manifest that the increments or decrements of the lines PC and pC are always equal; and therefore if each of the several motions of the bodies in the places P and p be resolved into two (by Cor. 2 of the Laws of Motion), one of which is directed towards the centre, or according to the lines PC, pC, and the other, transverse to the former, hath a direction perpendicular to the lines PC and pC; the motions towards the centre will be equal, and the transverse motion of the body p will be to the transverse motion of the body P as the angular motion of the line pC to the angular motion of the line PC; that is, as the angle VCp to the angle VCP. Therefore, at the same time that the body P, by both its motions, comes to the point K, the body p, having an equal motion towards the centre, will be equally moved from p towards C; and therefore that time being expired, it will be found somewhere in the line mkr, which, passing through the point k, is perpendicular to the line pC; and by its transverse motion will acquire a distance from the line *p*C, that will be to the distance which the other body P acquires from the line PC as the transverse motion of the body *p* to the transverse motion of the other body P. Therefore since *kr* is equal to the distance which the body P acquires from the line PC, and *mr* is to *kr* as the angle VC*p* to the angle VCP, that is, as the transverse motion of the body *p* to the transverse motion of the body P, it is manifest that the body *p*, at the expiration of that time, will be found in the place *m*. These things will be so, if the bodies *p* and P are equally moved in the directions of the lines *p*C and PC, and are therefore urged with equal forces in those directions. But if we take an angle *p*C*n* that is to the angle *p*C*k* as the angle VC*p* to the angle VCP, and *n*C be equal to *k*C, in that case the body *p* at the expiration of the time will really be in *n*; and is therefore urged with a greater force than the body P, if the angle *n*C*p* is greater than the angle *k*C*p*, that is, if the orbit *upk*, move either *in consequentia*, or *in antecedentia*, with a celerity greater than the double of that with which the line CP moves *in consequentia*; and with a less force if the orbit moves slower *in antecedentia*. And the difference of the forces will be as the interval *mn* of the places through which the body would be carried by the action of that difference in that given space of time. About the centre C with the interval C*n* or C*k* suppose a circle described cutting the lines *mr*, *mn* produced in *s* and *t*, and the rectangle $mn \times mt$ will be equal to the rectangle $mk \times ms$, and therefore *mn* will be equal to $\frac{mk \times ms}{mt}$. But since the triangles *p*C*k*, *p*C*n*, in a given time, are of a given magnitude, *kr* and *mr*, and their difference *mk*, and their sum *ms*, are reciprocally as the altitude *p*C, and therefore the rectangle $mk \times ms$ is reciprocally as the square of the altitude *p*C. But, moreover, *mt* is directly as $\frac{1}{2}mt$, that is, as the altitude *p*C. These are the first ratios of the nascent lines; and hence $\frac{mk \times ms}{mt}$, that is, the nascent lineola *mn*, and the difference of the forces proportional thereto, are reciprocally as the cube of the altitude *p*C. Q.E.D.

Cor. 1. Hence the difference of the forces in the places P and *p*, or K and *k*, is to the force with which a body may revolve with a circular motion from R to K, in the same time that the body P in an immovable orb describes the arc PK, as the nascent line *mn* to the versed sine of the nascent arc RK, that is, as $\frac{mk \times ms}{mt}$ to $\frac{rk^2}{2kC}$, or as $mk \times ms$ to the square of *rk*; that is, if we take given quantities F and G in the same ratio to one another as the angle VCP bears to the angle VC*p*, as GG — FF to FF. And, therefore, if from the centre C, with any distance CP or C*p*, there be described a circular sector equal to the whole area VPC, which the body revolving in an immovable orbit has by a radius drawn to the centre described in any certain time, the difference of the forces, with which the body P revolves in an immovable orbit, and the body p in a movable orbit, will be to the centripetal force, with which another body by a radius drawn to the centre can uniformly describe that sector in the same time as the area VPC is described, as GG — FF to FF. For that sector and the area pCk are to one another as the times in which they are described.

COR. 2. If the orbit VPK be an ellipsis, having its focus C, and its highest apsis V, and we suppose the the ellipsis upk similar and equal to it, so that pC may be always equal to PC, and the angle VCp be to the angle VCP in the given ratio of G to F; and for the altitude PC or pC we put A, and 2R for the latus rectum of the ellipsis, the force with which a body may be made to revolve in a movable ellipsis will be as $\frac{FF}{AA} + \frac{RGG - RFF}{A^3}$, and *vice versa*. Let the force with which a body may

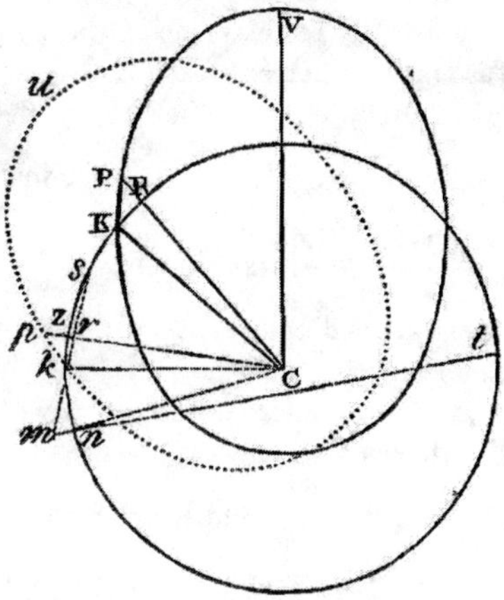

revolve in an immovable ellipsis be expressed by the quantity $\frac{FF}{AA}$, and the force in V will be $\frac{FF}{CV^2}$. But the force with which a body may revolve in a circle at the distance CV, with the same velocity as a body revolving in an ellipsis has in V, is to the force with which a body revolving in an ellipsis is acted upon in the apsis V, as half the latus rectum of the ellipsis to the semi-diameter CV of the circle, and therefore is as $\frac{RFF}{CV^3}$; and the force which is to this, as GG — FF to FF, is as $\frac{RGG - RFF}{CV^3}$: and this force (by Cor. 1 of this Prop.) is the difference of the forces in V, with which the body P revolves in the immovable ellipsis VPK, and the body p in the movable ellipsis upk. Therefore since by this Prop. that difference at any other altitude A is to itself at the altitude CV as $\frac{1}{A^3}$ to $\frac{1}{CV^3}$, the same difference in every altitude A will be as $\frac{RGG - RFF}{A^3}$. Therefore to the force $\frac{FF}{AA}$, by which the body may revolve in an immovable ellipsis VPK, add the excess $\frac{\text{RGG}-\text{RFF}}{\text{A}^3}$, and the sum will be the whole force $\frac{\text{FF}}{\text{AA}} + \frac{\text{RGG}-\text{RFF}}{\text{A}^3}$ by which a body may revolve in the same time in the movable ellipsis *upk*.

Cor. 3. In the same manner it will be found, that, if the immovable orbit VPK be an ellipsis having its centre in the centre of the forces C, and there be supposed a movable ellipsis *upk*, similar, equal, and concentrical to it; and 2R be the principal latus rectum of that ellipsis, and 2T the latus transversum, or greater axis; and the angle V*Cp* be continually to the angle VCP as G to F; the forces with which bodies may revolve in the immovable and movable ellipsis, in equal times, will be as $\frac{\text{FFA}}{\text{T}^3}$ and $\frac{\text{FFA}}{\text{T}^3} + \frac{\text{RGG}-\text{RFF}}{\text{A}^3}$ respectively.

Cor. 4. And universally, if the greatest altitude CV of the body be called T, and the radius of the curvature which the orbit VPK has in V, that is, the radius of a circle equally curve, be called R, and the centripetal force with which a body may revolve in any immovable trajectory VPK at the place V be called $\frac{\text{VFF}}{\text{TT}}$, and in other places P be indefinitely styled X; and the altitude CP be called A, and G be taken to F in the given ratio of the angle V*Cp* to the angle VCP; the centripetal force with which the same body will perform the same motions in the same time, in the same trajectory *upk* revolving with a circular motion, will be as the sum of the forces $\text{X} + \frac{\text{VRGG}-\text{VRFF}}{\text{A}^3}$.

Cor. 5. Therefore the motion of a body in an immovable orbit being given, its angular motion round the centre of the forces may be increased or diminished in a given ratio; and thence new immovable orbits may be found in which bodies may revolve with new centripetal forces.

Cor. 6. Therefore if there be erected the line VP of an indeterminate length, perpendicular to the line CV given by position, and CP be drawn, and *Cp* equal to it, making the angle V*Cp* having a given ratio to the angle VCP, the force with which a body may revolve in the curve line V*pk*, which the point *p* is continually describing, will be reciprocally as the cube of the altitude *Cp*. For the body P, by its *vis inertiæ* alone, no other force impelling it, will proceed uniformly in the right line VP. Add, then, a force tending to the centre C reciprocally as the cube of the altitude CP or *Cp*, and (by what was just demonstrated) the

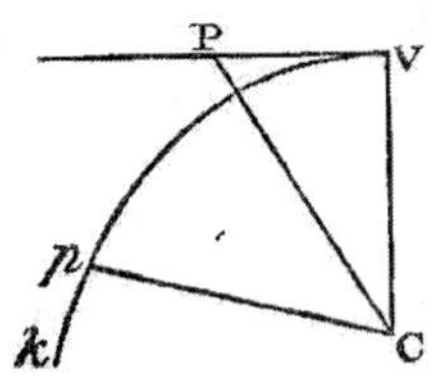

body will deflect from the rectilinear motion into the curve line V*pk*. But this curve V*pk* is the same with the curve VPQ found in Cor. 3, Prop. XLI, in which, I said, bodies attracted with such forces would ascend obliquely.

PROPOSITION XLV. PROBLEM XXXI.

To find the motion of the apsides in orbits approaching very near to circles.

This problem is solved arithmetically by reducing the orbit, which a body revolving in a movable ellipsis (as in Cor. 2 and 3 of the above Prop.) describes in an immovable plane, to the figure of the orbit whose apsides are required; and then seeking the apsides of the orbit which that body describes in an immovable plane. But orbits acquire the same figure, if the centripetal forces with which they are described, compared between themselves, are made proportional at equal altitudes. Let the point V be the highest apsis, and write T for the greatest altitude CV, A for any other altitude CP or C*p*, and X for the difference of the altitudes CV — CP; and the force with which a body moves in an ellipsis revolving about its focus C (as in Cor. 2), and which in Cor. 2 was as $\frac{FF}{AA} + \frac{RGG - RFF}{A^3}$, that is as, $\frac{FFA + RGG - RFF}{A^3}$, by substituting T — X for A, will become as $\frac{RGG - RFF + TFF - FFX}{A^3}$. In like manner any other centripetal force is to be reduced to a fraction whose denominator is A^3, and the numerators are to be made analogous by collating together the homologous terms. This will be made plainer by Examples.

EXAMPLE 1. Let us suppose the centripetal force to be uniform, and therefore as $\frac{A^3}{A^3}$ or, writing T — X for A in the numerator, as $\frac{T^3 - 3TTX + 3TXX - X^3}{A^3}$. Then collating together the correspondent terms of the numerators, that is, those that consist of given quantities, with those of given quantities, and those of quantities not given with those of quantities not given, it will become RGG — RFF + TFF to T^3 as — FFX to $3TTX + 3TXX - X^3$, or as —FF to —3TT + 3TX — XX. Now since the orbit is supposed extremely near to a circle, let it coincide with a circle; and because in that case R and T become equal, and X is infinitely diminished, the last ratios will be, as RGG to T^2, so — FF to — 3TT, or as GG to TT, so FF to 3TT; and again, as GG to FF, so TT to 3TT, that is, as 1 to 3; and therefore G is to F, that is, the angle VC*p* to the angle VCP, as 1 to $\sqrt{3}$. Therefore since the body, in an immovable ellipsis, in descending from the upper to the lower apsis, describes an angle, if I may so speak, of 180 deg., the other body in a movable ellipsis, and therefore in the immovable orbit we are treating of, will in its descent from the upper to the lower apsis, describe an angle VCp of $\frac{180}{\sqrt{3}}$ deg. And this comes to pass by reason of the likeness of this orbit which a body acted upon by an uniform centripetal force describes, and of that orbit which a body performing its circuits in a revolving ellipsis will describe in a quiescent plane. By this collation of the terms, these orbits are made similar; not universally, indeed, but then only when they approach very near to a circular figure. A body, therefore revolving with an uniform centripetal force in an orbit nearly circular, will always describe an angle of $\frac{180}{\sqrt{3}}$ deg., or 103 deg., 55 m., 23 sec., at the centre; moving from the upper apsis to the lower apsis when it has once described that angle, and thence returning to the upper apsis when it has described that angle again; and so on *in infinitum.*

EXAM. 2. Suppose the centripetal force to be as any power of the altitude A, as, for example, A^{n-3}, or $\frac{A^n}{A^3}$; where $n - 3$ and n signify any indices of powers whatever, whether integers or fractions, rational or surd, affirmative or negative. That numerator A^n or $\overline{T-X}|^n$ being reduced to an indeterminate series by my method of converging series, will become $T^n - nXT^{n-1} + \frac{nn-n}{2} XXT^{n-2}$, &c. And conferring these terms with the terms of the other numerator RGG − RFF + TFF − FFX, it becomes as RGG − RFF + TFF to T^n, so − FF to $-nT^{n-1} + \frac{nn-n}{2} XT^{n-2}$, &c. And taking the last ratios where the orbits approach to circles, it becomes as RGG to T^n, so − FF to $-nT^{n-1}$, or as GG to T^{n-1}, so FF to nT^{n-1}; and again, GG to FF, so T^{n-1} to nT^{n-1}, that is, as 1 to n; and therefore G is to F, that is the angle VCp to the angle VCP, as 1 to $\sqrt{n}$. Therefore since the angle VCP, described in the descent of the body from the upper apsis to the lower apsis in an ellipsis, is of 180 deg., the angle VCp, described in the descent of the body from the upper apsis to the lower apsis in an orbit nearly circular which a body describes with a centripetal force proportional to the power A^{n-3}, will be equal to an angle of $\frac{180}{\sqrt{n}}$ deg., and this angle being repeated, the body will return from the lower to the upper apsis, and so on *in infinitum.* As if the centripetal force be as the distance of the body from the centre, that is, as A, or $\frac{A^4}{A^3}$, n will be equal to 4, and $\sqrt{n}$ equal to 2; and therefore the angle between the upper and the lower apsis will be equal to $\frac{180}{2}$ deg., or 90 deg. Therefore the body having performed a fourth part of one revolution, will arrive at the lower apsis, and having performed another fourth part, will arrive at the upper apsis, and so on by turns *in infinitum.* This appears also from Prop. X. For a body acted on by this centripetal force will revolve in an immovable ellipsis, whose centre is the centre of force. If the centripetal force is reciprocally as the distance, that is, directly as $\frac{1}{A}$ or $\frac{A^2}{A^3}$, n will be equal to 2; and therefore the angle between the upper and lower apsis will be $\frac{180}{\sqrt{2}}$ deg., or 127 deg., 16 min., 45 sec.; and therefore a body revolving with such a force, will by a perpetual repetition of this angle, move alternately from the upper to the lower and from the lower to the upper apsis for ever. So, also, if the centripetal force be reciprocally as the biquadrate root of the eleventh power of the altitude, that is, reciprocally as $A^{\frac{11}{4}}$, and, therefore, directly as $\frac{1}{A^{\frac{11}{4}}}$, or as $\frac{A^{\frac{1}{4}}}{A^3}$, n will be equal to $\frac{1}{4}$, and $\frac{180}{\sqrt{n}}$ deg. will be equal to 360 deg.; and therefore the body parting from the upper apsis, and from thence perpetually descending, will arrive at the lower apsis when it has completed one entire revolution; and thence ascending perpetually, when it has completed another entire revolution, it will arrive again at the upper apsis; and so alternately for ever.

EXAM. 3. Taking m and n for any indices of the powers of the altitude, and b and c for any given numbers, suppose the centripetal force to be as $\frac{bA^m + cA^n}{A^3}$, that is, as $\frac{b \text{ into } \overline{T-X}|^m + c \text{ into } \overline{T-X}|^n}{A^3}$ or (by the method of converging series above-mentioned) as

$$\frac{bT^m + cT^n - mbXT^{m-1} - ncXT^{n-1} + \frac{mm-m}{2} bXXT^{m-2} + \frac{nn-n}{2} cXXT^{n-2}, \&c.}{A^3}$$

and comparing the terms of the numerators, there will arise RGG — RFF + TFF to $bT^m + cT^n$ as $-$ FF to $-mbT^{m-1} - ncT^{n-1} + \frac{mm-m}{2} bXT^{m-2} + \frac{nn-n}{2} cXT^{n-2}$, &c. And taking the last ratios that arise when the orbits come to a circular form, there will come forth GG to $bT^{m-1} + cT^{n-1}$ as FF to $mbT^{m-1} + ncT^{n-1}$; and again, GG to FF as $bT^{m-1} + cT^{n-1}$ to $mbT^{n-1} + ncT^{n-1}$. This proportion, by expressing the greatest altitude CV or T arithmetically by unity, becomes, GG to FF as $b + c$ to $mb + nc$, and therefore as 1 to $\frac{mb + nc}{b + c}$. Whence G becomes to F, that is, the angle VCp to the angle VCP, as 1 to $\sqrt{\frac{mb + nc}{b + c}}$. And therefore since the angle VCP between the upper and the lower apsis, in an immovable ellipsis, is of 180 deg., the angle VCp between the same apsides in an orbit which a body describes with a centripetal force, that is, as $\frac{bA^m + cA^n}{A^3}$, will be equal to an angle of $180\sqrt{\frac{b + c}{mb + nc}}$ deg. And by the same reasoning, if the centripetal force be as $\frac{bA^m - cA^n}{A^3}$, the angle between the apsides will be found equal to $180\sqrt{\frac{b - c}{mb - nc}}$ deg. After the same manner the Problem is solved in more difficult cases. The quantity to which the centripetal force is proportional must always be resolved into a converging series whose denominator is A^3. Then the given part of the numerator arising from that operation is to be supposed in the same ratio to that part of it which is not given, as the given part of this numerator RGG — RFF + TFF — FFX is to that part of the same numerator which is not given. And taking away the superfluous quantities, and writing unity for T, the proportion of G to F is obtained.

COR. 1. Hence if the centripetal force be as any power of the altitude, that power may be found from the motion of the apsides; and so contrariwise. That is, if the whole angular motion, with which the body returns to the same apsis, be to the angular motion of one revolution, or 360 deg., as any number as m to another as n, and the altitude called A; the force will be as the power $A^{\frac{nn}{mm} - 3}$ of the altitude A; the index of which power is $\frac{nn}{mm} - 3$. This appears by the second example. Hence it is plain that the force in its recess from the centre cannot decrease in a greater than a triplicate ratio of the altitude. A body revolving with such a force, and parting from the apsis, if it once begins to descend, can never arrive at the lower apsis or least altitude, but will descend to the centre, describing the curve line treated of in Cor. 3, Prop. XLI. But if it should, at its parting from the lower apsis, begin to ascend never so little, it will ascend *in infinitum*, and never come to the upper apsis; but will describe the curve line spoken of in the same Cor., and Cor. 6, Prop. XLIV. So that where the force in its recess from the centre decreases in a greater than a triplicate ratio of the altitude, the body at its parting from the apsis, will either descend to the centre, or ascend *in infinitum*, according as it descends or ascends at the beginning of its motion. But if the force in its recess from the centre either decreases in a less than a triplicate ratio of the altitude, or increases in any ratio of the altitude whatsoever, the body will never descend to the centre, but will at some time arrive at the lower apsis; and, on the contrary, if the body alternately ascending and descending from one apsis to another never comes to the centre, then either the force increases in the recess from the centre, or it decreases in a less than a triplicate ratio of the altitude; and the sooner the body returns from one apsis to another, the farther is the ratio of the forces from the triplicate ratio. As if the body should return to and from the upper apsis by an alternate descent and ascent in 8 revolutions, or in 4, or 2, or $1\frac{1}{2}$; that is, if m should be to n as 8, or 4, or 2, or $1\frac{1}{2}$ to 1, and therefore $\frac{nn}{mm} - 3$, be $\frac{1}{64} - 3$, or $\frac{1}{16} - 3$, or $\frac{1}{4} - 3$, or $\frac{4}{9} - 3$; then the force will be as $A^{\frac{1}{64} - 3}$, or $A^{\frac{1}{16} - 3}$, or $A^{\frac{1}{4} - 3}$, or $A^{\frac{4}{9} - 3}$; that is, it will be reciprocally as $A^{3 - \frac{1}{64}}$, or $A^{3 - \frac{1}{16}}$, or $A^{3 - \frac{1}{4}}$, or $A^{3 - \frac{4}{9}}$. If the body after each revolution returns to the same apsis, and the apsis remains unmoved, then m will be to n as 1 to 1, and therefore $A^{\frac{nn}{mm} - 3}$ will be equal to A^{-2}, or $\frac{1}{AA}$; and therefore the decrease of the forces will be in a duplicate ratio of the altitude; as was demonstrated above. If the body in three fourth parts, or two thirds, or one third, or one fourth part of an entire revolution, return to the same apsis; m will be to n as $\frac{3}{4}$ or $\frac{2}{3}$ or $\frac{1}{3}$ or $\frac{1}{4}$ to 1, and therefore $A^{\frac{nn}{mm} - 3}$ is equal to $A^{\frac{16}{9} - 3}$, or $A^{\frac{9}{4} - 3}$ or $A^{9 - 3}$, or $A^{16 - 3}$; and therefore the force is either reciprocally as $A^{\frac{11}{9}}$ or $A^{\frac{3}{4}}$, or directly as A^{6} or A^{13}. Lastly if the body in its progress from the upper apsis to the same upper apsis again, goes over one entire revolution and three deg. more, and therefore that apsis in each revolution of the body moves three deg. *in consequentia*; then m will be to n as 363 deg. to 360 deg. or as 121 to 120, and therefore $A^{\frac{nn}{mm} - 3}$ will be equal to $A^{-\frac{29523}{14641}}$, and therefore the centripetal force will be reciprocally as $A^{\frac{29523}{14641}}$, or reciprocally as $A^{2\frac{4}{243}}$ very nearly. Therefore the centripetal force decreases in a ratio something greater than the duplicate; but approaching $59\frac{3}{4}$ times nearer to the duplicate than the triplicate.

COR. 2. Hence also if a body, urged by a centripetal force which is reciprocally as the square of the altitude, revolves in an ellipsis whose focus is in the centre of the forces; and a new and foreign force should be added to or subducted from this centripetal force, the motion of the apsides arising from that foreign force may (by the third Example) be known; and so on the contrary. As if the force with which the body revolves in the ellipsis be as $\frac{1}{AA}$; and the foreign force subducted as cA, and therefore the remaining force as $\frac{A - cA^4}{A^3}$; then (by the third Example) b will be equal to 1, m equal to 1, and n equal to 4; and therefore the angle of revolution between the apsides is equal to $180\sqrt{\frac{1-c}{1-4c}}$ deg. Suppose that foreign force to be 357.45 parts less than the other force with which the body revolves in the ellipsis; that is, c to be $\frac{100}{35745}$; A or T being equal to 1; and then $180\sqrt{\frac{1-c}{1-4c}}$ will be $180\sqrt{\frac{35645}{35345}}$ or 180.7623, that is, 180 deg., 45 min., 44 sec. Therefore the body, parting from the upper apsis, will arrive at the lower apsis with an angular motion of 180 deg., 45 min., 44 sec, and this angular motion being repeated, will return to the upper apsis; and therefore the upper apsis in each revolution will go forward 1 deg., 31 min., 28 sec. The apsis of the moon is about twice as swift.

So much for the motion of bodies in orbits whose planes pass through the centre of force. It now remains to determine those motions in eccentrical planes. For those authors who treat of the motion of heavy bodies used to consider the ascent and descent of such bodies, not only in a perpendicular direction, but at all degrees of obliquity upon any given planes; and for the same reason we are to consider in this place the motions of bodies tending to centres by means of any forces whatsoever, when those bodies move in eccentrical planes. These planes are supposed to be perfectly smooth and polished, so as not to retard the motion of the bodies in the least. Moreover, in these demonstrations, instead of the planes upon which those bodies roll or slide, and which are therefore tangent planes to the bodies, I shall use planes parallel to them, in which the centres of the bodies move, and by that motion describe orbits. And by the same method I afterwards determine the motions of bodies performed in curve superficies.

SECTION X.

Of the motion of bodies in given superficies, and of the reciprocal motion of funependulous bodies.

PROPOSITION XLVI. PROBLEM XXXII.

Any kind of centripetal force being supposed, and the centre of force, and any plane whatsoever in which the body revolves, being given, and the quadratures of curvilinear figures being allowed; it is required to determine the motion of a body going off from a given place, with a given velocity, in the direction of a given right line in that plane.

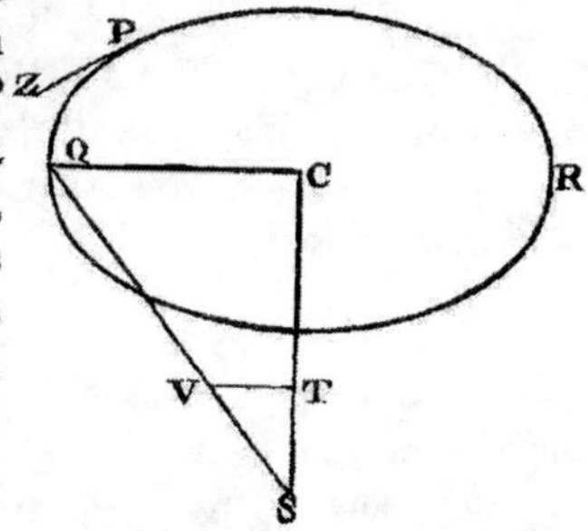

Let S be the centre of force, SC the least distance of that centre from the given plane, P a body issuing from the place P in the direction of the right line PZ, Q the same body revolving in its trajectory, and PQR the trajectory itself which is required to be found, described in that given plane. Join CQ, QS, and if in QS we take SV proportional to the centripetal force with which the body is attracted towards the centre S, and draw VT parallel to CQ, and meeting SC in T; then will the force SV be resolved into two (by Cor. 2, of the Laws of Motion), the force ST, and the force TV; of which ST attracting the body in the direction of a line perpendicular to that plane, does not at all change its motion in that plane. But the action of the other force TV, coinciding with the position of the plane itself, attracts the body directly towards the given point C in that plane; and therefore causes the body to move in this plane in the same manner as if the force ST were taken away, and the body were to revolve in free space about the centre C by means of the force TV alone. But there being given the centripetal force TV with which the body Q revolves in free space about the given centre C, there is given (by Prop. XLII) the trajectory PQR which the body describes; the place Q, in which the body will be found at any given time; and, lastly, the velocity of the body in that place Q. And so è contra. Q.E.I.

PROPOSITION XLVII. THEOREM XV.

Supposing the centripetal force to be proportional to the distance of the body from the centre; all bodies revolving in any planes whatsoever will describe ellipses, and complete their revolutions in equal times; and those which move in right lines, running backwards and forwards alternately, will complete their several periods of going and returning in the same times.

For letting all things stand as in the foregoing Proposition, the force SV, with which the body Q revolving in any plane PQR is attracted towards the centre S, is as the distance SQ; and therefore because SV and SQ, TV and CQ are proportional, the force TV with which the body is attracted towards the given point C in the plane of the orbit is as the distance CQ. Therefore the forces with which bodies found in the plane PQR are attracted towards the point C, are in proportion to the distances equal to the forces with which the same bodies are attracted every way towards the centre S; and therefore the bodies will move in the same times, and in the same figures, in any plane PQR about the point C, as they would do in free spaces about the centre S; and therefore (by Cor. 2, Prop. X, and Cor. 2, Prop. XXXVIII.) they will in equal times either describe ellipses in that plane about the centre C, or move to and fro in right lines passing through the centre C in that plane; completing the same periods of time in all cases. Q.E.D.

SCHOLIUM.

The ascent and descent of bodies in curve superficies has a near relation to these motions we have been speaking of. Imagine curve lines to be described on any plane, and to revolve about any given axes passing through the centre of force, and by that revolution to describe curve superficies; and that the bodies move in such sort that their centres may be always found in those superficies. If those bodies reciprocate to and fro with an oblique ascent and descent, their motions will be performed in planes passing through the axis, and therefore in the curve lines, by whose revolution those curve superficies were generated. In those cases, therefore, it will be sufficient to consider the motion in those curve lines.

PROPOSITION XLVIII. THEOREM XVI.

If a wheel stands upon the outside of a globe at right angles thereto, and revolving about its own axis goes forward in a great circle, the length of the curvilinear path which any point, given in the perimeter of the wheel, hath described since the time that it touched the globe (which curvilinear path we may call the cycloid or epicycloid), will be to double the versed sine of half the arc which since that time has touched the globe in passing over it, as the sum of the diameters of the globe and the wheel to the semi-diameter of the globe.

PROPOSITION XLIX. THEOREM XVII.

If a wheel stand upon the inside of a concave globe at right angles thereto, and revolving about its own axis go forward in one of the great circles of the globe, the length of the curvilinear path which any point, given in the perimeter of the wheel, hath described since it touched the globe, will be to the double of the versed sine of half the arc which in all that time has touched the globe in passing over it, as the difference of the diameters of the globe and the wheel to the semi-diameter of the globe.

Let ABL be the globe, C its centre, BPV the wheel insisting thereon, E the centre of the wheel, B the point of contact, and P the given point in the perimeter of the wheel. Imagine this wheel to proceed in the great circle ABL from A through B towards L, and in its progress to revolve in such a manner that the arcs AB, PB may be always equal one to the other, and the given point P in the perimeter of the wheel may describe in the

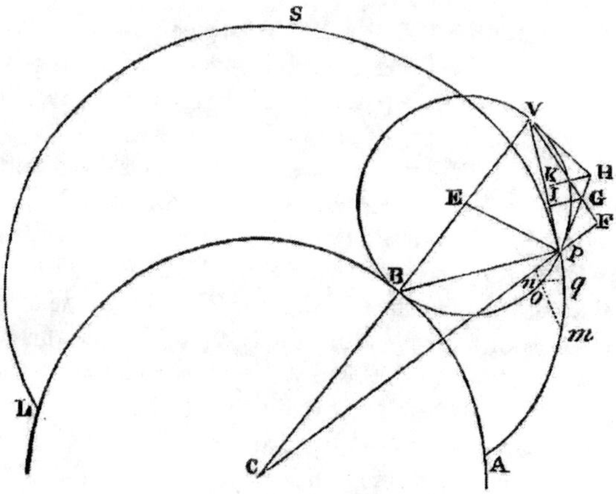

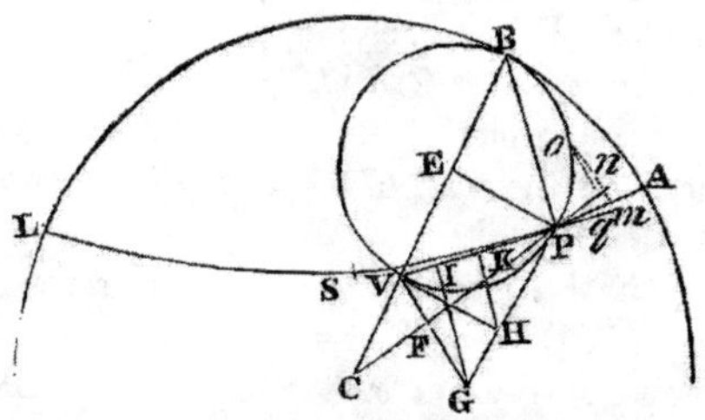

mean time the curvilinear path AP. Let AP be the whole curvilinear path described since the wheel touched the globe in A, and the length of this path AP will be to twice the versed sine of the arc $\frac{1}{2}$PB as 2CE to CB. For let the right line CE (produced if need be) meet the wheel in V, and join CP, BP, EP, VP; produce CP, and let fall thereon the perpendicular VF. Let PH, VH, meeting in H, touch the circle in P and V, and let PH cut VF in G, and to VP let fall the perpendiculars GI, HK. From the centre C with any interval let there be described the circle *nom*, cutting the right line CP in *n*, the perimeter of the wheel BP in *o*, and the curvilinear path AP in *m*; and from the centre V with the interval V*o* let there be described a circle cutting VP produced in *q*.

Because the wheel in its progress always revolves about the point of contact B, it is manifest that the right line BP is perpendicular to that curve line AP which the point P of the wheel describes, and therefore that the right line VP will touch this curve in the point P. Let the radius of the circle *nom* be gradually increased or diminished so that at last it become equal to the distance CP; and by reason of the similitude of the evanescent figure P*nomq*, and the figure PFGVI, the ultimate ratio of the evanescent lineolæ P*m*, P*n*, P*o*, P*q*, that is, the ratio of the momentary mutations of the curve AP, the right line CP, the circular arc BP, and the right line VP, will be the same as of the lines PV, PF, PG, PI, respectively. But since VF is perpendicular to CF, and VH to CV, and therefore the angles HVG, VCF equal; and the angle VHG (because the angles of the quadrilateral figure HVEP are right in V and P) is equal to the angle CEP, the triangles VHG, CEP will be similar; and thence it will come to pass that as EP is to CE so is HG to HV or HP, and so KI to KP, and by composition or division as CB to CE so is PI to PK, and doubling the consequents as CB to 2CE so PI to PV, and so is P*q* to P*m*. Therefore the decrement of the line VP, that is, the increment of the line BV—VP to the increment of the curve line AP is in a given ratio of CB to 2CE, and therefore (by Cor. Lem. IV) the lengths BV—VP and AP, generated by those increments, are in the same ratio. But if BV be radius, VP is the cosine of the angle BVP or ½BEP, and therefore BV—VP is the versed sine of the same angle, and therefore in this wheel, whose radius is ½BV, BV—VP will be double the versed sine of the arc ½BP. Therefore AP is to double the versed sine of the arc ½BP as 2CE to CB. Q.E.D.

The line AP in the former of these Propositions we shall name the cycloid without the globe, the other in the latter Proposition the cycloid within the globe, for distinction sake.

Cor. 1. Hence if there be described the entire cycloid ASL, and the same be bisected in S, the length of the part PS will be to the length PV (which is the double of the sine of the angle VBP, when EB is radius) as 2CE to CB, and therefore in a given ratio.

Cor. 2. And the length of the semi-perimeter of the cycloid AS will be equal to a right line which is to the diameter of the wheel BV as 2CE to CB.

PROPOSITION L. PROBLEM XXXIII.

To cause a pendulous body to oscillate in a given cycloid.

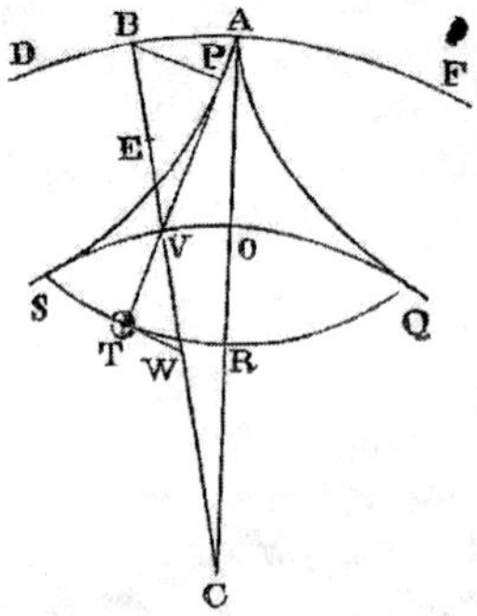

Let there be given within the globe QVS described with the centre C, the cycloid QRS, bisected in R, and meeting the superficies of the globe with its extreme points Q and S on either hand. Let there be drawn CR bisecting the arc QS in O, and let it be produced to A in such sort that CA may be to CO as CO to CR. About the centre C, with the interval CA, let there be described an exterior globe DAF; and within this globe, by a wheel whose diameter is AO, let there be described two semi-cycloids AQ, AS, touching the interior globe in Q and S, and meeting the exterior globe in A. From that point A, with a thread APT in length equal to the line AR, let the body T depend, and oscillate in such manner between the two semi-cycloids AQ, AS, that, as often as the pendulum parts from the perpendicular AR, the upper part of the thread AP may be applied to that semi-cycloid APS towards which the motion tends, and fold itself round that curve line, as if it were some solid obstacle, the remaining part of the same thread PT which has not yet touched the semi-cycloid continuing straight. Then will the weight T oscillate in the given cycloid QRS. Q.E.F.

For let the thread PT meet the cycloid QRS in T, and the circle QOS in V, and let CV be drawn; and to the rectilinear part of the thread PT from the extreme points P and T let there be erected the perpendiculars BP, TW, meeting the right line CV in B and W. It is evident, from the construction and generation of the similar figures AS, SR, that those perpendiculars PB, TW, cut off from CV the lengths VB, VW equal the diameters of the wheels OA, OR. Therefore TP is to VP (which is double the sine of the angle VBP when $\frac{1}{2}$BV is radius) as BW to BV, or AO + OR to AO, that is (since CA and CO, CO and CR, and by division AO and OR are proportional), as CA + CO to CA, or, if BV be bisected in E, as 2CE to CB. Therefore (by Cor. 1, Prop. XLIX), the length of the rectilinear part of the thread PT is always equal to the arc of the cycloid PS, and the whole thread APT is always equal to the half of the cycloid APS, that is (by Cor. 2, Prop. XLIX), to the length AR. And therefore contrariwise, if the string remain always equal to the length AR, the point T will always move in the given cycloid QRS. Q.E.D.

COR. The string AR is equal to the semi-cycloid AS, and therefore has the same ratio to AC the semi-diameter of the exterior globe as the like semi-cycloid SR has to CO the semi-diameter of the interior globe.

PROPOSITION LI. THEOREM XVIII.

If a centripetal force tending on all sides to the centre C *of a globe, be in all places as the distance of the place from the centre, and by this force alone acting upon it, the body* T *oscillate (in the manner above described) in the perimeter of the cycloid* QRS; *I say, that all the oscillations, how unequal soever in themselves, will be performed in equal times.*

For upon the tangent TW infinitely produced let fall the perpendicular CX, and join CT. Because the centripetal force with which the body T is impelled towards C is as the distance CT, let this (by Cor. 2, of the Laws) be resolved into the parts CX, TX, of which CX impelling the body directly from P stretches the thread PT, and by the resistance the thread makes to it is totally employed, producing no other effect; but the other part TX, impelling the body transversely or towards X, directly accelerates the motion in the cycloid. Then it is plain that the acceleration of the body, proportional to this accelerating force, will be every

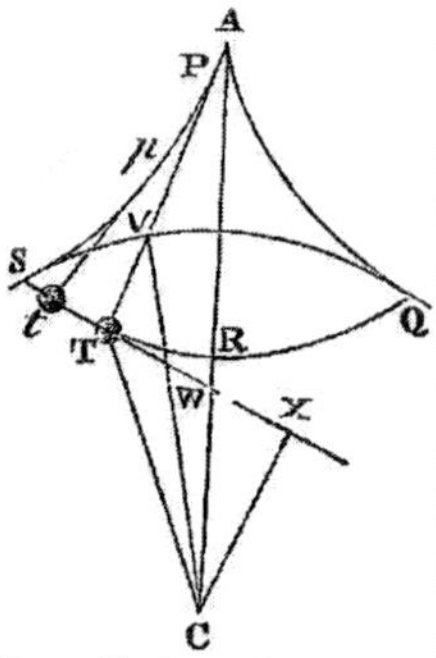

moment as the length TX, that is (because CV, WV, and TX, TW proportional to them are given), as the length TW, that is (by Cor. 1, Prop. XLIX), as the length of the arc of the cycloid TR. If therefore two pendulums APT, *Apt*, be unequally drawn aside from the perpendicular AR, and let fall together, their accelerations will be always as the arcs to be described TR, *t*R. But the parts described at the beginning of the motion are as the accelerations, that is, as the wholes that are to be described at the beginning, and therefore the parts which remain to be described, and the subsequent accelerations proportional to those parts, are also as the wholes, and so on. Therefore the accelerations, and consequently the velocities generated, and the parts described with those velocities, and the parts to be described, are always as the wholes; and therefore the parts to be described preserving a given ratio to each other will vanish together, that is, the two bodies oscillating will arrive together at the perpendicular AR. And since on the other hand the ascent of the pendulums from the lowest place R through the same cycloidal arcs with a retrograde motion, is retarded in the several places they pass through by the same forces by which their descent was accelerated; it is plain that the velocities of their ascent and descent through the same arcs are equal, and consequently performed in equal times; and, therefore, since the two parts of the cycloid RS and RQ lying on either side of the perpendicular are similar and equal, the two pendulums will perform as well the wholes as the halves of their oscillations in the same times. Q.E.D.

Cor. The force with which the body T is accelerated or retarded in any place T of the cycloid, is to the whole weight of the same body in the highest place S or Q as the arc of the cycloid TR is to the arc SR or QR.

PROPOSITION LII. PROBLEM XXXIV.

To define the velocities of the pendulums in the several places, and the times in which both the entire oscillations, and the several parts of them are performed.

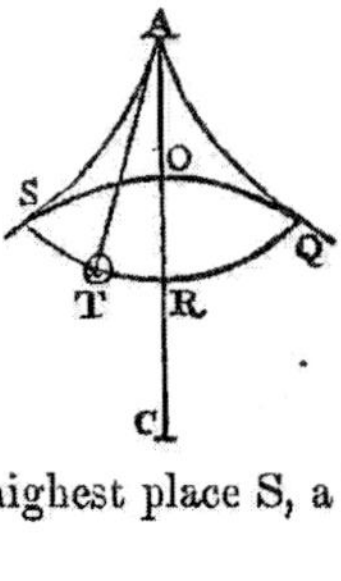

About any centre G, with the interval GH equal to the arc of the cycloid RS, describe a semi-circle HKM bisected by the semi-diameter GK. And if a centripetal force proportional to the distance of the places from the centre tend to the centre G, and it be in the perimeter HIK equal to the centripetal force in the perimeter of the globe QOS tending towards its centre, and at the same time that the pendulum T is let fall from the highest place S, a body, as L, is let fall from H to G; then because the

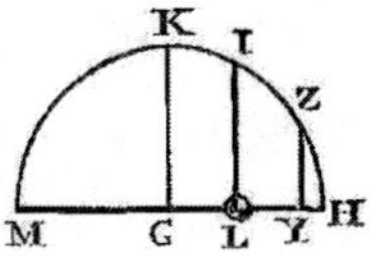

forces which act upon the bodies are equal at the beginning, and always proportional to the spaces to be described TR, LG, and therefore if TR and LG are equal, are also equal in the places T and L, it is plain that those bodies describe at the beginning equal spaces ST, HL, and therefore are still acted upon equally, and continue to describe equal spaces. Therefore by Prop. XXXVIII, the time in which the body describes the arc ST is to the time of one oscillation, as the arc HI the time in which the body H arrives at L, to the semi-periphery HKM, the time in which the body H will come to M. And the velocity of the pendulous body in the place T is to its velocity in the lowest place R, that is, the velocity of the body H in the place L to its velocity in the place G, or the momentary increment of the line HL to the momentary increment of the line HG (the arcs HI, HK increasing with an equable flux) as the ordinate LI to the radius GK, or as $\sqrt{SR^2 - TR^2}$ to SR. Hence, since in unequal oscillations there are described in equal time arcs proportional to the entire arcs of the oscillations, there are obtained from the times given, both the velocities and the arcs described in all the oscillations universally. Which was first required.

Let now any pendulous bodies oscillate in different cycloids described within different globes, whose absolute forces are also different; and if the absolute force of any globe QOS be called V, the accelerative force with which the pendulum is acted on in the circumference of this globe, when it begins to move directly towards its centre, will be as the distance of the pendulous body from that centre and the absolute force of the globe conjunctly, that is, as $CO \times V$. Therefore the lineola HY, which is as this accelerated force $CO \times V$, will be described in a given time; and if there be erected the perpendicular YZ meeting the circumference in Z, the nascent arc HZ will denote that given time. But that nascent arc HZ is in the subduplicate ratio of the rectangle GHY, and therefore as $\sqrt{GH \times CO \times V}$ Whence the time of an entire oscillation in the cycloid QRS (it being as the semi-periphery HKM, which denotes that entire oscillation, directly; and as the arc HZ which in like manner denotes a given time inversely) will be as GH directly and $\sqrt{GH \times CO \times V}$ inversely; that is, because GH and SR are equal, as $\sqrt{\frac{SR}{CO \times V}}$, or (by Cor. Prop. L,) as $\sqrt{\frac{AR}{AC \times V}}$. Therefore the oscillations in all globes and cycloids, performed with what absolute forces soever, are in a ratio compounded of the subduplicate ratio of the length of the string directly, and the subduplicate ratio of the distance between the point of suspension and the centre of the globe inversely, and the subduplicate ratio of the absolute force of the globe inversely also. Q.E.I.

COR. 1. Hence also the times of oscillating, falling, and revolving bodies may be compared among themselves. For if the diameter of the wheel with which the cycloid is described within the globe is supposed equal to the semi-diameter of the globe, the cycloid will become a right line passing through the centre of the globe, and the oscillation will be changed into a descent and subsequent ascent in that right line. Whence there is given both the time of the descent from any place to the centre, and the time equal to it in which the body revolving uniformly about the centre of the globe at any distance describes an arc of a quadrant For this time (by Case 2) is to the time of half the oscillation in any cycloid QRS as 1 to $\sqrt{\frac{AR}{AC}}$.

COR. 2. Hence also follow what Sir *Christopher Wren* and M. *Huygens* have discovered concerning the vulgar cycloid. For if the diameter of the globe be infinitely increased, its sphærical superficies will be changed into a plane, and the centripetal force will act uniformly in the direction of lines perpendicular to that plane, and this cycloid of our's will become the same with the common cycloid. But in that case the length of the arc of the cycloid between that plane and the describing point will become equal to four times the versed sine of half the arc of the wheel between the same plane and the describing point, as was discovered by Sir *Christopher Wren.* And a pendulum between two such cycloids will oscillate in a similar and equal cycloid in equal times, as M. *Huygens* demonstrated. The descent of heavy bodies also in the time of one oscillation will be the same as M. *Huygens* exhibited.

The propositions here demonstrated are adapted to the true constitution of the Earth, in so far as wheels moving in any of its great circles will describe, by the motions of nails fixed in their perimeters, cycloids without the globe; and pendulums, in mines and deep caverns of the Earth, must oscillate in cycloids within the globe, that those oscillations may be performed in equal times. For gravity (as will be shewn in the third book) decreases in its progress from the superficies of the Earth; upwards in a duplicate ratio of the distances from the centre of the Earth; downwards in a simple ratio of the same.

PROPOSITION LIII. PROBLEM XXXV.

Granting the quadratures of curvilinear figures, it is required to find the forces with which bodies moving in given curve lines may always perform their oscillations in equal times.

Let the body T oscillate in any curve line STRQ, whose axis is AR passing through the centre of force C. Draw TX touching that curve in any place of the body T, and in that tangent TX take TY equal to the arc TR. The length of that arc is known from the common methods used

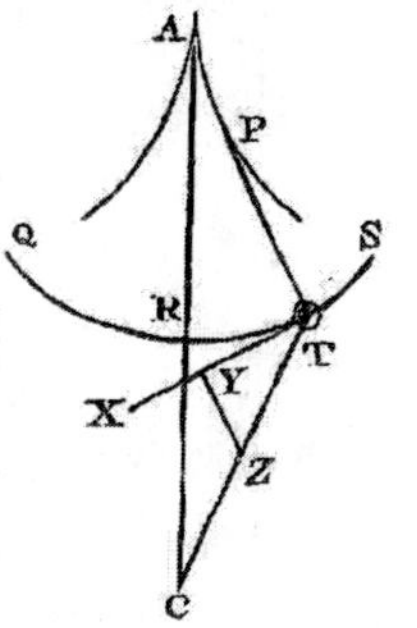

for the quadratures of figures. From the point Y draw the right line YZ perpendicular to the tangent. Draw CT meeting that perpendicular in Z, and the centripetal force will be proportional to the right line TZ. Q.E.I.

For if the force with which the body is attracted from T towards C be expressed by the right line TZ taken proportional to it, that force will be resolved into two forces TY, YZ, of which YZ drawing the body in the direction of the length of the thread PT, does not at all change its motion; whereas the other force TY directly accelerates or retards its motion in the curve STRQ. Wherefore since that force is as the space to be described TR, the accelerations or retardations of the body in describing two proportional parts (a greater and a less) of two oscillations, will be always as those parts, and therefore will cause those parts to be described together. But bodies which continually describe together parts proportional to the wholes, will describe the wholes together also. Q.E.D.

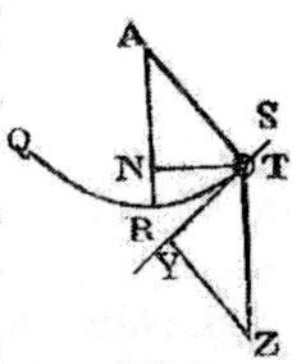

COR. 1. Hence if the body T, hanging by a rectilinear thread AT from the centre A, describe the circular arc STRQ, and in the mean time be acted on by any force tending downwards with parallel directions, which is to the uniform force of gravity as the arc TR to its sine TN, the times of the several oscillations will be equal. For because TZ, AR are parallel, the triangles ATN, ZTY are similar; and therefore TZ will be to AT as TY to TN; that is, if the uniform force of gravity be expressed by the given length AT, the force TZ, by which the oscillations become isochronous, will be to the force of gravity AT, as the arc TR equal to TY is to TN the sine of that arc.

COR. 2. And therefore in clocks, if forces were impressed by some machine upon the pendulum which preserves the motion, and so compounded with the force of gravity that the whole force tending downwards should be always as a line produced by applying the rectangle under the arc TR and the radius AR to the sine TN, all the oscillations will become isochronous.

PROPOSITION LIV. PROBLEM XXXVI.

Granting the quadratures of curvilinear figures, it is required to find the times in which bodies by means of any centripetal force will descend or ascend in any curve lines described in a plane passing through the centre of force.

Let the body descend from any place S, and move in any curve ST*t*R given in a plane passing through the centre of force C. Join CS, and let

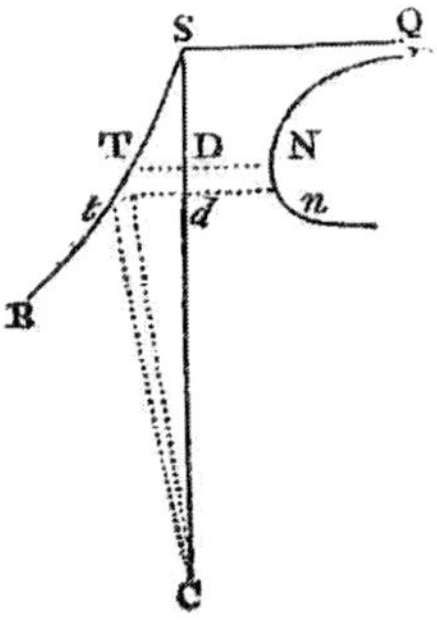

it be divided into innumerable equal parts, and let D*d* be one of those parts. From the centre C, with the intervals CD, C*d*, let the circles DT, *dt* be described, meeting the curve line ST*t*R in T and *t*. And because the law of centripetal force is given, and also the altitude CS from which the body at first fell, there will be given the velocity of the body in any other altitude CT (by Prop. XXXIX). But the time in which the body describes the lineola T*t* is as the length of that lineola, that is, as the secant of the angle *t*TC directly, and the velocity inversely. Let the ordinate DN, proportional to this time, be made perpendicular to the right line CS at the point D, and because D*d* is given, the rectangle D*d* × DN, that is, the area DN*nd*, will be proportional to the same time. Therefore if PN*n* be a curve line in which the point N is perpetually found, and its asymptote be the right line SQ standing upon the line CS at right angles, the area SQPND will be proportional to the time in which the body in its descent hath described the line ST; and therefore that area being found, the time is also given. Q.E.I.

PROPOSITION LV. THEOREM XIX.

If a body move in any curve superficies, whose axis passes through the centre of force, and from the body a perpendicular be let fall upon the axis; and a line parallel and equal thereto be drawn from any given point of the axis; I say, that this parallel line will describe an area proportional to the time.

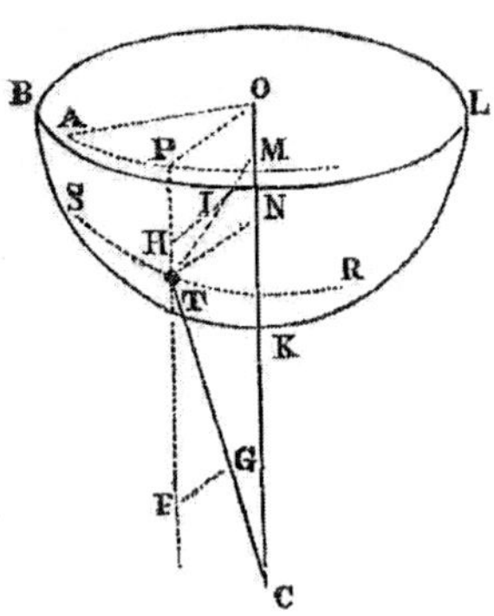

Let BKL be a curve superficies, T a body revolving in it, STR a trajectory which the body describes in the same, S the beginning of the trajectory, OMK the axis of the curve superficies, TN a right line let fall perpendicularly from the body to the axis; OP a line parallel and equal thereto drawn from the given point O in the axis; AP the orthographic projection of the trajectory described by the point P in the plane AOP in which the revolving line OP is found; A the beginning of that projection, answering to the point S; TC a right line drawn from the body to the centre; TG a part thereof proportional to the centripetal force with which the body tends towards the centre C; TM a right line perpendicular to the curve superficies; TI a part thereof proportional to the force of pressure with which the body urges the superficies, and therefore with which it is again repelled by the superficies towards M; PTF a right line parallel to the axis and passing through the body, and GF, IH right lines let fall perpendicularly from the points G and I upon that parallel PHTF. I say, now, that the area AOP, described by the radius OP from the beginning of the motion, is proportional to the time. For the force TG (by Cor. 2, of the Laws of Motion) is resolved into the forces TF, FG; and the force TI into the forces TH, HI; but the forces TF, TH, acting in the direction of the line PF perpendicular to the plane AOP, introduce no change in the motion of the body but in a direction perpendicular to that plane. Therefore its motion, so far as it has the same direction with the position of the plane, that is, the motion of the point P, by which the projection AP of the trajectory is described in that plane, is the same as if the forces TF, TH were taken away, and the body were acted on by the forces FG, HI alone; that is, the same as if the body were to describe in the plane AOP the curve AP by means of a centripetal force tending to the centre O, and equal to the sum of the forces FG and HI. But with such a force as that (by Prop. 1) the area AOP will be described proportional to the time. Q.E.D.

Cor. By the same reasoning, if a body, acted on by forces tending to two or more centres in any the same right line CO, should describe in a free space any curve line ST, the area AOP would be always proportional to the time.

PROPOSITION LVI. PROBLEM XXXVII.

Granting the quadratures of curvilinear figures, and supposing that there are given both the law of centripetal force tending to a given centre, and the curve superficies whose axis passes through that centre; it is required to find the trajectory which a body will describe in that superficies, when going off from a given place with a given velocity, and in a given direction in that superficies.

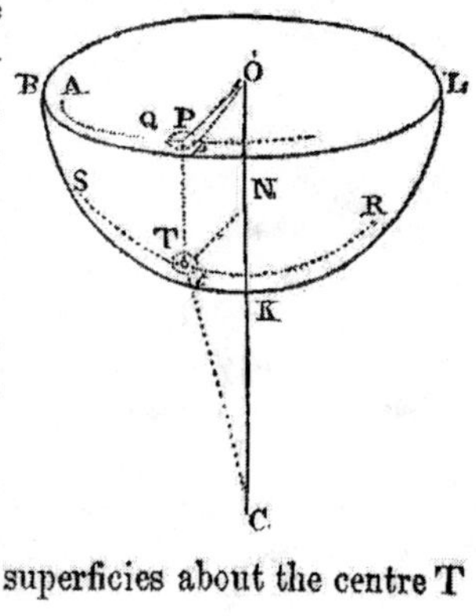

The last construction remaining, let the body T go from the given place S, in the direction of a line given by position, and turn into the trajectory sought STR, whose orthographic projection in the plane BDO is AP. And from the given velocity of the body in the altitude SC, its velocity in any other altitude TC will be also given. With that velocity, in a given moment of time, let the body describe the particle Tt of its trajectory, and let Pp be the projection of that particle described in the plane AOP. Join Op, and a little circle being described upon the curve superficies about the centre T with the interval T*t* let the projection of that little circle in the plane AOP be the ellipsis *p*Q. And because the magnitude of that little circle T*t*, and TN or PO its distance from the axis CO is also given, the ellipsis *p*Q will be given both in kind and magnitude, as also its position to the right line PO. And since the area PO*p* is proportional to the time, and therefore given because the time is given, the angle PO*p* will be given. And thence will be given *p* the common intersection of the ellipsis and the right line O*p*, together with the angle OP*p*, in which the projection AP*p* of the trajectory cuts the line OP. But from thence (by conferring Prop. XLI, with its 2d Cor.) the manner of determining the curve AP*p* easily appears. Then from the several points P of that projection erecting to the plane AOP, the perpendiculars PT meeting the curve superficies in T, there will be given the several points T of the trajectory. Q.E.I.

SECTION XI.

Of the motions of bodies tending to each other with centripetal forces.

I have hitherto been treating of the attractions of bodies towards an immovable centre; though very probably there is no such thing existent in nature. For attractions are made towards bodies, and the actions of the bodies attracted and attracting are always reciprocal and equal, by Law III; so that if there are two bodies, neither the attracted nor the attracting body is truly at rest, but both (by Cor. 4, of the Laws of Motion), being as it were mutually attracted, revolve about a common centre of gravity. And if there be more bodies, which are either attracted by one single one which is attracted by them again, or which all of them, attract each other mutually, these bodies will be so moved among themselves, as that their common centre of gravity will either be at rest, or move uniformly forward in a right line. I shall therefore at present go on to treat of the motion of bodies mutually attracting each other; considering the centripetal forces as attractions; though perhaps in a physical strictness they may more truly be called impulses. But these propositions are to be considered as purely mathematical; and therefore, laying aside all physical considerations, I make use of a familiar way of speaking, to make myself the more easily understood by a mathematical reader.

PROPOSITION LVII. THEOREM XX.

Two bodies attracting each other mutually describe similar figures about their common centre of gravity, and about each other mutually.

For the distances of the bodies from their common centre of gravity are reciprocally as the bodies; and therefore in a given ratio to each other; and thence, by composition of ratios, in a given ratio to the whole distance between the bodies. Now these distances revolve about their common term with an equable angular motion, because lying in the same right line they never change their inclination to each other mutually. But right lines that are in a given ratio to each other, and revolve about their terms with an equal angular motion, describe upon planes, which either rest with those terms, or move with any motion not angular, figures entirely similar round those terms. Therefore the figures described by the revolution of these distances are similar. Q.E.D.

PROPOSITION LVIII. THEOREM XXI.

If two bodies attract each other mutually with forces of any kind, and in the mean time revolve about the common centre of gravity; I say, that, by the same forces, there may be described round either body unmoved a figure similar and equal to the figures which the bodies so moving describe round each other mutually.

Let the bodies S and P revolve about their common centre of gravity C, proceeding from S to T, and from P to Q. From the given point *s* let

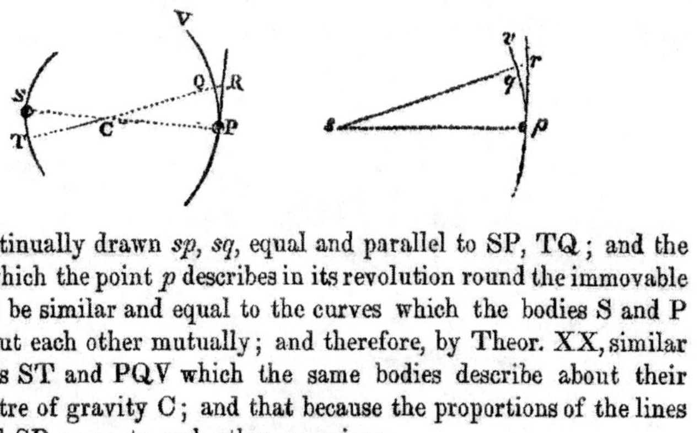

there be continually drawn *sp*, *sq*, equal and parallel to SP, TQ; and the curve *pqv*, which the point *p* describes in its revolution round the immovable point *s*, will be similar and equal to the curves which the bodies S and P describe about each other mutually; and therefore, by Theor. XX, similar to the curves ST and PQV which the same bodies describe about their common centre of gravity C; and that because the proportions of the lines SC, CP, and SP or *sp*, to each other, are given.

CASE 1. The common centre of gravity C (by Cor. 4, of the Laws of Motion) is either at rest, or moves uniformly in a right line. Let us first suppose it at rest, and in *s* and *p* let there be placed two bodies, one immovable in *s*, the other movable in *p*, similar and equal to the bodies S and P. Then let the right lines PR and *pr* touch the curves PQ and *pq* in P and *p*, and produce CQ and *sq* to R and *r*. And because the figures CPRQ, *sprq* are similar, RQ will be to *rq* as CP to *sp*, and therefore in a given ratio. Hence if the force with which the body P is attracted towards the body S, and by consequence towards the intermediate point the centre C, were to the force with which the body *p* is attracted towards the centre *s*, in the same given ratio, these forces would in equal times attract the bodies from the tangents PR, *pr* to the arcs PQ, *pq*, through the intervals proportional to them RQ, *rq*; and therefore this last force (tending to *s*) would make the body *p* revolve in the curve *pqv*, which would becom similar to the curve PQV, in which the first force obliges the body P to revolve; and their revolutions would be completed in the same times. But because those forces are not to each other in the ratio of CP to *sp*, but (by reason of the similarity and equality of the bodies S and *s*, P and *p*, and the equality of the distances SP, *sp*) mutually equal, the bodies in equal times will be equally drawn from the tangents; and therefore that the body *p* may be attracted through the greater interval *rq*, there is required a greater time, which will be in the subduplicate ratio of the intervals; because, by Lemma X, the spaces described at the very beginning of the motion are in a duplicate ratio of the times. Suppose, then the velocity of the body *p* to be to the velocity of the body P in a subduplicate ratio of the distance *sp* to the distance CP, so that the arcs *pq*, PQ, which are in a simple proportion to each other, may be described in times that are in a subduplicate ratio of the distances; and the bodies P, *p*, always attracted by equal forces, will describe round the quiescent centres C and *s* similar figures PQV, *pqv*, the latter of which *pqv* is similar and equal to the figure which the body P describes round the movable body S. Q.E.D.

Case 2. Suppose now that the common centre of gravity, together with the space in which the bodies are moved among themselves, proceeds uniformly in a right line; and (by Cor. 6, of the Laws of Motion) all the motions in this space will be performed in the same manner as before; and therefore the bodies will describe mutually about each other the same figures as before, which will be therefore similar and equal to the figure *pqv*. Q.E.D.

Cor. 1. Hence two bodies attracting each other with forces proportional to their distance, describe (by Prop. X) both round their common centre of gravity, and round each other mutually concentrical ellipses; and, *vice versa*, if such figures are described, the forces are proportional to the distances.

Cor. 2. And two bodies, whose forces are reciprocally proportional to the square of their distance, describe (by Prop. XI, XII, XIII), both round their common centre of gravity, and round each other mutually, conic sections having their focus in the centre about which the figures are described. And, *vice versa*, if such figures are described, the centripetal forces are reciprocally proportional to the squares of the distance.

Cor. 3. Any two bodies revolving round their common centre of gravity describe areas proportional to the times, by radii drawn both to that centre and to each other mutually.

PROPOSITION LIX., THEOREM XXII.

The periodic time of two bodies S *and* P *revolving round their common centre of gravity* C, *is to the periodic time of one of the bodies* P *revolving round the other* S *remaining unmoved, and describing a figure similar and equal to those which the bodies describe about each other mutually, in a subduplicate ratio of the other body* S *to the sum of the bodies* S + P.

For, by the demonstration of the last Proposition, the times in which any similar arcs PQ and *pq* are described are in a subduplicate ratio of the distances CP and SP, or *sp*, that is, in a subduplicate ratio of the body S to the sum of the bodies S + P. And by composition of ratios, the sums of the times in which all the similar arcs PQ and *pq* are described, that is, the whole times in which the whole similar figures are described are in the same subduplicate ratio. Q.E.D.

PROPOSITION LX. THEOREM XXIII.

If two bodies S *and* P, *attracting each other with forces reciprocally proportional to the squares of their distance, revolve about their common centre of gravity; I say, that the principal axis of the ellipsis which either of the bodies, as* P, *describes by this motion about the other* S, *will be to the principal axis of the ellipsis, which the same body* P *may describe in the same periodical time about the other body* S *quiescent, as the sum of the two bodies* S + P *to the first of two mean proportionals between that sum and the other body* S.

For if the ellipses described were equal to each other, their periodic times by the last Theorem would be in a subduplicate ratio of the body S to the sum of the bodies S + P. Let the periodic time in the latter ellipsis be diminished in that ratio, and the periodic times will become equal; but, by Prop. XV, the principal axis of the ellipsis will be diminished in a ratio sesquiplicate to the former ratio; that is, in a ratio to which the ratio of S to S + P is triplicate; and therefore that axis will be to the principal axis of the other ellipsis as the first of two mean proportionals between S + P and S to S + P. And inversely the principal axis of the ellipsis described about the movable body will be to the principal axis of that described round the immovable as S + P to the first of two mean proportionals between S + P and S. Q.E.D.

PROPOSITION LXI. THEOREM XXIV.

If two bodies attracting each other with any kind of forces, and not otherwise agitated or obstructed, are moved in any manner whatsoever, those motions will be the same as if they did not at all attract each other mutually, but were both attracted with the same forces by a third body placed in their common centre of gravity; and the law of the attracting forces will be the same in respect of the distance of the bodies from the common centre, as in respect of the distance between the two bodies.

For those forces with which the bodies attract each other mutually, by tending to the bodies, tend also to the common centre of gravity lying directly between them; and therefore are the same as if they proceeded from an intermediate body. Q.E.D.

And because there is given the ratio of the distance of either body from that common centre to the distance between the two bodies, there is given, of course, the ratio of any power of one distance to the same power of the other distance; and also the ratio of any quantity derived in any manner from one of the distances compounded any how with given quantities, to another quantity derived in like manner from the other distance, and as many given quantities having that given ratio of the distances to the first. Therefore if the force with which one body is attracted by another be directly or inversely as the distance of the bodies from each other, or as any power of that distance; or, lastly, as any quantity derived after any manner from that distance compounded with given quantities; then will the same force with which the same body is attracted to the common centre of gravity be in like manner directly or inversely as the distance of the attracted body from the common centre, or as any power of that distance; or, lastly, as a quantity derived in like sort from that distance compounded with analogous given quantities. That is, the law of attracting force will be the same with respect to both distances. Q.E.D.

PROPOSITION LXII. PROBLEM XXXVIII.

To determine the motions of two bodies which attract each other with forces reciprocally proportional to the squares of the distance between them, and are let fall from given places.

The bodies, by the last Theorem, will be moved in the same manner as if they were attracted by a third placed in the common centre of their gravity; and by the hypothesis that centre will be quiescent at the beginning of their motion, and therefore (by Cor. 4, of the Laws of Motion) will be always quiescent. The motions of the bodies are therefore to be determined (by Prob. XXV) in the same manner as if they were impelled by forces tending to that centre; and then we shall have the motions of the bodies attracting each other mutually. Q.E.I.

PROPOSITION LXIII. PROBLEM XXXIX.

To determine the motions of two bodies attracting each other with forces reciprocally proportional to the squares of their distance, and going off from given places in given directions with given velocities.

The motions of the bodies at the beginning being given, there is given also the uniform motion of the common centre of gravity, and the motion of the space which moves along with this centre uniformly in a right line, and also the very first, or beginning motions of the bodies in respect of this space. Then (by Cor. 5, of the Laws, and the last Theorem) the subsequent motions will be performed in the same manner in that space, as if that space together with the common centre of gravity were at rest, and as if the bodies did not attract each other, but were attracted by a third body placed in that centre. The motion therefore in this movable space of each body going off from a given place, in a given direction, with a given velocity, and acted upon by a centripetal force tending to that centre, is to be determined by Prob. IX and XXVI, and at the same time will be obtained the motion of the other round the same centre. With this motion compound the uniform progressive motion of the entire system of the space and the bodies revolving in it, and there will be obtained the absolute motion of the bodies in immovable space. Q.E.I.

PROPOSITION LXIV. PROBLEM XL.

Supposing forces with which bodies mutually attract each other to increase in a simple ratio of their distances from the centres; it is required to find the motions of several bodies among themselves.

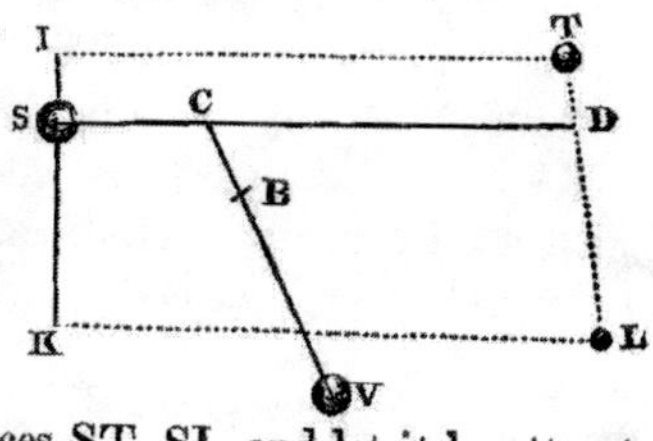

Suppose the first two bodies T and L to have their common centre of gravity in D. These, by Cor. 1, Theor. XXI, will describe ellipses having their centres in D, the magnitudes of which ellipses are known by Prob. V.

Let now a third body S attract the two former T and L with the accelerative forces ST, SL, and let it be attracted again by them. The force ST (by Cor. 2, of the Laws of Motion) is resolved into the forces SD, DT; and the force SL into the forces SD and DL. Now the forces DT, DL, which are as their sum TL, and therefore as the accelerative forces with which the bodies T and L attract each other mutually, added to the forces of the bodies T and L, the first to the first, and the last to the last, compose forces proportional to the distances DT and DL as before, but only greater than those former forces; and therefore (by Cor. 1, Prop. X, and Cor. 1, and 8, Prop. IV) they will cause those bodies to describe ellipses as before, but with a swifter motion. The remaining accelerative forces SD and DL, by the motive forces SD × T and SD × L, which are as the bodies attracting those bodies equally and in the direction of the lines TI, LK parallel to DS, do not at all change their situations with respect to one another, but cause them equally to approach to the line IK; which must be imagined drawn through the middle of the body S, and perpendicular to the line DS. But that approach to the line

IK will be hindered by causing the system of the bodies T and L on one side, and the body S on the other, with proper velocities, to revolve round the common centre of gravity C. With such a motion the body S, because the sum of the motive forces SD × T and SD × L is proportional to the distance CS, tends to the centre C, will describe an ellipsis round the same centre C; and the point D, because the lines CS and CD are proportional, will describe a like ellipsis over against it. But the bodies T and L, attracted by the motive forces SD × T and SD × L, the first by the first, and the last by the last, equally and in the direction of the parallel lines TI and LK, as was said before, will (by Cor. 5 and 6, of the Laws of Motion) continue to describe their ellipses round the movable centre D, as before. Q.E.I.

Let there be added a fourth body V, and, by the like reasoning, it will be demonstrated that this body and the point C will describe ellipses about the common centre of gravity B; the motions of the bodies T, L, and S round the centres D and C remaining the same as before; but accelerated. And by the same method one may add yet more bodies at pleasure. Q.E.I.

This would be the case, though the bodies T and L attract each other mutually with accelerative forces either greater or less than those with which they attract the other bodies in proportion to their distance. Let all the mutual accelerative attractions be to each other as the distances multiplied into the attracting bodies; and from what has gone before it will easily be concluded that all the bodies will describe different ellipses with equal periodical times about their common centre of gravity B, in an immovable plane. Q.E.I.

PROPOSITION LXV. THEOREM XXV.

Bodies, whose forces decrease in a duplicate ratio of their distances from their centres, may move among themselves in ellipses; and by radii drawn to the foci may describe areas proportional to the times very nearly.

In the last Proposition we demonstrated that case in which the motions will be performed exactly in ellipses. The more distant the law of the forces is from the law in that case, the more will the bodies disturb each other's motions; neither is it possible that bodies attracting each other mutually according to the law supposed in this Proposition should move exactly in ellipses, unless by keeping a certain proportion of distances from each other. However, in the following cases the orbits will not much differ from ellipses.

Case 1. Imagine several lesser bodies to revolve about some very great one at different distances from it, and suppose absolute forces tending to every one of the bodies proportional to each. And because (by Cor. 4, of the Laws) the common centre of gravity of them all is either at rest, or moves uniformly forward in a right line, suppose the lesser bodies so small that the great body may be never at a sensible distance from that centre; and then the great body will, without any sensible error, be either at rest, or move uniformly forward in a right line; and the lesser will revolve about that great one in ellipses, and by radii drawn thereto will describe areas proportional to the times; if we except the errors that may be introduced by the receding of the great body from the common centre of gravity, or by the mutual actions of the lesser bodies upon each other. But the lesser bodies may be so far diminished, as that this recess and the mutual actions of the bodies on each other may become less than any assignable; and therefore so as that the orbits may become ellipses, and the areas answer to the times, without any error that is not less than any assignable. Q.E.O.

CASE 2. Let us imagine a system of lesser bodies revolving about a very great one in the manner just described, or any other system of two bodies revolving about each other to be moving uniformly forward in a right line, and in the mean time to be impelled sideways by the force of another vastly greater body situate at a great distance. And because the equal accelerative forces with which the bodies are impelled in parallel directions do not change the situation of the bodies with respect to each other, but only oblige the whole system to change its place while the parts still retain their motions among themselves, it is manifest that no change in those motions of the attracted bodies can arise from their attractions towards the greater, unless by the inequality of the accelerative attractions, or by the inclinations of the lines towards each other, in whose directions the attractions are made. Suppose, therefore, all the accelerative attractions made towards the great body to be among themselves as the squares of the distances reciprocally; and then, by increasing the distance of the great body till the differences of the right lines drawn from that to the others in respect of their length, and the inclinations of those lines to each other, be less than any given, the motions of the parts of the system will continue without errors that are not less than any given. And because, by the small distance of those parts from each other, the whole system is attracted as if it were but one body, it will therefore be moved by this attraction as if it were one body; that is, its centre of gravity will describe about the great body one of the conic sections (that is, a parabola or hyperbola when the attraction is but languid, and an ellipsis when it is more vigorous); and by radii drawn thereto, it will describe areas proportional to the times, without any errors but those which arise from the distances of the parts, which are by the supposition exceedingly small, and may be diminished at pleasure. Q.E.O.

By a like reasoning one may proceed to more compounded cases *in infinitum.*

COR. 1. In the second Case, the nearer the very great body approaches to the system of two or more revolving bodies, the greater will the perturbation be of the motions of the parts of the system among themselves; because the inclinations of the lines drawn from that great body to those parts become greater; and the inequality of the proportion is also greater.

Cor. 2. But the perturbation will be greatest of all, if we suppose the accelerative attractions of the parts of the system towards the greatest body of all are not to each other reciprocally as the squares of the distances from that great body; especially if the inequality of this proportion be greater than the inequality of the proportion of the distances from the great body. For if the accelerative force, acting in parallel directions and equally, causes no perturbation in the motions of the parts of the system, it must of course, when it acts unequally, cause a perturbation somewhere, which will be greater or less as the inequality is greater or less. The excess of the greater impulses acting upon some bodies, and not acting upon others, must necessarily change their situation among themselves. And this perturbation, added to the perturbation arising from the inequality and inclination of the lines, makes the whole perturbation greater.

Cor. 3. Hence if the parts of this system move in ellipses or circles without any remarkable perturbation, it is manifest that, if they are at all impelled by accelerative forces tending to any other bodies, the impulse is very weak, or else is impressed very near equally and in parallel directions upon all of them.

PROPOSITION LXVI. THEOREM XXVI.

If three bodies whose forces decrease in a duplicate ratio of the distances attract each other mutually; and the accelerative attractions of any two towards the third be between themselves reciprocally as the squares of the distances; and the two least revolve about the greatest; I say, that the interior of the two revolving bodies will, by radii drawn to the innermost and greatest, describe round that body areas more proportional to the times, and a figure more approaching to that of an ellipsis having its focus in the point of concourse of the radii, if that great body be agitated by those attractions, than it would do if that great body were not attracted at all by the lesser, but remained at rest; or than it would if that great body were very much more or very much less attracted, or very much more or very much less agitated, by the attractions.

This appears plainly enough from the demonstration of the second Corollary of the foregoing Proposition; but it may be made out after this manner by a way of reasoning more distinct and more universally convincing.

Case 1. Let the lesser bodies P and S revolve in the same plane about the greatest body T, the body P describing the interior orbit PAB, and S the exterior orbit ESE. Let SK be the mean distance of the bodies P and S; and let the accelerative attraction of the body P towards S, at that mean distance, be expressed by that line SK. Make SL to SK as the

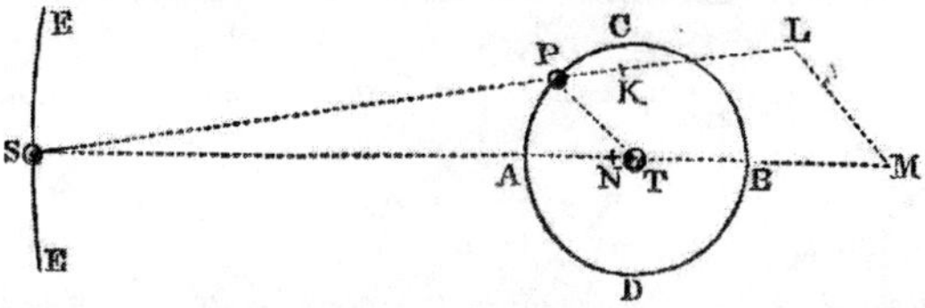

square of SK to the square of SP, and SL will be the accelerative attraction of the body P towards S at any distance SP. Join PT, and draw LM parallel to it meeting ST in M; and the attraction SL will be resolved (by Cor. 2, of the Laws of Motion) into the attractions SM, LM. And so the body P will be urged with a threefold accelerative force. One of these forces tends towards T, and arises from the mutual attraction of the bodies T and P. By this force alone the body P would describe round the body T, by the radius PT, areas proportional to the times, and an ellipsis whose focus is in the centre of the body T; and this it would do whether the body T remained unmoved, or whether it were agitated by that attraction. This appears from Prop. XI, and Cor. 2 and 3 of Theor. XXI. The other force is that of the attraction LM, which, because it tends from P to T, will be superadded to and coincide with the former force; and cause the areas to be still proportional to the times, by Cor. 3, Theor. XXI. But because it is not reciprocally proportional to the square of the distance PT, it will compose, when added to the former, a force varying from that proportion; which variation will be the greater by how much the proportion of this force to the former is greater, *cæteris paribus*. Therefore, since by Prop. XI, and by Cor. 2, Theor. XXI, the force with which the ellipsis is described about the focus T ought to be directed to that focus, and to be reciprocally proportional to the square of the distance PT, that compounded force varying from that proportion will make the orbit PAB vary from the figure of an ellipsis that has its focus in the point T; and so much the more by how much the variation from that proportion is greater; and by consequence by how much the proportion of the second force LM to the first force is greater, *cæteris paribus*. But now the third force SM, attracting the body P in a direction parallel to ST, composes with the other forces a new force which is no longer directed from P to T; and which varies so much more from this direction by how much the proportion of this third force to the other forces is greater, *cæteris paribus*; and therefore causes the body P to describe, by the radius TP, areas no longer proportional to the times; and therefore makes the variation from that proportionality so much greater by how much the proportion of this force to the others is greater. But this third force will increase the variation of the orbit PAB from the elliptical figure before-mentioned upon two accounts; first because that force is not directed from P to T; and, secondly, because it is not reciprocally proportional to the square of the distance PT. These things being premised, it is manifest that the areas are then most nearly proportional to the times, when that third force is the least possible, the rest preserving their former quantity; and that the orbit PAB does then approach nearest to the elliptical figure above-mentioned, when both the second and third, but especially the third force, is the least possible; the first force remaining in its former quantity.

Let the accelerative attraction of the body T towards S be expressed by the line SN; then if the accelerative attractions SM and SN were equal, these, attracting the bodies T and P equally and in parallel directions would not at all change their situation with respect to each other. The motions of the bodies between themselves would be the same in that case as if those attractions did not act at all, by Cor. 6, of the Laws of Motion. And, by a like reasoning, if the attraction SN is less than the attraction SM, it will take away out of the attraction SM the part SN, so that there will remain only the part (of the attraction) MN to disturb the proportionality of the areas and times, and the elliptical figure of the orbit. And in like manner if the attraction SN be greater than the attraction SM, the perturbation of the orbit and proportion will be produced by the difference MN alone. After this manner the attraction SN reduces always the attraction SM to the attraction MN, the first and second attractions remaining perfectly unchanged; and therefore the areas and times come then nearest to proportionality, and the orbit PAB to the above-mentioned elliptical figure, when the attraction MN is either none, or the least that is possible; that is, when the accelerative attractions of the bodies P and T approach as near as possible to equality; that is, when the attraction SN is neither none at all, nor less than the least of all the attractions SM, but is, as it were, a mean between the greatest and least of all those attractions SM, that is, not much greater nor much less than the attraction SK. Q.E.D.

Case 2. Let now the lesser bodies P, S, revolve about a greater T in different planes; and the force LM, acting in the direction of the line PT situate in the plane of the orbit PAB, will have the same effect as before; neither will it draw the body P from the plane of its orbit. But the other force NM acting in the direction of a line parallel to ST (and which, therefore, when the body S is without the line of the nodes is inclined to the plane of the orbit PAB), besides the perturbation of the motion just now spoken of as to longitude, introduces another perturbation also as to latitude, attracting the body P out of the plane of its orbit. And this perturbation, in any given situation of the bodies P and T to each other, will be as the generating force MN; and therefore becomes least when the force MN is least, that is (as was just now shewn), where the attraction SN is not much greater nor much less than the attraction SK. Q.E.D.

COR. 1. Hence it may be easily collected, that if several less bodies P, S, R, &c., revolve about a very great body T, the motion of the innermost revolving body P will be least disturbed by the attractions of the others, when the great body is as well attracted and agitated by the rest (according to the ratio of the accelerative forces) as the rest are by each other mutually.

COR. 2. In a system of three bodies, T, P, S, if the accelerative attractions of any two of them towards a third be to each other reciprocally as the squares of the distances, the body P, by the radius PT, will describe its area about the body T swifter near the conjunction A and the opposition B than it will near the quadratures C and D. For every force with which the body P is acted on and the body T is not, and which does not act in the direction of the line PT, does either accelerate or retard the description of the area, according as it is directed, whether *in consequentia* or *in antecedentia*. Such is the force NM. This force in the passage of the body P from C to A is directed *in consequentia* to its motion, and therefore accelerates it; then as far as D *in antecedentia*, and retards the motion; then *in consequentia* as far as B; and lastly *in antecedentia* as it moves from B to C.

COR. 3. And from the same reasoning it appears that the body P *cæteris paribus*, moves more swiftly in the conjunction and opposition than in the quadratures.

COR. 4. The orbit of the body P, *cæteris paribus*, is more curve at the quadratures than at the conjunction and opposition. For the swifter bodies move, the less they deflect from a rectilinear path. And besides the force KL, or NM, at the conjunction and opposition, is contrary to the force with which the body T attracts the body P, and therefore diminishes that force; but the body P will deflect the less from a rectilinear path the less it is impelled towards the body T.

COR. 5. Hence the body P, *cæteris paribus*, goes farther from the body T at the quadratures than at the conjunction and opposition. This is said,

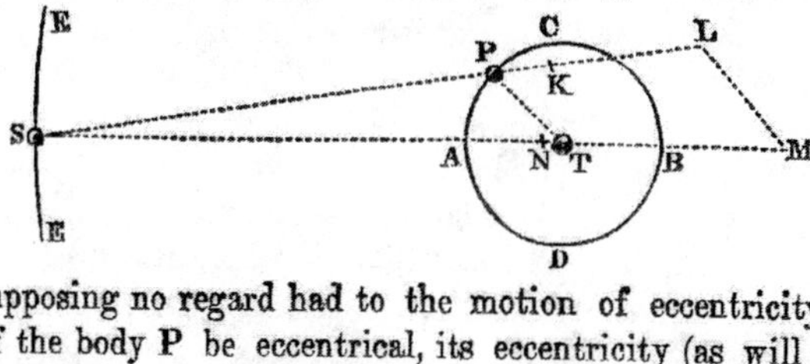

however, supposing no regard had to the motion of eccentricity. For if the orbit of the body P be eccentrical, its eccentricity (as will be shewn presently by Cor. 9) will be greatest when the apsides are in the syzygies; and thence it may sometimes come to pass that the body P, in its near approach to the farther apsis, may go farther from the body T at the syzygies than at the quadratures.

COR. 6. Because the centripetal force of the central body T, by which the body P is retained in its orbit, is increased at the quadratures by the addition caused by the force LM, and diminished at the syzygies by the subduction caused by the force KL, and, because the force KL is greater than LM, it is more diminished than increased; and, moreover, since that centripetal force (by Cor. 2, Prop. IV) is in a ratio compounded of the simple ratio of the radius TP directly, and the duplicate ratio of the periodical time inversely; it is plain that this compounded ratio is diminished by the action of the force KL; and therefore that the periodical time, supposing the radius of the orbit PT to remain the same, will be increased, and that in the subduplicate of that ratio in which the centripetal force is diminished; and, therefore, supposing this radius increased or diminished, the periodical time will be increased more or diminished less than in the sesquiplicate ratio of this radius, by Cor. 6, Prop. IV. If that force of the central body should gradually decay, the body P being less and less attracted would go farther and farther from the centre T; and, on the contrary, if it were increased, it would draw nearer to it. Therefore if the action of the distant body S, by which that force is diminished, were to increase and decrease by turns, the radius TP will be also increased and diminshed by turns; and the periodical time will be increased and diminished in a ratio compounded of the sesquiplicate ratio of the radius, and of the subduplicate of that ratio in which the centripetal force of the central body T is diminished or increased, by the increase or decrease of the action of the distant body S.

Cor. 7. It also follows, from what was before laid down, that the axis of the ellipsis described by the body P, or the line of the apsides, does as to its angular motion go forwards and backwards by turns, but more forwards than backwards, and by the excess of its direct motion is in the whole carried forwards. For the force with which the body P is urged to the body T at the quadratures, where the force MN vanishes, is compounded of the force LM and the centripetal force with which the body T attracts the body P. The first force LM, if the distance PT be increased, is increased in nearly the same proportion with that distance, and the other force decreases in the duplicate ratio of the distance; and therefore the sum of these two forces decreases in a less than the duplicate ratio of the distance PT; and therefore, by Cor. 1, Prop. XLV, will make the line of the apsides, or, which is the same thing, the upper apsis, to go backward. But at the conjunction and opposition the force with which the body P is urged towards the body T is the difference of the force KL, and of the force with which the body T attracts the body P; and that difference, because the force KL is very nearly increased in the ratio of the distance PT, decreases in more than the duplicate ratio of the distance PT; and therefore, by Cor. 1, Prop. XLV, causes the line of the apsides to go forwards. In the places between the syzygies and the quadratures, the motion of the line of the apsides depends upon both of these causes conjunctly, so that it either goes forwards or backwards in proportion to the excess of one of these causes above the other. Therefore since the force KL in the syzygies is almost twice as great as the force LM in the quadratures, the excess will be on the side of the force KL, and by consequence the line of the apsides will be carried forwards. The truth of this and the foregoing

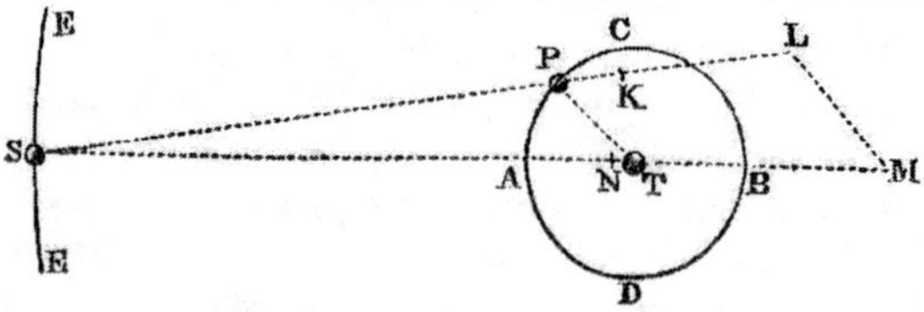

Corollary will be more easily understood by conceiving the system of the two bodies T and P to be surrounded on every side by several bodies S, S, S, &c., disposed about the orbit ESE. For by the actions of these bodies the action of the body T will be diminished on every side, and decrease in more than a duplicate ratio of the distance.

COR. 8. But since the progress or regress of the apsides depends upon the decrease of the centripetal force, that is, upon its being in a greater or less ratio than the duplicate ratio of the distance TP, in the passage of the body from the lower apsis to the upper; and upon a like increase in its return to the lower apsis again; and therefore becomes greatest where the proportion of the force at the upper apsis to the force at the lower apsis recedes farthest from the duplicate ratio of the distances inversely; it is plain, that, when the apsides are in the syzygies, they will, by reason of the subducting force KL or NM — LM, go forward more swiftly; and in the quadratures by the additional force LM go backward more slowly. Because the velocity of the progress or slowness of the regress is continued for a long time; this inequality becomes exceedingly great.

COR. 9. If a body is obliged, by a force reciprocally proportional to the square of its distance from any centre, to revolve in an ellipsis round that centre; and afterwards in its descent from the upper apsis to the lower apsis, that force by a perpetual accession of new force is increased in more than a duplicate ratio of the diminished distance; it is manifest that the body, being impelled always towards the centre by the perpetual accession of this new force, will incline more towards that centre than if it were urged by that force alone which decreases in a duplicate ratio of the diminished distance, and therefore will describe an orbit interior to that elliptical orbit, and at the lower apsis approaching nearer to the centre than before. Therefore the orbit by the accession of this new force will become more eccentrical. If now, while the body is returning from the lower to the upper apsis, it should decrease by the same degrees by which it increases before the body would return to its first distance; and therefore if the force decreases in a yet greater ratio, the body, being now less attracted than before, will ascend to a still greater distance, and so the eccentricity of the orbit will be increased still more. Therefore if the ratio of the increase and decrease of the centripetal force be augmented each revolution, the eccentricity will be augmented also; and, on the contrary, if that ratio decrease, it will be diminished.

Now, therefore, in the system of the bodies T, P, S, when the apsides of the orbit PAB are in the quadratures, the ratio of that increase and decrease is least of all, and becomes greatest when the apsides are in the syzygies. If the apsides are placed in the quadratures, the ratio near the apsides is less, and near the syzygies greater, than the duplicate ratio of the distances; and from that greater ratio arises a direct motion of the line of the apsides, as was just now said. But if we consider the ratio of the whole increase or decrease in the progress between the apsides, this is less than the duplicate ratio of the distances. The force in the lower is to the force in the upper apsis in less than a duplicate ratio of the distance of the upper apsis from the focus of the ellipsis to the distance of the lower apsis from the same focus; and, contrariwise, when the apsides are placed in the syzygies, the force in the lower apsis is to the force in the upper apsis in a greater than a duplicate ratio of the distances. For the forces LM in the quadratures added to the forces of the body T compose forces in a less ratio; and the forces KL in the syzygies subducted from the forces of the body T, leave the forces in a greater ratio. Therefore the ratio of the whole increase and decrease in the passage between the apsides is least at the quadratures and greatest at the syzygies; and therefore in the passage of the apsides from the quadratures to the syzygies it is continually augmented, and increases the eccentricity of the ellipsis; and in the passage from the syzygies to the quadratures it is perpetually decreasing, and diminishes the eccentricity.

Cor. 10. That we may give an account of the errors as to latitude, let us suppose the plane of the orbit EST to remain immovable; and from the cause of the errors above explained, it is manifest, that, of the two forces NM, ML, which are the only and entire cause of them, the force ML acting always in the plane of the orbit PAB never disturbs the motions as to latitude; and that the force NM, when the nodes are in the syzygies, acting also in the same plane of the orbit, does not at that time affect those motions. But when the nodes are in the quadratures, it disturbs them very much, and, attracting the body P perpetually out of the plane of its orbit, it diminishes the inclination of the plane in the passage of the body from the quadratures to the syzygies, and again increases the same in the passage from the syzygies to the quadratures. Hence it comes to pass that when the body is in the syzygies, the inclination is then least of all, and returns to the first magnitude nearly, when the body arrives at the next node. But if the nodes are situate at the octants after the quadratures, that is, between C and A, D and B, it will appear, from

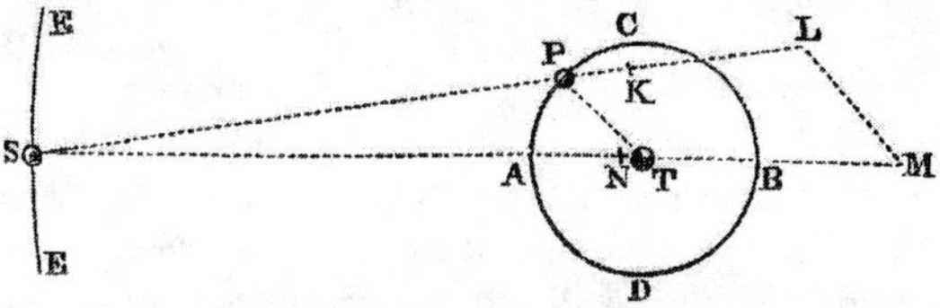

what was just now shewn, that in the passage of the body P from either node to the ninetieth degree from thence, the inclination of the plane is perpetually diminished; then, in the passage through the next 45 degrees to the next quadrature, the inclination is increased; and afterwards, again, in its passage through another 45 degrees to the next node, it is diminished. Therefore the inclination is more diminished than increased, and is therefore always less in the subsequent node than in the preceding one. And, by a like reasoning, the inclination is more increased than diminished when the nodes are in the other octants between A and D, B and C. The inclination, therefore, is the greatest of all when the nodes are in the syzygies. In their passage from the syzygies to the quadratures the inclination is diminished at each appulse of the body to the nodes; and becomes least of all when the nodes are in the quadratures, and the body in the syzygies; then it increases by the same degrees by which it decreased before; and, when the nodes come to the next syzygies, returns to its former magnitude.

COR. 11. Because when the nodes are in the quadratures the body P is perpetually attracted from the plane of its orbit; and because this attraction is made towards S in its passage from the node C through the conjunction A to the node D; and to the contrary part in its passage from the node D through the opposition B to the node C; it is manifest that, in its motion from the node C, the body recedes continually from the former plane CD of its orbit till it comes to the next node; and therefore at that node, being now at its greatest distance from the first plane CD, it will pass through the plane of the orbit EST not in D, the other node of that plane, but in a point that lies nearer to the body S, which therefore becomes a new place of the node *in antecedentia* to its former place. And, by a like reasoning, the nodes will continue to recede in their passage from this node to the next. The nodes, therefore, when situate in the quadratures, recede perpetually; and at the syzygies, where no perturbation can be produced in the motion as to latitude, are quiescent: in the intermediate places they partake of both conditions, and recede more slowly; and, therefore, being always either retrograde or stationary, they will be carried backwards, or *in antecedentia*, each revolution.

COR. 12. All the errors described in these corrollaries are a little greater at the conjunction of the bodies P, S, than at their opposition; because the generating forces NM and ML are greater.

Cor. 13. And since the causes and proportions of the errors and variations mentioned in these Corollaries do not depend upon the magnitude of the body S, it follows that all things before demonstrated will happen, if the magnitude of the body S be imagined so great as that the system of the two bodies P and T may revolve about it. And from this increase of the body S, and the consequent increase of its centripetal force, from which the errors of the body P arise, it will follow that all these errors, at equal distances, will be greater in this case, than in the other where the body S revolves about the system of the bodies P and T.

Cor. 14. But since the forces NM, ML, when the body S is exceedingly distant, are very nearly as the force SK and the ratio PT to ST conjunctly; that is, if both the distance PT, and the absolute force of the body S be given, as ST^3 reciprocally; and since those forces NM, ML are the causes of all the errors and effects treated of in the foregoing Corollaries; it is manifest that all those effects, if the system of bodies T and P continue as before, and only the distance ST and the absolute force of the body S be changed, will be very nearly in a ratio compounded of the direct ratio of the absolute force of the body S, and the triplicate inverse ratio of the distance ST. Hence if the system of bodies T and P revolve about a distant body S, those forces NM, ML, and their effects, will be (by Cor. 2 and 6, Prop IV) reciprocally in a duplicate ratio of the periodical time. And thence, also, if the magnitude of the body S be proportional to its absolute force, those forces NM, ML, and their effects, will be directly as the cube of the apparent diameter of the distant body S viewed from T, and so *vice versa.* For these ratios are the same as the compounded ratio above mentioned.

Cor. 15. And because if the orbits ESE and PAB, retaining their figure, proportions, and inclination to each other, should alter their magnitude; and the forces of the bodies S and T should either remain, or be changed in any given ratio; these forces (that is, the force of the body T, which obliges the body P to deflect from a rectilinear course into the orbit PAB, and the force of the body S, which causes the body P to deviate from that orbit) would act always in the same manner, and in the same proportion; it follows, that all the effects will be similar and proportional, and the times of those effects proportional also; that is, that all the linear errors will be as the diameters of the orbits, the angular errors the same as before; and the times of similar linear errors, or equal angular errors, as the periodical times of the orbits.

Cor. 16. Therefore if the figures of the orbits and their inclination to each other be given, and the magnitudes, forces, and distances of the bodies be any how changed, we may, from the errors and times of those errors in one case, collect very nearly the errors and times of the errors in any other case. But this may be done more expeditiously by the following method. The forces NM, ML, other things remaining unaltered, are as the radius TP; and their periodical effects (by Cor. 2, Lem. X) are as the forces and the square of the periodical time of the body P conjunctly. These are the linear errors of the body P; and hence the angular errors as they appear from the centre T (that is, the motion of the apsides and of the nodes, and all the apparent errors as to longitude and latitude) are in each revolution of the body P as the square of the time of the revolution, very nearly. Let these ratios be compounded with the ratios in Cor. 14, and in any system of bodies T, P, S, where P revolves about T very near to it, and T revolves about S at a great distance, the angular errors of the body P, observed from the centre T, will be in each revolution of the body P as the square of the periodical time of the body P directly, and the square of the periodical time of the body T inversely. And therefore the mean motion of the line of the apsides will be in a given ratio to the mean motion of the nodes; and both those motions will be as the periodical time of the body P directly, and the square of the periodical time of the body T inversely. The increase or diminution of the eccentricity and inclination of the orbit PAB makes no sensible variation in the motions of the apsides and nodes, unless that increase or diminution be very great indeed.

COR. 17. Since the line LM becomes sometimes greater and sometimes less than the radius PT, let the mean quantity of the force LM be expressed

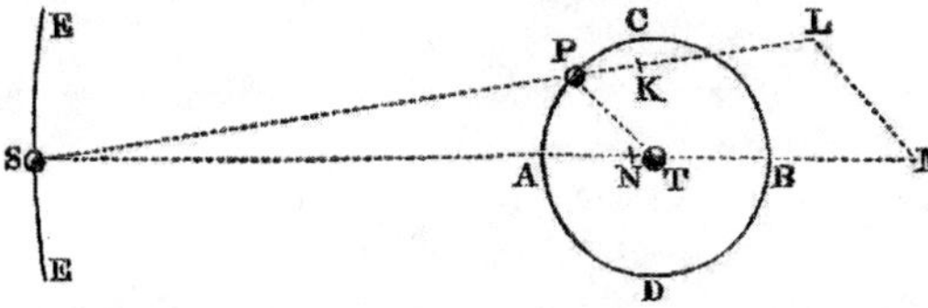

by that radius PT; and then that mean force will be to the mean force SK or SN (which may be also expressed by ST) as the length PT to the length ST. But the mean force SN or ST, by which the body T is retained in the orbit it describes about S, is to the force with which the body P is retained in its orbit about T in a ratio compounded of the ratio of the radius ST to the radius PT, and the duplicate ratio of the periodical time of the body P about T to the periodical time of the body T about S. And, *ex æquo*, the mean force LM is to the force by which the body P is retained in its orbit about T (or by which the same body P might revolve at the distance PT in the same periodical time about any immovable point T) in the same duplicate ratio of the periodical times. The periodical times therefore being given, together with the distance PT, the mean force LM is also given; and that force being given, there is given also the force MN, very nearly, by the analogy of the lines PT and MN.

Cor. 18. By the same laws by which the body P revolves about the body T, let us suppose many fluid bodies to move round T at equal distances from it; and to be so numerous, that they may all become contiguous to each other, so as to form a fluid annulus, or ring, of a round figure, and concentrical to the body T; and the several parts of this annulus, performing their motions by the same law as the body P, will draw nearer to the body T, and move swifter in the conjunction and opposition of themselves and the body S, than in the quadratures. And the nodes of this annulus, or its intersections with the plane of the orbit of the body S or T, will rest at the syzygies; but out of the syzygies they will be carried backward, or *in antecedentia;* with the greatest swiftness in the quadratures, and more slowly in other places. The inclination of this annulus also will vary, and its axis will oscillate each revolution, and when the revolution is completed will return to its former situation, except only that it will be carried round a little by the præcession of the nodes.

Cor. 19. Suppose now the sphærical body T, consisting of some matter not fluid, to be enlarged, and to extend itself on every side as far as that annulus, and that a channel were cut all round its circumference containing water; and that this sphere revolves uniformly about its own axis in the same periodical time. This water being accelerated and retarded by turns (as in the last Corollary), will be swifter at the syzygies, and slower at the quadratures, than the surface of the globe, and so will ebb and flow in its channel after the manner of the sea. If the attraction of the body S were taken away, the water would acquire no motion of flux and reflux by revolving round the quiescent centre of the globe. The case is the same of a globe moving uniformly forwards in a right line, and in the mean time revolving about its centre (by Cor. 5 of the Laws of Motion), and of a globe uniformly attracted from its rectilinear course (by Cor. 6, of the same Laws). But let the body S come to act upon it, and by its unequable attraction the water will receive this new motion; for there will be a stronger attraction upon that part of the water that is nearest to the body, and a weaker upon that part which is more remote. And the force LM will attract the water downwards at the quadratures, and depress it as far as the syzygies; and the force KL will attract it upwards in the syzygies, and withhold its descent, and make it rise as far as the quadratures; except only in so far as the motion of flux and reflux may be directed by the channel of the water, and be a little retarded by friction.

Cor. 20. If, now, the annulus becomes hard, and the globe is diminished, the motion of flux and reflux will cease; but the oscillating motion of the inclination and the præcession of the nodes will remain. Let the globe have the same axis with the annulus, and perform its revolutions in the same times, and at its surface touch the annulus within, and adhere to it; then the globe partaking of the motion of the annulus, this whole compages will oscillate, and the nodes will go backward, for the globe, as we shall shew presently, is perfectly indifferent to the receiving of all impressions. The greatest angle of the inclination of the annulus single is when the nodes are in the syzygies. Thence in the progress of the nodes to the quadratures, it endeavours to diminish its inclination, and by that endeavour impresses a motion upon the whole globe. The globe retains this motion impressed, till the annulus by a contrary endeavour destroys that motion, and impresses a new motion in a contrary direction. And by this means the greatest motion of the decreasing inclination happens when the nodes are in the quadratures, and the least angle of inclination in the octants

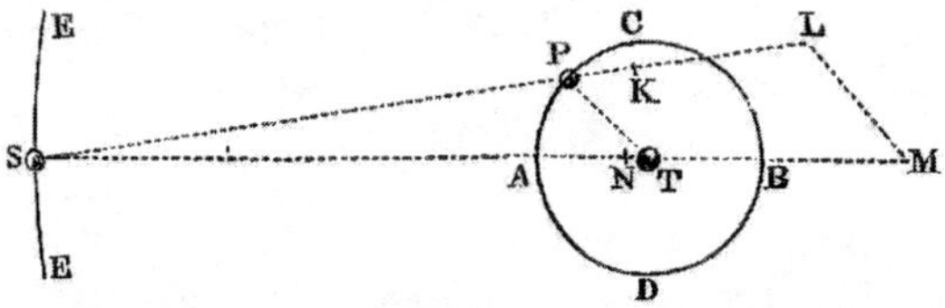

after the quadratures; and, again, the greatest motion of reclination happens when the nodes are in the syzygies; and the greatest angle of reclination in the octants following. And the case is the same of a globe without this annulus, if it be a little higher or a little denser in the equatorial than in the polar regions; for the excess of that matter in the regions near the equator supplies the place of the annulus. And though we should suppose the centripetal force of this globe to be any how increased, so that all its parts were to tend downwards, as the parts of our earth gravitate to the centre, yet the phænomena of this and the preceding Corollary would scarce be altered; except that the places of the greatest and least height of the water will be different; for the water is now no longer sustained and kept in its orbit by its centrifugal force, but by the channel in which it flows. And, besides, the force LM attracts the water downwards most in the quadratures, and the force KL or NM — LM attracts it upwards most in the syzygies. And these forces conjoined cease to attract the water downwards, and begin to attract it upwards in the octants before the syzygies; and cease to attract the water upwards, and begin to attract the water downwards in the octants after the syzygies. And thence the greatest height of the water may happen about the octants after the syzygies; and the least height about the octants after the quadratures; excepting only so far as the motion of ascent or descent impressed by these forces may by the *vis insita* of the water continue a little longer, or be stopped a little sooner by impediments in its channel.

COR. 21. For the same reason that redundant matter in the equatorial regions of a globe causes the nodes to go backwards, and therefore by the increase of that matter that retrogradation is increased, by the diminution is diminished, and by the removal quite ceases; it follows, that, if more than that redundant matter be taken away, that is, if the globe be either more depressed, or of a more rare consistence near the equator than near the poles, there will arise a motion of the nodes *in consequentia*.

Cor. 22. And thence from the motion of the nodes is known the constitution of the globe. That is, if the globe retains unalterably the same poles, and the motion (of the nodes) be *in antecedentia*, there is a redundance of the matter near the equator; but if *in consequentia*, a deficiency. Suppose a uniform and exactly sphærical globe to be first at rest in a free space; then by some impulse made obliquely upon its superficies to be driven from its place, and to receive a motion partly circular and partly right forward. Because this globe is perfectly indifferent to all the axes that pass through its centre, nor has a greater propensity to one axis or to one situation of the axis than to any other, it is manifest that by its own force it will never change its axis, or the inclination of it. Let now this globe be impelled obliquely by a new impulse in the same part of its superficies as before; and since the effect of an impulse is not at all changed by its coming sooner or later, it is manifest that these two impulses, successively impressed, will produce the same motion as if they were impressed at the same time; that is, the same motion as if the globe had been impelled by a simple force compounded of them both (by Cor. 2, of the Laws), that is, a simple motion about an axis of a given inclination. And the case is the same if the second impulse were made upon any other place of the equator of the first motion; and also if the first impulse were made upon any place in the equator of the motion which would be generated by the second impulse alone; and therefore, also, when both impulses are made in any places whatsoever; for these impulses will generate the same circular motion as if they were impressed together, and at once, in the place of the intersections of the equators of those motions, which would be generated by each of them separately. Therefore, a homogeneous and perfect globe will not retain several distinct motions, but will unite all those that are impressed on it, and reduce them into one; revolving, as far as in it lies, always with a simple and uniform motion about one single given axis, with an inclination perpetually invariable. And the inclination of the axis, or the velocity of the rotation, will not be changed by centripetal force. For if the globe be supposed to be divided into two hemispheres, by any plane whatsoever passing through its own centre, and the centre to which the force is directed, that force will always urge each hemisphere equally; and therefore will not incline the globe any way as to its motion round its own axis. But let there be added any where between the pole and the equator a heap of new matter like a mountain, and this, by its perpetual endeavour to recede from the centre of its motion, will disturb the motion of the globe, and cause its poles to wander about its superficies, describing circles about themselves and their opposite points. Neither can this enormous evagation of the poles be corrected, unless by placing that mountain either in one of the poles; in which case, by Cor. 21, the nodes of the equator will go forwards; or in the equatorial regions, in which case, by Cor. 20, the nodes will go backwards; or, lastly, by adding on the other side of the axis a new quantity of matter, by which the mountain may be balanced in its motion; and then the nodes will either go forwards or backwards, as the mountain and this newly added matter happen to be nearer to the pole or to the equator.

PROPOSITION LXVII. THEOREM XXVII.

The same laws of attraction being supposed, I say, that the exterior body S *does, by radii drawn to the point* O, *the common centre of gravity of the interior bodies* P *and* T, *describe round that centre areas more proportional to the times, and an orbit more approaching to the form of an ellipsis having its focus in that centre, than it can describe round the innermost and greatest body* T *by radii drawn to that body.*

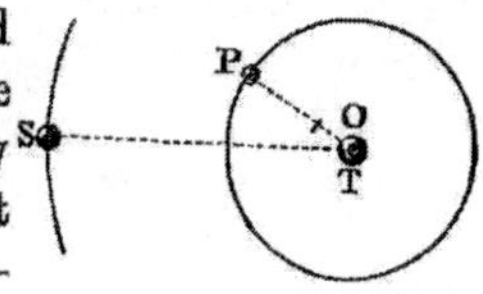

For the attractions of the body S towards T and P compose its absolute attraction, which is more directed towards O, the common centre of gravity of the bodies T and P, than it is to the greatest body T; and which is more in a reciprocal proportion to the square of the distance SO, than it is to the square of the distance ST; as will easily appear by a little consideration.

PROPOSITION LXVIII. THEOREM XXVIII.

The same laws of attraction supposed, I say, that the exterior body S *will, by radii drawn to* O, *the common centre of gravity of the interior bodies P and T, describe round that centre areas more proportional to the times, and an orbit more approaching to the form of an ellipsis having its focus in that centre, if the innermost and greatest body be agitated by these attractions as well as the rest, than it would do if that body were either at rest as not attracted, or were much more or much less attracted, or much more or much less agitated.*

This may be demonstrated after the same manner as Prop. LXVI, but by a more prolix reasoning, which I therefore pass over. It will be sufficient to consider it after this manner. From the demonstration of the last Proposition it is plain, that the centre, towards which the body S is urged by the two forces conjunctly, is very near to the common centre of gravity of those two other bodies. If this centre were to coincide with that common centre, and moreover the common centre of gravity of all the three bodies were at rest, the body S on one side, and the common centre of gravity of the other two bodies on the other side, would describe true ellipses about that quiescent common centre. This appears from Cor. 2, Prop. LVIII, compared with what was demonstrated in Prop. LXIV, and LXV. Now this accurate elliptical motion will be disturbed a little by the distance of the centre of the two bodies from the centre towards which the third body S is attracted. Let there be added, moreover, a motion to the common centre of the three, and the perturbation will be increased yet

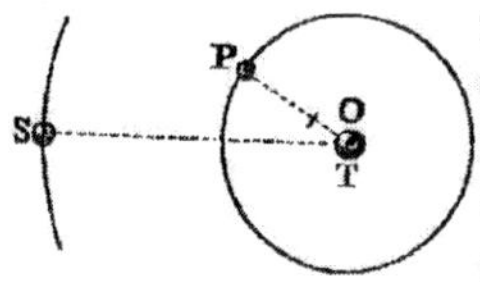

more. Therefore the perturbation is least when the common centre of the three bodies is at rest; that is, when the innermost and greatest body T is attracted according to the same law as the rest are; and is always greatest when the common centre of the three, by the diminution of the motion of the body T, begins to be moved, and is more and more agitated.

Cor. And hence if more lesser bodies revolve about the great one, it may easily be inferred that the orbits described will approach nearer to ellipses; and the descriptions of areas will be more nearly equable, if all the bodies mutually attract and agitate each other with accelerative forces that are as their absolute forces directly, and the squares of the distances inversely; and if the focus of each orbit be placed in the common centre of gravity of all the interior bodies (that is, if the focus of the first and innermost orbit be placed in the centre of gravity of the greatest and innermost body; the focus of the second orbit in the common centre of gravity of the two innermost bodies; the focus of the third orbit in the common centre of gravity of the three innermost; and so on), than if the innermost body were at rest, and was made the common focus of all the orbits.

PROPOSITION LXIX. THEOREM XXIX.

In a system of several bodies A, B, C, D, *&c., if any one of those bodies, as* A, *attract all the rest,* B, C, D, *&c., with accelerative forces that are reciprocally as the squares of the distances from the attracting body; and another body, as* B, *attracts also the rest,* A, C, D, *&c., with forces that are reciprocally as the squares of the distances from the attracting body; the absolute forces of the attracting bodies* A *and* B *will be to each other as those very bodies* A *and* B *to which those forces belong.*

For the accelerative attractions of all the bodies B, C, D, towards A, are by the supposition equal to each other at equal distances; and in like manner the accelerative attractions of all the bodies towards B are also equal to each other at equal distances. But the absolute attractive force of the body A is to the absolute attractive force of the body B as the accelerative attraction of all the bodies towards A to the accelerative attraction of all the bodies towards B at equal distances; and so is also the accelerative attraction of the body B towards A to the accelerative attraction of the body A towards B. But the accelerative attraction of the body B towards A is to the accelerative attraction of the body A towards B as the mass of the body A to the mass of the body B; because the motive forces which (by the 2d, 7th, and 8th Definition) are as the accelerative forces and the bodies attracted conjunctly are here equal to one another by the third Law. Therefore the absolute attractive force of the body A is to the absolute attractive force of the body B as the mass of the body A to the mass of the body B. Q.E.D.

COR. 1. Therefore if each of the bodies of the system A, B, C, D, &c. does singly attract all the rest with accelerative forces that are reciprocally as the squares of the distances from the attracting body, the absolute forces of all those bodies will be to each other as the bodies themselves.

COR. 2. By a like reasoning, if each of the bodies of the system A, B, C, D, &c., do singly attract all the rest with accelerative forces, which are either reciprocally or directly in the ratio of any power whatever of the distances from the attracting body; or which are defined by the distances from each of the attracting bodies according to any common law; it is plain that the absolute forces of those bodies are as the bodies themselves.

COR. 3. In a system of bodies whose forces decrease in the duplicate ratio of the distances, if the lesser revolve about one very great one in ellipses, having their common focus in the centre of that great body, and of a figure exceedingly accurate; and moreover by radii drawn to that great body describe areas proportional to the times exactly; the absolute forces of those bodies to each other will be either accurately or very nearly in the ratio of the bodies. And so on the contrary. This appears from Cor. of Prop. XLVIII, compared with the first Corollary of this Prop.

SCHOLIUM.

These Propositions naturally lead us to the analogy there is between centripetal forces, and the central bodies to which those forces used to be directed; for it is reasonable to suppose that forces which are directed to bodies should depend upon the nature and quantity of those bodies, as we see they do in magnetical experiments. And when such cases occur, we are to compute the attractions of the bodies by assigning to each of their particles its proper force, and then collecting the sum of them all. I here use the word attraction in general for any endeavour, of what kind soever, made by bodies to approach to each other; whether that endeavour arise from the action of the bodies themselves, as tending mutually to or agitating each other by spirits emitted; or whether it arises from the action of the æther or of the air, or of any medium whatsoever, whether corporeal or incorporeal, any how impelling bodies placed therein towards each other. In the same general sense I use the word impulse, not defining in this treatise the species or physical qualities of forces, but investigating the quantities and mathematical proportions of them; as I observed before in the Definitions. In mathematics we are to investigate the quantities of forces with their proportions consequent upon any conditions supposed; then, when we enter upon physics, we compare those proportions with the phænomena of Nature, that we may know what conditions of those forces answer to the several kinds of attractive bodies. And this preparation being made, we argue more safely concerning the physical species, causes, and proportions of the forces. Let us see, then, with what forces sphærical bodies consisting of particles endued with attractive powers in the manner above spoken of must act mutually upon one another; and what kind of motions will follow from thence.

SECTION XII.

Of the attractive forces of sphærical bodies.

PROPOSITION LXX. THEOREM XXX.

If to every point of a sphærical surface there tend equal centripetal forces decreasing in the duplicate ratio of the distances from those points; I say, that a corpuscle placed within that superficies will not be attracted by those forces any way.

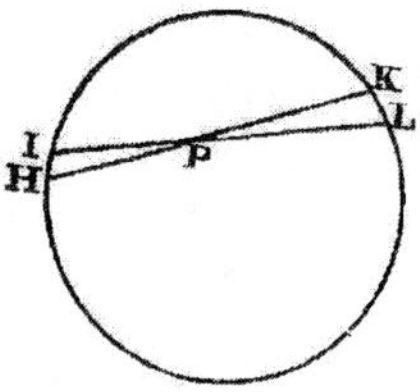

Let HIKL, be that sphærical superficies, and P a corpuscle placed within. Through P let there be drawn to this superficies to two lines HK, IL, intercepting very small arcs HI, KL; and because (by Cor. 3, Lem. VII) the triangles HPI, LPK are alike, those arcs will be proportional to the distances HP, LP; and any particles at HI and KL of the sphærical superficies, terminated by right lines passing through P, will be in the duplicate ratio of those distances. Therefore the forces of these particles exerted upon the body P are equal between themselves. For the forces are as the particles directly, and the squares of the distances inversely. And these two ratios compose the ratio of equality. The attractions therefore, being made equally towards contrary parts, destroy each other. And by a like reasoning all the attractions through the whole sphærical superficies are destroyed by contrary attractions. Therefore the body P will not be any way impelled by those attractions. Q.E.D.

PROPOSITION LXXI. THEOREM XXXI.

The same things supposed as above, I say, that a corpuscle placed without the sphærical superficies is attracted towards the centre of the sphere with a force reciprocally proportional to the square of its distance from that centre.

Let AHKB, *ahkb*, be two equal sphærical superficies described about the centre S, *s*; their diameters AB, *ab*; and let P and *p* be two corpuscles situate without the spheres in those diameters produced. Let there

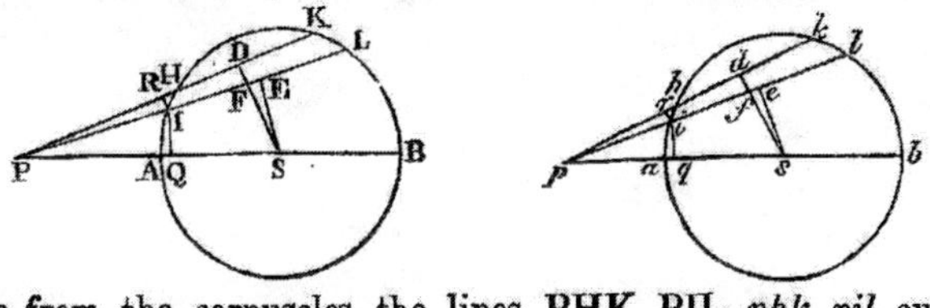

be drawn from the corpuscles the lines PHK, PIL, *phk*, *pil*, cutting off from the great circles AHB, *ahb*, the equal arcs HK, *hk*, IL, *il*; and to those lines let fall the perpendiculars SD, *sd*, SE, *se*, IR, *ir*; of which let SD, *sd*, cut PL, *pl*, in F and *f*. Let fall also to the diameters the perpendiculars IQ, *iq*. Let now the angles DPE, *dpe*, vanish; and because DS and *ds*, ES and *es* are equal, the lines PE, PF, and *pe*, *pf*, and the lineolæ DF, *df* may be taken for equal; because their last ratio, when the angles DPE, *dpe* vanish together, is the ratio of equality. These things then supposed, it will be, as PI to PF so is RI to DF, and as *pf* to *pi* so is *df* or DF to *ri*; and, *ex æquo*, as PI $\times$ *pf* to PF $\times$ *pi* so is RI to *ri*, that is (by Cor. 3, Lem. VII), so is the arc IH to the arc *ih*. Again, PI is to PS as IQ to SE, and *ps* to *pi* as *se* or SE to *iq*; and, *ex æquo*, PI $\times$ *ps* to PS $\times$ *pi* as IQ to *iq*. And compounding the ratios $PI^2 \times pf \times ps$ is to $pi^2 \times PF \times PS$, as IH $\times$ IQ to *ih* $\times$ *iq*; that is, as the circular superficies which is described by the arc IH, as the semi-circle AKB revolves about the diameter AB, is to the circular superficies described by the arc *ih* as the semi-circle *akb* revolves about the diameter *ab*. And the forces with which these superficies attract the corpuscles P and *p* in the direction of lines tending to those superficies are by the hypothesis as the superficies themselves directly, and the squares of the distances of the superficies from those corpuscles inversely; that is, as $pf \times ps$ to $PF \times PS$. And these forces again are to the oblique parts of them which (by the resolution of forces as in Cor. 2, of the Laws) tend to the centres in the directions of the lines PS, *ps*, as PI to PQ, and *pi* to *pq*; that is (because of the like triangles PIQ and PSF, *piq* and *psf*), as PS to PF and *ps* to *pf*. Thence *ex æquo*, the attraction of the corpuscle P towards S is to the attraction of the corpuscle *p* towards *s* as $\frac{PF \times pf \times ps}{PS}$ is to $\frac{pf \times PF \times PS}{ps}$, that is, as ps^2 to PS^2. And, by a like reasoning, the forces with which the superficies described by the revolution of the arcs KL, *kl* attract those corpuscles, will be as ps^2 to PS^2. And in the same ratio will be the forces of all the circular superficies into which each of the sphærical superficies may be divided by taking *sd* always equal to SD, and *se* equal to SE. And therefore, by composition, the forces of the entire sphærical superficies exerted upon those corpuscles will be in the same ratio. Q.E.D.

PROPOSITION LXXII. THEOREM XXXII.

If to the several points of a sphere there tend equal centripetal forces decreasing in a duplicate ratio of the distances from those points; and there be given both the density of the sphere and the ratio of the diameter of the sphere to the distance of the corpuscle from its centre; I say, that the force with which the corpuscle is attracted is proportional to the semi-diameter of the sphere.

For conceive two corpuscles to be severally attracted by two spheres, one by one, the other by the other, and their distances from the centres of the spheres to be proportional to the diameters of the spheres respectively; and the spheres to be resolved into like particles, disposed in a like situation to the corpuscles. Then the attractions of one corpuscle towards the several particles of one sphere will be to the attractions of the other towards as many analogous particles of the other sphere in a ratio compounded of the ratio of the particles directly, and the duplicate ratio of the distances inversely. But the particles are as the spheres, that is, in a triplicate ratio of the diameters, and the distances are as the diameters; and the first ratio directly with the last ratio taken twice inversely, becomes the ratio of diameter to diameter. Q.E.D.

Cor. 1. Hence if corpuscles revolve in circles about spheres composed of matter equally attracting, and the distances from the centres of the spheres be proportional to their diameters, the periodic times will be equal.

Cor. 2. And, *vice versa*, if the periodic times are equal, the distances will be proportional to the diameters. These two Corollaries appear from Cor. 3, Prop. IV.

Cor. 3. If to the several points of any two solids whatever, of like figure and equal density, there tend equal centripetal forces decreasing in a duplicate ratio of the distances from those points, the forces, with which corpuscles placed in a like situation to those two solids will be attracted by them, will be to each other as the diameters of the solids.

PROPOSITION LXXIII. THEOREM XXXIII.

If to the several points of a given sphere there tend equal centripetal forces decreasing in a duplicate ratio of the distances from the points; I say, that a corpuscle placed within the sphere is attracted by a force proportional to its distance from the centre.

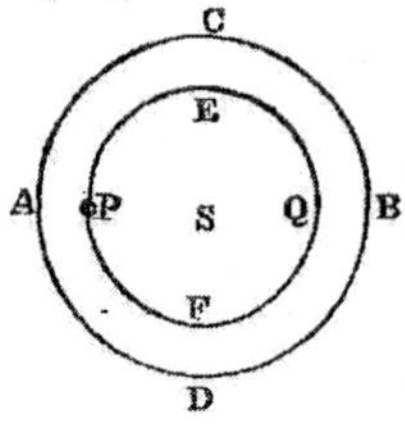

In the sphere ABCD, described about the centre S, let there be placed the corpuscle P; and about the same centre S, with the interval SP, conceive described an interior sphere PEQF. It is plain (by Prop. LXX) that the concentric sphærical superficies, of which the difference AEBF of the spheres is composed, have no effect at all upon the body P, their attractions being destroyed by contrary attractions. There remains, therefore, only the attraction of the interior sphere PEQF. And (by Prop. LXXII) this is as the distance PS. Q.E.D.

SCHOLIUM.

By the superficies of which I here imagine the solids composed, I do not mean superficies purely mathematical, but orbs so extremely thin, that their thickness is as nothing; that is, the evanescent orbs of which the sphere will at last consist, when the number of the orbs is increased, and their thickness diminished without end. In like manner, by the points of which lines, surfaces, and solids are said to be composed, are to be understood equal particles, whose magnitude is perfectly inconsiderable.

PROPOSITION LXXIV. THEOREM XXXIV.

The same things supposed, I say, that a corpuscle situate without the sphere is attracted with a force reciprocally proportional to the square of its distance from the centre.

For suppose the sphere to be divided into innumerable concentric sphærical superficies, and the attractions of the corpuscle arising from the several superficies will be reciprocally proportional to the square of the distance of the corpuscle from the centre of the sphere (by Prop. LXXI). And, by composition, the sum of those attractions, that is, the attraction of the corpuscle towards the entire sphere, will be in the same ratio. Q.E.D.

COR. 1. Hence the attractions of homogeneous spheres at equal distances from the centres will be as the spheres themselves. For (by Prop. LXXII) if the distances be proportional to the diameters of the spheres, the forces will be as the diameters. Let the greater distance be diminished in that ratio; and the distances now being equal, the attraction will be increased in the duplicate of that ratio; and therefore will be to the other attraction in the triplicate of that ratio; that is, in the ratio of the spheres.

COR. 2. At any distances whatever the attractions are as the spheres applied to the squares of the distances.

COR. 3. If a corpuscle placed without an homogeneous sphere is attracted by a force reciprocally proportional to the square of its distance from the centre, and the sphere consists of attractive particles, the force of every particle will decrease in a duplicate ratio of the distance from each particle.

PROPOSITION LXXV. THEOREM XXXV.

If to the several points of a given sphere there tend equal centripetal forces decreasing in a duplicate ratio of the distances from the points; I say, that another similar sphere will be attracted by it with a force reciprocally proportional to the square of the distance of the centres.

For the attraction of every particle is reciprocally as the square of its distance from the centre of the attracting sphere (by Prop. LXXIV), and is therefore the same as if that whole attracting force issued from one single corpuscle placed in the centre of this sphere. But this attraction is as great as on the other hand the attraction of the same corpuscle would be, if that were itself attracted by the several particles of the attracted sphere with the same force with which they are attracted by it. But that attraction of the corpuscle would be (by Prop. LXXIV) reciprocally proportional to the square of its distance from the centre of the sphere; therefore the attraction of the sphere, equal thereto, is also in the same ratio. Q.E.D.

Cor. 1. The attractions of spheres towards other homogeneous spheres are as the attracting spheres applied to the squares of the distances of their centres from the centres of those which they attract.

Cor. 2. The case is the same when the attracted sphere does also attract. For the several points of the one attract the several points of the other with the same force with which they themselves are attracted by the others again; and therefore since in all attractions (by Law III) the attracted and attracting point are both equally acted on, the force will be doubled by their mutual attractions, the proportions remaining.

Cor. 3. Those several truths demonstrated above concerning the motion of bodies about the focus of the conic sections will take place when an attracting sphere is placed in the focus, and the bodies move without the sphere.

Cor. 4. Those things which were demonstrated before of the motion of bodies about the centre of the conic sections take place when the motions are performed within the sphere.

PROPOSITION LXXVI. THEOREM XXXVI.

If spheres be however dissimilar (as to density of matter and attractive force) in the same ratio onward from the centre to the circumference; but every where similar, at every given distance from the centre, on all sides round about; and the attractive force of every point decreases in the duplicate ratio of the distance of the body attracted; I say, that the whole force with which one of these spheres attracts the other will be reciprocally proportional to the square of the distance of the centres.

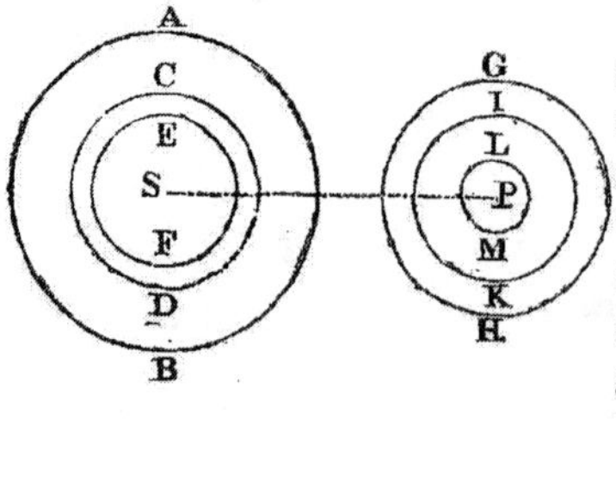

Imagine several concentric similar spheres, AB, CD, EF, &c., the innermost of which added to the outermost may compose a matter more dense towards the centre, or subducted from them may leave the same more lax and rare. Then, by Prop. LXXV, these spheres will attract other similar concentric spheres GH, IK, LM, &c., each the other, with forces reciprocally proportional to the square of the distance SP. And, by composition or division, the sum of all those forces, or the excess of any of them above the others; that is, the entire force with which the whole sphere AB (composed of any concentric spheres or of their differences) will attract the whole sphere GH (composed of any concentric spheres or their differences) in the same ratio. Let the number of the concentric spheres be increased *in infinitum*, so that the density of the matter together with the attractive force may, in the progress from the circumference to the centre, increase or decrease according to any given law; and by the addition of matter not attractive, let the deficient density be supplied, that so the spheres may acquire any form desired; and the force with which one of these attracts the other will be still, by the former reasoning, in the same ratio of the square of the distance inversely. Q.E.D.

COR. 1. Hence if many spheres of this kind, similar in all respects, attract each other mutually, the accelerative attractions of each to each, at any equal distances of the centres, will be as the attracting spheres.

COR. 2. And at any unequal distances, as the attracting spheres applied to the squares of the distances between the centres.

COR. 3. The motive attractions, or the weights of the spheres towards one another, will be at equal distances of the centres as the attracting and attracted spheres conjunctly; that is, as the products arising from multiplying the spheres into each other.

COR. 4. And at unequal distances, as those products directly, and the squares of the distances between the centres inversely.

COR. 5. These proportions take place also when the attraction arises from the attractive virtue of both spheres mutually exerted upon each other. For the attraction is only doubled by the conjunction of the forces, the proportions remaining as before.

COR. 6. If spheres of this kind revolve about others at rest, each about each; and the distances between the centres of the quiescent and revolving bodies are proportional to the diameters of the quiescent bodies; the periodic times will be equal.

COR. 7. And, again, if the periodic times are equal, the distances will be proportional to the diameters.

COR. 8. All those truths above demonstrated, relating to the motions of bodies about the foci of conic sections, will take place when an attracting sphere, of any form and condition like that above described, is placed in the focus.

COR. 9. And also when the revolving bodies are also attracting spheres of any condition like that above described.

PROPOSITION LXXVII. THEOREM XXXVII.

If to the several points of spheres there tend centripetal forces proportional to the distances of the points from the attracted bodies; I say, that the compounded force with which two spheres attract each other mutually is as the distance between the centres of the spheres.

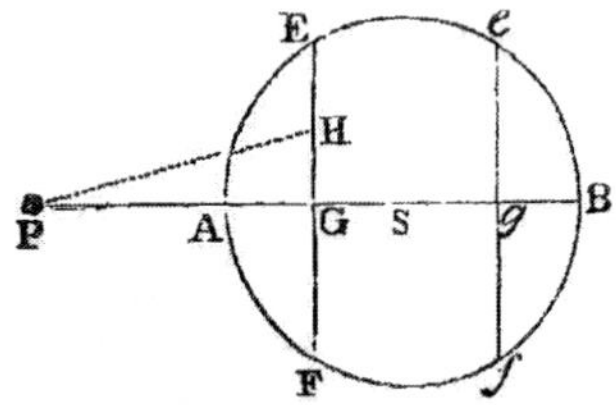

Case 1. Let AEBF be a sphere; S its centre; P a corpuscle attracted; PASB the axis of the sphere passing through the centre of the corpuscle; EF, *ef* two planes cutting the sphere, and perpendicular to the axis, and equi-distant, one on one side, the other on the other, from the centre of the sphere; G and *g* the intersections of the planes and the axis; and H any point in the plane EF. The centripetal force of the point H upon the corpuscle P, exerted in the direction of the line PH, is as the distance PH; and (by Cor. 2, of the Laws) the same exerted in the direction of the line PG, or towards the centre S, is as the length PG. Therefore the force of all the points in the plane EF (that is, of that whole plane) by which the corpuscle P is attracted towards the centre S is as the distance PG multiplied by the number of those points, that is, as the solid contained under that plane EF and the distance PG. And in like manner the force of the plane *ef*, by which the corpuscle P is attracted towards the centre S, is as that plane drawn into its distance P*g*, or as the equal plane EF drawn into that distance P*g*; and the sum of the forces of both planes as the plane EF drawn into the sum of the distances PG + P*g*, that is, as that plane drawn into twice the distance PS of the centre and the corpuscle; that is, as twice the plane EF drawn into the distance PS, or as the sum of the equal planes EF + *ef* drawn into the same distance. And, by a like reasoning, the forces of all the planes in the whole sphere, equi-distant on each side from the centre of the sphere, are as the sum of those planes drawn into the distance PS, that is, as the whole sphere and the distance PS conjunctly. Q.E.D.

Case 2. Let now the corpuscle P attract the sphere AEBF. And, by the same reasoning, it will appear that the force with which the sphere is attracted is as the distance PS. Q.E.D.

Case 3. Imagine another sphere composed of innumerable corpuscles P; and because the force with which every corpuscle is attracted is as the distance of the corpuscle from the centre of the first sphere, and as the same sphere conjunctly, and is therefore the same as if it all proceeded from a single corpuscle situate in the centre of the sphere, the entire force with which all the corpuscles in the second sphere are attracted, that is, with which that whole sphere is attracted, will be the same as if that sphere were attracted by a force issuing from a single corpuscle in the centre of the first sphere; and is therefore proportional to the distance between the centres of the spheres. Q.E.D.

CASE 4. Let the spheres attract each other mutually, and the force will be doubled, but the proportion will remain. Q.E.D.

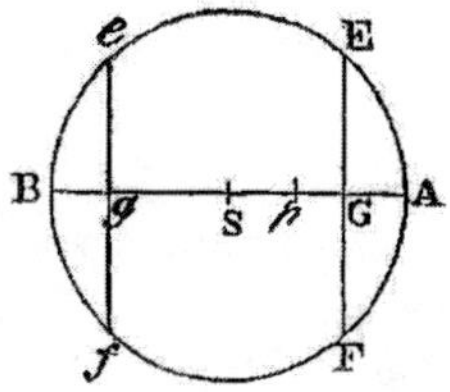

CASE 5. Let the corpuscle p be placed within the sphere AEBF; and because the force of the plane ef upon the corpuscle is as the solid contained under that plane and the distance pg; and the contrary force of the plane EF as the solid contained under that plane and the distance pG; the force compounded of both will be as the difference of the solids, that is, as the sum of the equal planes drawn into half the difference of the distances; that is, as that sum drawn into pS, the distance of the corpuscle from the centre of the sphere. And, by a like reasoning, the attraction of all the planes EF, ef, throughout the whole sphere, that is, the attraction of the whole sphere, is conjunctly as the sum of all the planes, or as the whole sphere, and as pS, the distance of the corpuscle from the centre of the sphere. Q.E.D.

CASE 6. And if there be composed a new sphere out of innumerable corpuscles such as p, situate within the first sphere AEBF, it may be proved, as before, that the attraction, whether single of one sphere towards the other, or mutual of both towards each other, will be as the distance pS of the centres. Q.E.D.

PROPOSITION LXXVIII. THEOREM XXXVIII.

If spheres in the progress from the centre to the circumference be however dissimilar and unequable, but similar on every side round about at all given distances from the centre; and the attractive force of every point be as the distance of the attracted body; I say, that the entire force with which two spheres of this kind attract each other mutually is proportional to the distance between the centres of the spheres.

This is demonstrated from the foregoing Proposition, in the same manner as Proposition LXXVI was demonstrated from Proposition LXXV.

COR. Those things that were above demonstrated in Prop. X and LXIV, of the motion of bodies round the centres of conic sections, take place when all the attractions are made by the force of sphærical bodies of the condition above described, and the attracted bodies are spheres of the same kind.

SCHOLIUM.

I have now explained the two principal cases of attractions; to wit, when the centripetal forces decrease in a duplicate ratio of the distances, or increase in a simple ratio of the distances, causing the bodies in both cases to revolve in conic sections, and composing sphærical bodies whose centripetal forces observe the same law of increase or decrease in the recess from the centre as the forces of the particles themselves do; which is very remarkable. It would be tedious to run over the other cases, whose conclusions are less elegant and important, so particularly as I have done these. I choose rather to comprehend and determine them all by one general method as follows.

LEMMA XXIX.

If about the centre S *there be described any circle as* AEB, *and about the centre* P *there be also described two circles* EF, ef, *cutting the first in* E *and* e, *and the line* PS *in* F *and* f; *and there be let fall to* PS *the perpendiculars* ED, ed; *I say, that if the distance of the arcs* EF, ef *be supposed to be infinitely diminished, the last ratio of the evanscent line* Dd *to the evanescent line* Ff *is the same as that of the line* PE *to the line* PS.

For if the line P*e* cut the arc EF in *q*; and the right line E*e*, which

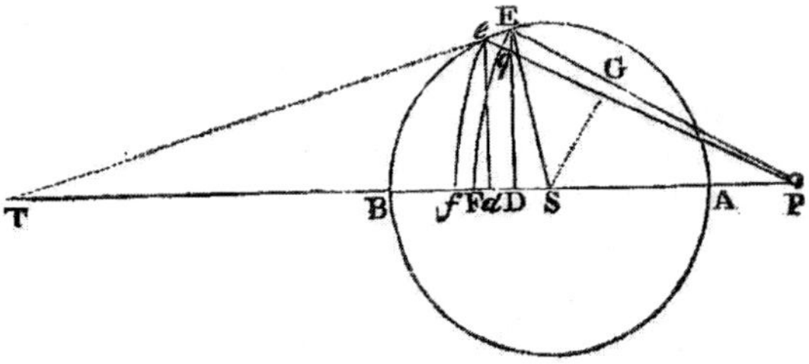

coincides with the evanescent arc E*e*, be produced, and meet the right line PS in T; and there be let fall from S to PE the perpendicular SG; then, because of the like triangles DTE, *d*T*e*, DES, it will be as D*d* to E*e* so DT to TE, or DE to ES; and because the triangles, E*eq*, ESG (by Lem. VIII, and Cor. 3, Lem. VII) are similar, it will be as E*e* to *eq* or F*f* so ES to SG; and, *ex æquo*, as D*d* to F*f* so DE to SG; that is (because of the similar triangles PDE, PGS), so is PE to PS. Q.E.D.

PROPOSITION LXXIX. THEOREM XXXIX.

Suppose a superficies as EFfe *to have its breadth infinitely diminished, and to be just vanishing; and that the same superficies by its revolution round the axis* PS *describes a sphærical concavo-convex solid, to the several equal particles of which there tend equal centripetal forces; I say, that the force with which that solid attracts a corpuscle situate in* P *is in a ratio compounded of the ratio of the solid* $DE^2 \times Ff$ *and the ratio of the force with which the given particle in the place* Ff *would attract the same corpuscle.*

For if we consider, first, the force of the sphærical superficies FE which

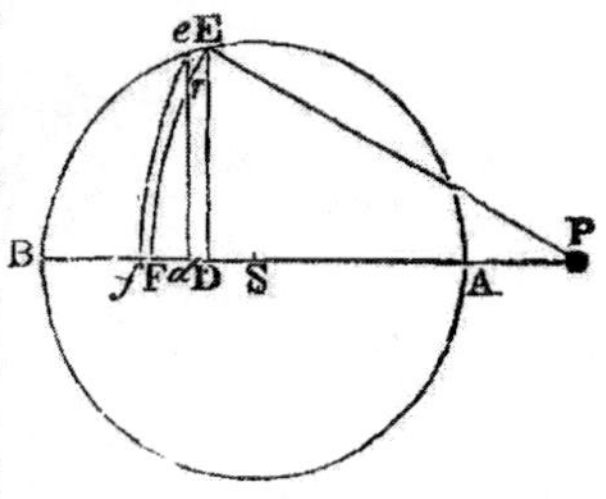

is generated by the revolution of the arc FE, and is cut any where, as in r, by the line de, the annular part of the superficies generated by the revolution of the arc rE will be as the lineola Dd, the radius of the sphere PE remaining the same; as *Archimedes* has demonstrated in his Book of the Sphere and Cylinder. And the force of this superficies exerted in the direction of the lines PE or Pr situate all round in the conical superficies, will be as this annular superficies itself; that is as the lineola Dd, or, which is the same, as the rectangle under the given radius PE of the sphere and the lineola Dd; but that force, exerted in the direction of the line PS tending to the centre S, will be less in the ratio PD to PE, and therefore will be as PD $\times$ Dd. Suppose now the line DF to be divided into innumerable little equal particles, each of which call Dd, and then the superficies FE will be divided into so many equal annuli, whose forces will be as the sum of all the rectangles PD $\times$ Dd, that is, as $\frac{1}{2}$PF^2 — $\frac{1}{2}$PD^2, and therefore as DE^2. Let now the superficies FE be drawn into the altitude Ff; and the force of the solid EFfe exerted upon the corpuscle P will be as DE^2 $\times$ Ff; that is, if the force be given which any given particle as Ff exerts upon the corpuscle P at the distance PF. But if that force be not given, the force of the solid EFfe will be as the solid DE^2 $\times$ Ff and that force not given, conjunctly. Q.E.D.

PROPOSITION LXXX. THEOREM XL.

If to the several equal parts of a sphere ABE *described about the centre* S *there tend equal centripetal forces; and from the several points* D *in the axis of the sphere* AB *in which a corpuscle, as* P, *is placed, there be erected the perpendiculars* DE *meeting the sphere in* E, *and if in those perpendiculars the lengths* DN *be taken as the quantity* $\frac{DE^2 \times PS}{PE}$, *and as the force which a particle of the sphere situate in the axis exerts at the distance* PE *upon the corpuscle* P *conjunctly; I say, that the whole force with which the corpuscle P is attracted towards the sphere is as the area* ANB, *comprehended under the axis of the sphere* AB, *and the curve line* ANB, *the locus of the point* N.

For supposing the construction in the last Lemma and Theorem to stand, conceive the axis of the sphere AB to be divided into innumerable equal particles Dd, and the whole sphere to be divided into so many sphærical concavo-convex laminæ EFfe; and erect the perpendicular dn. By the last Theorem, the force with which the laminæ EFfe attracts the corpuscle P is as DE^2 $\times$ Ff and the force of one particle exerted at the

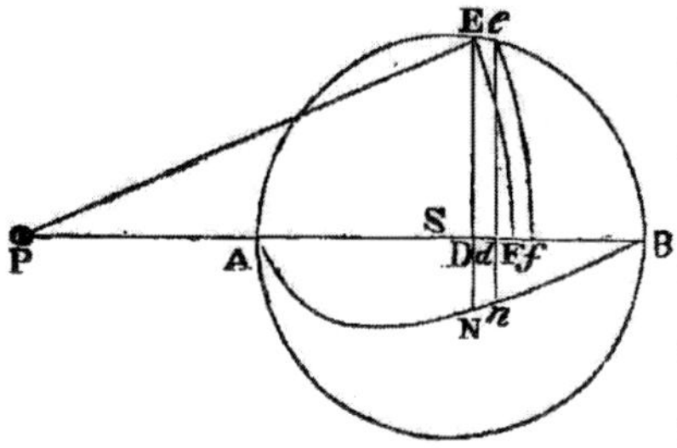

distance PE or PF, conjunctly. But (by the last Lemma) Dd is to Ff as PE to PS, and therefore Ff is equal to $\frac{PS \times Dd}{PE}$; and $DE^2 \times Ff$ is equal to $Dd \times \frac{DE^2 \times PS}{PE}$; and therefore the force of the lamina EFfe is as $Dd \times \frac{DE^2 \times PS}{PE}$ and the force of a particle exerted at the distance PF conjunctly; that is, by the supposition, as DN × Dd, or as the evanescent area DNnd. Therefore the forces of all the laminæ exerted upon the corpuscle P are as all the areas DNnd, that is, the whole force of the sphere will be as the whole area ANB. Q.E.D.

Cor. 1. Hence if the centripetal force tending to the several particles remain always the same at all distances, and DN be made as $\frac{DE^2 \times PS}{PE}$, the whole force with which the corpuscle is attracted by the sphere is as the area ANB.

Cor. 2. If the centripetal force of the particles be reciprocally as the distance of the corpuscle attracted by it, and DN be made as $\frac{DE^2 \times PS}{PE^2}$, the force with which the corpuscle P is attracted by the whole sphere will be as the area ANB.

Cor. 3. If the centripetal force of the particles be reciprocally as the cube of the distance of the corpuscle attracted by it, and DN be made as $\frac{DE^2 \times PS}{PE^4}$, the force with which the corpuscle is attracted by the whole sphere will be as the area ANB.

Cor. 4. And universally if the centripetal force tending to the several particles of the sphere be supposed to be reciprocally as the quantity V; and DN be made as $\frac{DE^2 \times PS}{PE \times V}$; the force with which a corpuscle is attracted by the whole sphere will be as the area ANB.

PROPOSITION LXXXI. PROBLEM XLI.

The things remaining as above, it is required to measure the area ANB.

From the point P let there be drawn the right line PH touching the sphere in H; and to the axis PAB, letting fall the perpendicular HI, bisect PI in L; and (by Prop. XII, Book II, Elem.) PE^2 is equal to

$PS^2 + SE^2 + 2PSD$. But because the triangles SPH, SHI are alike, SE^2 or SH^2 is equal to the rectangle PSI. Therefore PE^2 is equal to the rectangle contained under PS and $PS + SI + 2SD$; that is, under PS and $2LS + 2SD$; that is, under PS and 2LD. Moreover DE^2 is equal to $SE^2 - SD^2$, or $SE^2 -$

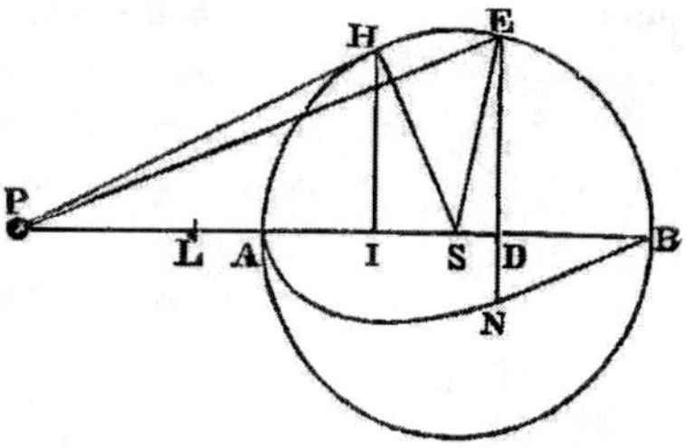

$LS^2 + 2SLD - LD^2$, that is, $2SLD - LD^2 - ALB$. For $LS^2 - SE^2$ or $LS^2 - SA^2$ (by Prop. VI, Book II, Elem.) is equal to the rectangle ALB. Therefore if instead of DE^2 we write $2SLD - LD^2 - ALB$, the quantity $\frac{DE^2 \times PS}{PE \times V}$, which (by Cor. 4 of the foregoing Prop.) is as the length of the ordinate DN, will now resolve itself into three parts $\frac{2SLD \times PS}{PE \times V} - \frac{LD^2 \times PS}{PE \times V} - \frac{ALB \times PS}{PE \times V}$; where if instead of V we write the inverse ratio of the centripetal force, and instead of PE the mean proportional between PS and 2LD, those three parts will become ordinates to so many curve lines, whose areas are discovered by the common methods. Q.E.D.

EXAMPLE 1. If the centripetal force tending to the several particles of the sphere be reciprocally as the distance; instead of V write PE the distance, then $2PS \times LD$ for PE^2; and DN will become as $SL - \frac{1}{2} LD - \frac{ALB}{2LD}$. Suppose DN equal to its double $2SL - LD - \frac{ALB}{LD}$; and 2SL the given part of the ordinate drawn into the length AB will describe the rectangular area $2SL \times AB$; and the indefinite part LD, drawn perpendicularly into the same length with a continued motion, in such sort as in its motion one way or another it may either by increasing or decreasing remain always equal to the length LD, will describe the area $\frac{LB^2 - LA^2}{2}$, that is, the area $SL \times AB$; which taken from the former area $2SL \times AB$, leaves the area $SL \times AB$. But the third part $\frac{ALB}{LD}$, drawn after the same manner with a continued motion perpendicularly into the same length, will describe the area of an hyperbola, which subducted from the area $SL \times AB$ will leave ANB the area sought. Whence arises this construction of the Problem. At the points, L, A, B, erect the perpendiculars Ll, Aa, Bb; making Aa equal to LB, and Bb equal to LA. Making Ll and LB asymptotes, describe through the points a, b,

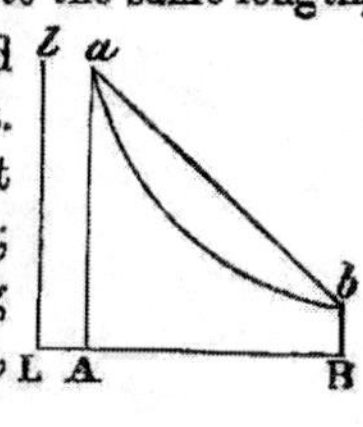

the hyperbolic curve *ab*. And the chord *ba* being drawn, will inclose the area *aba* equal to the area sought ANB.

Example 2. If the centripetal force tending to the several particles of the sphere be reciprocally as the cube of the distance, or (which is the same thing) as that cube applied to any given plane; write $\frac{PE^3}{2AS^2}$ for V, and $2PS \times LD$ for PE^2; and DN will become as $\frac{SL \times AS^2}{PS \times LD} - \frac{AS^2}{2PS} - \frac{ALB \times AS^2}{2PS \times LD^2}$, that is (because PS, AS, SI are continually proportional), as $\frac{LSI}{LD} - \frac{1}{2}SI - \frac{ALB \times SI}{2LD^2}$. If we draw then these three parts into the length AB, the first $\frac{LSI}{LD}$ will generate the area of an hyperbola; the second $\frac{1}{2}SI$ the area $\frac{1}{2}AB \times SI$; the third $\frac{ALB \times SI}{2LD^2}$ the area $\frac{ALB \times SI}{2LA} - \frac{ALB \times SI}{2LB}$, that is, $\frac{1}{2}AB \times SI$. From the first subduct the sum of the second and third, and there will remain ANB, the area sought. Whence arises this construction of the problem. At the points L, A, S, B, erect

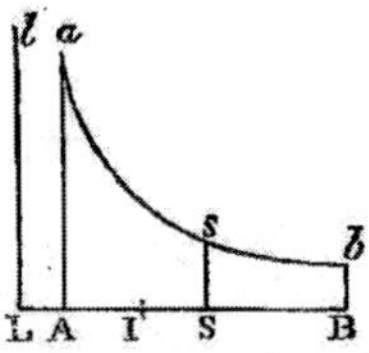

the perpendiculars *Ll* *Aa* *Ss*, *Bb*, of which suppose *Ss* equal to SI; and through the point *s*, to the asymptotes *Ll*, LB, describe the hyperbola *asb* meeting the perpendiculars A*a*, B*b*, in *a* and *b*; and the rectangle 2ASI, subducted from the hyberbolic area A*asb*B, will leave ANB the area sought.

Example 3. If the centripetal force tending to the several particles of the spheres decrease in a quadruplicate ratio of the distance from the particles; write $\frac{PE^4}{2AS^3}$ for V, then $\sqrt{2PS + LD}$ for PE, and DN will become as $\frac{SI^2 \times SL}{\sqrt{2SI}} \times \frac{1}{\sqrt{LD^3}} - \frac{SI^2}{2\sqrt{2SI}} \times \frac{1}{\sqrt{LD}} - \frac{SI^2 \times ALB}{2\sqrt{2SI}} \times \frac{1}{\sqrt{LD^5}}$. These three parts drawn into the length AB, produce so many areas, viz.

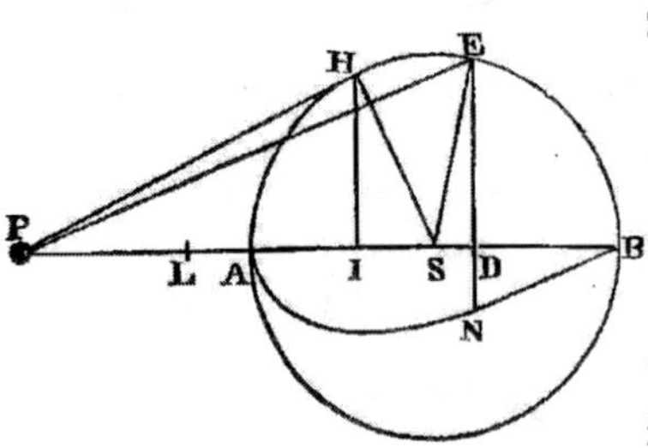

$\frac{2SI^2 \times SL}{\sqrt{2SI}}$ into $\frac{1}{\sqrt{LA}} - \frac{1}{\sqrt{LB}}$; $\frac{SI^2}{\sqrt{2SI}}$ into $\sqrt{LB} - \sqrt{LA}$; and $\frac{SI^2 \times ALB}{3\sqrt{2SI}}$ into $\frac{1}{\sqrt{LA^3}} - \frac{1}{\sqrt{LB^3}}$. And these after due reduction come forth $\frac{2SI^2 \times SL}{LI}$, SI^2, and SI^2 +

$\frac{2SI^3}{3LI}$. And these by subducting the last from the first, become $\frac{4SI^3}{3LI}$. Therefore the entire force with which the corpuscle P is attracted towards the centre of the sphere is as $\frac{SI^3}{PI}$, that is, reciprocally as $PS^3 \times PI$. Q.E.I.

By the same method one may determine the attraction of a corpuscle situate within the sphere, but more expeditiously by the following Theorem.

PROPOSITION LXXXII. THEOREM XLI.

In a sphere described about the centre S *with the interval* SA, *if there be taken* SI, SA, SP *continually proportional; I say, that the attraction of a corpuscle within the sphere in any place* I *is to its attraction without the sphere in the place* P *in a ratio compounded of the subduplicate ratio of* IS, PS, *the distances from the centre, and the subduplicate ratio of the centripetal forces tending to the centre in those places* P *and* I.

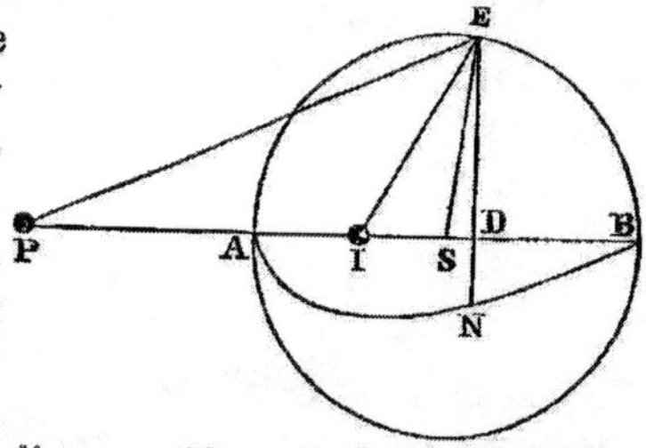

As if the centripetal forces of the particles of the sphere be reciprocally as the distances of the corpuscle attracted by them; the force with which the corpuscle situate in I is attracted by the entire sphere will be to the force with which it is attracted in P in a ratio compounded of the subduplicate ratio of the distance SI to the distance SP, and the subduplicate ratio of the centripetal force in the place I arising from any particle in the centre to the centripetal force in the place P arising from the same particle in the centre; that is, in the subduplicate ratio of the distances SI, SP to each other reciprocally. These two subduplicate ratios compose the ratio of equality, and therefore the attractions in I and P produced by the whole sphere are equal. By the like calculation, if the forces of the particles of the sphere are reciprocally in a duplicate ratio of the distances, it will be found that the attraction in I is to the attraction in P as the distance SP to the semi-diameter SA of the sphere. If those forces are reciprocally in a triplicate ratio of the distances, the attractions in I and P will be to each other as SP^2 to SA^2; if in a quadruplicate ratio, as SP^3 to SA^3. Therefore since the attraction in P was found in this last case to be reciprocally as $PS^3 \times PI$, the attraction in I will be reciprocally as $SA^3 \times PI$, that is, because SA^3 is given reciprocally as PI. And the progression is the same *in infinitum*. The demonstration of this Theorem is as follows:

The things remaining as above constructed, and a corpuscle being in any place P, the ordinate DN was found to be as $\frac{DE^2 \times PS}{PE \times V}$. Therefore if IE be drawn, that ordinate for any other place of the corpuscle, as I, will become (*mutatis mutandis*) as $\frac{DE^2 \times IS}{IE \times V}$. Suppose the centripetal forces flowing from any point of the sphere, as E, to be to each other at the distances IE and PE as PE^n to IE^n (where the number n denotes the index of the powers of PE and IE), and those ordinates will become as $\frac{DE^2 \times PS}{PE \times PE^n}$ and $\frac{DE^2 \times IS}{IE \times IE^n}$ whose ratio to each other is as $PS \times IE \times IE^n$ to $IS \times PE \times PE^n$. Because SI, SE, SP are in continued proportion, the triangles SPE, SEI are alike; and thence IE is to PE as IS to SE or SA. For the ratio of IE to PE write the ratio of IS to SA; and the ratio of the ordinates becomes that of $PS \times IE^n$ to $SA \times PE^n$. But the ratio of PS to SA is subduplicate of that of the distances PS, SI; and the ratio of IE^n to PE^n (because IE is to PE as IS to SA) is subduplicate of that of the forces at the distances PS, IS. Therefore the ordinates, and consequently the areas which the ordinates describe, and the attractions proportional to them, are in a ratio compounded of those subduplicate ratios. Q.E.D.

PROPOSITION LXXXIII. PROBLEM XLII.

To find the force with which a corpuscle placed in the centre of a sphere is attracted towards any segment of that sphere whatsoever.

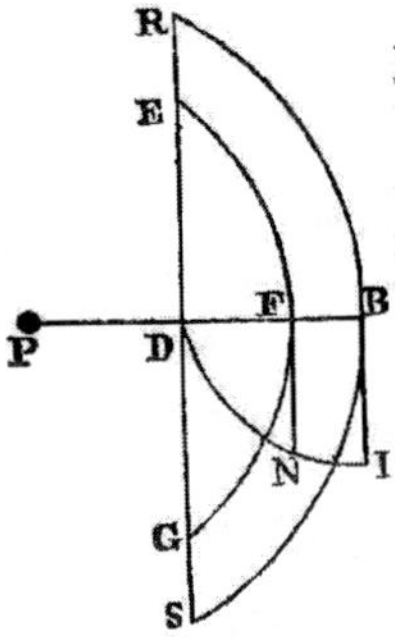

Let P be a body in the centre of that sphere and RBSD a segment thereof contained under the plane RDS, and the sphærical superficies RBS. Let DB be cut in F by a sphærical superficies EFG described from the centre P, and let the segment be divided into the parts BREFGS, FEDG. Let us suppose that segment to be not a purely mathematical but a physical superficies, having some, but a perfectly inconsiderable thickness. Let that thickness be called O, and (by what *Archimedes* has demonstrated) that superficies will be as $PF \times DF \times O$. Let us suppose besides the attractive forces of the particles of the sphere to be reciprocally as that power of the distances, of which n is index; and the force with which the superficies EFG attracts the body P will be (by Prop. LXXIX) as $\frac{DE^2 \times O}{PF^n}$, that is, as $\frac{2DF \times O}{PF^{n-1}} - \frac{DF^2 \times O}{PF^n}$. Let the perpendicular FN drawn into

O be proportional to this quantity; and the curvilinear area BDI, which the ordinate FN, drawn through the length DB with a continued motion will describe, will be as the whole force with which the whole segment RBSD attracts the body P. Q.E.I.

PROPOSITION LXXXIV. PROBLEM XLIII.

To find the force with which a corpuscle, placed without the centre of a sphere in the axis of any segment, is attracted by that segment.

Let the body P placed in the axis ADB of the segment EBK be attracted by that segment. About the centre P, with the interval PE, let the sphærical superficies EFK be described; and let it divide the segment into two parts EBKFE and EFKDE. Find the force of the first of those parts by Prop. LXXXI, and the force of the latter part by Prop. LXXXIII, and the sum of the forces will be the force of the whole segment EBKDE. Q.E.I.

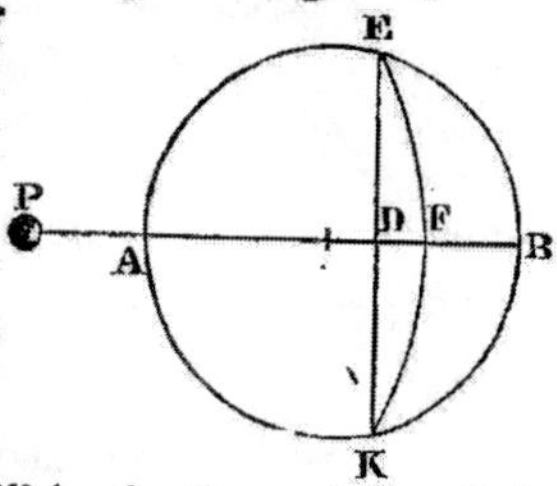

SCHOLIUM.

The attractions of sphærical bodies being now explained, it comes next in order to treat of the laws of attraction in other bodies consisting in like manner of attractive particles; but to treat of them particularly is not necessary to my design. It will be sufficient to subjoin some general propositions relating to the forces of such bodies, and the motions thence arising, because the knowledge of these will be of some little use in philosophical inquiries.

SECTION XIII.

Of the attractive forces of bodies which are not of a sphærical figure.

PROPOSITION LXXXV. THEOREM XLII.

If a body be attracted by another, and its attraction be vastly stronger when it is contiguous to the attracting body than when they are separated from one another by a very small interval; the forces of the particles of the attracting body decrease, in the recess of the body attracted, in more than a duplicate ratio of the distance of the particles.

For if the forces decrease in a duplicate ratio of the distances from the particles, the attraction towards a sphærical body being (by Prop. LXXIV) reciprocally as the square of the distance of the attracted body from the centre of the sphere, will not be sensibly increased by the contact, and it will be still less increased by it, if the attraction, in the recess of the body attracted, decreases in a still less proportion. The proposition, therefore, is evident concerning attractive spheres. And the case is the same of concave sphærical orbs attracting external bodies. And much more does it appear in orbs that attract bodies placed within them, because there the attractions diffused through the cavities of those orbs are (by Prop. LXX) destroyed by contrary attractions, and therefore have no effect even in the place of contact. Now if from these spheres and sphærical orbs we take away any parts remote from the place of contact, and add new parts any where at pleasure, we may change the figures of the attractive bodies at pleasure; but the parts added or taken away, being remote from the place of contact, will cause no remarkable excess of the attraction arising from the contact of the two bodies. Therefore the proposition holds good in bodies of all figures. Q.E.D.

PROPOSITION LXXXVI. THEOREM XLIII.

If the forces of the particles of which an attractive body is composed decrease, in the recess of the attractive body, in a triplicate or more than a triplicate ratio of the distance from the particles, the attraction will be vastly stronger in the point of contact than when the attracting and attracted bodies are separated from each other, though by never so small an interval.

For that the attraction is infinitely increased when the attracted corpuscle comes to touch an attracting sphere of this kind, appears, by the solution of Problem XLI, exhibited in the second and third Examples. The same will also appear (by comparing those Examples and Theorem XLI together) of attractions of bodies made towards concavo-convex orbs, whether the attracted bodies be placed without the orbs, or in the cavities within them. And by adding to or taking from those spheres and orbs any attractive matter any where without the place of contact, so that the attractive bodies may receive any assigned figure, the Proposition will hold good of all bodies universally. Q.E.D.

PROPOSITION LXXXVII. THEOREM XLIV.

If two bodies similar to each other, and consisting of matter equally attractive, attract separately two corpuscles proportional to those bodies, and in a like situation to them, the accelerative attractions of the corpuscles towards the entire bodies will be as the accelerative attractions of the corpuscles towards particles of the bodies proportional to the wholes, and alike situated in them.

For if the bodies are divided into particles proportional to the wholes, and alike situated in them, it will be, as the attraction towards any particle of one of the bodies to the attraction towards the correspondent particle in the other body, so are the attractions towards the several particles of the first body, to the attractions towards the several correspondent particles of the other body; and, by composition, so is the attraction towards the first whole body to the attraction towards the second whole body. Q.E.D.

COR. 1. Therefore if, as the distances of the corpuscles attracted increase, the attractive forces of the particles decrease in the ratio of any power of the distances, the accelerative attractions towards the whole bodies will be as the bodies directly, and those powers of the distances inversely. As if the forces of the particles decrease in a duplicate ratio of the distances from the corpuscles attracted, and the bodies are as A^3 and B^3, and therefore both the cubic sides of the bodies, and the distance of the attracted corpuscles from the bodies, are as A and B; the accelerative attractions towards the bodies will be as $\frac{A^3}{A^2}$ and $\frac{B^3}{B^2}$, that is, as A and B the cubic sides of those bodies. If the forces of the particles decrease in a triplicate ratio of the distances from the attracted corpuscles, the accelerative attractions towards the whole bodies will be as $\frac{A^3}{A^3}$ and $\frac{B^3}{B^3}$, that is, equal. If the forces decrease in a quadruplicate ratio, the attractions towards the bodies will be as $\frac{A^3}{A^4}$ and $\frac{B^3}{B^4}$, that is, reciprocally as the cubic sides A and B. And so in other cases.

COR. 2. Hence, on the other hand, from the forces with which like bodies attract corpuscles similarly situated, may be collected the ratio of the decrease of the attractive forces of the particles as the attracted corpuscle recedes from them; if so be that decrease is directly or inversely in any ratio of the distances.

PROPOSITION LXXXVIII. THEOREM XLV.

If the attractive forces of the equal particles of any body be as the distance of the places from the particles, the force of the whole body will tend to its centre of gravity; and will be the same with the force of a globe, consisting of similar and equal matter, and having its centre in the centre of gravity.

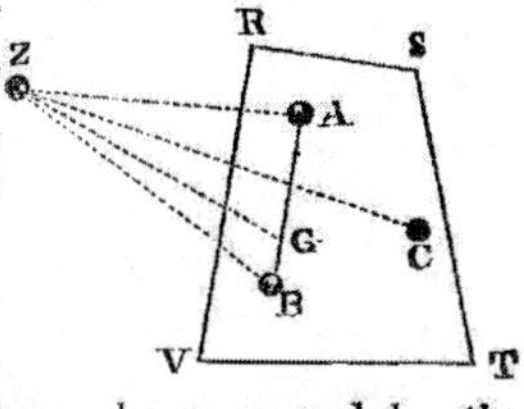

Let the particles A, B, of the body RSTV attract any corpuscle Z with forces which, supposing the particles to be equal between themselves, are as the distances AZ, BZ; but, if they are supposed unequal, are as those particles and their distances AZ, BZ, conjunctly, or (if I may so speak) as those particles drawn into their distances AZ, BZ respectively. And let those forces be expressed by the contents under A × AZ, and B × BZ. Join AB, and let it be cut in G, so that AG may be to BG as the particle B to the particle A; and G will be the common centre of gravity of the particles A and B. The force A × AZ will (by Cor. 2, of the Laws) be resolved into the forces A × GZ and A × AG; and the force B × BZ into the forces B × GZ and B × BG. Now the forces A × AG and B × BG, because A is proportional to B, and BG to AG, are equal, and therefore having contrary directions destroy one another. There remain then the forces A × GZ and B × GZ. These tend from Z towards the centre G, and compose the force $\overline{A + B}$ × GZ; that is, the same force as if the attractive particles A and B were placed in their common centre of gravity G, composing there a little globe.

By the same reasoning, if there be added a third particle C, and the force of it be compounded with the force $\overline{A + B}$ × GZ tending to the centre G, the force thence arising will tend to the common centre of gravity of that globe in G and of the particle C; that is, to the common centre of gravity of the three particles A, B, C; and will be the same as if that globe and the particle C were placed in that common centre composing a greater globe there; and so we may go on *in infinitum.* Therefore the whole force of all the particles of any body whatever RSTV is the same as if that body, without removing its centre of gravity, were to put on the form of a globe. Q.E.D.

Cor. Hence the motion of the attracted body Z will be the same as if the attracting body RSTV were sphærical; and therefore if that attracting body be either at rest, or proceed uniformly in a right line, the body attracted will move in an ellipsis having its centre in the centre of gravity of the attracting body.

PROPOSITION LXXXIX. THEOREM XLVI.

If there be several bodies consisting of equal particles whose forces are as the distances of the places from each, the force compounded of all the forces by which any corpuscle is attracted will tend to the common centre of gravity of the attracting bodies; and will be the same as if those attracting bodies, preserving their common centre of gravity, should unite there, and be formed into a globe.

This is demonstrated after the same manner as the foregoing Proposition.

Cor. Therefore the motion of the attracted body will be the same as if the attracting bodies, preserving their common centre of gravity, should unite there, and be formed into a globe. And, therefore, if the common centre of gravity of the attracting bodies be either at rest, or proceed uniformly in a right line, the attracted body will move in an ellipsis having its centre in the common centre of gravity of the attracting bodies.

PROPOSITION XC. PROBLEM XLIV.

If to the several points of any circle there tend equal centripetal forces, increasing or decreasing in any ratio of the distances; it is required to find the force with which a corpuscle is attracted, that is, situate any where in a right line which stands at right angles to the plane of the circle at its centre.

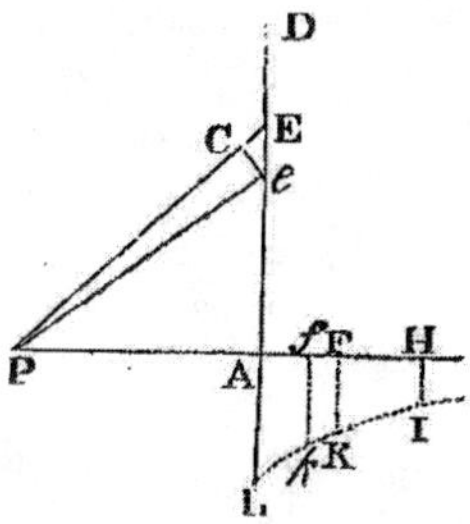

Suppose a circle to be described about the centre A with any interval AD in a plane to which the right line AP is perpendicular; and let it be required to find the force with which a corpuscle P is attracted towards the same. From any point E of the circle, to the attracted corpuscle P, let there be drawn the right line PE. In the right line PA take PF equal to PE, and make a perpendicular FK, erected at F, to be as the force with which the point E attracts the corpuscle P. And let the curve line IKL be the locus of the point K. Let that curve meet the plane of the circle in L. In PA take PH equal to PD, and erect the perpendicular HI meeting that curve in I; and the attraction of the corpuscle P towards the circle will be as the area AHIL drawn into the altitude AP. Q.E.I.

For let there be taken in AE a very small line E*e*. Join P*e*, and in PE, PA take PC, P*f* equal to P*e*. And because the force, with which any point E of the annulus described about the centre A with the interval AE in the aforesaid plane attracts to itself the body P, is supposed to be as FK; and, therefore, the force with which that point attracts the body P towards A is as $\frac{AP \times FK}{PE}$; and the force with which the whole annulus attracts the body P towards A is as the annulus and $\frac{AP \times FK}{PE}$ conjunctly; and that annulus also is as the rectangle under the radius AE and the breadth E*e*, and this rectangle (because PE and AE, E*e* and CE are proportional) is equal to the rectangle PE × CE or PE × F*f*; the force with which that annulus attracts the body P towards A will be as PE × F*f* and $\frac{AP \times FK}{PE}$ conjunctly; that is, as the content under F*f* × FK × AP, or as the area FK*kf* drawn into AP. And therefore the sum of the forces with which all the annuli, in the circle described about the centre A with the interval AD, attract the body P towards A, is as the whole area AHIKL drawn into AP. Q.E.D.

COR. 1. Hence if the forces of the points decrease in the duplicate ratio of the distances, that is, if FK be as $\frac{1}{PF^2}$, and therefore the area AHIKL as $\frac{1}{PA} - \frac{1}{PH}$; the attraction of the corpuscle P towards the circle will be as $1 - \frac{PA}{PH}$; that is, as $\frac{AH}{PH}$.

Cor. 2. And universally if the forces of the points at the distances D be reciprocally as any power D^n of the distances; that is, if FK be as $\frac{1}{D^n}$, and therefore the area AHIKL as $\frac{1}{PA^{n-1}} - \frac{1}{PH^{n-1}}$; the attraction of the corpuscle P towards the circle will be as $\frac{1}{PA^{n-2}} - \frac{PA}{PH^{n-1}}$.

Cor. 3. And if the diameter of the circle be increased *in infinitum*, and the number n be greater than unity; the attraction of the corpuscle P towards the whole infinite plane will be reciprocally as PA^{n-2}, because the other term $\frac{PA}{PH^{n-1}}$ vanishes.

PROPOSITION XCI. PROBLEM XLV.

To find the attraction of a corpuscle situate in the axis of a round solid, to whose several points there tend equal centripetal forces decreasing in any ratio of the distances whatsoever.

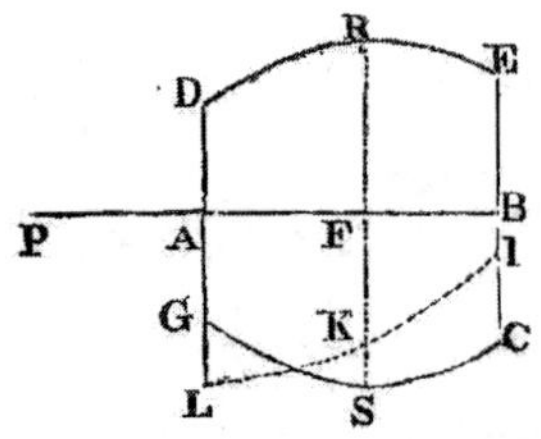

Let the corpuscle P, situate in the axis AB of the solid DECG, be attracted towards that solid. Let the solid be cut by any circle as RFS, perpendicular to the axis; and in its semi-diameter FS, in any plane PALKB passing through the axis, let there be taken (by Prop. XC) the length FK proportional to the force with which the corpuscle P is attracted towards that circle. Let the locus of the point K be the curve line LKI, meeting the planes of the outermost circles AL and BI in L and I; and the attraction of the corpuscle P towards the solid will be as the area LABI. Q.E.I.

Cor. 1. Hence if the solid be a cylinder described by the parallelogram ADEB revolved about the axis AB, and the centripetal forces tending to the several points be reciprocally as the squares of the distances from the points; the attraction of the corpuscle P towards this cylinder will be as AB — PE + PD. For the ordinate FK (by Cor. 1, Prop. XC) will be as $1 - \frac{PF}{PR}$. The part 1 of this quantity, drawn into the length AB, describes the area $1 \times AB$; and the other part $\frac{PF}{PR}$, drawn into the length PB describes the area 1 into $\overline{PE - AD}$ (as may be easily shewn from the quadrature of the curve LKI); and, in like manner, the same part drawn into the length PA describes the area 1 into $\overline{PD - AD}$, and drawn into AB, the difference of PB and PA, describes 1 into $\overline{PE - PD}$, the difference of the areas. From the first content $1 \times AB$ take away the last content 1 into $\overline{PE - PD}$, and there will remain the area LABI equal to 1 into $\overline{AB - PE + PD}$. Therefore the force, being proportional to this area, is as $AB - PE + PD$.

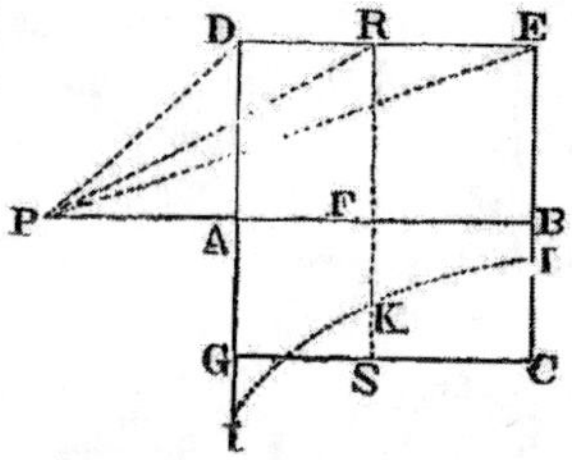

COR. 2. Hence also is known the force by which a spheroid AGBC attracts any body P situate externally in its axis AB. Let NKRM be a conic section whose ordinate ER perpendicular to PE may be always equal to the length of the line PD, continually drawn to the point D in which that ordinate cuts the spheroid. From the vertices A, B, of the spheriod, let there be erected to its axis AB the perpendiculars AK, BM, respectively equal to AP, BP, and therefore meeting the conic section in K and M; and join KM cutting off from it the segment KMRK. Let S be the centre of the spheroid, and SC its greatest semi-diameter; and the force with which the spheroid attracts the body P will be to the force with which a sphere described with the diameter AB attracts the same body as $\frac{AS \times CS^2 - PS \times KMRK}{PS^2 + CS^2 - AS^2}$ is to $\frac{AS^3}{3PS^2}$. And by a calculation founded on the same principles may be found the forces of the segments of the spheroid.

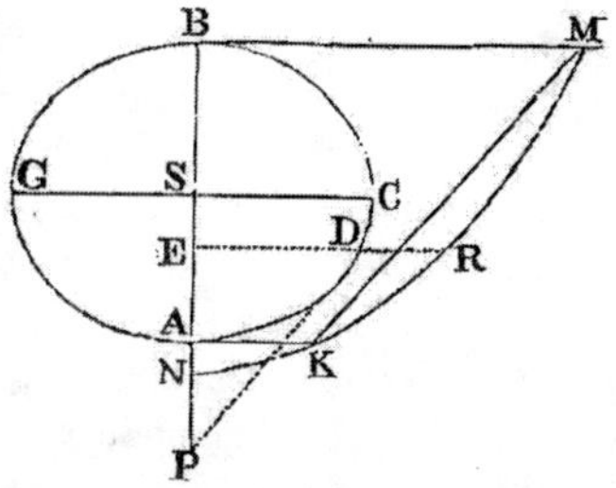

COR. 3. If the corpuscle be placed within the spheroid and in its axis, the attraction will be as its distance from the centre. This may be easily collected from the following reasoning, whether the particle be in the axis or in any other given diameter. Let AGOF be an attracting spheroid, S its centre, and P the body attracted. Through the body P let there be drawn the semi-diameter SPA, and two right lines DE, FG meeting the spheroid in D and E, F and G; and let PCM, HLN be the superficies of

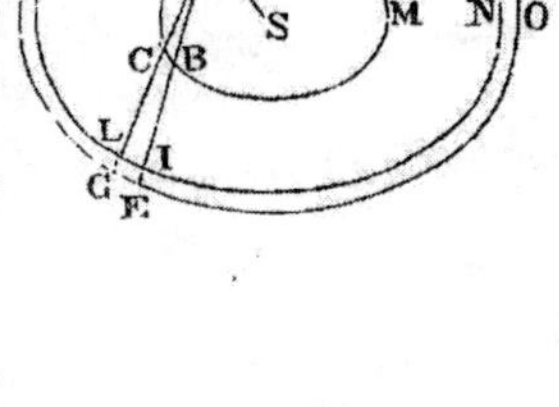

two interior spheroids similar and concentrical to the exterior, the first of which passes through the body P, and cuts the right lines DE, FG in B and C; and the latter cuts the same right lines in H and I, K and L. Let the spheroids have all one common axis, and the parts of the right lines intercepted on both sides DP and BE, FP and CG, DH and IE, FK and LG, will be mutually equal; because the right lines DE, PB, and HI, are bisected in the same point, as are also the right lines FG, PC, and KL. Conceive now DPF, EPG to represent opposite cones described with the infinitely small vertical angles DPF, EPG, and the lines DH, EI to be infinitely small also. Then the particles of the cones DHKF, GLIE, cut off by the spheroidical superficies, by reason of the equality of the lines DH and EI, will be to one another as the squares of the distances from the body P, and will therefore attract that corpuscle equally. And by a like reasoning if the spaces DPF, EGCB be divided into particles by the superficies of innumerable similar spheroids concentric to the former and having one common axis, all these particles will equally attract on both sides the body P towards contrary parts. Therefore the forces of the cone DPF, and of the conic segment EGCB, are equal, and by their contrariety destroy each other. And the case is the same of the forces of all the matter that lies without the interior spheroid PCBM. Therefore the body P is attracted by the interior spheroid PCBM alone, and therefore (by Cor. 3, Prop. LXXII) its attraction is to the force with which the body A is attracted by the whole spheroid AGOD as the distance PS to the distance AS. Q.E.D.

PROPOSITION XCII. PROBLEM XLVI.

An attracting body being given, it is required to find the ratio of the decrease of the centripetal forces tending to its several points.

The body given must be formed into a sphere, a cylinder, or some regular figure, whose law of attraction answering to any ratio of decrease may be found by Prop. LXXX, LXXXI, and XCI. Then, by experiments, the force of the attractions must be found at several distances, and the law of attraction towards the whole, made known by that means, will give the ratio of the decrease of the forces of the several parts; which was to be found.

PROPOSITION XCIII. THEOREM XLVII.

If a solid be plane on one side, and infinitely extended on all other sides, and consist of equal particles equally attractive, whose forces decrease, in the recess from the solid, in the ratio of any power greater than the square of the distances; and a corpuscle placed towards either part of the plane is attracted by the force of the whole solid; I say, that the attractive force of the whole solid, in the recess from its plane superfi-

cies, will decrease in the ratio of a power whose side is the distance of the corpuscle from the plane, and its index less by 3 than the index of the power of the distances.

Case 1. Let LGl be the plane by which the solid is terminated. Let the solid lie on that hand of the plane that is towards I, and let it be resolved into innumerable planes mHM, nIN, oKO, &c., parallel to GL. And first let the attracted body C be placed without the solid. Let there be drawn CGHI perpendicular to those innumerable planes,

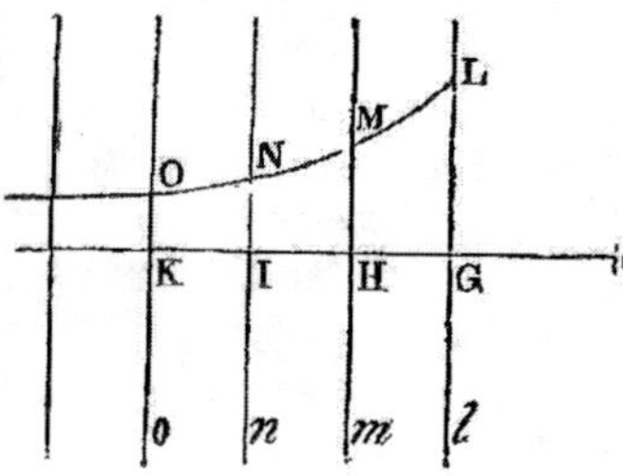

and let the attractive forces of the points of the solid decrease in the ratio of a power of the distances whose index is the number n not less than 3. Therefore (by Cor. 3, Prop. XC) the force with which any plane mHM attracts the point C is reciprocally as CH^{n-2}. In the plane mHM take the length HM reciprocally proportional to CH^{n-2}, and that force will be as HM. In like manner in the several planes lGL, nIN, oKO, &c., take the lengths GL, IN, KO, &c., reciprocally proportional to CG^{n-2}, CI^{n-2}, CK^{n-2}, &c., and the forces of those planes will be as the lengths so taken, and therefore the sum of the forces as the sum of the lengths, that is, the force of the whole solid as the area GLOK produced infinitely towards OK. But that area (by the known methods of quadratures) is reciprocally as CG^{n-3}, and therefore the force of the whole solid is reciprocally as CG^{n-3}. Q.E.D.

Case 2. Let the corpuscle C be now placed on that hand of the plane lGL that is within the solid, and take the distance CK equal to the distance CG. And the part of the solid LGloKO terminated by the parallel planes lGL, oKO, will attract the corpuscle C, situate in the middle, neither one way nor another, the contrary actions of the opposite points destroying one another by reason of

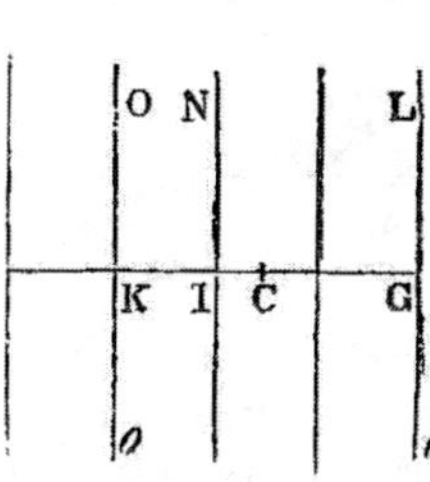

their equality. Therefore the corpuscle C is attracted by the force only of the solid situate beyond the plane OK. But this force (by Case 1) is reciprocally as CK^{n-3}, that is, (because CG, CK are equal) reciprocally as CG^{n-3}. Q.E.D.

Cor. 1. Hence if the solid LGIN be terminated on each side by two infinite parallel planes LG, IN, its attractive force is known, subducting from the attractive force of the whole infinite solid LGKO the attractive force of the more distant part NIKO infinitely produced towards KO.

Cor. 2. If the more distant part of this solid be rejected, because its attraction compared with the attraction of the nearer part is inconsiderable, the attraction of that nearer part will, as the distance increases, decrease nearly in the ratio of the power CG^{n-3}.

Cor. 3. And hence if any finite body, plane on one side, attract a corpuscle situate over against the middle of that plane, and the distance between the corpuscle and the plane compared with the dimensions of the attracting body be extremely small; and the attracting body consist of homogeneous particles, whose attractive forces decrease in the ratio of any power of the distances greater than the quadruplicate; the attractive force of the whole body will decrease very nearly in the ratio of a power whose side is that very small distance, and the index less by 3 than the index of the former power. This assertion does not hold good, however, of a body consisting of particles whose attractive forces decrease in the ratio of the triplicate power of the distances; because, in that case, the attraction of the remoter part of the infinite body in the second Corollary is always infinitely greater than the attraction of the nearer part.

SCHOLIUM.

If a body is attracted perpendicularly towards a given plane, and from the law of attraction given, the motion of the body be required; the Problem will be solved by seeking (by Prop. XXXIX) the motion of the body descending in a right line towards that plane, and (by Cor. 2, of the Laws) compounding that motion with an uniform motion performed in the direction of lines parallel to that plane. And, on the contrary, if there be required the law of the attraction tending towards the plane in perpendicular directions, by which the body may be caused to move in any given curve line, the Problem will be solved by working after the manner of the third Problem.

But the operations may be contracted by resolving the ordinates into converging series. As if to a base A the length B be ordinately applied in any given angle, and that length be as any power of the base $A^{\frac{m}{n}}$; and there be sought the force with which a body, either attracted towards the base or driven from it in the direction of that ordinate, may be caused to move in the curve line which that ordinate always describes with its superior extremity; I suppose the base to be increased by a very small part O, and I resolve the ordinate $\overline{A+O}|^{\frac{m}{n}}$ into an infinite series $A^{\frac{m}{n}} + \frac{m}{n}OA^{\frac{m-n}{n}} + \frac{mm-mn}{2nn}OOA^{\frac{m-2n}{n}}$ &c., and I suppose the force proportional to the term of this series in which O is of two dimensions, that is, to the term $\frac{mm-mn}{2nn}OOA^{\frac{m-2n}{n}}$. Therefore the force sought is as

$\frac{mm - mn}{nn} A^{\frac{m-2n}{n}}$, or, which is the same thing, as $\frac{mm - mn}{nn} B^{\frac{m-2n}{m}}$. As if the ordinate describe a parabola, m being $= 2$, and $n = 1$, the force will be as the given quantity $2B^{\circ}$, and therefore is given. Therefore with a given force the body will move in a parabola, as *Galileo* has demonstrated. If the ordinate describe an hyperbola, m being $= 0 - 1$, and $n = 1$, the force will be as $2A^{-3}$ or $2B^{3}$; and therefore a force which is as the cube of the ordinate will cause the body to move in an hyperbola. But leaving this kind of propositions, I shall go on to some others relating to motion which I have not yet touched upon.

SECTION XIV.

Of the motion of very small bodies when agitated by centripetal forces tending to the several parts of any very great body.

PROPOSITION XCIV. THEOREM XLVIII.

If two similar mediums be separated from each other by a space terminated on both sides by parallel planes, and a body in its passage through that space be attracted or impelled perpendicularly towards either of those mediums, and not agitated or hindered by any other force; and the attraction be every where the same at equal distances from either plane, taken towards the same hand of the plane; I say, that the sine of incidence upon either plane will be to the sine of emergence from the other plane in a given ratio.

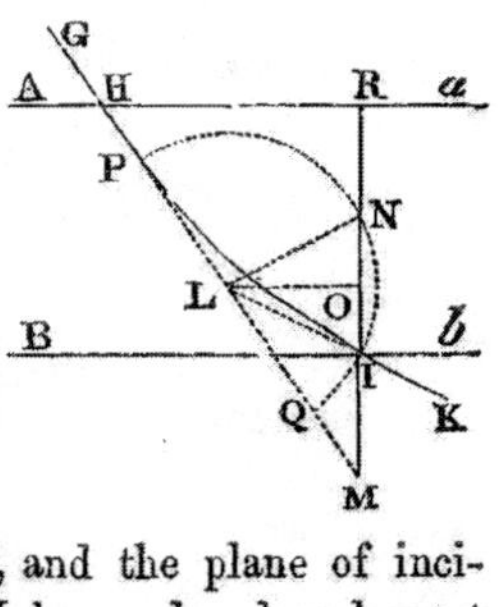

CASE 1. Let Aa and Bb be two parallel planes, and let the body light upon the first plane Aa in the direction of the line GH, and in its whole passage through the intermediate space let it be attracted or impelled towards the medium of incidence, and by that action let it be made to describe a curve line HI, and let it emerge in the direction of the line IK. Let there be erected IM perpendicular to Bb the plane of emergence, and meeting the line of incidence GH prolonged in M, and the plane of incidence Aa in R; and let the line of emergence KI be produced and meet HM in L. About the centre L, with the interval LI, let a circle be described cutting both HM in P and Q, and MI produced in N; and, first, if the attraction or impulse be supposed uniform, the curve HI (by what *Galileo* has demonstrated) be a parabola, whose property is that of a rec-

tangle under its given latus rectum and the line IM is equal to the square of HM; and moreover the line HM will be bisected in L. Whence if to MI there be let fall the perpendicular LO, MO, OR will be equal; and adding the equal lines ON, OI, the wholes MN, IR will be equal also. Therefore since IR is given, MN is also given, and the rectangle NMI is to the rectangle under the latus rectum and IM, that is, to HM^2 in a given ratio. But the rectangle NMI is equal to the rectangle PMQ, that is, to the difference of the squares ML^2, and PL^2 or LI^2; and HM^2 hath a given ratio to its fourth part ML^2; therefore the ratio of $ML^2—LI^2$ to ML^2 is given, and by conversion the ratio of LI^2 to ML^2, and its subduplicate, the ratio of LI to ML. But in every triangle, as LMI, the sines of the angles are proportional to the opposite sides. Therefore the ratio of the sine of the angle of incidence LMR to the sine of the angle of emergence LIR is given. Q.E.D.

Case 2. Let now the body pass successively through several spaces terminated with parallel planes *AabB*, *BbcC*, &c., and let it be acted on by a force which is uniform in each of them separately, but different in the different spaces; and by what was just demonstrated, the sine of the angle of incidence on the first plane *Aa* is to the sine of emergence from the second plane *Bb* in a given ratio; and this sine of incidence upon the second plane *Bb* will be to the sine of emergence from the third plane *Cc* in a given ratio; and this sine to the sine of emergence from the fourth plane *Dd* in a given ratio; and so on *in infinitum;* and, by equality, the sine of incidence on the first plane to the sine of emergence from the last plane in a given ratio. Let now the intervals of the planes be diminished, and their number be infinitely increased, so that the action of attraction or impulse, exerted according to any assigned law, may become continual, and the ratio of the sine of incidence on the first plane to the sine of emergence from the last plane being all along given, will be given then also. Q.E.D.

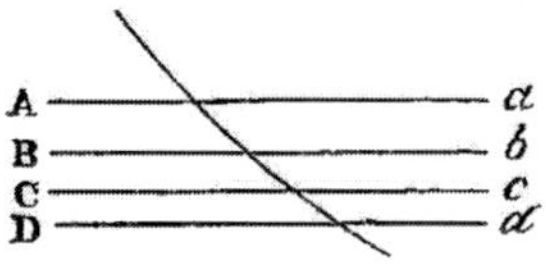

PROPOSITION XCV. THEOREM XLIX.

The same things being supposed, I say, that the velocity of the body before its incidence is to its velocity after emergence as the sine of emergence to the sine of incidence.

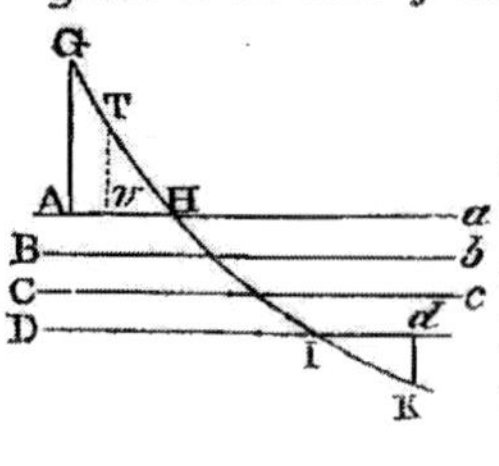

Make AH and *Id* equal, and erect the perpendiculars AG, *d*K meeting the lines of incidence and emergence GH, IK, in G and K. In GH take TH equal to IK, and to the plane *Aa* let fall a perpendicular T*v*. And (by Cor. 2 of the Laws of Motion) let the motion of the body be resolved into two, one perpendicular to the planes

Aa, Bb, Cc, &c, and another parallel to them. The force of attraction or impulse, acting in directions perpendicular to those planes, does not at all alter the motion in parallel directions; and therefore the body proceeding with this motion will in equal times go through those equal parallel intervals that lie between the line AG and the point H, and between the point I and the line dK; that is, they will describe the lines GH, IK in equal times. Therefore the velocity before incidence is to the velocity after emergence as GH to IK or TH, that is, as AH or Id to vH, that is (supposing TH or IK radius), as the sine of emergence to the sine of incidence. Q.E.D.

PROPOSITION XCVI. THEOREM L.

The same things being supposed, and that the motion before incidence is swifter than afterwards; I say, that if the line of incidence be inclined continually, the body will be at last reflected, and the angle of reflexion will be equal to the angle of incidence.

For conceive the body passing between the parallel planes Aa, Bb, Cc, &c., to describe parabolic arcs as above; and let those arcs be HP, PQ, QR, &c. And let the obliquity of the line of incidence GH to the first plane Aa be such

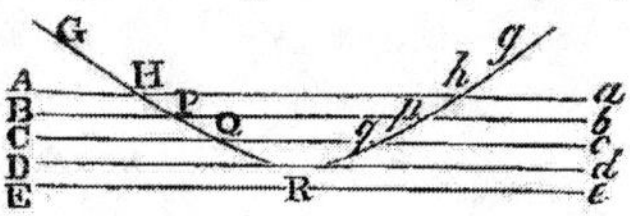

that the sine of incidence may be to the radius of the circle whose sine it is, in the same ratio which the same sine of incidence hath to the sine of emergence from the plane Dd into the space DdeE; and because the sine of emergence is now become equal to radius, the angle of emergence will be a right one, and therefore the line of emergence will coincide with the plane Dd. Let the body come to this plane in the point R; and because the line of emergence coincides with that plane, it is manifest that the body can proceed no farther towards the plane Ee. But neither can it proceed in the line of emergence Rd; because it is perpetually attracted or impelled towards the medium of incidence. It will return, therefore, between the planes Cc, Dd, describing an arc of a parabola QRq, whose principal vertex (by what *Galileo* has demonstrated) is in R, cutting the plane Cc in the same angle at q, that it did before at Q; then going on in the parabolic arcs qp, ph, &c., similar and equal to the former arcs QP, PH, &c., it will cut the rest of the planes in the same angles at p, h, &c., as it did before in P, H, &c., and will emerge at last with the same obliquity at h with which it first impinged on that plane at H. Conceive now the intervals of the planes Aa, Bb, Cc, Dd, Ee, &c., to be infinitely diminished, and the number infinitely increased, so that the action of attraction or impulse, exerted according to any assigned law, may become continual; and, the angle of emergence remaining all along equal to the angle of incidence, will be equal to the same also at last. Q.E.D.

SCHOLIUM.

These attractions bear a great resemblance to the reflexions and refractions of light made in a given ratio of the secants, as was discovered by *Snellius*; and consequently in a given ratio of the sines, as was exhibited by *Des Cartes*. For it is now certain from the phenomena of *Jupiter's* satellites, confirmed by the observations of different astronomers, that light is propagated in succession, and requires about seven or eight minutes to travel from the sun to the earth. Moreover, the rays of light that are in our air (as lately was discovered by *Grimaldus*, by the admission of light into a dark room through a small hole, which I have also tried) in their passage near the angles of bodies, whether transparent or opaque (such as the circular and rectangular edges of gold, silver and brass coins, or of knives, or broken pieces of stone or glass), are bent or inflected round those bodies as if they were attracted to them; and those rays which in their passage come nearest to the bodies are the most inflected, as if they were most attracted; which thing I myself have also carefully observed. And those which pass at greater distances are less inflected; and those at still greater distances are a little inflected the contrary way, and form three fringes of colours. In the figure *s* represents the edge of a knife, or any

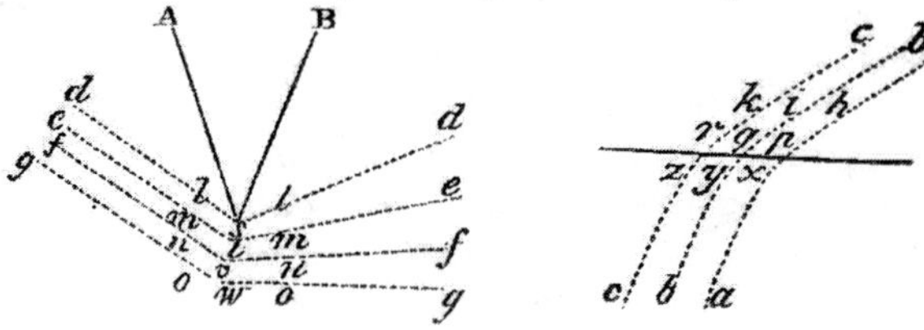

kind of wedge A*s*B; and *gowog*, *fnunf*, *emtme*, *dlsld*, are rays inflected towards the knife in the arcs *owo*, *nvn*, *mtm*, *lsl*; which inflection is greater or less according to their distance from the knife. Now since this inflection of the rays is performed in the air without the knife, it follows that the rays which fall upon the knife are first inflected in the air before they touch the knife. And the case is the same of the rays falling upon glass. The refraction, therefore, is made not in the point of incidence, but gradually, by a continual inflection of the rays; which is done partly in the air before they touch the glass, partly (if I mistake not) within the glass, after they have entered it; as is represented in the rays *ckzc*, *biyb*, *ahxa*, falling upon *r*, *q*, *p*, and inflected between *k* and *z*, *i* and *y*, *h* and *x*. Therefore because of the analogy there is between the propagation of the rays of light and the motion of bodies, I thought it not amiss to add the following Propositions for optical uses; not at all considering the nature of the rays of light, or inquiring whether they are bodies or not; but only determining the trajectories of bodies which are extremely like the trajectories of the rays.

PROPOSITION XCVII. PROBLEM XLVII.

Supposing the sine of incidence upon any superficies to be in a given ratio to the sine of emergence; and that the inflection of the paths of those bodies near that superficies is performed in a very short space, which may be considered as a point; it is required to determine such a superficies as may cause all the corpuscles issuing from any one given place to converge to another given place.

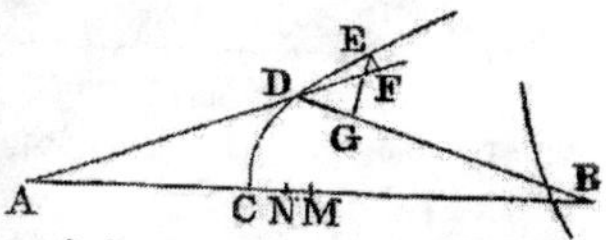

Let A be the place from whence the corpuscles diverge; B the place to which they should converge; CDE the curve line which by its revolution round the axis AB describes the superficies sought; D, E, any two points of that curve; and EF, EG, perpendiculars let fall on the paths of the bodies AD, DB. Let the point D approach to and coalesce with the point E; and the ultimate ratio of the line DF by which AD is increased, to the line DG by which DB is diminished, will be the same as that of the sine of incidence to the sine of emergence. Therefore the ratio of the increment of the line AD to the decrement of the line DB is given; and therefore if in the axis AB there be taken any where the point C through which the curve CDE must pass, and CM the increment of AC be taken in that given ratio to CN the decrement of BC, and from the centres A, B, with the intervals AM, BN, there be described two circles cutting each other in D; that point D will touch the curve sought CDE, and, by touching it any where at pleasure, will determine that curve. Q.E.I.

COR. 1. By causing the point A or B to go off sometimes *in infinitum*, and sometimes to move towards other parts of the point C, will be obtained all those figures which *Cartesius* has exhibited in his Optics and Geometry relating to refractions. The invention of which *Cartesius* having thought fit to conceal, is here laid open in this Proposition.

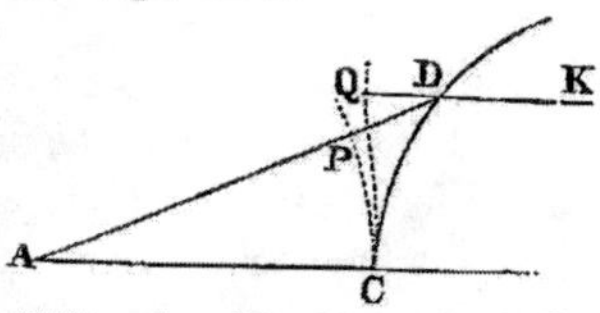

COR. 2. If a body lighting on any superficies CD in the direction of a right line AD, drawn according to any law, should emerge in the direction of another right line DK; and from the point C there be drawn curve lines CP, CQ, always perpendicular to AD, DK; the increments of the lines PD, QD, and therefore the lines themselves PD, QD, generated by those increments, will be as the sines of incidence and emergence to each other, and *è contra.*

PROPOSITION XCVIII. PROBLEM XLVIII.

The same things supposed; if round the axis AB *any attractive superficies be described as* CD, *regular or irregular, through which the bodies issuing from the given place* A *must pass; it is required to find a second attractive superficies* EF, *which may make those bodies converge to a given place* B.

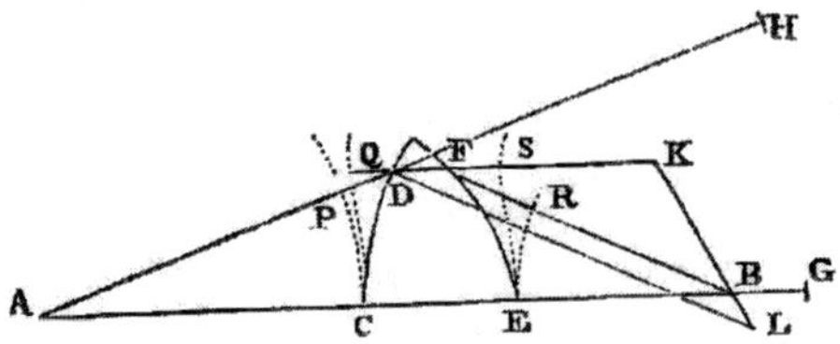

Let a line joining AB cut the first superficies in C and the second in E, the point D being taken any how at pleasure. And supposing the sine of incidence on the first superficies to the sine of emergence from the same, and the sine of emergence from the second superficies to the sine of incidence on the same, to be as any given quantity M to another given quantity N; then produce AB to G, so that BG may be to CE as M — N to N; and AD to H, so that AH may be equal to AG; and DF to K, so that DK may be to DH as N to M. Join KB, and about the centre D with the interval DH describe a circle meeting KB produced in L, and draw BF parallel to DL; and the point F will touch the line EF, which, being turned round the axis AB, will describe the superficies sought. Q.E.F.

For conceive the lines CP, CQ to be every where perpendicular to AD, DF, and the lines ER, ES to FB, FD respectively, and therefore QS to be always equal to CE; and (by Cor. 2, Prop. XCVII) PD will be to QD as M to N, and therefore as DL to DK, or FB to FK; and by division as DL — FB or PH — PD — FB to FD or FQ — QD; and by composition as PH — FB to FQ, that is (because PH and CG, QS and CE, are equal), as CE + BG — FR to CE — FS. But (because BG is to CE as M — N to N) it comes to pass also that CE + BG is to CE as M to N; and therefore, by division, FR is to FS as M to N; and therefore (by Cor. 2, Prop. XCVII) the superficies EF compels a body, falling upon it in the direction DF, to go on in the line FR to the place B. Q.E.D.

SCHOLIUM.

In the same manner one may go on to three or more superficies. But of all figures the sphærical is the most proper for optical uses. If the object glasses of telescopes were made of two glasses of a sphærical figure, containing water between them, it is not unlikely that the errors of the refractions made in the extreme parts of the superficies of the glasses may be accurately enough corrected by the refractions of the water. Such object glasses are to be preferred before elliptic and hyperbolic glasses, not only because they may be formed with more ease and accuracy, but because the pencils of rays situate without the axis of the glass would be more accurately refracted by them. But the different refrangibility of different rays is the real obstacle that hinders optics from being made perfect by sphærical or any other figures. Unless the errors thence arising can be corrected, all the labour spent in correcting the others is quite thrown away.

BOOK II.

BOOK II.

OF THE MOTION OF BODIES.

SECTION I.

Of the motion of bodies that are resisted in the ratio of the velocity.

PROPOSITION I. THEOREM I.

If a body is resisted in the ratio of its velocity, the motion lost by resistance is as the space gone over in its motion.

For since the motion lost in each equal particle of time is as the velocity, that is, as the particle of space gone over, then, by composition, the motion lost in the whole time will be as the whole space gone over. Q.E.D.

COR. Therefore if the body, destitute of all gravity, move by its innate force only in free spaces, and there be given both its whole motion at the beginning, and also the motion remaining after some part of the way is gone over, there will be given also the whole space which the body can describe in an infinite time. For that space will be to the space now described as the whole motion at the beginning is to the part lost of that motion.

LEMMA I.

Quantities proportional to their differences are continually proportional.

Let A be to A — B as B to B — C and C to C — D, &c., and, by conversion, A will be to B as B to C and C to D, &c. Q.E.D.

PROPOSITION II. THEOREM II.

If a body is resisted in the ratio of its velocity, and moves, by its vis insita *only, through a similar medium, and the times be taken equal, the velocities in the beginning of each of the times are in a geometrical progression, and the spaces described in each of the times are as the velocities.*

CASE 1. Let the time be divided into equal particles; and if at the very beginning of each particle we suppose the resistance to act with one single impulse which is as the velocity, the decrement of the velocity in each of the particles of time will be as the same velocity. Therefore the velocities are proportional to their differences, and therefore (by Lem. 1, Book II) continually proportional. Therefore if out of an equal number of particles there be compounded any equal portions of time, the velocities at the beginning of those times will be as terms in a continued progression, which are taken by intervals, omitting every where an equal number of intermediate terms. But the ratios of these terms are compounded of the equal ratios of the intermediate terms equally repeated, and therefore are equal. Therefore the velocities, being proportional to those terms, are in geometrical progression. Let those equal particles of time be diminished, and their number increased *in infinitum*, so that the impulse of resistance may become continual; and the velocities at the beginnings of equal times, always continually proportional, will be also in this case continually proportional. Q.E.D.

Case 2. And, by division, the differences of the velocities, that is, the parts of the velocities lost in each of the times, are as the wholes; but the spaces described in each of the times are as the lost parts of the velocities (by Prop. 1, Book I), and therefore are also as the wholes. Q.E.D.

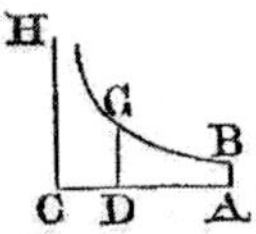

Corol. Hence if to the rectangular asymptotes AC, CH, the hyperbola BG is described, and AB, DG be drawn perpendicular to the asymptote AC, and both the velocity of the body, and the resistance of the medium, at the very beginning of the motion, be expressed by any given line AC, and, after some time is elapsed, by the indefinite line DC; the time may be expressed by the area ABGD, and the space described in that time by the line AD. For if that area, by the motion of the point D, be uniformly increased in the same manner as the time, the right line DC will decrease in a geometrical ratio in the same manner as the velocity; and the parts of the right line AC, described in equal times, will decrease in the same ratio.

PROPOSITION III. PROBLEM I.

To define the motion of a body which, in a similar medium, ascends or descends in a right line, and is resisted in the ratio of its velocity, and acted upon by an uniform force of gravity.

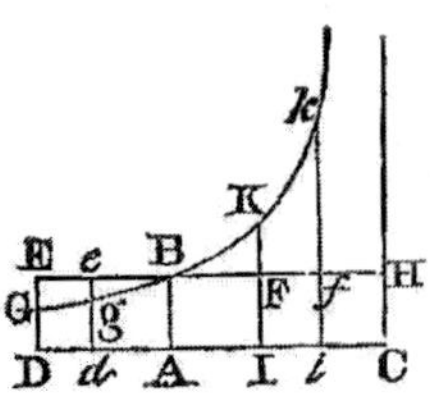

The body ascending, let the gravity be expounded by any given rectangle BACH; and the resistance of the medium, at the beginning of the ascent, by the rectangle BADE, taken on the contrary side of the right line AB. Through the point B, with the rectangular asymptotes AC, CH, describe an hyperbola, cutting the perpendiculars DE, *de*, in

G, *g*; and the body ascending will in the time DG*gd* describe the space EG*ge*; in the time DGBA, the space of the whole ascent EGB; in the time ABKI, the space of descent BFK; and in the time IK*ki* the space of descent KF*fk*; and the velocities of the bodies (proportional to the resistance of the medium) in these periods of time will be ABED, AB*ed*, O, ABFI, AB*fi* respectively; and the greatest velocity which the body can acquire by descending will be BACH.

For let the rectangle BACH be resolved into innumerable rectangles A*k*, K*l*, L*m*, M*n*, &c., which shall be as the increments of the velocities produced in so many equal times; then will 0, A*k*, A*l*, A*m*, A*n*, &c., be as the whole velocities, and therefore (by supposition) as the resistances of the medium in the beginning of each of the equal times. Make AC to AK, or ABHC to AB*k*K, as the force of gravity to the resistance in the beginning of the second time; then from the force of gravity subduct the resistances, and ABHC, K*k*HC, L*l*HC, M*m*HC, &c., will be as the absolute forces with which the body is acted upon in the beginning of each of the times, and therefore (by Law I) as the increments of the velocities, that is, as the rectangles A*k*, K*l*, L*m*, M*n*, &c., and therefore (by Lem. 1, Book II) in a geometrical progression. Therefore, if the right lines K*k*, L*l*, M*m*, N*n*, &c., are produced so as to meet the hyperbola in *q*, *r*, *s*, *t*, &c., the areas AB*q*K, K*qr*L, L*rs*M, M*st*N, &c., will be equal, and therefore analogous to the equal times and equal gravitating forces. But the area AB*q*K (by Corol. 3, Lem. VII and VIII, Book I) is to the area B*kq* as K*q* to $\frac{1}{2}$*kq*, or AC to $\frac{1}{2}$AK, that is, as the force of gravity to the resistance in the middle of the first time. And by the like reasoning, the areas *q*KL*r*, *r*LM*s*, *s*MN*t*, &c., are to the areas *qklr*, *rlms*, *smnt*, &c., as the gravitating forces to the resistances in the middle of the second, third, fourth time, and so on. Therefore since the equal areas BAK*q*, *q*KL*r*, *r*LM*s*, *s*MN*t*, &c., are analogous to the gravitating forces, the areas B*kq*, *qklr*, *rlms*, *smnt*, &c., will be analogous to the resistances in the middle of each of the times, that is (by supposition), to the velocities, and so to the spaces described. Take the sums of the analogous quantities, and the areas B*kq*, B*lr*, B*ms*, B*nt*, &c., will be analogous to the whole spaces described; and also the areas AB*q*K, AB*r*L, AB*s*M, AB*t*N, &c., to the times. Therefore the body, in descending, will in any time AB*r*L describe the space B*lr*, and in the time L*rt*N the space *rlnt*. Q.E.D. And the like demonstration holds in ascending motion.

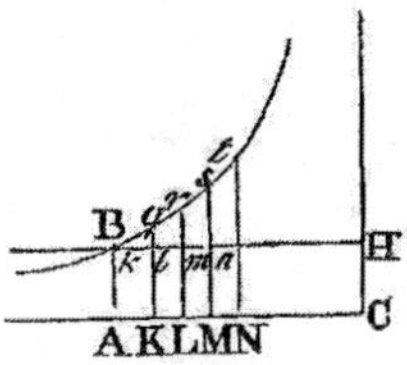

COROL. 1. Therefore the greatest velocity that the body can acquire by falling is to the velocity acquired in any given time as the given force of gravity which perpetually acts upon it to the resisting force which opposes it at the end of that time.

COROL. 2. But the time being augmented in an arithmetical progression, the sum of that greatest velocity and the velocity in the ascent, and also their difference in the descent, decreases in a geometrical progression.

COROL. 3. Also the differences of the spaces, which are described in equal differences of the times, decrease in the same geometrical progression.

COROL. 4. The space described by the body is the difference of two spaces, whereof one is as the time taken from the beginning of the descent, and the other as the velocity; which [spaces] also at the beginning of the descent are equal among themselves.

PROPOSITION IV. PROBLEM II.

Supposing the force of gravity in any similar medium to be uniform, and to tend perpendicularly to the plane of the horizon; to define the motion of a projectile therein, which suffers resistance proportional to its velocity.

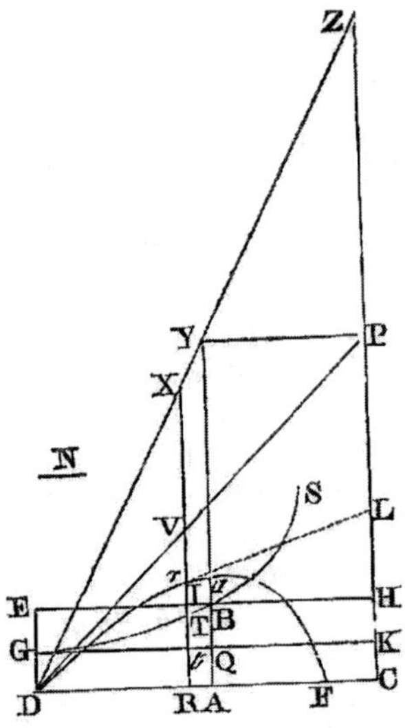

Let the projectile go from any place D in the direction of any right line DP, and let its velocity at the beginning of the motion be expounded by the length DP. From the point P let fall the perpendicular PC on the horizontal line DC, and cut DC in A, so that DA may be to AC as the resistance of the medium arising from the motion upwards at the beginning to the force of gravity; or (which comes to the same) so that the rectangle under DA and DP may be to that under AC and CP as the whole resistance at the beginning of the motion to the force of gravity. With the asymptotes DC, CP describe any hyperbola GTBS cutting the perpendiculars DG, AB in G and B; complete the parallelogram DGKC, and let its side GK cut AB in Q. Take a line N in the same ratio to QB as DC is in to CP; and from any point R of the right line DC erect RT perpendicular to it, meeting the hyperbola in T, and the right lines EH, GK, DP in I, t, and V; in that perpendicular take Vr equal to $\frac{t\text{GT}}{\text{N}}$, or which is the same thing, take Rr equal to $\frac{\text{GTIE}}{\text{N}}$; and the projectile in the time DRTG will arrive at the point r, describing the curve line DraF, the locus of the point r; thence it will come to its greatest height a in the perpendicular AB; and afterwards ever approach to the asymptote PC. And its velocity in any point r will be as the tangent rL to the curve. Q.E.I.

For N is to QB as DC to CP or DR to RV, and therefore RV is equal to $\frac{DR \times QB}{N}$, and Rr (that is, RV — Vr, or $\frac{DR \times QB - tGT}{N}$) is equal to $\frac{DR \times AB - RDGT}{N}$. Now let the time be expounded by the area RDGT and (by Laws, Cor. 2), distinguish the motion of the body into two others, one of ascent, the other lateral. And since the resistance is as the motion, let that also be distinguished into two parts proportional and contrary to the parts of the motion: and therefore the length described by the lateral motion will be (by Prop. II, Book II) as the line DR, and the height (by Prop. III, Book II) as the area DR × AB — RDGT, that is, as the line Rr. But in the very beginning of the motion the area RDGT is equal to the rectangle DR × AQ, and therefore that line Rr (or $\frac{DR \times AB - DR \times AQ}{N}$) will then be to DR as AB — AQ or QB to N, that is, as CP to DC; and therefore as the motion upwards to the motion lengthwise at the beginning. Since, therefore, Rr is always as the height, and DR always as the length, and Rr is to DR at the beginning as the height to the length, it follows, that Rr is always to DR as the height to the length; and therefore that the body will move in the line DraF, which is the locus of the point r. Q.E.D.

COR. 1. Therefore Rr is equal to $\frac{DR \times AB}{N} - \frac{RDGT}{N}$; and therefore if RT be produced to X so that RX may be equal to $\frac{DR \times AB}{N}$, that is, if the parallelogram ACPY be completed, and DY cutting CP in Z be drawn, and RT be produced till it meets DY in X; Xr will be equal to $\frac{RDGT}{N}$, and therefore proportional to the time.

COR. 2. Whence if innumerable lines CR, or, which is the same, innumerable lines ZX, be taken in a geometrical progression, there will be as many lines Xr in an arithmetical progression. And hence the curve DraF is easily delineated by the table of logarithms.

COR. 3. If a parabola be constructed to the vertex D, and the diameter DG produced downwards, and its latus rectum is to 2 DP as the whole resistance at the beginning of the motion to the gravitating force, the velocity with which the body ought to go from the place D, in the direction of the right line DP, so as in an uniform resisting medium to describe the curve DraF, will be the same as that with which it ought to go from the same place D, in the direction of the same right line DP, so as to describe

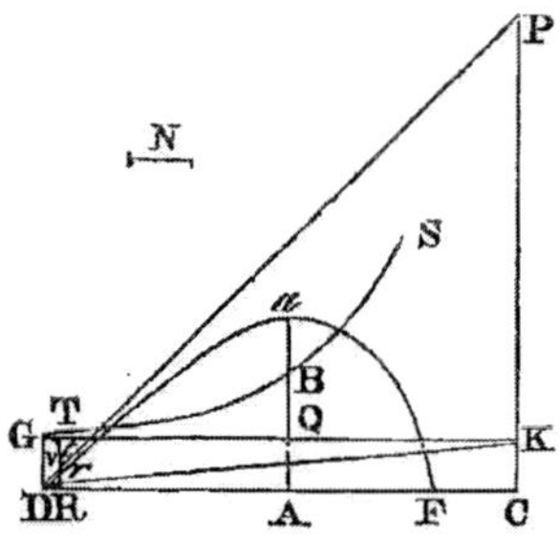

a parabola in a non-resisting medium. For the latus rectum of this parabola, at the very beginning of the motion, is $\frac{DV^2}{Vr}$; and Vr is $\frac{tGT}{N}$ or $\frac{DR \times Tt}{2N}$. But a right line, which, if drawn, would touch the hyperbola GTS in G, is parallel to DK, and therefore Tt is $\frac{CK \times DR}{DC}$, and N is $\frac{QB \times DC}{CP}$. And therefore Vr is equal to $\frac{DR^2 \times CK \times CP}{2DC^2 \times QB}$, that is (because DR and DC, DV and DP are proportionals), to $\frac{DV^2 \times CK \times CP}{2DP^2 \times QB}$; and the latus rectum $\frac{DV^2}{Vr}$ comes out $\frac{2DP^2 \times QB}{CK \times CP}$, that is (because QB and CK, DA and AC are proportional), $\frac{2DP^2 \times DA}{AC \times CP}$, and therefore is to 2DP as DP $\times$ DA to CP $\times$ AC; that is, as the resistance to the gravity. Q.E.D.

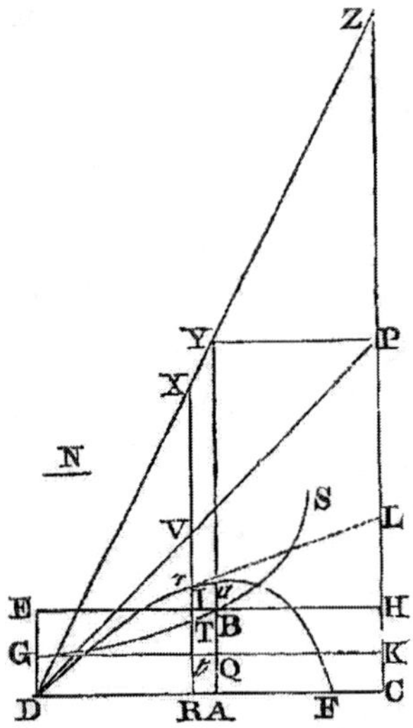

COR. 4. Hence if a body be projected from any place D with a given velocity, in the direction of a right line DP given by position, and the resistance of the medium, at the beginning of the motion, be given, the curve DraF, which that body will describe, may be found. For the velocity being given, the latus rectum of the parabola is given, as is well known. And taking 2DP to that latus rectum, as the force of gravity to the resisting force, DP is also given. Then cutting DC in A, so that CP $\times$ AC may be to DP $\times$ DA in the same ratio of the gravity to the resistance, the point A will be given. And hence the curve DraF is also given.

COR. 5. And, on the contrary, if the curve DraF be given, there will be given both the velocity of the body and the resistance of the medium in each of the places r. For the ratio of CP $\times$ AC to DP $\times$ DA being given, there is given both the resistance of the medium at the beginning of the motion, and the latus rectum of the parabola; and thence the velocity at the beginning of the motion is given also. Then from the length of the tangent

L there is given both the velocity proportional to it, and the resistance proportional to the velocity in any place r.

COR. 6. But since the length 2DP is to the latus rectum of the parabola as the gravity to the resistance in D; and, from the velocity augmented, the resistance is augmented in the same ratio, but the latus rectum of the parabola is augmented in the duplicate of that ratio, it is plain that the length 2DP is augmented in that simple ratio only; and is therefore always proportional to the velocity; nor will it be augmented or diminished by the change of the angle CDP, unless the velocity be also changed.

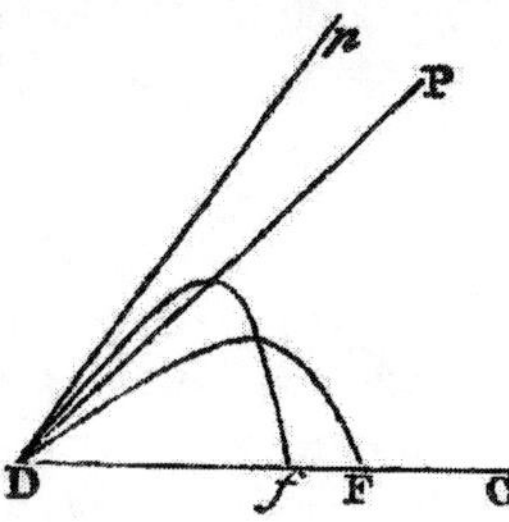

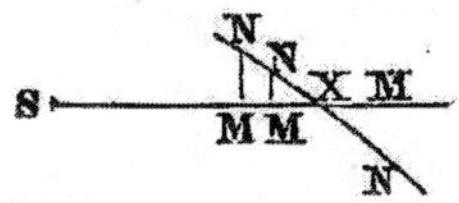

COR. 7. Hence appears the method of determining the curve D*ra*F nearly from the phenomena, and thence collecting the resistance and velocity with which the body is projected. Let two similar and equal bodies be projected with the same velocity, from the place D, in different angles CDP, CD*p*; and let the places F, *f*, where they fall upon the horizontal plane DC, be known. Then taking any length for DP or D*p* suppose the resistance in D to be to the gravity in any ratio whatsoever, and let that ratio be expounded by any length SM. Then, by computation, from that assumed length DP, find the lengths DF, D*f*; and from the ratio $\frac{Ff}{DF}$, found by calculation, subduct the same ratio as found by experiment; and let the difference be expounded by the perpendicular MN. Repeat the same a second and a third time, by assuming always a new ratio SM of the resistance to the gravity, and collecting a new difference MN. Draw the affirmative differences on one side of the right line SM, and the negative on the other side; and through the points N, N, N, draw a regular curve NNN, cutting the right line SMMM in X, and SX will be the true ratio of the resistance to the gravity, which was to be found. From this ratio the length DF is to be collected by calculation; and a length, which is to the assumed length DP as the length DF known by experiment to the length DF just now found, will be the true length DP. This being known, you will have both the curve line D*ra*F which the body describes, and also the velocity and resistance of the body in each place.

SCHOLIUM.

But, yet, that the resistance of bodies is in the ratio of the velocity, is more a mathematical hypothesis than a physical one. In mediums void of all tenacity, the resistances made to bodies are in the duplicate ratio of the velocities. For by the action of a swifter body, a greater motion in proportion to a greater velocity is communicated to the same quantity of the medium in a less time; and in an equal time, by reason of a greater quantity of the disturbed medium, a motion is communicated in the duplicate ratio greater; and the resistance (by Law II and III) is as the motion communicated. Let us, therefore, see what motions arise from this law of resistance.

SECTION II.

Of the motion of bodies that are resisted in the duplicate ratio of their velocities.

PROPOSITION V. THEOREM III.

If a body is resisted in the duplicate ratio of its velocity, and moves by its innate force only through a similar medium; and the times be taken in a geometrical progression, proceeding from less to greater terms: I say, that the velocities at the beginning of each of the times are in the same geometrical progression inversely; and that the spaces are equal, which are described in each of the times.

For since the resistance of the medium is proportional to the square of the velocity, and the decrement of the velocity is proportional to the resistance: if the time be divided into innumerable equal particles, the squares of the velocities at the beginning of each of the times will be proportional to the differences of the same velocities. Let those particles of time be AK, KL, LM, &c., taken in the right line CD; and erect the perpendiculars AB, K*k*, L*l*, M*m*, &c., meeting the hyperbola B*klm*G, described with the centre C, and the rectangular asymptotes CD, CH, in B, *k*, *l*, *m*, &c.; then AB will be to K*k* as CK to CA, and, by division, AB — K*k* to K*k* as AK to CA, and alternately, AB — K*k* to AK as K*k* to CA; and therefore as AB × K*k* to AB × CA.

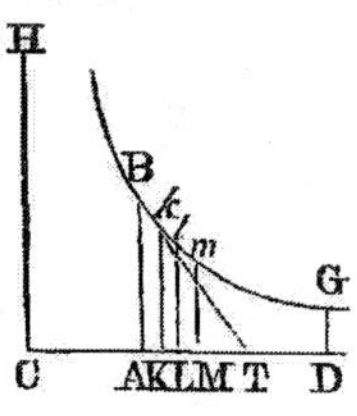

Therefore since AK and AB × CA are given, AB — K*k* will be as AB × K*k*; and, lastly, when AB and K*k* coincide, as AB². And, by the like reasoning, K*k*—L*l*, L*l*—M*m*, &c., will be as K*k*², L*l*², &c. Therefore the squares of the lines AB, K*k*, L*l*, M*m*, &c., are as their differences; and, therefore, since the squares of the velocities were shewn above to be as their differences, the progression of both will be alike. This being demonstrated it follows also that the areas described by these lines are in a like progression with the spaces described by these velocities. Therefore if the velocity at the beginning of the first time AK be expounded by the line AB, and the velocity at the beginning of the second time KL by the line K*k*, and the length described in the first time by the area AK*k*B, all the following velocities will be expounded by the following lines L*l*, M*m*, &c., and the lengths described, by the areas K*l*, L*m*, &c. And, by composition, if the whole time be expounded by AM, the sum of its parts, the whole length described will be expounded by AM*m*B the sum of its parts. Now conceive the time AM to be divided into the parts AK, KL, LM, &c., so that CA, CK, CL, CM, &c. may be in a geometrical progression; and those parts will be in the same progression, and the velocities AB, K*k*, L*l*, M*m*, &c., will be in the same progression inversely, and the spaces described A*k*, K*l*, L*m*, &c., will be equal. Q.E.D.

COR. 1. Hence it appears, that if the time be expounded by any part AD of the asymptote, and the velocity in the beginning of the time by the ordinate AB, the velocity at the end of the time will be expounded by the ordinate DG; and the whole space described by the adjacent hyperbolic area ABGD; and the space which any body can describe in the same time AD, with the first velocity AB, in a non-resisting medium, by the rectangle AB × AD.

COR. 2. Hence the space described in a resisting medium is given, by taking it to the space described with the uniform velocity AB in a non-resisting medium, as the hyperbolic area ABGD to the rectangle AB × AD.

COR. 3. The resistance of the medium is also given, by making it equal, in the very beginning of the motion, to an uniform centripetal force, which could generate, in a body falling through a non-resisting medium, the velocity AB in the time AC. For if BT be drawn touching the hyperbola in B, and meeting the asymptote in T, the right line AT will be equal to AC, and will express the time in which the first resistance, uniformly continued, may take away the whole velocity AB.

COR. 4. And thence is also given the proportion of this resistance to the force of gravity, or any other given centripetal force.

COR. 5. And, *vice versa*, if there is given the proportion of the resistance to any given centripetal force, the time AC is also given, in which a centripetal force equal to the resistance may generate any velocity as AB; and thence is given the point B, through which the hyperbola, having CH, CD for its asymptotes, is to be described; as also the space ABGD, which a body, by beginning its motion with that velocity AB, can describe in any time AD, in a similar resisting medium.

PROPOSITION VI. THEOREM IV.

Homogeneous and equal spherical bodies, opposed by resistances that are in the duplicate ratio of the velocities, and moving on by their innate force only, will, in times which are reciprocally as the velocities at the beginning, describe equal spaces, and lose parts of their velocities proportional to the wholes.

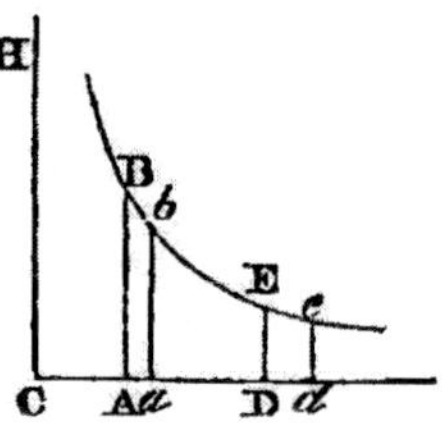

To the rectangular asymptotes CD, CH describe any hyperbola B*b*E*e*, cutting the perpendiculars AB, *ab*, DE, *de* in B, *b*, E, *e*; let the initial velocities be expounded by the perpendiculars AB, DE, and the times by the lines A*a*, D*d*. Therefore as A*a* is to D*d*, so (by the hypothesis) is DE to AB, and so (from the nature of the hyperbola) is CA to CD; and, by composition, so is C*a* to C*d*. Therefore the areas AB*ba*, DE*ed*, that is, the spaces described, are equal among themselves, and the first velocities AB, DE are proportional to the last *ab*, *de*; and therefore, by division, proportional to the parts of the velocities lost, AB — *ab*, DE — *de*. Q.E.D.

PROPOSITION VII. THEOREM V.

If spherical bodies are resisted in the duplicate ratio of their velocities, in times which are as the first motions directly, and the first resistances inversely, they will lose parts of their motions proportional to the wholes, and will describe spaces proportional to those times and the first velocities conjunctly.

For the parts of the motions lost are as the resistances and times conjunctly. Therefore, that those parts may be proportional to the wholes, the resistance and time conjunctly ought to be as the motion. Therefore the time will be as the motion directly and the resistance inversely. Wherefore the particles of the times being taken in that ratio, the bodies will always lose parts of their motions proportional to the wholes, and therefore will retain velocities always proportional to their first velocities. And because of the given ratio of the velocities, they will always describe spaces which are as the first velocities and the times conjunctly. Q.E.D.

Cor. 1. Therefore if bodies equally swift are resisted in a duplicate ratio of their diameters, homogeneous globes moving with any velocities whatsoever, by describing spaces proportional to their diameters, will lose parts of their motions proportional to the wholes. For the motion of each globe will be as its velocity and mass conjunctly, that is, as the velocity and the cube of its diameter; the resistance (by supposition) will be as the square of the diameter and the square of the velocity conjunctly; and the time (by this proposition) is in the former ratio directly, and in the latter inversely, that is, as the diameter directly and the velocity inversely; and therefore the space, which is proportional to the time and velocity is as the diameter.

Cor. 2. If bodies equally swift are resisted in a sesquiplicate ratio of their diameters, homogeneous globes, moving with any velocities whatsoever, by describing spaces that are in a sesquiplicate ratio of the diameters, will lose parts of their motions proportional to the wholes.

COR. 3. And universally, if equally swift bodies are resisted in the ratio of any power of the diameters, the spaces, in which homogeneous globes, moving with any velocity whatsoever, will lose parts of their motions proportional to the wholes, will be as the cubes of the diameters applied to that power. Let those diameters be D and E; and if the resistances, where the velocities are supposed equal, are as D^n and E^n; the spaces in which the globes, moving with any velocities whatsoever, will lose parts of their motions proportional to the wholes, will be as D^{3-n} and E^{3-n}. And therefore homogeneous globes, in describing spaces proportional to D^{3-n} and E^{3-n}, will retain their velocities in the same ratio to one another as at the beginning.

COR. 4. Now if the globes are not homogeneous, the space described by the denser globe must be augmented in the ratio of the density. For the motion, with an equal velocity, is greater in the ratio of the density, and the time (by this Prop.) is augmented in the ratio of motion directly, and the space described in the ratio of the time.

COR. 5. And if the globes move in different mediums, the space, in a medium which, *cæteris paribus*, resists the most, must be diminished in the ratio of the greater resistance. For the time (by this Prop.) will be diminished in the ratio of the augmented resistance, and the space in the ratio of the time.

LEMMA II.

The moment of any genitum *is equal to the moments of each of the generating sides drawn into the indices of the powers of those sides, and into their co-efficients continually.*

I call any quantity a *genitum* which is not made by addition or subduction of divers parts, but is generated or produced in arithmetic by the multiplication, division, or extraction of the root of any terms whatsoever; in geometry by the invention of contents and sides, or of the extremes and means of proportionals. Quantities of this kind are products, quotients, roots, rectangles, squares, cubes, square and cubic sides, and the like. These quantities I here consider as variable and indetermined, and increasing or decreasing, as it were, by a perpetual motion or flux; and I understand their momentaneous increments or decrements by the name of moments; so that the increments may be esteemed as added or affirmative moments; and the decrements as subducted or negative ones. But take care not to look upon finite particles as such. Finite particles are not moments, but the very quantities generated by the moments. We are to conceive them as the just nascent principles of finite magnitudes. Nor do we in this Lemma regard the magnitude of the moments, but their first proportion, as nascent. It will be the same thing, if, instead of moments, we use either the velocities of the increments and decrements (which may also be called the motions, mutations, and fluxions of quantities), or any finite quantities proportional to those velocities. The co-efficient of any generating side is the quantity which arises by applying the genitum to that side.

Wherefore the sense of the Lemma is, that if the moments of any quantities A, B, C, &c., increasing or decreasing by a perpetual flux, or the velocities of the mutations which are proportional to them, be called a, b, c, &c., the moment or mutation of the generated rectangle AB will be $aB + bA$; the moment of the generated content ABC will be $aBC + bAC + cAB$; and the moments of the generated powers A^2, A^3, A^4, $A^{\frac{1}{2}}$, $A^{\frac{3}{2}}$, $A^{\frac{1}{3}}$, $A^{\frac{2}{3}}$, A^{-1}, A^{-2}, $A^{-\frac{1}{2}}$ will be $2aA$, $3aA^2$, $4aA^3$, $\frac{1}{2}aA^{-\frac{1}{2}}$, $\frac{3}{2}aA^{\frac{1}{2}}$, $\frac{1}{3}aA^{-\frac{3}{3}}$, $\frac{2}{3}aA^{-\frac{1}{3}}$, $-aA^{-2}$, $-2aA^{-3}$, $-\frac{1}{2}aA^{-\frac{3}{2}}$ respectively; and, in general, that the moment of any power $A^{\frac{n}{m}}$, will be $\frac{n}{m}aA^{\frac{n-m}{m}}$. Also, that the moment of the generated quantity A^2B will be $2aAB + bA^2$; the moment of the generated quantity $A^3B^4C^2$ will be $3aA^2B^4C^2 + 4bA^3B^3C^2 + 2cA^3B^4C$; and the moment of the generated quantity $\frac{A^3}{B^2}$ or A^3B^{-2} will be $3aA^2B^{-2} - 2bA^3B^{-3}$; and so on. The Lemma is thus demonstrated.

Case 1. Any rectangle, as AB, augmented by a perpetual flux, when, as yet, there wanted of the sides A and B half their moments $\frac{1}{2}a$ and $\frac{1}{2}b$, was $A - \frac{1}{2}a$ into $B - \frac{1}{2}b$, or $AB - \frac{1}{2}a\,B - \frac{1}{2}b\,A + \frac{1}{4}ab$; but as soon as the sides A and B are augmented by the other half moments, the rectangle becomes $A + \frac{1}{2}a$ into $B + \frac{1}{2}b$, or $AB + \frac{1}{2}a\,B + \frac{1}{2}b\,A + \frac{1}{4}ab$. From this rectangle subduct the former rectangle, and there will remain the excess $aB + bA$. Therefore with the whole increments a and b of the sides, the increment $aB + bA$ of the rectangle is generated. Q.E.D.

Case 2. Suppose AB always equal to G, and then the moment of the content ABC or GC (by Case 1) will be $gC + cG$, that is (putting AB and $aB + bA$ for G and g), $aBC + bAC + cAB$. And the reasoning is the same for contents under ever so many sides. Q.E.D.

Case 3. Suppose the sides A, B, and C, to be always equal among themselves; and the moment $aB + bA$, of A^2, that is, of the rectangle AB, will be $2aA$; and the moment $aBC + bAC + cAB$ of A^3, that is, of the content ABC, will be $3aA^2$. And by the same reasoning the moment of any power A^n is naA^{n-1}. Q.E.D

Case 4. Therefore since $\frac{1}{A}$ into A is 1, the moment of $\frac{1}{A}$ drawn into

A, together with $\frac{1}{A}$ drawn into a, will be the moment of 1, that is, nothing. Therefore the moment of $\frac{1}{A}$, or of A^{-1}, is $\frac{-a}{A^2}$. And generally since $\frac{1}{A^n}$ into A^n is 1, the moment of $\frac{1}{A^n}$drawn into A^n together with $\frac{1}{A^n}$ into naA^{n-1} will be nothing. And, therefore, the moment of $\frac{1}{A^n}$ or A^{-n} will be $-\frac{na}{A^{n+1}}$. Q.E.D.

CASE 5. And since $A^{\frac{1}{2}}$ into $A^{\frac{1}{2}}$ is A, the moment of $A^{\frac{1}{2}}$ drawn into $2A^{\frac{1}{2}}$ will be a (by Case 3); and, therefore, the moment of $A^{\frac{1}{2}}$ will be $\frac{a}{2A^{\frac{1}{2}}}$ or $\frac{1}{2}aA^{-\frac{1}{2}}$. And, generally, putting $A^{\frac{m}{n}}$ equal to B, then A^m will be equal to B^n, and therefore maA^{m-1} equal to nbB^{n-1}, and maA^{-1} equal to nbB^{-1}, or $nbA^{-\frac{m}{n}}$; and therefore $\frac{m}{n}aA^{\frac{m-n}{n}}$ is equal to b, that is, equal to the moment of $A^{\frac{m}{n}}$. Q.E.D.

CASE 6. Therefore the moment of any generated quantity A^mB^n is the moment of A^m drawn into B^n, together with the moment of B^n drawn into A^m, that is, $maA^{m-1}B^n + nbB^{n-1}A^m$; and that whether the indices m and n of the powers be whole numbers or fractions, affirmative or negative. And the reasoning is the same for contents under more powers. Q.E.D.

COR. 1. Hence in quantities continually proportional, if one term is given, the moments of the rest of the terms will be as the same terms multiplied by the number of intervals between them and the given term. Let A, B, C, D, E, F, be continually proportional; then if the term C is given, the moments of the rest of the terms will be among themselves as —2A, —B, D, 2E, 3F.

COR. 2. And if in four proportionals the two means are given, the moments of the extremes will be as those extremes. The same is to be understood of the sides of any given rectangle.

COR. 3. And if the sum or difference of two squares is given, the moments of the sides will be reciprocally as the sides.

SCHOLIUM.

In a letter of mine to Mr. *J. Collins*, dated *December* 10, 1672, having described a method of tangents, which I suspected to be the same with *Slusius's* method, which at that time was not made public, I subjoined these words: *This is one particular, or rather a Corollary, of a general me-*

thod, which extends itself, without any troublesome calculation, not only to the drawing of tangents to any curve lines, whether geometrical or mechanical, or any how respecting right lines or other curves, but also to the resolving other abstruser kinds of problems about the crookedness, areas, lengths, centres of gravity of curves, &c.; nor is it (as Hudden's *method* de Maximis & Minimis) *limited to equations which are free from surd quantities. This method I have interwoven with that other of working in equations, by reducing them to infinite series.* So far that letter. And these last words relate to a treatise I composed on that subject in the year 1671. The foundation of that general method is contained in the preceding Lemma.

PROPOSITION VIII. THEOREM VI.

If a body in an uniform medium, being uniformly acted upon by the force of gravity, ascends or descends in a right line; and the whole space described be distinguished into equal parts, and in the beginning of each of the parts (by adding or subducting the resisting force of the medium to or from the force of gravity, when the body ascends or descends) you collect the absolute forces; I say, that those absolute forces are in a geometrical progression.

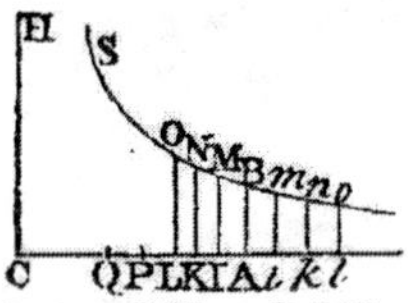

For let the force of gravity be expounded by the given line AC; the force of resistance by the indefinite line AK; the absolute force in the descent of the body by the difference KC; the velocity of the body by a line AP, which shall be a mean proportional between AK and AC, and therefore in a subduplicate ratio of the resistance; the increment of the resistance made in a given particle of time by the lineola KL, and the contemporaneous increment of the velocity by the lineola PQ; and with the centre C, and rectangular asymptotes CA, CH, describe any hyperbola BNS meeting the erected perpendiculars AB, KN, LO in B, N and O. Because AK is as AP^2, the moment KL of the one will be as the moment 2APQ of the other, that is, as $AP \times KC$; for the increment PQ of the velocity is (by Law II) proportional to the generating force KC. Let the ratio of KL be compounded with the ratio KN, and the rectangle $KL \times KN$ will become as $AP \times KC \times KN$; that is (because the rectangle $KC \times KN$ is given), as AP. But the ultimate ratio of the hyperbolic area KNOL to the rectangle $KL \times KN$ becomes, when the points K and L coincide, the ratio of equality. Therefore that hyperbolic evanescent area is as AP. Therefore the whole hyperbolic area ABOL is composed of particles KNOL which are always proportional to the velocity AP; and therefore is itself proportional to the space described with that velocity. Let that area be now divided into equal parts, as ABMI, IMNK, KNOL, &c., and the absolute forces AC, IC, KC, LC, &c., will be in a geometrical progression. Q.E.D. And by a like reasoning, in the ascent of the body, taking, on the contrary side of the point A, the equal areas AB*mi*, *imnk*, *knol*, &c., it will appear that the absolute forces AC, *i*C, *k*C, *l*C, &c., are continually proportional. Therefore if all the spaces in the ascent and descent are taken equal, all the absolute forces *l*C, *k*C, *i*C, AC, IC, KC, LC, &c., will be continually proportional. Q.E.D.

COR. 1. Hence if the space described be expounded by the hyperbolic area ABNK, the force of gravity, the velocity of the body, and the resistance of the medium, may be expounded by the lines AC, AP, and AK respectively; and *vice versa.*

COR. 2. And the greatest velocity which the body can ever acquire in an infinite descent will be expounded by the line AC.

COR. 3. Therefore if the resistance of the medium answering to any given velocity be known, the greatest velocity will be found, by taking it to that given velocity in a ratio subduplicate of the ratio which the force of gravity bears to that known resistance of the medium.

PROPOSITION IX. THEOREM VII.

Supposing what is above demonstrated, I say, that if the tangents of the angles of the sector of a circle, and of an hyperbola, be taken proportional to the velocities, the radius being of a fit magnitude, all the time of the ascent to the highest place will be as the sector of the circle, and all the time of descending from the highest place as the sector of the hyperbola.

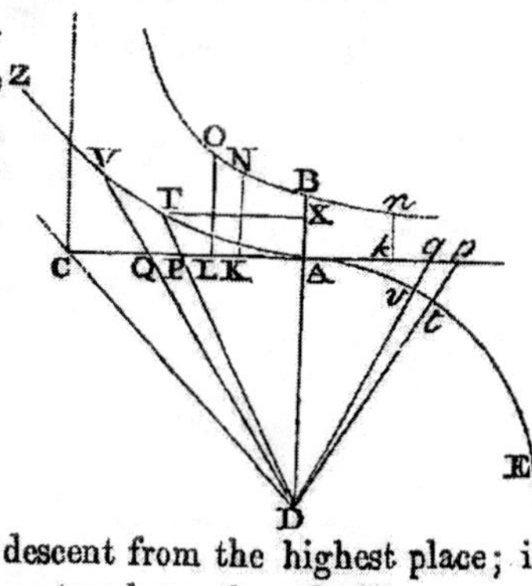

To the right line AC, which expresses the force of gravity, let AD be drawn perpendicular and equal. From the centre D with the semi-diameter AD describe as well the quadrant A*t*E of a circle, as the rectangular hyperbola AVZ, whose axis is AK, principal vertex A, and asymptote DC. Let D*p*, DP be drawn; and the circular sector A*t*D will be as all the time of the ascent to the highest place; and the hyperbolic sector ATD as all the time of descent from the highest place; if so be that the tangents A*p*, AP of those sectors be as the velocities.

CASE 1. Draw D*vq* cutting off the moments or least particles *t*D*v* and *q*D*p*, described in the same time, of the sector AD*t* and of the triangle AD*p*. Since those particles (because of the common angle D) are in a duplicate ratio of the sides, the particle *t*D*v* will be as $\frac{q\mathrm{D}p \times t\mathrm{D}^2}{p\mathrm{D}^2}$, that is (because tD is given), as $\frac{qDp}{pD^2}$. But pD^2 is $AD^2 + Ap^2$, that is, $AD^2 + AD \times Ak$, or $AD \times Ck$; and qDp is $\frac{1}{2} AD \times pq$. Therefore tDv, the particle of the sector, is as $\frac{pq}{Ck}$; that is, as the least decrement pq of the velocity directly, and the force Ck which diminishes the velocity, inversely; and therefore as the particle of time answering to the decrement of the velocity. And, by composition, the sum of all the particles tDv in the sector ADt will be as the sum of the particles of time answering to each of the lost particles pq of the decreasing velocity Ap, till that velocity, being diminished into nothing, vanishes; that is, the whole sector ADt is as the whole time of ascent to the highest place. Q.E.D.

CASE 2. Draw DQV cutting off the least particles TDV and PDQ of the sector DAV, and of the triangle DAQ; and these particles will be to each other as DT^2 to DP^2, that is (if TX and AP are parallel), as DX^2 to DA^2 or TX^2 to AP^2; and, by division, as $DX^2 - TX^2$ to $DA^2 - AP^2$. But, from the nature of the hyperbola, $DX^2 - TX^2$ is AD^2; and, by the supposition, AP^2 is $AD \times AK$. Therefore the particles are to each other as AD^2 to $AD^2 - AD \times AK$; that is, as AD to AD — AK or AC to CK: and therefore the particle TDV of the sector is $\frac{PDQ \times AC}{CK}$; and therefore (because AC and AD are given) as $\frac{PQ}{CK}$; that is, as the increment of the velocity directly, and as the force generating the increment inversely; and therefore as the particle of the time answering to the increment. And, by composition, the sum of the particles of time, in which all the particles PQ of the velocity AP are generated, will be as the sum of the particles of the sector ATD; that is, the whole time will be as the whole sector. Q.E.D.

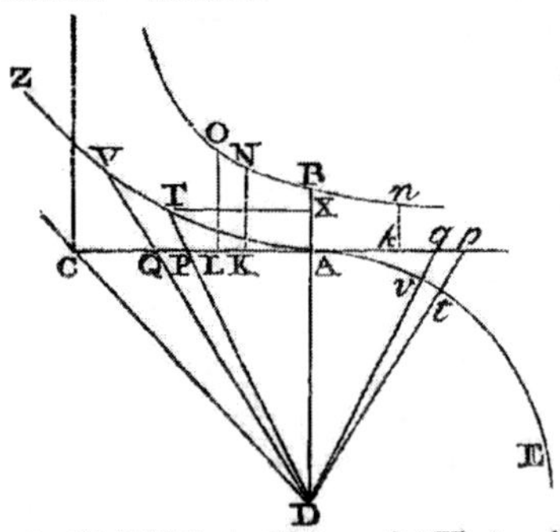

COR. 1. Hence if AB be equal to a fourth part of AC, the space which a body will describe by falling in any time will be to the space which the body could describe, by moving uniformly on in the same time with its greatest velocity AC, as the area ABNK, which expresses the space described in falling to the area ATD, which expresses the time. For since AC is to AP as AP to AK, then (by Cor. 1, Lem. II, of this Book) LK is to PQ as 2AK to AP, that is, as 2AP to AC, and thence LK is to $\frac{1}{2}$PQ as AP to $\frac{1}{4}$AC or AB; and KN is to AC or AD as AB to

. CK; and therefore, *ex æquo*, LKNO to DPQ as AP to CK. But DPQ was to DTV as CK to AC. Therefore, *ex æquo*, LKNO is to DTV as AP to AC; that is, as the velocity of the falling body to the greatest velocity which the body by falling can acquire. Since, therefore, the moments LKNO and DTV of the areas ABNK and ATD are as the velocities, all the parts of those areas generated in the same time will be as the spaces described in the same time; and therefore the whole areas ABNK and ADT, generated from the beginning, will be as the whole spaces described from the beginning of the descent. Q.E.D.

Cor. 2. The same is true also of the space described in the ascent. That is to say, that all that space is to the space described in the same time, with the uniform velocity AC, as the area AB*nk* is to the sector AD*t*.

Cor. 3. The velocity of the body, falling in the time ATD, is to the velocity which it would acquire in the same time in a non-resisting space, as the triangle APD to the hyperbolic sector ATD. For the velocity in a non-resisting medium would be as the time ATD, and in a resisting medium is as AP, that is, as the triangle APD. And those velocities, at the beginning of the descent, are equal among themselves, as well as those areas ATD, APD.

Cor. 4. By the same argument, the velocity in the ascent is to the velocity with which the body in the same time, in a non-resisting space, would lose all its motion of ascent, as the triangle A*p*D to the circular sector A*t*D; or as the right line A*p* to the arc A*t*.

Cor. 5. Therefore the time in which a body, by falling in a resisting medium, would acquire the velocity AP, is to the time in which it would acquire its greatest velocity AC, by falling in a non-resisting space, as the sector ADT to the triangle ADC: and the time in which it would lose its velocity A*p*, by ascending in a resisting medium, is to the time in which it would lose the same velocity by ascending in a non-resisting space, as the arc A*t* to its tangent A*p*.

Cor. 6. Hence from the given time there is given the space described in the ascent or descent. For the greatest velocity of a body descending *in infinitum* is given (by Corol. 2 and 3, Theor. VI, of this Book); and thence the time is given in which a body would acquire that velocity by falling in a non-resisting space. And taking the sector ADT or AD*t* to the triangle ADC in the ratio of the given time to the time just now found, there will be given both the velocity AP or A*p*, and the area ABNK or AB*nk*, which is to the sector ADT, or AD*t*, as the space sought to the space which would, in the given time, be uniformly described with that greatest velocity found just before.

Cor. 7. And by going backward, from the given space of ascent or descent AB*nk* or ABNK, there will be given the time AD*t* or ADT.

PROPOSITION X. PROBLEM III.

Suppose the uniform force of gravity to tend directly to the plane of the horizon, and the resistance to be as the density of the medium and the square of the velocity conjunctly: it is proposed to find the density of the medium in each place, which shall make the body move in any given curve line; the velocity of the body and the resistance of the medium in each place.

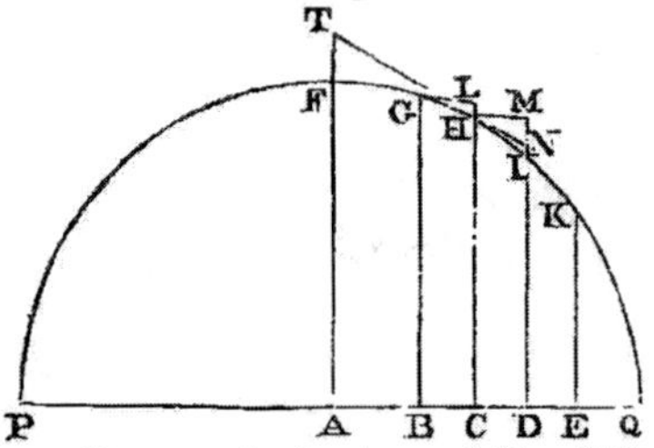

Let PQ be a plane perpendicular to the plane of the scheme itself; PFHQ a curve line meeting that plane in the points P and Q; G, H, I, K four places of the body going on in this curve from F to Q; and GB, HC, ID, KE four parallel ordinates let fall from these points to the horizon, and standing on the horizontal line PQ at the points B, C, D, E; and let the distances BC, CD, DE, of the ordinates be equal among themselves. From the points G and H let the right lines GL, HN, be drawn touching the curve in G and H, and meeting the ordinates CH, DI, produced upwards, in L and N: and complete the parallelogram HCDM. And the times in which the body describes the arcs GH, HI, will be in a subduplicate ratio of the altitudes LH, NI, which the bodies would describe in those times, by falling from the tangents; and the velocities will be as the lengths described GH, HI directly, and the times inversely. Let the times be expounded by T and t, and the velocities by $\frac{\text{GH}}{\text{T}}$ and $\frac{\text{HI}}{t}$; and the decrement of the velocity produced in the time t will be expounded by $\frac{\text{GH}}{\text{T}} - \frac{\text{HI}}{t}$. This decrement arises from the resistance which retards the body, and from the gravity which accelerates it. Gravity, in a falling body, which in its fall describes the space NI, produces a velocity with which it would be able to describe twice that space in the same time, as *Galileo* has demonstrated; that is, the velocity $\frac{2\text{NI}}{t}$: but if the body describes the arc HI, it augments that arc only by the length HI — HN or $\frac{\text{MI} \times \text{NI}}{\text{HI}}$; and therefore generates only the velocity $\frac{2\text{MI} \times \text{NI}}{t \times \text{HI}}$. Let this velocity be added to the before-mentioned decrement, and we shall have the decrement of the velocity arising from the resistance alone, that is, $\frac{\text{GH}}{\text{T}} - \frac{\text{HI}}{t} + \frac{2\text{MI} \times \text{NI}}{t \times \text{HI}}$.

Therefore since, in the same time, the action of gravity generates, in a falling body, the velocity $\frac{2NI}{t}$, the resistance will be to the gravity as $\frac{GH}{T} - \frac{HI}{t} + \frac{2MI \times NI}{t \times HI}$ to $\frac{2NI}{t}$ or as $\frac{t \times GH}{T} - HI + \frac{2MI \times NI}{HI}$ to 2NI.

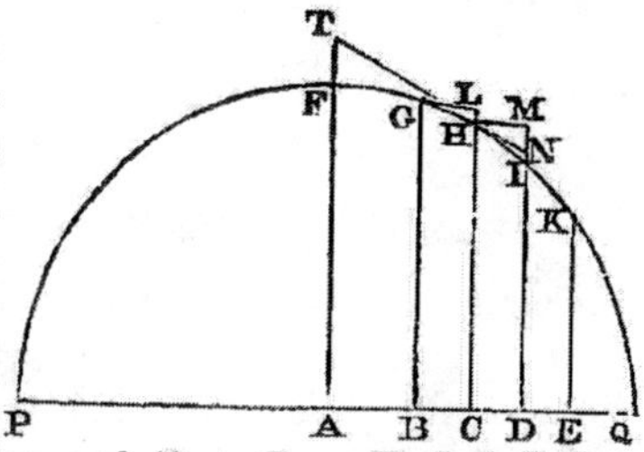

Now for the abscissas CB, CD, CE, put $-o$, o, $2o$. For the ordinate CH put P; and for MI put any series $Qo + Ro^2 + So^3 +$, &c. And all the terms of the series after the first, that is, $Ro^2 + So^3 +$, &c., will be NI; and the ordinates DI, EK, and BG will be $P - Qo - Ro^2 - So^3 -$, &c., $P - 2Qo - 4Ro^2 - 8So^3 -$, &c., and $P + Qo - Ro^2 + So^3 -$, &c., respectively. And by squaring the differences of the ordinates BG — CH and CH — DI, and to the squares thence produced adding the squares of BC and CD themselves, you will have $oo + QQoo - 2QRo^3 +$, &c., and $oo + QQoo + 2QRo^3 +$, &c., the squares of the arcs GH, HI; whose roots $o\sqrt{1+QQ} - \frac{QRoo}{\sqrt{1+QQ}}$, and $o\sqrt{1+QQ} + \frac{QRoo}{\sqrt{1+QQ}}$ are the arcs GH and HI. Moreover, if from the ordinate CH there be subducted half the sum of the ordinates BG and DI, and from the ordinate DI there be subducted half the sum of the ordinates CH and EK, there will remain Roo and $Roo + 3So^3$, the versed sines of the arcs GI and HK. And these are proportional to the lineolæ LH and NI, and therefore in the duplicate ratio of the infinitely small times T and t: and thence the ratio $\frac{t}{T}$ is $\sqrt{\frac{R+3So}{R}}$ or $\frac{R+\frac{3}{2}So}{R}$; and $\frac{t \times GH}{T} - HI + \frac{2MI \times NI}{HI}$, by substituting the values of $\frac{t}{T}$, GH, HI, MI and NI just found, becomes $\frac{3Soo}{2R}\sqrt{1+QQ}$. And since 2NI is $2Roo$, the resistance will be now to the gravity as $\frac{3Soo}{2R}\sqrt{1+QQ}$ to $2Roo$, that is, as $3S\sqrt{1+QQ}$ to 4RR.

And the velocity will be such, that a body going off therewith from any place H, in the direction of the tangent HN, would describe, in vacuo, a parabola, whose diameter is HC, and its latus rectum $\frac{HN^2}{NI}$ or $\frac{1+QQ}{R}$.

And the resistance is as the density of the medium and the square of the velocity conjunctly; and therefore the density of the medium is as the resistance directly, and the square of the velocity inversely; that is, as

$\frac{3S\sqrt{1+QQ}}{4RR}$ directly and $\frac{1+QQ}{R}$ inversely; that is, as $\frac{S}{R\sqrt{1+QQ}}$. Q.E.I.

Cor. 1. If the tangent HN be produced both ways, so as to meet any ordinate AF in T $\frac{HT}{AC}$ will be equal to $\sqrt{1+QQ}$, and therefore in what has gone before may be put for $\sqrt{1+QQ}$. By this means the resistance will be to the gravity as $3S \times HT$ to $4RR \times AC$; the velocity will be as $\frac{HT}{AC\sqrt{R}}$, and the density of the medium will be as $\frac{S \times AC}{R \times HT}$.

Cor. 2. And hence, if the curve line PFHQ be defined by the relation between the base or abscissa AC and the ordinate CH, as is usual, and the value of the ordinate be resolved into a converging series, the Problem will be expeditiously solved by the first terms of the series; as in the following examples.

Example 1. Let the line PFHQ be a semi-circle described upon the diameter PQ, to find the density of the medium that shall make a projectile move in that line.

Bisect the diameter PQ in A; and call AQ, n; AC, a; CH, e; and CD, o; then DI^2 or $AQ^2 - AD^2 = nn - aa - 2ao - oo$, or $ee - 2ao - oo$; and the root being extracted by our method, will give $DI = e - \frac{ao}{e} - \frac{oo}{2e} - \frac{aaoo}{2e^3} - \frac{ao^3}{2e^3} - \frac{a^3o^3}{2e^5} -$, &c. Here put nn for $ee + aa$, and DI will become $= e - \frac{ao}{e} - \frac{nnoo}{2e^3} - \frac{anno^3}{2e^5} -$, &c.

Such series I distinguish into successive terms after this manner: I call that the first term in which the infinitely small quantity o is not found; the second, in which that quantity is of one dimension only; the third, in which it arises to two dimensions; the fourth, in which it is of three; and so *ad infinitum*. And the first term, which here is e, will always denote the length of the ordinate CH, standing at the beginning of the indefinite quantity o. The second term, which here is $\frac{ao}{e}$, will denote the difference between CH and DN; that is, the lineola MN which is cut off by completing the parallelogram HCDM; and therefore always determines the position of the tangent HN; as, in this case, by taking MN to HM as $\frac{ao}{e}$ to o, or a to e. The third term, which here is $\frac{nnoo}{2e^3}$, will represent the lineola IN, which lies between the tangent and the curve; and therefore determines the angle of contact IHN, or the curvature which the curve line has in H. If that lineola IN is of a finite magnitude, it will be expressed by the third term, together with those that follow *in infinitum*. But if that lineola be diminished *in infinitum*, the terms following become infinitely less than the third term, and therefore may be neglected. The fourth term determines the variation of the curvature; the fifth, the variation of the variation; and so on. Whence, by the way, appears no contemptible use of these series in the solution of problems that depend upon tangents, and the curvature of curves.

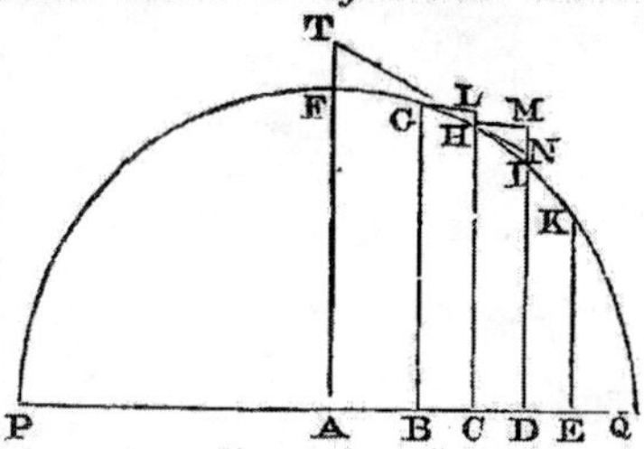

Now compare the series $e - \frac{ao}{e} - \frac{nnoo}{2e^3} - \frac{anno^3}{2e^5}$ — &c., with the series $P - Qo - Roo - So^3$ — &c., and for P, Q, R and S, put e, $\frac{a}{e}$, $\frac{nn}{2e^3}$ and $\frac{ann}{2e^5}$, and for $\sqrt{1 + QQ}$ put $\sqrt{1 + \frac{aa}{ee}}$ or $\frac{n}{e}$; and the density of the medium will come out as $\frac{a}{ne}$; that is (because n is given), as $\frac{a}{e}$ or $\frac{AC}{CH}$, that is, as that length of the tangent HT, which is terminated at the semi-diameter AF standing perpendicularly on PQ: and the resistance will be to the gravity as $3a$ to $2n$, that is, as 3AC to the diameter PQ of the circle; and the velocity will be as $\sqrt{CH}$. Therefore if the body goes from the place F, with a due velocity, in the direction of a line parallel to PQ, and the density of the medium in each of the places H is as the length of the tangent HT, and the resistance also in any place H is to the force of gravity as 3AC to PQ, that body will describe the quadrant FHQ of a circle. Q.E.I.

But if the same body should go from the place P, in the direction of a line perpendicular to PQ, and should begin to move in an arc of the semicircle PFQ, we must take AC or a on the contrary side of the centre A; and therefore its sign must be changed, and we must put $-a$ for $+a$. Then the density of the medium would come out as $-\frac{a}{e}$. But nature does not admit of a negative density, that is, a density which accelerates the motion of bodies; and therefore it cannot naturally come to pass that a body by ascending from P should describe the quadrant PF of a circle. To produce such an effect, a body ought to be accelerated by an impelling medium, and not impeded by a resisting one.

EXAMPLE 2. Let the line PFQ be a parabola, having its axis AF perpendicular to the horizon PQ, to find the density of the medium, which will make a projectile move in that line.

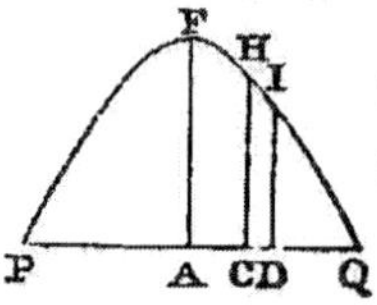

From the nature of the parabola, the rectangle PDQ is equal to the rectangle under the ordinate DI and some given right line; that is, if that right line be called b; PC, a; PQ, c; CH, e; and CD, o; the rectangle $a + o$ into $c - a - o$ or $ac - aa - 2ao + co - oo$, is equal to the rectangle b into DI, and therefore DI is equal to $\frac{ac - aa}{b} + \frac{c - 2a}{b} o - \frac{oo}{b}$. Now the second term $\frac{c - 2a}{b} o$ of this series is to be put for Qo, and the third term $\frac{oo}{b}$ for Roo. But since there are no more terms, the co-efficient S of the fourth term will vanish; and therefore the quantity $\frac{S}{R\sqrt{1 + QQ}}$, to which the density of the medium is proportional, will be nothing. Therefore, where the medium is of no density, the projectile will move in a parabola; as *Galileo* hath heretofore demonstrated. Q.E.I.

Example 3. Let the line AGK be an hyperbola, having its asymptote NX perpendicular to the horizontal plane AK, to find the density of the medium that will make a projectile move in that line.

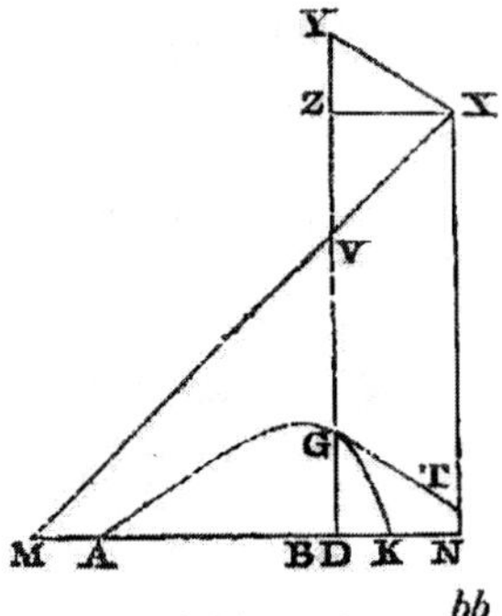

Let MX be the other asymptote, meeting the ordinate DG produced in V; and from the nature of the hyperbola, the rectangle of XV into VG will be given. There is also given the ratio of DN to VX, and therefore the rectangle of DN into VG is given. Let that be bb: and, completing the parallelogram DNXZ, let BN be called a; BD, o; NX, c; and let the given ratio of VZ to ZX or DN be $\frac{m}{n}$. Then DN will be equal to $a - o$, VG equal to $\frac{bb}{a - o}$, VZ equal to $\frac{m}{n} \times \overline{a - o}$, and GD or NX $-$VZ$-$VG equal to $c - \frac{m}{n} a + \frac{m}{n} o - \frac{bb}{a - o}$. Let the term $\frac{bb}{a - o}$ be resolved into the converging series $\frac{bb}{a} + \frac{bb}{aa} o + \frac{bb}{a^3} oo + \frac{bb}{a^4} o^3$, &c., and GD will become equal to $c - \frac{m}{n} a - \frac{bb}{a} + \frac{m}{n} o - \frac{bb}{aa} o - \frac{bb}{a^3} o^2 - \frac{bb}{a^4} o^3$

&c. The second term $\frac{m}{n}o - \frac{bb}{aa}o$ of this series is to be used for Qo; the third $\frac{bb}{a^3}o^2$, with its sign changed for Ro^2; and the fourth $\frac{bb}{a^4}o^3$, with its sign changed also for So^3, and their coefficients $\frac{m}{n} - \frac{bb}{aa}$, $\frac{bb}{a^3}$ and $\frac{bb}{a^4}$ are to be put for Q, R, and S in the former rule. Which being done, the density of the medium will come out as $\dfrac{\frac{bb}{a^4}}{\frac{bb}{a^3}\sqrt{1 + \frac{mm}{nn} - \frac{2mbb}{naa} + \frac{b^4}{a^4}}}$ or $\dfrac{1}{\sqrt{aa + \frac{mm}{nn}aa - \frac{2mbb}{n} + \frac{b^4}{aa}}}$, that is, if in VZ you take VY equal to VG, as $\frac{1}{XY}$. For aa and $\frac{m^2}{n^2}a^2 - \frac{2mbb}{n} + \frac{b^4}{aa}$ are the squares of XZ and ZY. But the ratio of the resistance to gravity is found to be that of 3XY to 2YG; and the velocity is that with which the body would describe a parabola, whose vertex is G, diameter DG, latus rectum $\frac{XY^2}{VG}$. Suppose, therefore, that the densities of the medium in each of the places G are reciprocally as the distances XY, and that the resistance in any place G is to the gravity as 3XY to 2YG; and a body let go from the place A, with a due velocity, will describe that hyperbola AGK. Q.E.I.

EXAMPLE 4. Suppose, indefinitely, the line AGK to be an hyperbola described with the centre X, and the asymptotes MX, NX, so that, having constructed the rectangle XZDN, whose side ZD cuts the hyperbola in G and its asymptote in V, VG may be reciprocally as any power DN^n of the line ZX or DN, whose index is the number n: to find the density of the medium in which a projected body will describe this curve.

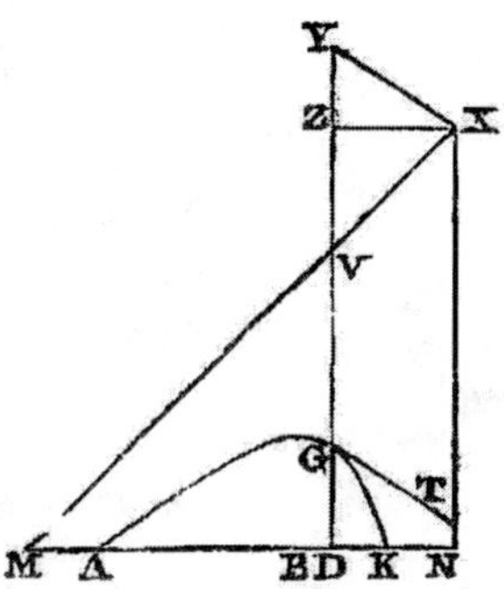

For BN, BD, NX, put A, O, C, respectively, and let VZ be to XZ or DN as d to e, and VG be equal to $\frac{bb}{DN^n}$; then DN will be equal to A — O, $VG = \frac{bb}{\overline{A-O}|^n}$, $VZ = \frac{d}{e}\overline{A-O}$, and GD or NX — VZ — VG equal to $C - \frac{d}{e}A + \frac{d}{e}O - \frac{bb}{\overline{A-O}|^n}$. Let the term $\frac{bb}{\overline{A - O}|^n}$ be resolved into an infinite series $\frac{bb}{A^n} + \frac{nbb}{A^{n+1}} \times O + \frac{nn+n}{2A^{n+2}} \times bb\, O^2 + \frac{n^3 + 3nn + 2n}{6A^{n+3}} \times bb\, O^3$, &c., and GD will be equal to $C - \frac{d}{e} A - \frac{bb}{A^n} + \frac{d}{e} O - \frac{nbb}{A^{n+1}} O - \frac{+nn+n}{2A^{n+2}} bb\, O^2 - \frac{+n^3+3nn+2n}{6A^{n+3}} bbO^3$, &c. The second term $\frac{d}{e} O - \frac{nbb}{A^{n+1}} O$ of this series is to be used for Qo, the third $\frac{nn+n}{2A^{n+2}} bbO^2$ for Roo, the fourth $\frac{n^3+3nn+2n}{6A^{n+3}} bbO^3$ for So^3. And thence the density of the medium $\frac{S}{R\sqrt{1+QQ}}$, in any place G, will be $\frac{n+2}{3\sqrt{A^2 + \frac{dd}{ee} A^2 - \frac{2dnbb}{eA^n} A + \frac{nnb^4}{A^{2n}}}}$, and therefore if in VZ you take VY equal to $n \times$ VG, that density is reciprocally as XY. For A^2 and $\frac{dd}{ee} A^2 - \frac{2dnbb}{eA^n} A + \frac{nnb^4}{A^{2n}}$ are the squares of XZ and ZY. But the resistance in the same place G is to the force of gravity as $3S \times \frac{XY}{A}$ to 4RR, that is, as XY to $\frac{2nn+2n}{n+2}$ VG. And the velocity there is the same wherewith the projected body would move in a parabola, whose vertex is G, diameter GD, and latus rectum $\frac{1+QQ}{R}$ or $\frac{2XY^2}{nn+n \times VG}$. Q.E.I.

SCHOLIUM.

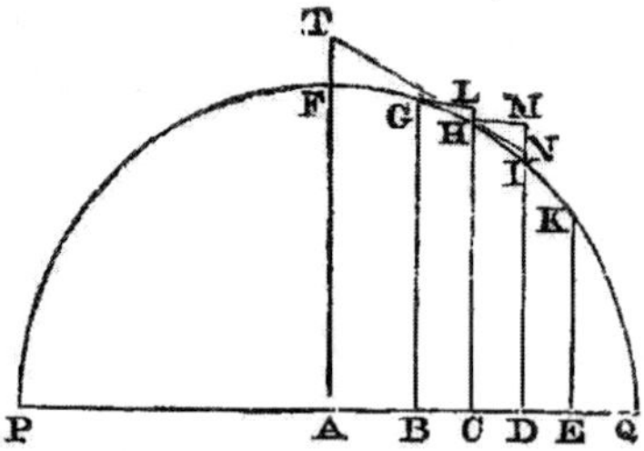

In the same manner that the density of the medium comes out to be as $\frac{S \times AC}{R \times HT}$, in Cor. 1, if the resistance is put as any power V^n of the velocity V, the density of the medium will come out to be as $\frac{S}{R^{\frac{4-n}{2}}} \times \overline{\frac{AC}{HT}}|^{n-1}$.

And therefore if a curve can be found, such that the ratio of $\frac{S}{R^{\frac{4-n}{2}}}$ to

$\left|\frac{HT}{AC}\right|^{n-1}$, or of $\frac{S^2}{R^{4-n}}$ to $\overline{1+QQ}|^{n-1}$ may be given; the body, in an uniform medium, whose resistance is as the power V^n of the velocity V, will move in this curve. But let us return to more simple curves.

Because there can be no motion in a parabola except in a non-resisting medium, but in the hyperbolas here described it is produced by a perpetual resistance; it is evident that the line which a projectile describes in an uniformly resisting medium approaches nearer to these hyperbolas than to a parabola. That line is certainly of the hyperbolic kind, but about the vertex it is more distant from the asymptotes, and in the parts remote from the vertex draws nearer to them than these hyperbolas here described. The difference, however, is not so great between the one and the other but that these latter may be commodiously enough used in practice instead of the former. And perhaps these may prove more useful than an hyperbola that is more accurate, and at the same time more compounded. They may be made use of, then, in this manner.

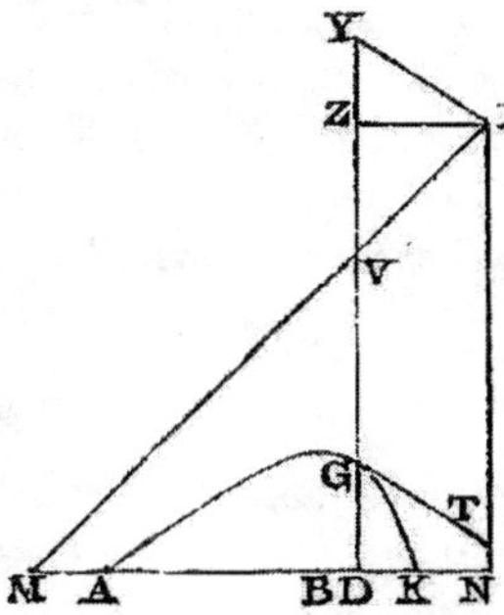

Complete the parallelogram XYGT, and the right line GT will touch the hyperbola in G, and therefore the density of the medium in G is reciprocally as the tangent GT, and the velocity there as $\sqrt{\frac{GT^2}{GV}}$; and the resistance is to the force of gravity as GT to $\frac{2nn+2n}{n+2} \times GV$.

Therefore if a body projected from the place A, in the direction of the right line AH, describes the hyperbola AGK and AH produced meets the asymptote NX in H, and AI drawn parallel to it meets the other asymptote MX in I; the density of the medium in A will be reciprocally as AH, and the velocity of the body as $\sqrt{\frac{AH^2}{AI}}$, and the resistance there to the force of gravity as AH to $\frac{2nn+2n}{n+2} \times AI$. Hence the following rules are deduced.

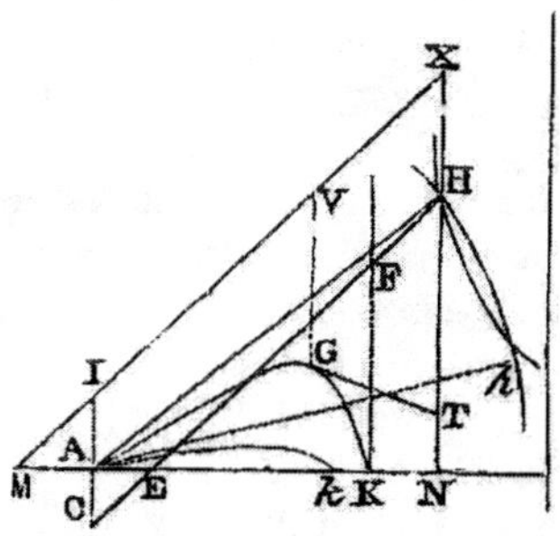

RULE 1. If the density of the medium at A, and the velocity with which the body is projected remain the same, and the angle NAH be changed; the lengths AH, AI, HX will remain. Therefore if those lengths, in any

one case, are found, the hyperbola may afterwards be easily determined from any given angle NAH.

Rule 2. If the angle NAH, and the density of the medium at A, remain the same, and the velocity with which the body is projected be changed, the length AH will continue the same; and AI will be changed in a duplicate ratio of the velocity reciprocally.

Rule 3. If the angle NAH, the velocity of the body at A, and the accelerative gravity remain the same, and the proportion of the resistance at A to the motive gravity be augmented in any ratio; the proportion of AH to AI will be augmented in the same ratio, the latus rectum of the above-mentioned parabola remaining the same, and also the length $\frac{AH^2}{AI}$ proportional to it; and therefore AH will be diminished in the same ratio, and AI will be diminished in the duplicate of that ratio. But the proportion of the resistance to the weight is augmented, when either the specific gravity is made less, the magnitude remaining equal, or when the density of the medium is made greater, or when, by diminishing the magnitude, the resistance becomes diminished in a less ratio than the weight.

Rule 4. Because the density of the medium is greater near the vertex of the hyperbola than it is in the place A, that a mean density may be preserved, the ratio of the least of the tangents GT to the tangent AH ought to be found, and the density in A augmented in a ratio a little greater than that of half the sum of those tangents to the least of the tangents GT.

Rule 5. If the lengths AH, AI are given, and the figure AGK is to be described, produce HN to X, so that HX may be to AI as $n + 1$ to 1; and with the centre X, and the asymptotes MX, NX, describe an hyperbola through the point A, such that AI may be to any of the lines VG as XV^n to XI^n.

Rule 6. By how much the greater the number n is, so much the more accurate are these hyperbolas in the ascent of the body from A, and less accurate in its descent to K; and the contrary. The conic hyperbola keeps a mean ratio between these, and is more simple than the rest. Therefore if the hyperbola be of this kind, and you are to find the point K, where the projected body falls upon any right line AN passing through the point A, let AN produced meet the asymptotes MX, NX in M and N, and take NK equal to AM.

Rule 7. And hence appears an expeditious method of determining this hyperbola from the phenomena. Let two similar and equal bodies be projected with the same velocity, in different angles HAK, hAk, and let them fall upon the plane of the horizon in K and k; and note the proportion of AK to Ak. Let it be as d to e. Then erecting a perpendicular AI of any length, assume any how the length AH or Ah, and thence graphically, or by scale and compass, collect the lengths AK, Ak (by Rule 6). If the ratio of AK to Ak be the same with that of d to e, the length of AH was

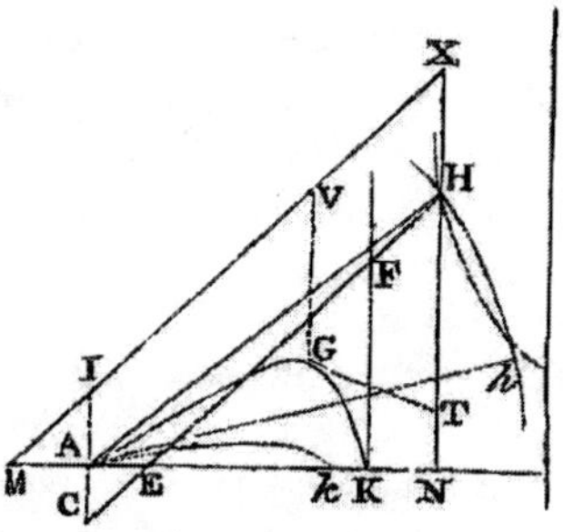

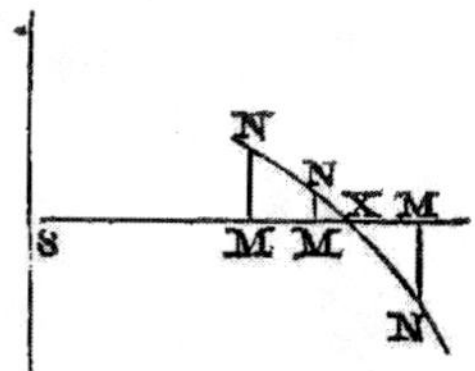

rightly assumed. If not, take on the indefinite right line SM, the length SM equal to the assumed AH; and erect a perpendicular MN equal to the difference $\frac{AK}{Ak} - \frac{d}{e}$ of the ratios drawn into any given right line. By the like method, from several assumed lengths AH, you may find several points N; and draw through them all a regular curve NNXN, cutting the right line SMMM in X. Lastly, assume AH equal to the abscissa SX, and thence find again the length AK; and the lengths, which are to the assumed length AI, and this last AH, as the length AK known by experiment, to the length AK last found, will be the true lengths AI and AH, which were to be found. But these being given, there will be given also the resisting force of the medium in the place A, it being to the force of gravity as AH to $\frac{4}{3}$AI. Let the density of the medium be increased by Rule 4, and if the resisting force just found be increased in the same ratio, it will become still more accurate.

RULE 8. The lengths AH, HX being found; let there be now required the position of the line AH, according to which a projectile thrown with that given velocity shall fall upon any point K. At the points A and K, erect the lines AC, KF perpendicular to the horizon; whereof let AC be drawn downwards, and be equal to AI or $\frac{1}{2}$HX. With the asymptotes AK, KF, describe an hyperbola, whose conjugate shall pass through the point C; and from the centre A, with the interval AH, describe a circle cutting that hyperbola in the point H; then the projectile thrown in the direction of the right line AH will fall upon the point K. Q.E.I. For the point H, because of the given length AH, must be somewhere in the circumference of the described circle. Draw CH meeting AK and KF in E and F; and because CH, MX are parallel, and AC, AI equal, AE will be equal to AM, and therefore also equal to KN. But CE is to AE as FH to KN, and therefore CE and FH are equal. Therefore the point H falls upon the hyperbolic curve described with the asymptotes AK, KF, whose conjugate passes through the point C; and is therefore found in the

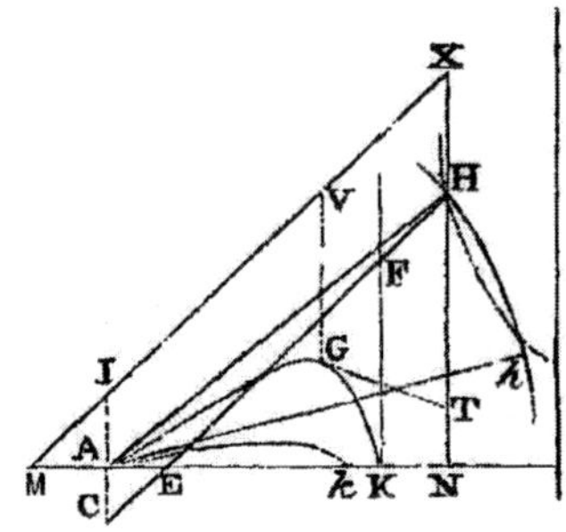

common intersection of this hyperbolic curve and the circumference of the described circle. Q.E.D. It is to be observed that this operation is the same, whether the right line AKN be parallel to the horizon, or inclined thereto in any angle; and that from two intersections H, h, there arise two angles NAH, NAh; and that in mechanical practice it is sufficient once to describe a circle, then to apply a ruler CH, of an indeterminate length, so to the point C, that its part FH, intercepted between the circle and the right line FK, may be equal to its part CE placed between the point C and the right line AK.

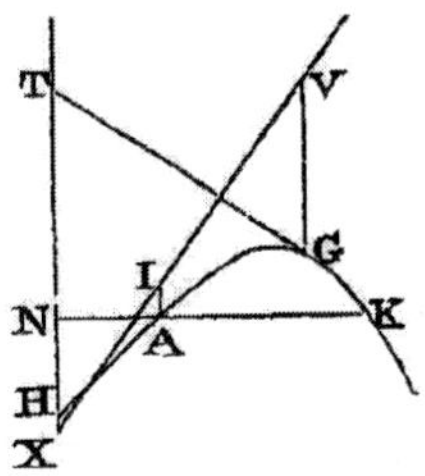

What has been said of hyperbolas may be easily applied to parabolas. For if a parabola be represented by XAGK, touched by a right line XV in the vertex X, and the ordinates IA, VG be as any powers XI^n, XV^n, of the abscissas XI, XV; draw XT, GT, AH, whereof let XT be parallel to VG, and let GT, AH touch the parabola in G and A: and a body projected from any place A, in the direction of the right line AH, with a due velocity, will describe this parabola, if the density of the medium in each of the places G be reciprocally as the tangent GT. In that case the velocity in G will be the same as would cause a body, moving in a non-resisting space, to describe a conic parabola, having G for its vertex, VG produced downwards for its diameter, and $\frac{2GT^2}{\overline{nn-n} \times VG}$ for its latus rectum. And the resisting force in G will be to the force of gravity as GT to $\frac{2nn-2n}{n-2}$ VG. Therefore if NAK represent an horizontal line, and both the density of the medium at A, and the velocity with which the body is projected, remaining the same, the angle NAH be any how altered, the lengths AH, AI, HX will remain; and thence will be given the vertex X of the parabola, and the position of the right line XI; and by taking VG to IA as XV^n to XI^n, there will be given all the points G of the parabola, through which the projectile will pass.

SECTION III.

Of the motions of bodies which are resisted partly in the ratio of the velocities, and partly in the duplicate of the same ratio.

PROPOSITION XI. THEOREM VIII.

If a body be resisted partly in the ratio and partly in the duplicate ratio of its velocity, and moves in a similar medium by its innate force only; and the times be taken in arithmetical progression; then quantities reciprocally proportional to the velocities, increased by a certain given quantity, will be in geometrical progression.

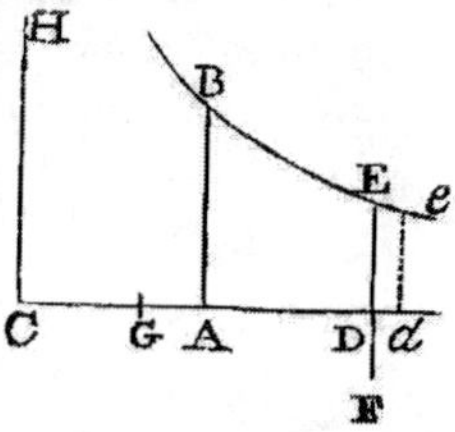

With the centre C, and the rectangular asymptotes CADd and CH, describe an hyperbola BEe, and let AB, DE, de, be parallel to the asymptote CH. In the asymptote CD let A, G be given points; and if the time be expounded by the hyperbolic area ABED uniformly increasing, I say, that the velocity may be expressed by the length DF, whose reciprocal GD, together with the given line CG, compose the length CD increasing in a geometrical progression.

For let the areola DEed be the least given increment of the time, and Dd will be reciprocally as DE, and therefore directly as CD. Therefore the decrement of $\frac{1}{\text{GD}}$, which (by Lem. II, Book II) is $\frac{\text{D}d}{\text{GD}^2}$, will be also as $\frac{\text{CD}}{\text{GD}^2}$ or $\frac{\text{CG}+\text{GD}}{\text{GD}^2}$, that is, as $\frac{1}{\text{GD}}+\frac{\text{CG}}{\text{GD}^2}$. Therefore the time ABED uniformly increasing by the addition of the given particles EDde, it follows that $\frac{1}{\text{GD}}$ decreases in the same ratio with the velocity. For the decrement of the velocity is as the resistance, that is (by the supposition), as the sum of two quantities, whereof one is as the velocity, and the other as the square of the velocity; and the decrement of $\frac{1}{\text{GD}}$ is as the sum of the quantities $\frac{1}{\text{GD}}$ and $\frac{\text{CG}}{\text{GD}^2}$, whereof the first is $\frac{1}{\text{GD}}$ itself, and the last $\frac{\text{CG}}{\text{GD}^2}$ is as $\frac{1}{\text{GD}^2}$: therefore $\frac{1}{\text{GD}}$ is as the velocity, the decrements of both being analogous. And if the quantity GD reciprocally proportional to $\frac{1}{\text{GD}}$, be augmented by the given quantity CG; the sum CD, the time ABED uniformly increasing, will increase in a geometrical progression. Q.E.D.

Cor. 1. Therefore, if, having the points A and G given, the time be expounded by the hyperbolic area ABED, the velocity may be expounded by $\frac{1}{GD}$ the reciprocal of GD.

Cor. 2. And by taking GA to GD as the reciprocal of the velocity at the beginning to the reciprocal of the velocity at the end of any time ABED, the point G will be found. And that point being found the velocity may be found from any other time given.

PROPOSITION XII. THEOREM IX.

The same things being supposed, I say, that if the spaces described are taken in arithmetical progression, the velocities augmented by a certain given quantity will be in geometrical progression.

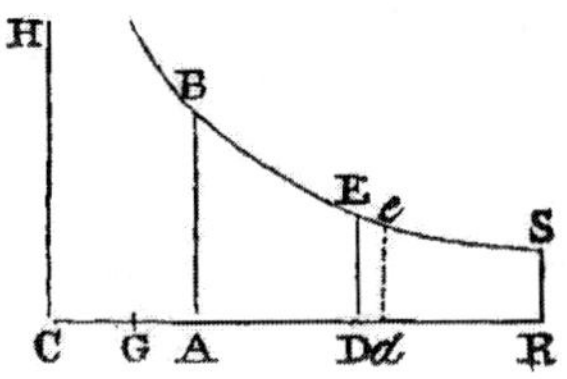

In the asymptote CD let there be given the point R, and, erecting the perpendicular RS meeting the hyperbola in S, let the space described be expounded by the hyperbolic area RSED; and the velocity will be as the length GD, which, together with the given line CG, composes a length CD decreasing in a geometrical progression, while the space RSED increases in an arithmetical progression.

For, because the increment EDde of the space is given, the lineola Dd, which is the decrement of GD, will be reciprocally as ED, and therefore directly as CD; that is, as the sum of the same GD and the given length CG. But the decrement of the velocity, in a time reciprocally proportional thereto, in which the given particle of space DdeE is described, is as the resistance and the time conjunctly, that is, directly as the sum of two quantities, whereof one is as the velocity, the other as the square of the velocity, and inversely as the velocity; and therefore directly as the sum of two quantities, one of which is given, the other is as the velocity. Therefore the decrement both of the velocity and the line GD is as a given quantity and a decreasing quantity conjunctly; and, because the decrements are analogous, the decreasing quantities will always be analogous; viz., the velocity, and the line GD. Q.E.D.

Cor. 1. If the velocity be expounded by the length GD, the space described will be as the hyperbolic area DESR.

Cor. 2. And if the point R be assumed any how, the point G will be found, by taking GR to GD as the velocity at the beginning to the velocity after any space RSED is described. The point G being given, the space is given from the given velocity: and the contrary.

Cor. 3. Whence since (by Prop. XI) the velocity is given from the given time, and (by this Prop.) the space is given from the given velocity; the space will be given from the given time: and the contrary.

PROPOSITION XIII. THEOREM X.

Supposing that a body attracted downwards by an uniform gravity ascends or descends in a right line; and that the same is resisted partly in the ratio of its velocity, and partly in the duplicate ratio thereof: I say, that, if right lines parallel to the diameters of a circle and an hyperbola be drawn through the ends of the conjugate diameters, and the velocities be as some segments of those parallels drawn from a given point, the times will be as the sectors of the areas cut off by right lines drawn from the centre to the ends of the segments; and the contrary.

CASE Suppose first that the body is ascending, and from the centre D, with any semi-diameter DB, describe a quadrant BETF of a circle, and through the end B of the semi-diameter DB draw the indefinite line BAP, parallel to the semi-diameter DF. In that line let there be given the point A, and take the segment AP proportional to the velocity. And since one part of the resistance is as the velocity, and another part as the square of the velocity, let the whole resistance be as $AP^2 + 2BAP$. Join DA, DP, cutting the circle in E and T, and let the gravity be expounded by DA^2, so that the gravity shall be to the resistance in P as DA^2 to $AP^2 + 2BAP$; and the time of the whole ascent will be as the sector EDT of the circle.

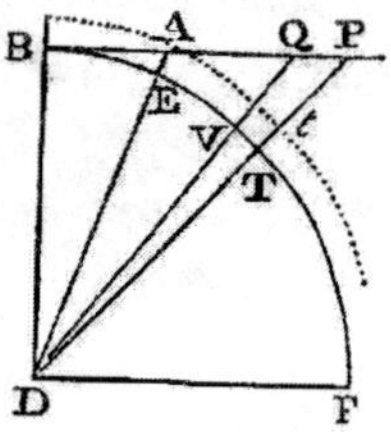

For draw DVQ, cutting off the moment PQ of the velocity AP, and the moment DTV of the sector DET answering to a given moment of time; and that decrement PQ of the velocity will be as the sum of the forces of gravity DA^2 and of resistance $AP^2 + 2BAP$, that is (by Prop. XII, Book II, Elem.), as DP^2. Then the area DPQ, which is proportional to PQ, is as DP^2, and the area DTV, which is to the area DPQ as DT^2 to DP^2, is as the given quantity DT^2. Therefore the area EDT decreases uniformly according to the rate of the future time, by subduction of given particles DTV, and is therefore proportional to the time of the whole ascent. Q.E.D.

CASE 2. If the velocity in the ascent of the body be expounded by the length AP as before, and the resistance be made as $AP^2 + 2BAP$, and if the force of gravity be less than can be expressed by DA^2; take BD of such a length, that $AB^2 - BD^2$ may be proportional to the gravity, and let DF be perpendicular and equal

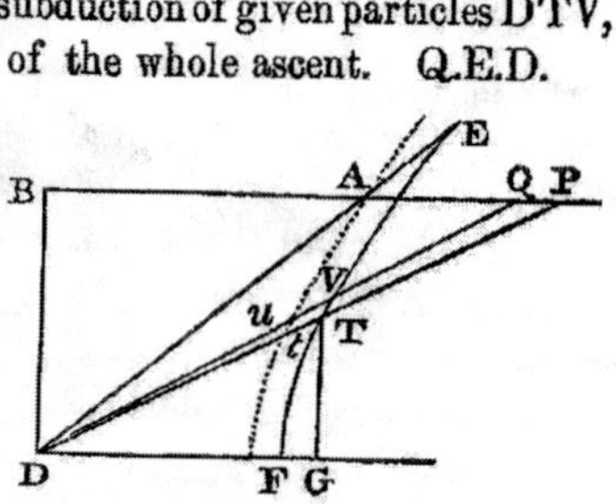

to DB, and through the vertex F describe the hyperbola FTVF, whose conjugate semi-diameters are DB and DF, and which cuts DA in E, and DP, DQ in T and V; and the time of the whole ascent will be as the hyperbolic sector TDE.

For the decrement PQ of the velocity, produced in a given particle of time, is as the sum of the resistance $AP^2 + 2BAP$ and of the gravity $AB^2 - BD^2$, that is, as $BP^2 - BD^2$. But the area DTV is to the area DPQ as DT^2 to DP^2; and, therefore, if GT be drawn perpendicular to DF, as GT^2 or $GD^2 - DF^2$ to BD^2, and as GD^2 to BP^2, and, by division, as DF^2 to $BP^2 - BD^2$. Therefore since the area DPQ is as PQ, that is, as $BP^2 - BD^2$, the area DTV will be as the given quantity DF^2. Therefore the area EDT decreases uniformly in each of the equal particles of time, by the subduction of so many given particles DTV, and therefore is proportional to the time. Q.E.D.

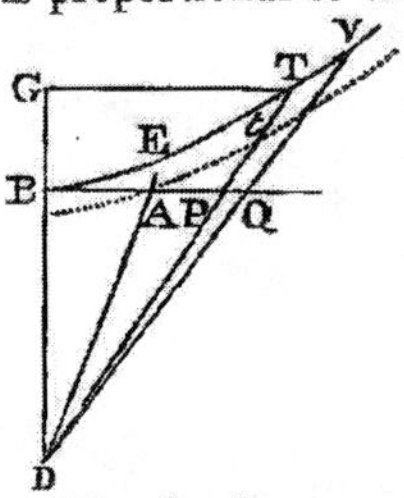

Case 3. Let AP be the velocity in the descent of the body, and $AP^2 + 2BAP$ the force of resistance, and $BD^2 - AB^2$ the force of gravity, the angle DBA being a right one. And if with the centre D, and the principal vertex B, there be described a rectangular hyperbola BETV cutting DA, DP, and DQ produced in E, T, and V; the sector DET of this hyperbola will be as the whole time of descent.

For the increment PQ of the velocity, and the area DPQ proportional to it, is as the excess of the gravity above the resistance, that is, as $BD^2 - AB^2 - 2BAP - AP^2$ or $BD^2 - BP^2$. And the area DTV is to the area DPQ as DT^2 to DP^2; and therefore as GT^2 or $GD^2 - BD^2$ to BP^2, and as GD^2 to BD^2, and, by division, as BD^2 to $BD^2 - BP^2$. Therefore since the area DPQ is as $BD^2 - BP^2$, the area DTV will be as the given quantity BD^2. Therefore the area EDT increases uniformly in the several equal particles of time by the addition of as many given particles DTV, and therefore is proportional to the time of the descent. Q.E.D.

Cor. If with the centre D and the semi-diameter DA there be drawn through the vertex A an arc A*t* similar to the arc ET, and similarly subtending the angle ADT, the velocity AP will be to the velocity which the body in the time EDT, in a non-resisting space, can lose in its ascent, or acquire in its descent, as the area of the triangle DAP to the area of the sector DA*t*; and therefore is given from the time given. For the velocity in a non-resisting medium is proportional to the time, and therefore to this sector; in a resisting medium, it is as the triangle; and in both mediums, where it is least, it approaches to the ratio of equality, as the sector and triangle do.

SCHOLIUM.

One may demonstrate also that case in the ascent of the body, where the force of gravity is less than can be expressed by DA^2 or $AB^2 + BD^2$, and greater than can be expressed by $AB^2 - DB^2$, and must be expressed by AB^2. But I hasten to other things.

PROPOSITION XIV. THEOREM XI.

The same things being supposed, I say, that the space described in the ascent or descent is as the difference of the area by which the time is expressed, and of some other area which is augmented or diminished in an arithmetical progression; if the forces compounded of the resistance and the gravity be taken in a geometrical progression.

Take AC (in these three figures) proportional to the gravity, and AK to the resistance; but take them on the same side of the point A, if the

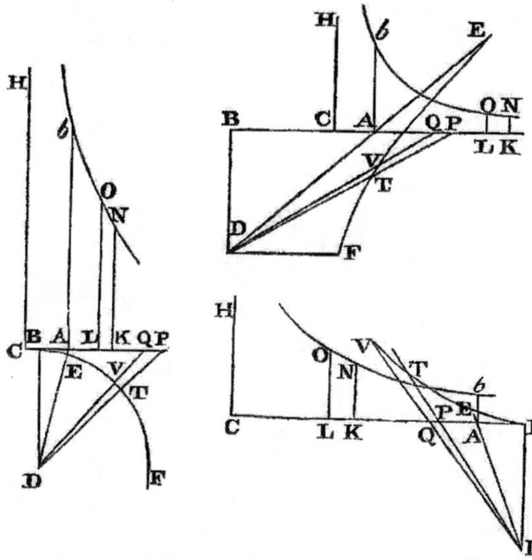

body is descending, otherwise on the contrary. Erect Ab, which make to DB as DB^2 to 4BAC: and to the rectangular asymptotes CK, CH, describe the hyperbola bN; and, erecting KN perpendicular to CK, the area AbNK will be augmented or diminished in an arithmetical progression, while the forces CK are taken in a geometrical progression. I say, therefore, that the distance of the body from its greatest altitude is as the excess of the area AbNK above the area DET.

For since AK is as the resistance, that is, as $AP^2 \times 2BAP$; assume any given quantity Z, and put AK equal to $\frac{AP^2 + 2BAP}{Z}$; then (by Lem.

II of this Book) the moment KL of AK will be equal to $\frac{2APQ + 2BA \times PQ}{Z}$ or $\frac{2BPQ}{Z}$, and the moment KLON of the area A*b*NK will be equal to $\frac{2BPQ \times LO}{Z}$ or $\frac{BPQ \times BD^3}{2Z \times CK \times AB}$.

Case 1. Now if the body ascends, and the gravity be as $AB^2 + BD^2$, BET being a circle, the line AC, which is proportional to the gravity, will be $\frac{AB^2 + BD^2}{Z}$, and DP^2 or $AP^2 + 2BAP + AB^2 + BD^2$ will be $AK \times Z + AC \times Z$ or $CK \times Z$; and therefore the area DTV will be to the area DPQ as DT^2 or DB^2 to $CK \times Z$.

Case 2. If the body ascends, and the gravity be as $AB^2 - BD^2$, the line AC will be $\frac{AB^2 - BD^2}{Z}$, and DT^2 will be to DP^2 as DF^2 or DB^2 to $BP^2 - BD^2$ or $AP^2 + 2BAP + AB^2 - BD^2$, that is, to $AK \times Z +$

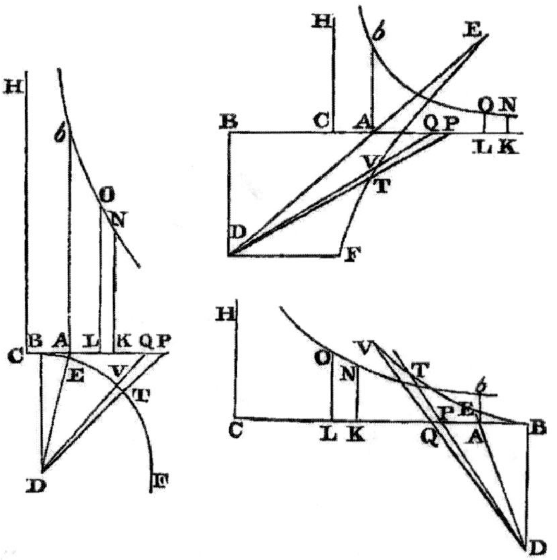

$AC \times Z$ or $CK \times Z$. And therefore the area DTV will be to the area DPQ as DB^2 to $CK \times Z$.

Case 3. And by the same reasoning, if the body descends, and therefore the gravity is as $BD^2 - AB^2$, and the line AC becomes equal to $\frac{BD^2 - AB^2}{Z}$; the area DTV will be to the area DPQ as DB^2 to $CK \times Z$: as above.

Since, therefore, these areas are always in this ratio, if for the area

DTV, by which the moment of the time, always equal to itself, is expressed, there be put any determinate rectangle, as $BD \times m$, the area DPQ, that is, $\frac{1}{2}BD \times PQ$, will be to $BD \times m$ as $CK \times Z$ to BD^3. And thence $PQ \times BD^3$ becomes equal to $2BD \times m \times CK \times Z$, and the moment KLON of the area A*b*NK, found before, becomes $\frac{BP \times BD \times m}{AB}$. From the area DET subduct its moment DTV or $BD \times m$, and there will remain $\frac{AP \times BD \times m}{AB}$. Therefore the difference of the moments, that is, the moment of the difference of the areas, is equal to $\frac{AP \times BD \times m}{AB}$; and therefore (because of the given quantity $\frac{BD \times m}{AB}$) as the velocity AP; that is, as the moment of the space which the body describes in its ascent or descent. And therefore the difference of the areas, and that space, increasing or decreasing by proportional moments, and beginning together or vanishing together, are proportional. Q.E.D.

COR. If the length, which arises by applying the area DET to the line BD, be called M; and another length V be taken in that ratio to the length M, which the line DA has to the line DE; the space which a body, in a resisting medium, describes in its whole ascent or descent, will be to the space which a body, in a non-resisting medium, falling from rest, can describe in the same time, as the difference of the aforesaid areas to $\frac{BD \times V^2}{AB}$; and therefore is given from the time given. For the space in a non-resisting medium is in a duplicate ratio of the time, or as V^2; and, because BD and AB are given, as $\frac{BD \times V^2}{AB}$. This area is equal to the area $\frac{DA^2 \times BD \times M^2}{DE^2 \times AB}$ and the moment of M is m; and therefor the moment of this area is $\frac{DA^2 \times BD \times 2M \times m}{DE^2 \times AB}$. But this moment is to the moment of the difference of the aforesaid areas DET and A*b*NK, viz., to $\frac{AP \times BD \times m}{AB}$, as $\frac{DA^2 \times BD \times M}{DE^2}$ to $\frac{1}{2}BD \times AP$, or as $\frac{DA^2}{DE^2}$ into DET to DAP; and, therefore, when the areas DET and DAP are least, in the ratio of equality. Therefore the area $\frac{BD \times V^2}{AB}$ and the difference of the areas DET and A*b*NK, when all these areas are least, have equal moments; and are therefore equal. Therefore since the velocities, and therefore also the spaces in both mediums described together, in the beginning of the descent, or the end of the ascent, approach to equality, and therefore are then one to another as the area $\frac{BD \times V^2}{AB}$, and the difference of the areas DET and A*b*NK; and moreover since the space, in a non-resisting medium, is perpetually as $\frac{BD \times V^2}{AB}$, and the space, in a resisting medium, is perpetually as the difference of the areas DET and A*b*NK; it necessarily follows, that the spaces, in both mediums, described in any equal times, are one to another as that area $\frac{BD \times V^2}{AB}$, and the difference of the areas DET and A*b*NK. Q.E.D.

SCHOLIUM.

The resistance of spherical bodies in fluids arises partly from the tenacity, partly from the attrition, and partly from the density of the medium. And that part of the resistance which arises from the density of the fluid is, as I said, in a duplicate ratio of the velocity; the other part, which arises from the tenacity of the fluid, is uniform, or as the moment of the time; and, therefore, we might now proceed to the motion of bodies, which are resisted partly by an uniform force, or in the ratio of the moments of the time, and partly in the duplicate ratio of the velocity. But it is sufficient to have cleared the way to this speculation in Prop. VIII and IX foregoing, and their Corollaries. For in those Propositions, instead of the uniform resistance made to an ascending body arising from its gravity, one may substitute the uniform resistance which arises from the tenacity of the medium, when the body moves by its *vis insita* alone; and when the body ascends in a right line, add this uniform resistance to the force of gravity, and subduct it when the body descends in a right line. One might also go on to the motion of bodies which are resisted in part uniformly, in part in the ratio of the velocity, and in part in the duplicate ratio of the same velocity. And I have opened a way to this in Prop. XIII and XIV foregoing, in which the uniform resistance arising from the tenacity of the medium may be substituted for the force of gravity, or be compounded with it as before. But I hasten to other things.

SECTION IV.

Of the circular motion of bodies in resisting mediums.

LEMMA III.

Let PQR *be a spiral cutting all the radii* SP, SQ, SR, *&c., in equal angles. Draw the right line* PT *touching the spiral in any point* P, *and cutting the radius* SQ *in* T; *draw* PO, QO *perpendicular to the spiral, and meeting in* O, *and join* SO. *I say, that if the points* P *and* Q *approach and coincide, the angle* PSO *will become a right angle, and the ultimate ratio of the rectangle* $TQ \times 2PS$ *to* PQ^2 *will be the ratio of equality.*

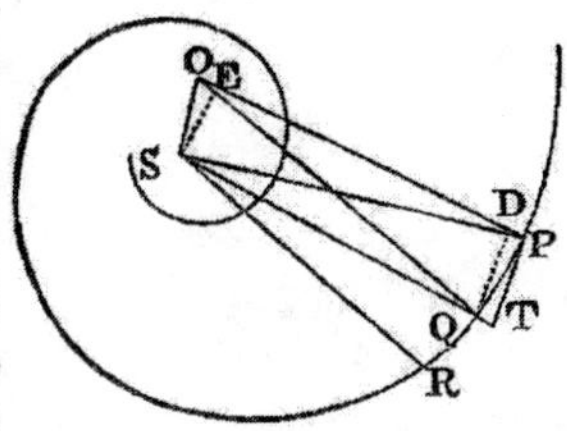

For from the right angles OPQ, OQR, subduct the equal angles SPQ, SQR, and there will remain the equal angles OPS, OQS. Therefore a circle which passes through the points OSP will pass also through the point Q. Let the points P and Q coincide, and this circle will touch the spiral in the place of coincidence PQ, and will therefore cut the right line OP perpendicularly. Therefore OP will become a diameter of this circle, and the angle OSP, being in a semi-circle, becomes a right one. Q.E.D.

Draw QD, SE perpendicular to OP, and the ultimate ratios of the lines will be as follows: TQ to PD as TS or PS to PE, or 2PO to 2PS; and PD to PQ as PQ to 2PO; and, *ex æquo perturbatè*, to TQ to PQ as PQ to 2PS. Whence PQ^2 becomes equal to $TQ \times 2PS$. Q.E.D.

PROPOSITION XV. THEOREM XII.

If the density of a medium in each place thereof be reciprocally as the distance of the places from an immovable centre, and the centripetal force be in the duplicate ratio of the density; I say, that a body may revolve in a spiral which cuts all the radii drawn from that centre in a given angle.

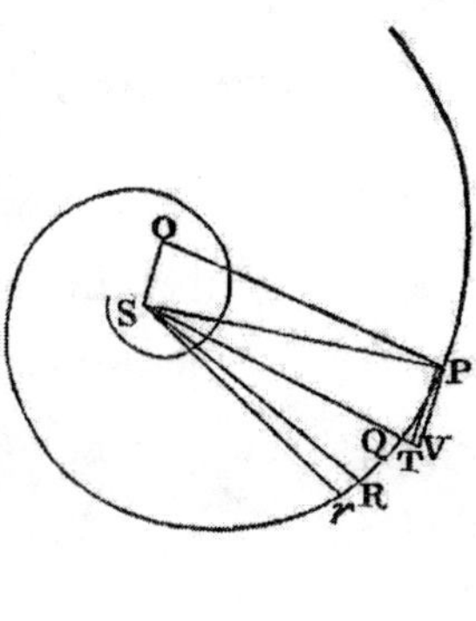

Suppose every thing to be as in the foregoing Lemma, and produce SQ to V so that SV may be equal to SP. In any time let a body, in a resisting medium, describe the least arc PQ, and in double the time the least arc PR; and the decrements of those arcs arising from the resistance, or their differences from the arcs which would be described in a non-resisting medium in the same times, will be to each other as the squares of the times in which they are generated; therefore the decrement of the arc PQ is the fourth part of the decrement of the arc PR. Whence also if the area QSr be taken equal to the area PSQ, the decrement of the arc PQ will be equal to half the lineola Rr; and therefore the force of resistance and the centripetal force are to each other as the lineola $\frac{1}{2}$Rr and TQ which they generate in the same time. Because the centripetal force with which the body is urged in P is reciprocally as SP^2, and (by Lem. X, Book I) the lineola TQ, which is generated by that force, is in a ratio compounded of the ratio of this force and the duplicate ratio of the time in which the arc PQ is described (for in this case I neglect the resistance, as being infinitely less than the centripetal force), it follows that $TQ \times SP^2$, that is (by the last Lemma), $\frac{1}{2}PQ^2 \times SP$, will be in a duplicate ratio of the time, and therefore the time is as $PQ \times \sqrt{SP}$; and the velocity of the body, with which the arc PQ is described in that time, as $\frac{PQ}{PQ \times \sqrt{SP}}$ or $\frac{1}{\sqrt{SP}}$, that is, in the subduplicate ratio of SP reciprocally. And, by a like reasoning, the velocity with which the arc QR is described, is in the subduplicate ratio of SQ reciprocally. Now those arcs PQ and QR are as the describing velocities to each other; that is, in the subduplicate ratio of SQ to SP, or as SQ to $\sqrt{SP \times SQ}$; and, because of the equal angles SPQ, SQr, and the equal areas PSQ, QSr, the arc PQ is to the arc Qr as SQ to SP. Take the differences of the proportional consequents, and the arc PQ will be to the arc Rr as SQ to $SP - \sqrt{SP \times SQ}$, or $\frac{1}{2}$VQ. For the points P and Q coinciding, the ultimate ratio of $SP - \sqrt{SP \times SQ}$ to $\frac{1}{2}$VQ is the ratio of equality. Because the decrement of the arc PQ arising from the resistance, or its double Rr, is as the resistance and the square of the time conjunctly, the resistance will be as $\frac{Rr}{PQ^2 \times SP}$. But PQ was to R$r$ as SQ to $\frac{1}{2}$VQ, and thence $\frac{Rr}{PQ^2 \times SP}$ becomes as $\frac{\frac{1}{2}VQ}{PQ \times SP \times SQ}$, or as $\frac{\frac{1}{2}OS}{OP \times SP^2}$. For the points P and Q coinciding, SP and SQ coincide also, and the angle PVQ becomes a right one; and, because of the similar triangles PVQ, PSO, PQ becomes to $\frac{1}{2}$VQ as OP to $\frac{1}{2}$OS. Therefore $\frac{OS}{OP \times SP^2}$ is as the resistance, that is, in the ratio of the density of the medium in P and the duplicate ratio of the velocity conjunctly. Subduct the duplicate ratio of the velocity, namely, the ratio $\frac{1}{SP}$, and there will remain the density of the medium in P, as $\frac{OS}{OP \times SP}$. Let the spiral be given, and, because of the given ratio of OS to OP, the density of the medium in P will be as $\frac{1}{SP}$. Therefore in a medium whose density is reciprocally as SP the distance from the centre, a body will revolve in this spiral. Q.E.D.

COR. 1. The velocity in any place P, is always the same wherewith a body in a non-resisting medium with the same centripetal force would revolve in a circle, at the same distance SP from the centre.

COR. 2. The density of the medium, if the distance SP be given, is as $\frac{OS}{OP}$, but if that distance is not given, as $\frac{OS}{OP \times SP}$. And thence a spiral may be fitted to any density of the medium.

COR. 3. The force of the resistance in any place P is to the centripetal force in the same place as $\frac{1}{2}$OS to OP. For those forces are to each other as $\frac{1}{2}Rr$ and TQ, or as $\frac{\frac{1}{4}VQ \times PQ}{SQ}$ and $\frac{\frac{1}{2}PQ^2}{SP}$, that is, as $\frac{1}{2}$VQ and PQ, or $\frac{1}{2}$OS and OP. The spiral therefore being given, there is given the proportion of the resistance to the centripetal force; and, *vice versa*, from that proportion given the spiral is given.

COR. 4. Therefore the body cannot revolve in this spiral, except where the force of resistance is less than half the centripetal force. Let the resistance be made equal to half the centripetal force, and the spiral will coincide with the right line PS, and in that right line the body will descend to the centre with a velocity that is to the velocity, with which it was proved before, in the case of the parabola (Theor. X, Book I), the descent would be made in a non-resisting medium, in the subduplicate ratio of unity to the number two. And the times of the descent will be here reciprocally as the velocities, and therefore given.

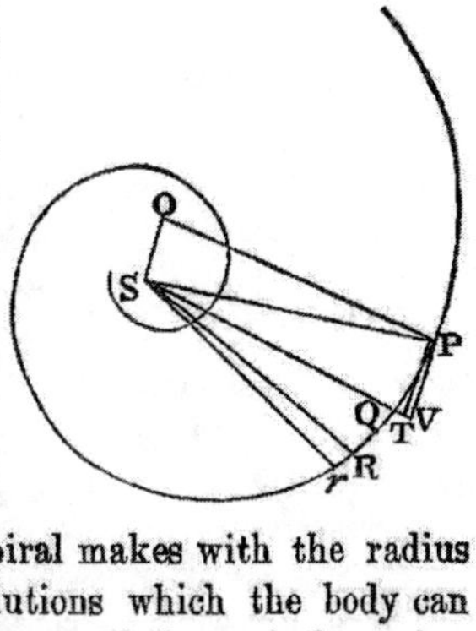

COR. 5. And because at equal distances from the centre the velocity is the same in the spiral PQR as it is in the right line SP, and the length of the spiral is to the length of the right line PS in a given ratio, namely, in the ratio of OP to OS; the time of the descent in the spiral will be to the time of the descent in the right line SP in the same given ratio, and therefore given.

COR. 6. If from the centre S, with any two given intervals, two circles are described; and these circles remaining, the angle which the spiral makes with the radius PS be any how changed; the number of revolutions which the body can complete in the space between the circumferences of those circles, going round in the spiral from one circumference to another, will be as $\frac{PS}{OS}$, or as the tangent of the angle which the spiral makes with the radius PS; and the time of the same revolutions will be as $\frac{OP}{OS}$, that is, as the secant of the same angle, or reciprocally as the density of the medium.

Cor. 7. If a body, in a medium whose density is reciprocally as the distances of places from the centre, revolves in any curve AEB about that centre, and cuts the first radius AS in the same angle in B as it did before in A, and that with a velocity that shall be to its first velocity in A reciprocally in a subduplicate ratio of the distances from the centre (that is, as AS to a mean proportional between AS and BS) that body will continue to describe innumerable similar revolutions BFC, CGD, &c., and by its intersections will distinguish the radius AS into parts AS, BS, CS, DS, &c., that are continually proportional. But the times of the revolutions will be as the perimeters of the orbits AEB, BFC, CGD, &c., directly, and the velocities at the beginnings A, B, C of those orbits inversely; that is as $AS^{\frac{3}{2}}$, $BS^{\frac{3}{2}}$, $CS^{\frac{3}{2}}$. And the whole time in which the body will arrive at the centre, will be to the time of the first revolution as the sum of all the continued proportionals $AS^{\frac{3}{2}}$, $BS^{\frac{3}{2}}$, $CS^{\frac{3}{2}}$, going on *ad infinitum*, to the first term $AS^{\frac{3}{2}}$; that is, as the first term $AS^{\frac{3}{2}}$ to the difference of the two first $AS^{\frac{3}{2}} - BS^{\frac{3}{2}}$, or as $\frac{2}{3}$ AS to AB very nearly. Whence the whole time may be easily found.

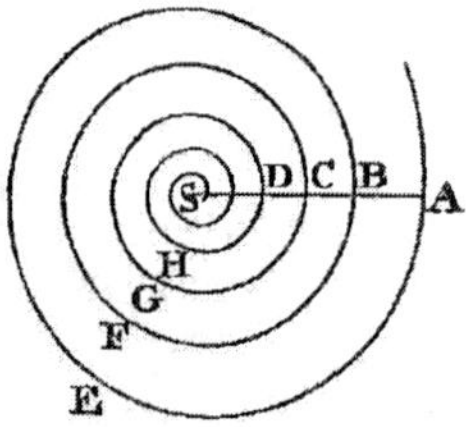

Cor. 8. From hence also may be deduced, near enough, the motions of bodies in mediums whose density is either uniform, or observes any other assigned law. From the centre S, with intervals SA, SB, SC, &c., continually proportional, describe as many circles; and suppose the time of the revolutions between the perimeters of any two of those circles, in the medium whereof we treated, to be to the time of the revolutions between the same in the medium proposed as the mean density of the proposed medium between those circles to the mean density of the medium whereof we treated, between the same circles, nearly: and that the secant of the angle in which the spiral above determined, in the medium whereof we treated, cuts the radius AS, is in the same ratio to the secant of the angle in which the new spiral, in the proposed medium, cuts the same radius: and also that the number of all the revolutions between the same two circles is nearly as the tangents of those angles. If this be done every where between every two circles, the motion will be continued through all the circles. And by this means one may without difficulty conceive at what rate and in what time bodies ought to revolve in any regular medium.

COR. 9. And although these motions becoming eccentrical should be performed in spirals approaching to an oval figure, yet, conceiving the several revolutions of those spirals to be at the same distances from each other, and to approach to the centre by the same degrees as the spiral above described, we may also understand how the motions of bodies may be performed in spirals of that kind.

PROPOSITION XVI. THEOREM XIII.

If the density of the medium in each of the places be reciprocally as the distance of the places from the immoveable centre, and the centripetal force be reciprocally as any power of the same distance, I say, that the body may revolve in a spiral intersecting all the radii drawn from that centre in a given angle.

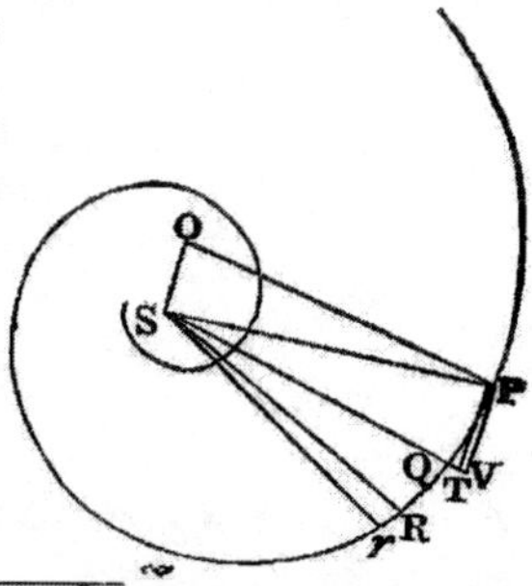

This is demonstrated in the same manner as the foregoing Proposition. For if the centripetal force in P be reciprocally as any power SP^{n+1} of the distance SP whose index is $n+1$; it will be collected, as above, that the time in which the body describes any arc PQ, will be as $PQ \times PS^{\frac{1}{2}n}$; and the resistance in P as $\frac{Rr}{PQ^2 \times SP^n}$, or as $\frac{\overline{1 - \frac{1}{2}n} \times VQ}{PQ \times SP^n \times SQ}$, and therefore as $\frac{\overline{1 - \frac{1}{2}n} \times OS}{OP \times SP^{n+1}}$, that is, (because $\frac{\overline{1 - \frac{1}{2}n} \times OS}{OP}$ is a given quantity), reciprocally as SP^{n+1}. And therefore, since the velocity is reciprocally as $SP^{\frac{1}{2}n}$, the density in P will be reciprocally as SP.

COR. 1. The resistance is to the centripetal force as $\overline{1 - \frac{1}{2}n} \times OS$ to OP.

COR. 2. If the centripetal force be reciprocally as SP^3, $1 - \frac{1}{2}n$ will be $= 0$; and therefore the resistance and density of the medium will be nothing, as in Prop. IX, Book I.

COR. 3. If the centripetal force be reciprocally as any power of the radius SP, whose index is greater than the number 3, the affirmative resistance will be changed into a negative.

SCHOLIUM.

This Proposition and the former, which relate to mediums of unequal density, are to be understood of the motion of bodies that are so small, that the greater density of the medium on one side of the body above that on the other is not to be considered. I suppose also the resistance, *cæteris paribus*, to be proportional to its density. Whence, in mediums whose force of resistance is not as the density, the density must be so much augmented or diminished, that either the excess of the resistance may be taken away, or the defect supplied.

PROPOSITION XVII. PROBLEM IV.

To find the centripetal force and the resisting force of the medium, by which a body, the law of the velocity being given, shall revolve in a given spiral.

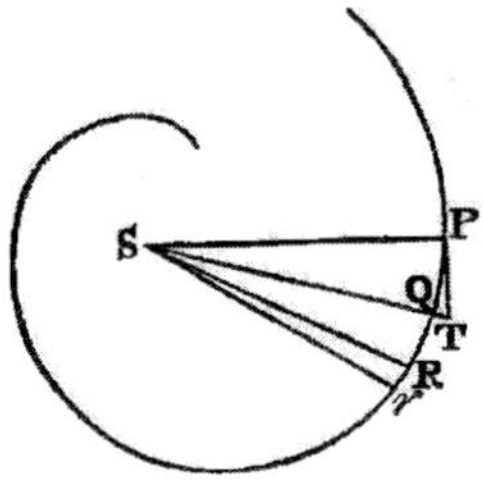

Let that spiral be PQR. From the velocity, with which the body goes over the very small arc PQ, the time will be given; and from the altitude TQ, which is as the centripetal force, and the square of the time, that force will be given. Then from the difference RS*r* of the areas PSQ and QSR described in equal particles of time, the retardation of the body will be given; and from the retardation will be found the resisting force and density of the medium.

PROPOSITION XVIII. PROBLEM V.

The law of centripetal force being given, to find the density of the medium in each of the places thereof, by which a body may describe a given spiral.

From the centripetal force the velocity in each place must be found; then from the retardation of the velocity the density of the medium is found, as in the foregoing Proposition.

But I have explained the method of managing these Problems in the tenth Proposition and second Lemma of this Book; and will no longer detain the reader in these perplexed disquisitions. I shall now add some things relating to the forces of progressive bodies, and to the density and resistance of those mediums in which the motions hitherto treated of, and those akin to them, are performed.

SECTION V.

Of the density and compression of fluids ; and of hydrostatics.

THE DEFINITION OF A FLUID.

A fluid is any body whose parts yield to any force impressed on it, and, by yielding, are easily moved among themselves.

PROPOSITION XIX. THEOREM XIV.

All the parts of a homogeneous and unmoved fluid included in any unmoved vessel, and compressed on every side (setting aside the consideration of condensation, gravity, and all centripetal forces), will be equally pressed on every side, and remain in their places without any motion arising from that pressure.

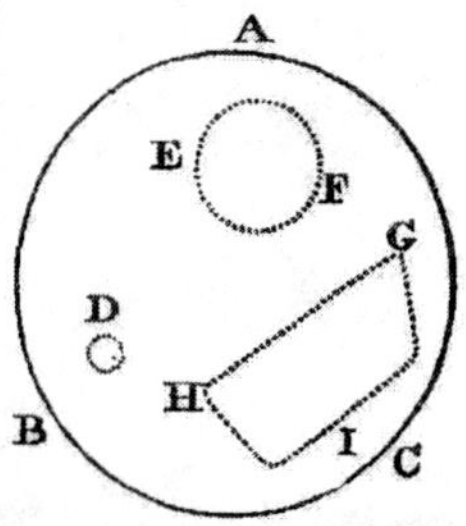

CASE 1. Let a fluid be included in the spherical vessel ABC, and uniformly compressed on every side: I say, that no part of it will be moved by that pressure. For if any part, as D, be moved, all such parts at the same distance from the centre on every side must necessarily be moved at the same time by a like motion; because the pressure of them all is similar and equal; and all other motion is excluded that does not arise from that pressure. But if these parts come all of them nearer to the centre, the fluid must be condensed towards the centre, contrary to the supposition. If they recede from it, the fluid must be condensed towards the circumference; which is also contrary to the supposition. Neither can they move in any one direction retaining their distance from the centre, because for the same reason, they may move in a contrary direction; but the same part cannot be moved contrary ways at the same time. Therefore no part of the fluid will be moved from its place. Q.E.D.

CASE 2. I say now, that all the spherical parts of this fluid are equally pressed on every side. For let EF be a spherical part of the fluid; if this be not pressed equally on every side, augment the lesser pressure till it be pressed equally on every side; and its parts (by Case 1) will remain in their places. But before the increase of the pressure, they would remain in their places (by Case 1); and by the addition of a new pressure they will be moved, by the definition of a fluid, from those places. Now these two conclusions contradict each other. Therefore it was false to say that the sphere EF was not pressed equally on every side. Q.E.D.

CASE 3. I say besides, that different spherical parts have equal pressures. For the contiguous spherical parts press each other mutually and equally in the point of contact (by Law III). But (by Case 2) they are pressed on every side with the same force. Therefore any two spherical parts not contiguous, since an intermediate spherical part can touch both, will be pressed with the same force. Q.E.D.

CASE 4. I say now, that all the parts of the fluid are every where pressed equally. For any two parts may be touched by spherical parts in any points whatever; and there they will equally press those spherical parts (by Case 3), and are reciprocally equally pressed by them (by Law III). Q.E.D.

CASE 5. Since, therefore, any part GHI of the fluid is inclosed by the rest of the fluid as in a vessel, and is equally pressed on every side; and also its parts equally press one another, and are at rest among themselves; it is manifest that all the parts of any fluid as GHI, which is pressed equally on every side, do press each other mutually and equally, and are at rest among themselves. Q.E.D.

CASE 6. Therefore if that fluid be included in a vessel of a yielding substance, or that is not rigid, and be not equally pressed on every side, the same will give way to a stronger pressure, by the Definition of fluidity.

CASE 7. And therefore, in an inflexible or rigid vessel, a fluid will not sustain a stronger pressure on one side than on the other, but will give way to it, and that in a moment of time; because the rigid side of the vessel does not follow the yielding liquor. But the fluid, by thus yielding, will press against the opposite side, and so the pressure will tend on every side to equality. And because the fluid, as soon as it endeavours to recede from the part that is most pressed, is withstood by the resistance of the vessel on the opposite side, the pressure will on every side be reduced to equality, in a moment of time, without any local motion: and from thence the parts of the fluid (by Case 5) will press each other mutually and equally, and be at rest among themselves. Q.E.D.

COR. Whence neither will a motion of the parts of the fluid among themselves be changed by a pressure communicated to the external superficies, except so far as either the figure of the superficies may be somewhere altered, or that all the parts of the fluid, by pressing one another more intensely or remissly, may slide with more or less difficulty among themselves.

PROPOSITION XX. THEOREM XV.

If all the parts of a spherical fluid, homogeneous at equal distances from the centre, lying on a spherical concentric bottom, gravitate towards the centre of the whole, the bottom will sustain the weight of a cylinder, whose base is equal to the superficies of the bottom, and whose altitude is the same with that of the incumbent fluid.

Let DHM be the superficies of the bottom, and AEI the upper superficies of the fluid. Let the fluid be distinguished into concentric orbs of equal thickness, by the innumerable spherical superficies BFK, CGL; and

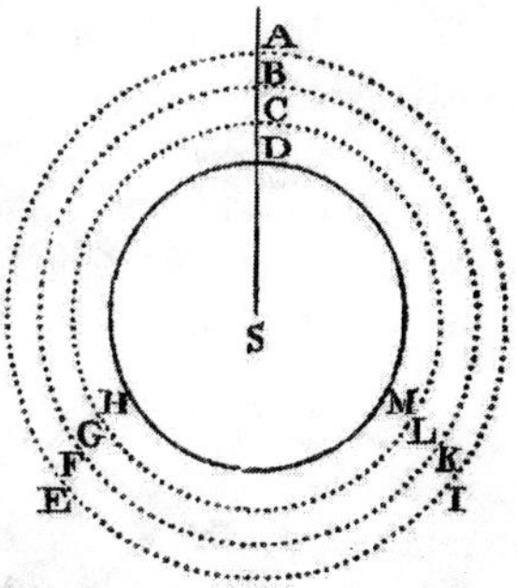

conceive the force of gravity to act only in the upper superficies of every orb, and the actions to be equal on the equal parts of all the superficies. Therefore the upper superficies AE is pressed by the single force of its own gravity, by which all the parts of the upper orb, and the second superficies BFK, will (by Prop. XIX), according to its measure, be equally pressed. The second superficies BFK is pressed likewise by the force of its own gravity, which, added to the former force, makes the pressure double. The third superficies CGL is, according to its measure, acted on by this pressure and the force of its own gravity besides, which makes its pressure triple. And in like manner the fourth superficies receives a quadruple pressure, the fifth superficies a quintuple, and so on. Therefore the pressure acting on every superficies is not as the solid quantity of the incumbent fluid, but as the number of the orbs reaching to the upper surface of the fluid; and is equal to the gravity of the lowest orb multiplied by the number of orbs: that is, to the gravity of a solid whose ultimate ratio to the cylinder above-mentioned (when the number of the orbs is increased and their thickness diminished, *ad infinitum*, so that the action of gravity from the lowest superficies to the uppermost may become continued) is the ratio of equality. Therefore the lowest superficies sustains the weight of the cylinder above determined. Q.E.D. And by a like reasoning the Proposition will be evident, where the gravity of the fluid decreases in any assigned ratio of the distance from the centre, and also where the fluid is more rare above and denser below. Q.E.D.

COR. 1. Therefore the bottom is not pressed by the whole weight of the incumbent fluid, but only sustains that part of it which is described in the Proposition; the rest of the weight being sustained archwise by the spherical figure of the fluid.

COR. 2. The quantity of the pressure is the same always at equal distances from the centre, whether the superficies pressed be parallel to the horizon, or perpendicular, or oblique; or whether the fluid, continued upwards from the compressed superficies, rises perpendicularly in a rectilinear direction, or creeps obliquely through crooked cavities and canals, whether those passages be regular or irregular, wide or narrow. That the pressure is not altered by any of these circumstances, may be collected by applying the demonstration of this Theorem to the several cases of fluids.

COR. 3. From the same demonstration it may also be collected (by Prop. XIX), that the parts of a heavy fluid acquire no motion among themselves by the pressure of the incumbent weight, except that motion which arises from condensation.

COR. 4. And therefore if another body of the same specific gravity, incapable of condensation, be immersed in this fluid, it will acquire no motion by the pressure of the incumbent weight: it will neither descend nor ascend, nor change its figure. If it be spherical, it will remain so, notwithstanding the pressure; if it be square, it will remain square; and that, whether it be soft or fluid; whether it swims freely in the fluid, or lies at the bottom. For any internal part of a fluid is in the same state with the submersed body; and the case of all submersed bodies that have the same magnitude, figure, and specific gravity, is alike. If a submersed body, retaining its weight, should dissolve and put on the form of a fluid, this body, if before it would have ascended, descended, or from any pressure assume a new figure, would now likewise ascend, descend, or put on a new figure; and that, because its gravity and the other causes of its motion remain. But (by Case 5, Prop. XIX) it would now be at rest, and retain its figure. Therefore also in the former case.

COR. 5. Therefore a body that is specifically heavier than a fluid contiguous to it will sink; and that which is specifically lighter will ascend, and attain so much motion and change of figure as that excess or defect of gravity is able to produce. For that excess or defect is the same thing as an impulse, by which a body, otherwise *in equilibrio* with the parts of the fluid, is acted on; and may be compared with the excess or defect of a weight in one of the scales of a balance.

COR. 6. Therefore bodies placed in fluids have a twofold gravity; the one true and absolute, the other apparent, vulgar, and comparative. Absolute gravity is the whole force with which the body tends downwards; relative and vulgar gravity is the excess of gravity with which the body tends downwards more than the ambient fluid. By the first kind of gravity the parts of all fluids and bodies gravitate in their proper places; and therefore their weights taken together compose the weight of the whole. For the whole taken together is heavy, as may be experienced in vessels full of liquor; and the weight of the whole is equal to the weights of all the parts, and is therefore composed of them. By the other kind of gravity bodies do not gravitate in their places; that is, compared with one another, they do not preponderate, but, hindering one another's endeavours to descend, remain in their proper places, as if they were not heavy. Those things which are in the air, and do not preponderate, are commonly looked on as not heavy. Those which do preponderate are commonly reckoned heavy, in as much as they are not sustained by the weight of the air. The common weights are nothing else but the excess of the true weights above the weight of the air. Hence also, vulgarly, those things are called light which are less heavy, and, by yielding to the preponderating air, mount upwards. But these are only comparatively light, and not truly so, because they descend *in vacuo*. Thus, in water, bodies which, by their greater or less gravity, descend or ascend, are comparatively and apparently heavy or light; and their comparative and apparent gravity or levity is the excess or defect by which their true gravity either exceeds the gravity of the water or is exceeded by it. But those things which neither by preponderating descend, nor, by yielding to the preponderating fluid, ascend, although by their true weight they do increase the weight of the whole, yet comparatively, and in the sense of the vulgar, they do not gravitate in the water. For these cases are alike demonstrated.

Cor. 7. These things which have been demonstrated concerning gravity take place in any other centripetal forces.

Cor. 8. Therefore if the medium in which any body moves be acted on either by its own gravity, or by any other centripetal force, and the body be urged more powerfully by the same force; the difference of the forces is that very motive force, which, in the foregoing Propositions, I have considered as a centripetal force. But if the body be more lightly urged by that force, the difference of the forces becomes a centrifugal force, and is to be considered as such.

Cor. 9. But since fluids by pressing the included bodies do not change their external figures, it appears also (by Cor. Prop. XIX) that they will not change the situation of their internal parts in relation to one another; and therefore if animals were immersed therein, and that all sensation did arise from the motion of their parts, the fluid will neither hurt the immersed bodies, nor excite any sensation, unless so far as those bodies may be condensed by the compression. And the case is the same of any system of bodies encompassed with a compressing fluid. All the parts of the system will be agitated with the same motions as if they were placed in a vacuum, and would only retain their comparative gravity; unless so far as the fluid may somewhat resist their motions, or be requisite to conglutinate them by compression.

PROPOSITION XXI. THEOREM XVI.

Let the density of any fluid be proportional to the compression, and its parts be attracted downwards by a centripetal force reciprocally proportional to the distances from the centre: I say, that, if those distances be taken continually proportional, the densities of the fluid at the same distances will be also continually proportional.

Let ATV denote the spherical bottom of the fluid, S the centre, SA, SB, SC, SD, SE, SF, &c., distances continually proportional. Erect the perpendiculars AH, BI, CK, DL, EM, FN, &c., which shall be as the densities of the medium in the places A, B, C, D, E, F; and the specific gravities in those places will be as $\frac{AH}{AS}, \frac{BI}{BS}, \frac{CK}{CS}$, &c., or, which is all one, as

$\frac{AH}{AB}, \frac{BI}{BC}, \frac{CK}{CD}$, &c. Suppose, first, these gravities to be uniformly continued from A to B, from B to C, from C to D, &c., the decrements in the points B, C, D, &c., being taken by steps. And these gravities drawn into the altitudes AB, BC, CD, &c., will give the pressures AH, BI, CK, &c., by which the bottom ATV is acted on (by Theor. XV). Therefore the particle A sustains all the pressures AH, BI, CK, DL, &c., proceeding *in infinitum;* and the particle B sustains the pressures of all but the first AH; and the particle C all but the two first AH, BI; and so on: and therefore the density AH of the first particle A is to the density BI of the second particle B as the sum of all AH + BI + CK + DL, *in infinitum*, to the sum of all BI + CK + DL, &c. And BI the density of the second particle B is to CK the density of the third C, as the sum of all BI + CK + DL, &c., to the sum of all CK + DL, &c. Therefore these sums are proportional to their differences AH, BI, CK, &c., and therefore continually proportional (by Lem. 1 of this Book); and therefore the differences AH, BI, CK, &c., proportional to the sums, are also continually proportional. Wherefore since the densities in the places A, B, C, &c., are as AH, BI, CK, &c., they will also be continually proportional. Proceed intermissively, and, *ex æquo*, at the distances SA, SC, SE, continually proportional, the densities AH, CK, EM will be continually proportional. And by the same reasoning, at any distances SA, SD, SG, continually proportional, the densities AH, DL, GO, will be continually proportional. Let now the points A, B, C, D, E, &c., coincide, so that the progression of the specific gravities from the bottom A to the top of the fluid may be made continual; and at any distances SA, SD, SG, continually proportional, the densities AH, DL, GO, being all along continually proportional, will still remain continually proportional. Q.E.D.

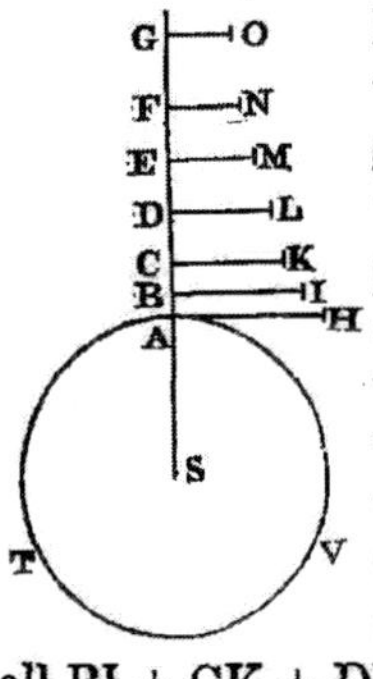

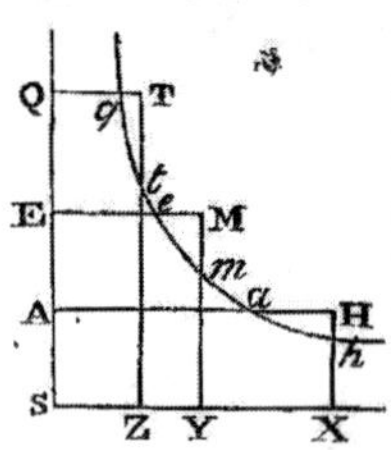

Cor. Hence if the density of the fluid in two places, as A and E, be given, its density in any other place Q may be collected. With the centre S, and the rectangular asymptotes SQ, SX, describe an hyperbola cutting the perpendiculars AH, EM, QT in *a*, *e*, and *q*, as also the perpendiculars HX, MY, TZ, let fall upon the asypmtote SX, in *h*, *m*, and *t*. Make the area Y*mt*Z to the given area Y*mh*X as the given area E*eq*Q to the given area E*ea*A; and the line Z*t* produced will cut off the line QT proportional to the density. For if the lines SA, SE, SQ are continually proportional, the areas E*eq*Q, E*ea*A will be equal, and thence the areas Y*mt*Z, X*hm*Y, proportional to them, will be also equal; and the lines SX, SY, SZ, that is, AH, EM, QT continually proportional, as they ought to be. And if the lines SA, SE, SQ, obtain any other order in the series of continued proportionals, the lines AH, EM, QT, because of the proportional hyperbolic areas, will obtain the same order in another series of quantities continually proportional.

PROPOSITION XXII. THEOREM XVII.

Let the density of any fluid be proportional to the compression, and its parts be attracted downwards by a gravitation reciprocally proportional to the squares of the distances from the centre: I say, that if the distances be taken in harmonic progression, the densities of the fluid at those distances will be in a geometrical progression.

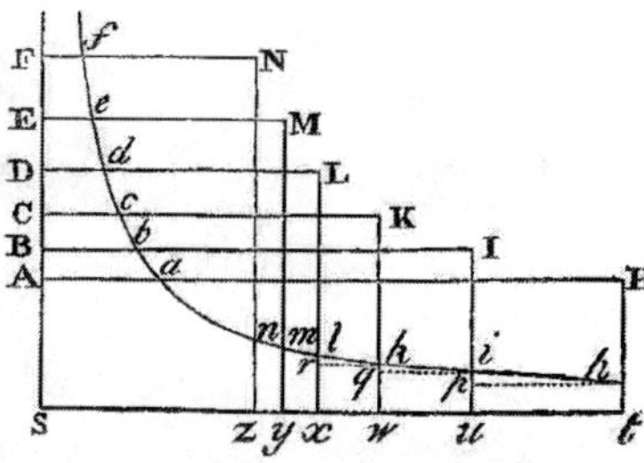

Let S denote the centre, and SA, SB, SC, SD, SE, the distances in geometrical progression. Erect the perpendiculars AH, BI, CK, &c., which shall be as the densities of the fluid in the places A, B, C, D, E, &c., and the specific gravities thereof in those places will be as $\frac{AH}{SA^2}, \frac{BI}{SB^2}, \frac{CK}{SC^2}$, &c. Suppose these gravities to be uniformly continued, the first from A to B, the second from B to C, the third from C to D, &c. And these drawn into the altitudes AB, BC, CD, DE, &c., or, which is the same thing, into the distances SA, SB, SC, &c., proportional to those altitudes, will give $\frac{AH}{SA}, \frac{BI}{SB}, \frac{CK}{SC}$, &c., the exponents of the pressures. Therefore since the densities are as the sums of those pressures, the differences AH — BI, BI — CK, &c., of the densities will be as the differences of those sums $\frac{AH}{SA}, \frac{BI}{SB}, \frac{CK}{SC}$, &c. With the centre S, and the asymptotes SA, S*x*, describe any hyperbola, cutting the perpendiculars AH, BI, CK, &c., in *a*, *b*, *c*, &c., and the perpendiculars H*t*, I*u*, K*w*, let fall upon the asymptote S*x*, in *h*, *i*, *k*; and the differences of the densities *tu*, *uw*, &c., will be as $\frac{AH}{SA}, \frac{BI}{SB}$, &c. And the rectangles $tu \times th$, $uw \times ui$, &c., or *tp*, *uq*, &c., as $\frac{AH \times th}{SA}, \frac{BI \times ui}{SB}$, &c., that is, as A*a*, B*b*, &c. For, by the nature of the hyperbola, SA is to AH or S*t* as *th* to A*a*, and therefore $\frac{AH \times th}{SA}$ is equal to A*a*. And, by a like reasoning, $\frac{BI \times ui}{SB}$ is equal to B*b*, &c. But A*a*, B*b*, C*c*, &c., are continually proportional, and therefore proportional to their differences A*a* — B*b*, B*b* — C*c*, &c., therefore the rectangles *tp*, *uq*, &c., are proportional to those differences; as also the sums of the rectangles *tp* + *uq*, or *tp* + *uq* + *wr* to the sums of the differences A*a* — C*c* or A*a* — D*d*. Suppose several of these terms, and the sum of all the differences, as A*a* — F*f*, will be proportional to the sum of all the rectangles, as *zthn*. Increase the number of terms, and diminish the distances of the points A, B, C, &c., *in infinitum*, and those rectangles will become equal to the hyperbolic area *zthn*, and therefore the difference A*a* — F*f* is proportional to this area. Take now any distances, as SA, SD, SF, in harmonic progression, and the differences A*a* — D*d*, D*d* — F*f* will be equal; and therefore the areas *thlx*, *xlnz*, proportional to those differences will be equal among themselves, and the densities S*t*, S*x*, S*z*, that is, AH, DL, FN, continually proportional. Q.E.D.

Cor. Hence if any two densities of the fluid, as AH and BI, be given, the area *thiu*, answering to their difference *tu*, will be given; and thence the density FN will be found at any height SF, by taking the area *thnz* to that given area *thiu* as the difference A*a* — F*f* to the difference A*a* — B*b*.

SCHOLIUM.

By a like reasoning it may be proved, that if the gravity of the particles of a fluid be diminished in a triplicate ratio of the distances from the centre; and the reciprocals of the squares of the distances SA, SB, SC, &c., (namely, $\frac{SA^3}{SA^2}, \frac{SA^3}{SB^2}, \frac{SA^3}{SC^2}$) be taken in an arithmetical progression, the densities AH, BI, CK, &c., will be in a geometrical progression. And if the gravity be diminished in a quadruplicate ratio of the distances, and the reciprocals of the cubes of the distances (as $\frac{SA^4}{SA^3}, \frac{SA^4}{SB^3}, \frac{SA^4}{SC^3}$, &c.,) be taken in arithmetical progression, the densities AH, BI, CK, &c., will be in geometrical progression. And so *in infinitum*. Again; if the gravity of the particles of the fluid be the same at all distances, and the distances be in arithmetical progression, the densities will be in a geometrical progression as Dr. *Halley* has found. If the gravity be as the distance, and the squares of the distances be in arithmetical progression, the densities will be in geometrical progression. And so *in infinitum*. These things will be so, when the density of the fluid condensed by compression is as the force of compression; or, which is the same thing, when the space possessed by the fluid is reciprocally as this force. Other laws of condensation may be supposed, as that the cube of the compressing force may be as the biquadrate of the density; or the triplicate ratio of the force the same with the quadruplicate ratio of the density: in which case, if the gravity be reciprocally as the square of the distance from the centre, the density will be reciprocally as the cube of the distance. Suppose that the cube of the compressing force be as the quadrato-cube of the density; and if the gravity be reciprocally as the square of the distance, the density will be reciprocally in a sesquiplicate ratio of the distance. Suppose the compressing force to be in a duplicate ratio of the density, and the gravity reciprocally in a duplicate ratio of the distance, and the density will be reciprocally as the distance. To run over all the cases that might be offered would be tedious. But as to our own air, this is certain from experiment, that its density is either accurately, or very nearly at least, as the compressing force; and therefore the density of the air in the atmosphere of the earth is as the weight of the whole incumbent air, that is, as the height of the mercury in the barometer.

PROPOSITION XXIII. THEOREM XVIII.

If a fluid be composed of particles mutually flying each other, and the density be as the compression, the centrifugal forces of the particles will be reciprocally proportional to the distances of their centres. And, vice versa, *particles flying each other, with forces that are reciprocally proportional to the distances of their centres, compose an elastic fluid, whose density is as the compression.*

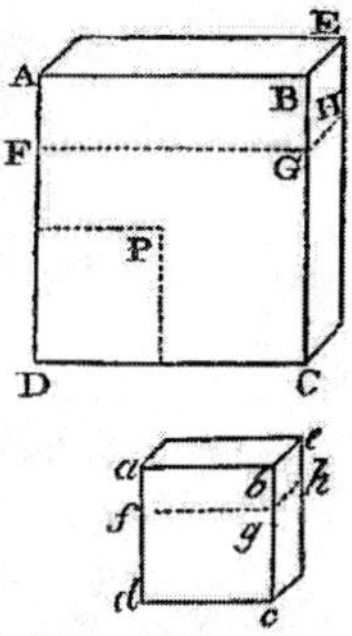

Let the fluid be supposed to be included in a cubic space ACE, and then to be reduced by compression into a lesser cubic space *ace*; and the distances of the particles retaining a like situation with respect to each other in both the spaces, will be as the sides AB, *ab* of the cubes; and the densities of the mediums will be reciprocally as the containing spaces AB^3, ab^3. In the plane side of the greater cube ABCD take the square DP equal to the plane side *db* of the lesser cube: and, by the supposition, the pressure with which the square DP urges the inclosed fluid will be to the pressure with which that square *db* urges the inclosed fluid as the densities of the mediums are to each other, that is, as ab^3 to AB^3. But the pressure with which the square DB urges the included fluid is to the pressure with which the square DP urges the same fluid as the square DB to the square DP, that is, as AB^2 to ab^2. Therefore, *ex æquo*, the pressure with which the square DB urges the fluid is to the pressure with which the square *db* urges the fluid as *ab* to AB. Let the planes FGH, *fgh*, be drawn through the middles of the two cubes, and divide the fluid into two parts. These parts will press each other mutually with the same forces with which they are themselves pressed by the planes AC, *ac*, that is, in the proportion of *ab* to AB: and therefore the centrifugal forces by which these pressures are sustained are in the same ratio. The number of the particles being equal, and the situation alike, in both cubes, the forces which all the particles exert, according to the planes FGH, *fgh*, upon all, are as the forces which each exerts on each. Therefore the forces which each exerts on each, according to the plane FGH in the greater cube, are to the forces which each exerts on each, according to the plane *fgh* in the lesser cube, as *ab* to AB, that is, reciprocally as the distances of the particles from each other. Q.E.D.

And, *vice versa*, if the forces of the single particles are reciprocally as the distances, that is, reciprocally as the sides of the cubes AB, *ab*; the sums of the forces will be in the same ratio, and the pressures of the sides DB, *db* as the sums of the forces; and the pressure of the square DP to the pressure of the side DB as ab^2 to AB^2. And, *ex æquo*, the pressure of the square DP to the pressure of the side *db* as ab^3 to AB^3; that is, the force of compression in the one to the force of compression in the other as the density in the former to the density in the latter. Q.E.D.

SCHOLIUM.

By a like reasoning, if the centrifugal forces of the particles are reciprocally in the duplicate ratio of the distances between the centres, the cubes of the compressing forces will be as the biquadrates of the densities. If the centrifugal forces be reciprocally in the triplicate or quadruplicate ratio of the distances, the cubes of the compressing forces will be as the quadrato-cubes, or cubo-cubes of the densities. And universally, if D be put for the distance, and E for the density of the compressed fluid, and the centrifugal forces be reciprocally as any power D^n of the distance, whose index is the number n, the compressing forces will be as the cube roots of the power E^{n+2}, whose index is the number $n + 2$; and the contrary. All these things are to be understood of particles whose centrifugal forces terminate in those particles that are next them, or are diffused not much further. We have an example of this in magnetical bodies. Their attractive virtue is terminated nearly in bodies of their own kind that are next them. The virtue of the magnet is contracted by the interposition of an iron plate, and is almost terminated at it: for bodies further off are not attracted by the magnet so much as by the iron plate. If in this manner particles repel others of their own kind that lie next them, but do not exert their virtue on the more remote, particles of this kind will compose such fluids as are treated of in this Proposition. If the virtue of any particle diffuse itself every way *in infinitum*, there will be required a greater force to produce an equal condensation of a greater quantity of the fluid. But whether elastic fluids do really consist of particles so repelling each other, is a physical question. We have here demonstrated mathematically the property of fluids consisting of particles of this kind, that hence philosophers may take occasion to discuss that question.

SECTION VI.

Of the motion and resistance of funependulous bodies.

PROPOSITION XXIV. THEOREM XIX.

The quantities of matter in funependulous bodies, whose centres of oscillation are equally distant from the centre of suspension, are in a ratio compounded of the ratio of the weights and the duplicate ratio of the times of the oscillations in vacuo.

For the velocity which a given force can generate in a given matter in a given time is as the force and the time directly, and the matter inversely. The greater the force or the time is, or the less the matter, the greater velocity will be generated. This is manifest from the second Law of Motion. Now if pendulums are of the same length, the motive forces in places equally distant from the perpendicular are as the weights: and therefore if two bodies by oscillating describe equal arcs, and those arcs are divided into equal parts; since the times in which the bodies describe each of the correspondent parts of the arcs are as the times of the whole oscillations, the velocities in the correspondent parts of the oscillations will be to each other as the motive forces and the whole times of the oscillations directly, and the quantities of matter reciprocally: and therefore the quantities of matter are as the forces and the times of the oscillations directly and the velocities reciprocally. But the velocities reciprocally are as the times, and therefore the times directly and the velocities reciprocally are as the squares of the times; and therefore the quantities of matter are as the motive forces and the squares of the times, that is, as the weights and the squares of the times. Q.E.D.

COR. 1. Therefore if the times are equal, the quantities of matter in each of the bodies are as the weights.

COR. 2. If the weights are equal, the quantities of matter will be as the squares of the times.

COR. 3. If the quantities of matter are equal, the weights will be reciprocally as the squares of the times.

COR. 4. Whence since the squares of the times, *cæteris paribus*, are as the lengths of the pendulums, therefore if both the times and quantities of matter are equal, the weights will be as the lengths of the pendulums.

Cor. 5. And universally, the quantity of matter in the pendulous body is as the weight and the square of the time directly, and the length of the pendulum inversely.

Cor. 6. But in a non-resisting medium, the quantity of matter in the pendulous body is as the comparative weight and the square of the time directly, and the length of the pendulum inversely. For the comparative weight is the motive force of the body in any heavy medium, as was shewn above; and therefore does the same thing in such a non-resisting medium as the absolute weight does in a vacuum.

Cor. 7. And hence appears a method both of comparing bodies one among another, as to the quantity of matter in each; and of comparing the weights of the same body in different places, to know the variation of its gravity. And by experiments made with the greatest accuracy, I have always found the quantity of matter in bodies to be proportional to their weight.

PROPOSITION XXV. THEOREM XX.

Funependulous bodies that are, in any medium, resisted in the ratio of the moments of time, and funependulous bodies that move in a non-resisting medium of the same specific gravity, perform their oscillations in a cycloid in the same time, and describe proportional parts of arcs together.

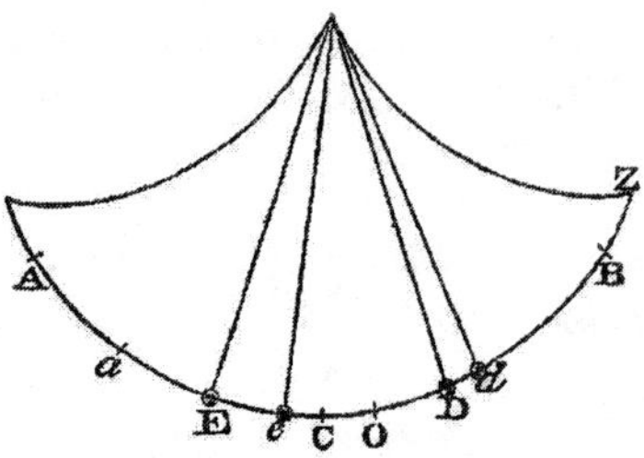

Let AB be an arc of a cycloid, which a body D, by vibrating in a non-resisting medium, shall describe in any time. Bisect that arc in C, so that C may be the lowest point thereof; and the accelerative force with which the body is urged in any place D, or *d* or E, will be as the length of the arc CD, or C*d*, or CE. Let that force be expressed by that same arc; and since the resistance is as the moment of the time, and therefore given, let it be expressed by the given part CO of the cycloidal arc, and take the arc O*d* in the same ratio to the arc CD that the arc OB has to the arc CB: and the force with which the body in *d* is urged in a resisting medium, being the excess of the force C*d* above the resistance CO, will be expressed by the arc O*d*, and will therefore be to the force with which the body D is urged in a non-resisting medium in the place D, as the arc O*d* to the arc CD; and therefore also in the place B, as the arc OB to the arc CB. Therefore if two bodies D, *d* go from the place B, and are urged by these forces; since the forces at the beginning are as the arc CB and OB, the first velocities and arcs first described will be in the same ratio. Let those arcs be BD and B*d*, and the remaining arcs

CD, *Od*, will be in the same ratio. Therefore the forces, being proportional to those arcs CD, *Od*, will remain in the same ratio as at the beginning, and therefore the bodies will continue describing together arcs in the same ratio. Therefore the forces and velocities and the remaining arcs CD, *Od*, will be always as the whole arcs CB, OB, and therefore those remaining arcs will be described together. Therefore the two bodies D and *d* will arrive together at the places C and O; that which moves in the non-resisting medium, at the place C, and the other, in the resisting medium, at the place O. Now since the velocities in C and O are as the arcs CB, OB, the arcs which the bodies describe when they go farther will be in the same ratio. Let those arcs be CE and O*e*. The force with which the body D in a non-resisting medium is retarded in E is as CE, and the force with which the body *d* in the resisting medium is retarded in *e*, is as the sum of the force C*e* and the resistance CO, that is, as O*e*; and therefore the forces with which the bodies are retarded are as the arcs CB, OB, proportional to the arcs CE, O*e*; and therefore the velocities, retarded in that given ratio, remain in the same given ratio. Therefore the velocities and the arcs described with those velocities are always to each other in that given ratio of the arcs CB and OB; and therefore if the entire arcs AB, *a*B are taken in the same ratio, the bodies D and *d* will describe those arcs together, and in the places A and *a* will lose all their motion together. Therefore the whole oscillations are isochronal, or are performed in equal times; and any parts of the arcs, as BD, B*d*, or BE, B*e*, that are described together, are proportional to the whole arcs BA, B*a*. Q.E.D.

COR. Therefore the swiftest motion in a resisting medium does not fall upon the lowest point C, but is found in that point O, in which the whole arc described B*a* is bisected. And the body, proceeding from thence to *a*, is retarded at the same rate with which it was accelerated before in its descent from B to O.

PROPOSITION XXVI. THEOREM XXI.

Funependulous bodies, that are resisted in the ratio of the velocity, have their oscillations in a cycloid isochronal.

FOR if two bodies, equally distant from their centres of suspension, describe, in oscillating, unequal arcs, and the velocities in the correspondent parts of the arcs be to each other as the whole arcs; the resistances, proportional to the velocities, will be also to each other as the same arcs. Therefore if these resistances be subducted from or added to the motive forces arising from gravity which are as the same arcs, the differences or sums will be to each other in the same ratio of the arcs; and since the increments and decrements of the velocities are as these differences or sums, the velocities will be always as the whole arcs; therefore if the velocities are in any one case as the whole arcs, they will remain always in the same ratio. But at the beginning of the motion, when the bodies begin to descend and describe those arcs, the forces, which at that time are proportional to the arcs, will generate velocities proportional to the arcs. Therefore the velocities will be always as the whole arcs to be described, and therefore those arcs will be described in the same time. Q.E.D.

PROPOSITION XXVII. THEOREM XXII.

If funependulous bodies are resisted in the duplicate ratio of their velocities, the differences between the times of the oscillations in a resisting medium, and the times of the oscillations in a non-resisting medium of the same specific gravity, will be proportional to the arcs described in oscillating nearly.

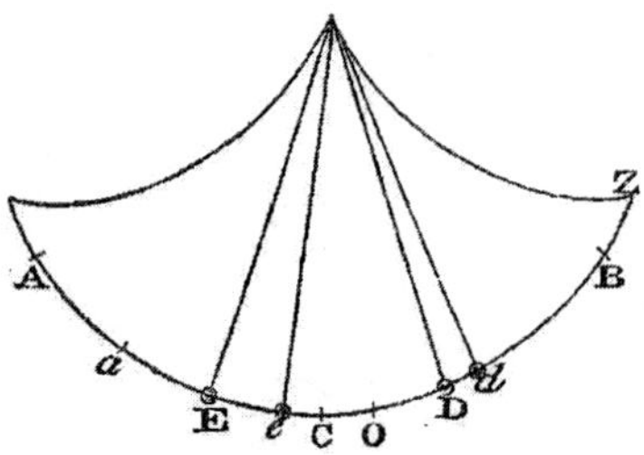

For let equal pendulums in a resisting medium describe the unequal arcs A, B; and the resistance of the body in the arc A will be to the resistance of the body in the correspondent part of the arc B in the duplicate ratio of the velocities, that is, as AA to BB nearly. If the resistance in the arc B were to the resistance in the arc A as AB to AA, the times in the arcs A and B would be equal (by the last Prop.) Therefore the resistance AA in the arc A, or AB in the arc B, causes the excess of the time in the arc A above the time in a non-resisting medium; and the resistance BB causes the excess of the time in the arc B above the time in a non-resisting medium. But those excesses are as the efficient forces AB and BB nearly, that is, as the arcs A and B. Q.E.D.

Cor. 1. Hence from the times of the oscillations in unequal arcs in a resisting medium, may be known the times of the oscillations in a non-resisting medium of the same specific gravity. For the difference of the times will be to the excess of the time in the lesser arc above the time in a non-resisting medium as the difference of the arcs to the lesser arc.

Cor. 2. The shorter oscillations are more isochronal, and very short ones are performed nearly in the same times as in a non-resisting medium. But the times of those which are performed in greater arcs are a little greater, because the resistance in the descent of the body, by which the time is prolonged, is greater, in proportion to the length described in the descent than the resistance in the subsequent ascent, by which the time is contracted. But the time of the oscillations, both short and long, seems to be prolonged in some measure by the motion of the medium. For retarded bodies are resisted somewhat less in proportion to the velocity, and accelerated bodies somewhat more than those that proceed uniformly forwards;

because the medium, by the motion it has received from the bodies, going forwards the same way with them, is more agitated in the former case, and less in the latter; and so conspires more or less with the bodies moved. Therefore it resists the pendulums in their descent more, and in their ascent less, than in proportion to the velocity; and these two causes concurring prolong the time.

PROPOSITION XXVIII. THEOREM XXIII.

If a funependulous body, oscillating in a cycloid, be resisted in the ratio of the moments of the time, its resistance will be to the force of gravity as the excess of the arc described in the whole descent above the arc described in the subsequent ascent to twice the length of the pendulum.

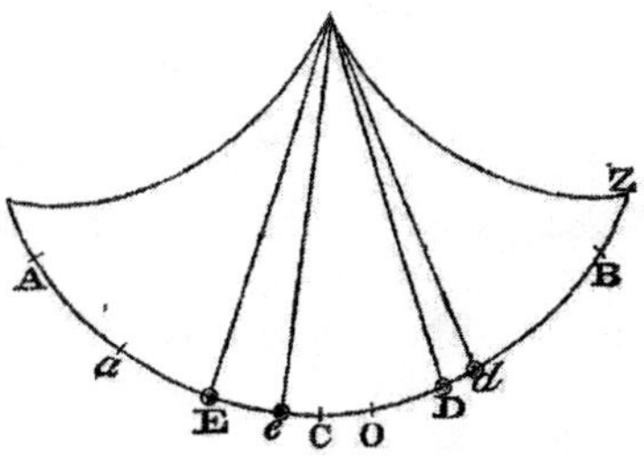

Let BC represent the arc described in the descent, C*a* the arc described in the ascent, and A*a* the difference of the arcs: and things remaining as they were constructed and demonstrated in Prop. XXV, the force with which the oscillating body is urged in any place D will be to the force of resistance as the arc CD to the arc CO, which is half of that difference A*a*. Therefore the force with which the oscillating body is urged at the beginning or the highest point of the cycloid, that is, the force of gravity, will be to the resistance as the arc of the cycloid, between that highest point and lowest point C, is to the arc CO; that is (doubling those arcs), as the whole cycloidal arc, or twice the length of the pendulum, to the arc A*a*. Q.E.D.

PROPOSITION XXIX. PROBLEM VI.

Supposing that a body oscillating in a cycloid is resisted in a duplicate ratio of the velocity: to find the resistance in each place.

Let B*a* be an arc described in one entire oscillation, C the lowest point

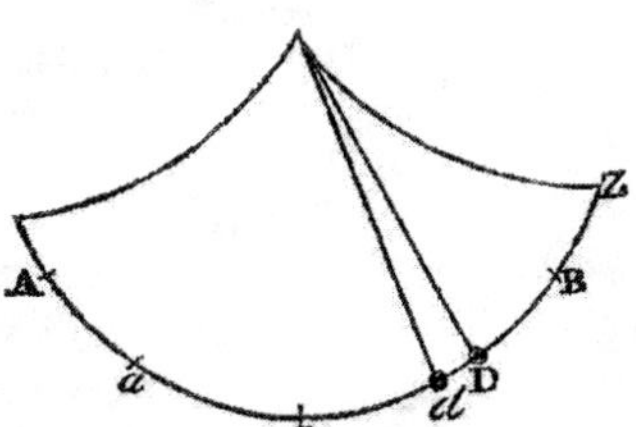

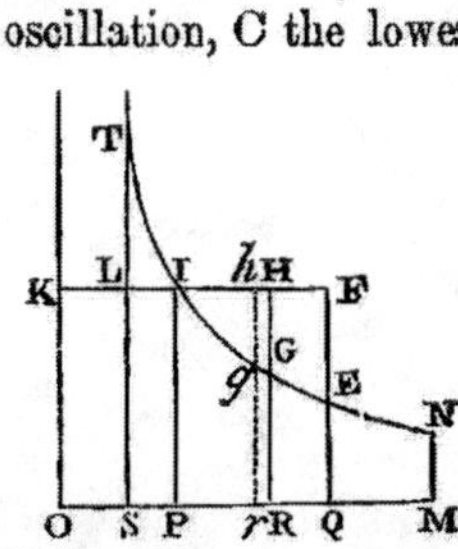

of the cycloid, and CZ half the whole cycloidal arc, equal to the length of the pendulum; and let it be required to find the resistance of the body in any place D. Cut the indefinite right line OQ in the points O, S, P, Q, so that (erecting the perpendiculars OK, ST, PI, QE, and with the centre O, and the aysmptotes OK, OQ, describing the hyperbola TIGE cutting the perpendiculars ST, PI, QE in T, I, and E, and through the point I drawing KF, parallel to the asymptote OQ, meeting the asymptote OK in K, and the perpendiculars ST and QE in L and F) the hyperbolic area PIEQ may be to the hyperbolic area PITS as the arc BC, described in the descent of the body, to the arc C*a* described in the ascent; and that the area IEF may be to the area ILT as OQ to OS. Then with the perpendicular MN cut off the hyperbolic area PINM, and let that area be to the hyperbolic area PIEQ as the arc CZ to the arc BC described in the descent. And if the perpendicular RG cut off the hyperbolic area PIGR, which shall be to the area PIEQ as any arc CD to the arc BC described in the whole descent, the resistance in any place D will be to the force of gravity as the area $\frac{OR}{OQ}$ IEF — IGH to the area PINM.

For since the forces arising from gravity with which the body is urged in the places Z, B, D, *a*, are as the arcs CZ, CB, CD, C*a* and those arcs are as the areas PINM, PIEQ, PIGR, PITS; let those areas be the exponents both of the arcs and of the forces respectively. Let D*d* be a very small space described by the body in its descent: and let it be expressed by the very small area RG*gr* comprehended between the parallels RG, *rg*; and produce *rg* to *h*, so that GH*hg* and RG*gr* may be the contemporaneous decrements of the areas IGH, PIGR. And the increment GH*hg* — $\frac{Rr}{OQ}$ IEF, or R*r* × HG — $\frac{Rr}{OQ}$ IEF, of the area $\frac{OR}{OQ}$ IEF — IGH will be to the decrement RG*gr*, or R*r* × RG, of the area PIGR, as HG — $\frac{IEF}{OQ}$ to RG; and therefore as OR × HG — $\frac{OR}{OQ}$ IEF to OR × GR or OP × PI, that is (because of the equal quantities OR × HG, OR × HR — OR × GR, ORHK — OPIK, PIHR and PIGR + IGH), as PIGR + IGH — $\frac{OR}{OQ}$ IEF to OPIK. Therefore if the area $\frac{OR}{OQ}$ IEF — IGH be called Y, and RG*gr* the decrement of the area PIGR be given, the increment of the area Y will be as PIGR — Y.

Then if V represent the force arising from the gravity, proportional to the arc CD to be described, by which the body is acted upon in D, and R be put for the resistance, V — R will be the whole force with which the body is urged in D. Therefore the increment of the velocity is as V — R and the particle of time in which it is generated conjunctly. But the velocity itself is as the contemporaneous increment of the space described directly and the same particle of time inversely. Therefore, since the resistance is, by the supposition, as the square of the velocity, the increment of the resistance will (by Lem. II) be as the velocity and the increment of the velocity conjunctly, that is, as the moment of the space and V — R conjunctly; and, therefore, if the moment of the space be given, as V — R; that is, if for the force V we put its exponent PIGR, and the resistance R be expressed by any other area Z, as PIGR — Z.

Therefore the area PIGR uniformly decreasing by the subduction of given moments, the area Y increases in proportion of PIGR — Y, and the area Z in proportion of PIGR — Z. And therefore if the areas Y and Z begin together, and at the beginning are equal, these, by the addition of equal moments, will continue to be equal; and in like manner decreasing by equal moments, will vanish together. And, *vice versa*, if they together begin and vanish, they will have equal moments and be always equal; and that, because if the resistance Z be augmented, the velocity together with the arc *Ca*, described in the ascent of the body, will be diminished; and the point in which all the motion together with the resistance ceases coming nearer to the point C, the resistance vanishes sooner than the area Y. And the contrary will happen when the resistance is diminished.

Now the area Z begins and ends where the resistance is nothing, that is, at the beginning of the motion where the arc CD is equal to the arc CB,

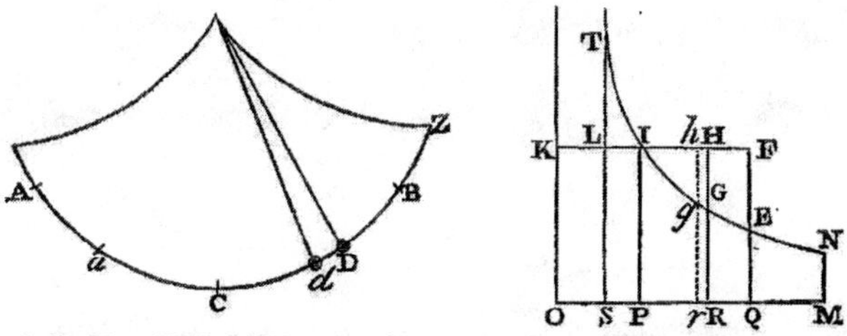

and the right line RG falls upon the right line QE; and at the end of the motion where the arc CD is equal to the arc *Ca*, and RG falls upon the right line ST. And the area Y or $\frac{OR}{OQ}$ IEF — IGH begins and ends also where the resistance is nothing, and therefore where $\frac{OR}{OQ}$ IEF and IGH are equal; that is (by the construction), where the right line RG falls successively upon the right lines QE and ST. Therefore those areas begin and vanish together, and are therefore always equal. Therefore the area $\frac{OR}{OQ}$ IEF — IGH is equal to the area Z, by which the resistance is expressed, and therefore is to the area PINM, by which the gravity is expressed, as the resistance to the gravity. Q.E.D.

Cor. 1. Therefore the resistance in the lowest place C is to the force of gravity as the area $\frac{OP}{OQ}$ IEF to the area PINM.

Cor. 2. But it becomes greatest where the area PIHR is to the area IEF as OR to OQ. For in that case its moment (that is, PIGR — Y) becomes nothing.

Cor. 3. Hence also may be known the velocity in each place, as being in the subduplicate ratio of the resistance, and at the beginning of the motion equal to the velocity of the body oscillating in the same cycloid without any resistance.

However, by reason of the difficulty of the calculation by which the resistance and the velocity are found by this Proposition, we have thought fit to subjoin the Proposition following.

PROPOSITION XXX. THEOREM XXIV.

If a right line aB *be equal to the arc of a cycloid which an oscillating body describes, and at each of its points* D *the perpendiculars* DK *be erected, which shall be to the length of the pendulum as the resistance of the body in the corresponding points of the arc to the force of gravity; I say, that the difference between the arc described in the whole descent and the arc described in the whole subsequent ascent drawn into half the sum of the same arcs will be equal to the area* BKa *which all those perpendiculars take up.*

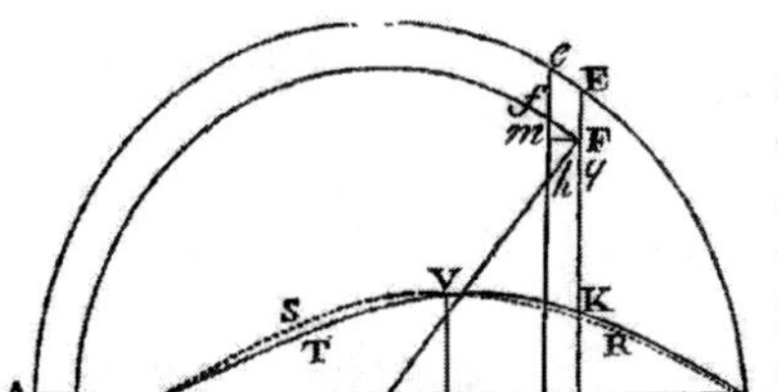

Let the arc of the cycloid, described in one entire oscillation, be expressed by the right line *a*B, equal to it, and the arc which would have been described *in vacuo* by the length AB. Bisect AB in C, and the point C will represent the lowest point of the cycloid, and CD will be as the force arising from gravity, with which the body in D is urged in the direction of the tangent of the cycloid, and will have the same ratio to the length of the pendulum as the force in D has to the force of gravity. Let that force, therefore, be expressed by that length CD, and the force of gravity by the length of the pendulum; and if in DE you take DK in the same ratio to the length of the pendulum as the resistance has to the gravity, DK will be the exponent of the resistance. From the centre C with the interval CA or CB describe a semi-circle BE*e*A. Let the body describe, in the least time, the space D*d;* and, erecting the perpendiculars DE, *de*, meeting the circumference in E and *e*, they will be as the velocities which the body descending *in vacuo* from the point B would acquire in the places D and *d*. This appears by Prop. LII, Book I. Let, therefore, these velocities be expressed by those perpendiculars DE, *de*; and let DF be the velocity which it acquires in D by falling from B in the resisting medium. And if from the centre C with the interval CF we describe the circle F*f*M meeting the right lines *de* and AB in *f* and M, then M will be the place to which it would thenceforward, without farther resistance, ascend, and *df* the velocity it would acquire in *d*. Whence, also, if F*g* represent the moment of the velocity which the body D, in describing the least space D*d*, loses by the resistance of the medium; and CN be taken equal to C*g*; then will N be the place to which the body, if it met no farther resistance, would thenceforward ascend, and MN will be the decrement of the ascent arising from the loss of that velocity. Draw F*m* perpendicular to *df*, and the decrement F*g* of the velocity DF generated by the resistance DK will be to the increment *fm* of the same velocity, generated by the force CD, as the generating force DK to the generating force CD. But because of the similar triangles F*mf*, F*hg*, FDC, *fm* is to F*m* or D*d* as CD to DF; and, *ex æquo*, F*g* to D*d* as DK to DF. Also F*h* is to F*g* as DF to CF; and, *ex æquo perturbatè*, F*h* or MN to D*d* as DK to CF or CM; and therefore the sum of all the MN × CM will be equal to the sum of all the D*d* × DK. At the moveable point M suppose always a rectangular ordinate erected equal to the indeterminate CM, which by a continual motion is drawn into the whole length A*a*; and the trapezium described by that motion, or its equal, the rectangle A*a* × $\frac{1}{2}a$B, will be equal to the sum of all the MN × CM, and therefore to the sum of all the D*d* × DK, that is, to the area BKVT*a*. Q.E.D.

COR. Hence from the law of resistance, and the difference A*a* of the arcs C*a*, CB, may be collected the proportion of the resistance to the gravity nearly.

For if the resistance DK be uniform, the figure BKT*a* will be a rectangle under B*a* and DK; and thence the rectangle under $\frac{1}{2}$B*a* and A*a* will be equal to the rectangle under B*a* and DK, and DK will be equal to $\frac{1}{2}$A*a*. Wherefore since DK is the exponent of the resistance, and the length of the pendulum the exponent of the gravity, the resistance will be to the gravity as $\frac{1}{2}$A*a* to the length of the pendulum; altogether as in Prop. XXVIII is demonstrated.

If the resistance be as the velocity, the figure BKT*a* will be nearly an ellipsis. For if a body, in a non-resisting medium, by one entire oscillation, should describe the length BA, the velocity in any place D would be as the ordinate DE of the circle described on the diameter AB. Therefore since B*a* in the resisting medium, and BA in the non-resisting one, are described nearly in the same times; and therefore the velocities in each of the points of B*a* are to the velocities in the correspondent points of the length BA nearly as B*a* is to BA, the velocity in the point D in the resisting medium will be as the ordinate of the circle or ellipsis described upon the diameter B*a*; and therefore the figure BKVT*a* will be nearly an ellipsis. Since the resistance is supposed proportional to the velocity, let OV be the exponent of the resistance in the middle point O; and an ellipsis BRVS*a* described with the centre O, and the semi-axes OB, OV, will be nearly equal to the figure BKVT*a*, and to its equal the rectangle A*a* × BO. Therefore A*a* × BO is to OV × BO as the area of this ellipsis to OV × BO; that is, A*a* is to OV as the area of the semi-circle to the square of the radius, or as 11 to 7 nearly; and, therefore, $\frac{7}{11}$A*a* is to the length of the pendulum as the resistance of the oscillating body in O to its gravity.

Now if the resistance DK be in the duplicate ratio of the velocity, the figure BKVT*a* will be almost a parabola having V for its vertex and OV for its axis, and therefore will be nearly equal to the rectangle under $\frac{2}{3}$B*a* and OV. Therefore the rectangle under $\frac{1}{2}$B*a* and A*a* is equal to the rectangle $\frac{2}{3}$B*a* × OV, and therefore OV is equal to $\frac{3}{4}$A*a*; and therefore the resistance in O made to the oscillating body is to its gravity as $\frac{3}{4}$A*a* to the length of the pendulum.

And I take these conclusions to be accurate enough for practical uses. For since an ellipsis or parabola BRVS*a* falls in with the figure BKVT*a* in the middle point V, that figure, if greater towards the part BRV or VS*a* than the other, is less towards the contrary part, and is therefore nearly equal to it.

PROPOSITION XXXI. THEOREM XXV.

If the resistance made to an oscillating body in each of the proportional parts of the arcs described be augmented or diminished in a given ratio, the difference between the arc described in the descent and the arc described in the subsequent ascent will be augmented or diminished in the same ratio.

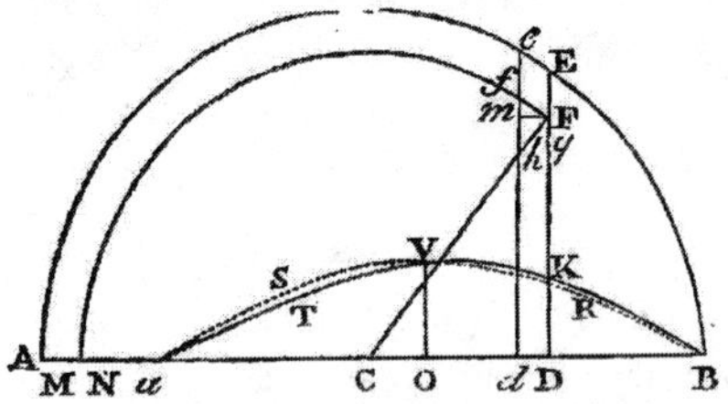

For that difference arises from the retardation of the pendulum by the resistance of the medium, and therefore is as the whole retardation and the retarding resistance proportional thereto. In the foregoing Proposition the rectangle under the right line $\frac{1}{2}a$B and the difference A*a* of the arcs CB, C*a*, was equal to the area BKT*a*. And that area, if the length *a*B remains, is augmented or diminished in the ratio of the ordinates DK; that is, in the ratio of the resistance and is therefore as the length *a*B and the resistance conjunctly. And therefore the rectangle under A*a* and $\frac{1}{2}a$B is as *a*B and the resistance conjunctly, and therefore A*a* is as the resistance. Q.E.D.

COR. 1. Hence if the resistance be as the velocity, the difference of the arcs in the same medium will be as the whole arc described: and the contrary.

COR. 2. If the resistance be in the duplicate ratio of the velocity, that difference will be in the duplicate ratio of the whole arc: and the contrary.

COR. 3. And universally, if the resistance be in the triplicate or any other ratio of the velocity, the difference will be in the same ratio of the whole arc: and the contrary.

COR. 4. If the resistance be partly in the simple ratio of the velocity, and partly in the duplicate ratio of the same, the difference will be partly in the ratio of the whole arc, and partly in the duplicate ratio of it: and the contrary. So that the law and ratio of the resistance will be the same for the velocity as the law and ratio of that difference for the length of the arc.

COR. 5. And therefore if a pendulum describe successively unequal arcs, and we can find the ratio of the increment or decrement of this difference for the length of the arc described, there will be had also the ratio of the increment or decrement of the resistance for a greater or less velocity.

GENERAL SCHOLIUM.

From these propositions we may find the resistance of mediums by pendulums oscillating therein. I found the resistance of the air by the following experiments. I suspended a wooden globe or ball weighing $57\frac{7}{22}$ ounces troy, its diameter $6\frac{7}{8}$ *London* inches, by a fine thread on a firm hook, so that the distance between the hook and the centre of oscillation of the globe was $10\frac{1}{2}$ feet. I marked on the thread a point 10 feet and 1 inch distant from the centre of suspension; and even with that point I placed a ruler divided into inches, by the help whereof I observed the lengths of the arcs described by the pendulum. Then I numbered the oscillations in which the globe would lose $\frac{1}{8}$ part of its motion. If the pendulum was drawn aside from the perpendicular to the distance of 2 inches, and thence let go, so that in its whole descent it described an arc of 2 inches, and in the first whole oscillation, compounded of the descent and subsequent ascent, an arc of almost 4 inches, the same in 164 oscillations lost $\frac{1}{8}$ part of its motion, so as in its last ascent to describe an arc of $1\frac{3}{4}$ inches. If in the first descent it described an arc of 4 inches, it lost $\frac{1}{8}$ part of its motion in 121 oscillations, so as in its last ascent to describe an arc of $3\frac{1}{2}$ inches. If in the first descent it described an arc of 8, 16, 32, or 64 inches, it lost $\frac{1}{8}$ part of its motion in 69, $35\frac{1}{2}$, $18\frac{1}{2}$, $9\frac{2}{3}$ oscillations, respectively. Therefore the difference between the arcs described in the first descent and the last ascent was in the 1st, 2d, 3d, 4th, 5th, 6th cases, $\frac{1}{4}$, $\frac{1}{2}$, 1, 2, 4, 8 inches respectively. Divide those differences by the number of oscillations in each case, and in one mean oscillation, wherein an arc of $3\frac{3}{4}$, $7\frac{1}{2}$, 15, 30,

60, 120 inches was described, the difference of the arcs described in the descent and subsequent ascent will be $\frac{1}{656}$, $\frac{1}{242}$, $\frac{1}{69}$, $\frac{4}{71}$, $\frac{8}{37}$, $\frac{24}{29}$ parts of an inch, respectively. But these differences in the greater oscillations are in the duplicate ratio of the arcs described nearly, but in lesser oscillátions something greater than in that ratio; and therefore (by Cor. 2, Prop. XXXI of this Book) the resistance of the globe, when it moves very swift, is in the duplicate ratio of the velocity, nearly; and when it moves slowly, somewhat greater than in that ratio.

Now let V represent the greatest velocity in any oscillation, and let A, B, and C be given quantities, and let us suppose the difference of the arcs to be $AV + BV^{\frac{3}{2}} + CV^2$. Since the greatest velocities are in the cycloid as $\frac{1}{2}$ the arcs described in oscillating, and in the circle as $\frac{1}{2}$ the chords of those arcs; and therefore in equal arcs are greater in the cycloid than in the circle in the ratio of $\frac{1}{2}$ the arcs to their chords; but the times in the circle are greater than in the cycloid, in a reciprocal ratio of the velocity; it is plain that the differences of the arcs (which are as the resistance and the square of the time conjunctly) are nearly the same in both curves: for in the cycloid those differences must be on the one hand augmented, with the resistance, in about the duplicate ratio of the arc to the chord, because of the velocity augmented in the simple ratio of the same; and on the other hand diminished, with the square of the time, in the same duplicate ratio. Therefore to reduce these observations to the cycloid, we must take the same differences of the arcs as were observed in the circle, and suppose the greatest velocities analogous to the half, or the whole arcs, that is, to the numbers $\frac{1}{2}$, 1, 2, 4, 8, 16. Therefore in the 2d, 4th, and 6th cases, put 1, 4, and 16 for V; and the difference of the arcs in the 2d case will become $\frac{\frac{1}{2}}{121} = A + B + C$; in the 4th case, $\frac{2}{35\frac{1}{2}} = 4A + 8B + 16C$; in the 6th case, $\frac{8}{9\frac{2}{3}} = 16A + 64B + 256C$. These equations reduced give $A = 0{,}0000916$, $B = 0{,}0010847$, and $C = 0{,}0029558$. Therefore the difference of the arcs is as $0{,}0000916V + 0{,}0010847V^{\frac{3}{2}} + 0{,}0029558V^2$: and therefore since (by Cor. Prop. XXX, applied to this case) the resistance of the globe in the middle of the arc described in oscillating, where the velocity is V, is to its weight as $\frac{7}{11}AV + \frac{7}{10}BV^{\frac{3}{2}} + \frac{3}{4}CV^2$ to the length of the pendulum, if for A, B, and C you put the numbers found, the resistance of the globe will be to its weight as $0{,}0000583V + 0{,}0007593V^{\frac{3}{2}} + 0{,}0022169V^2$ to the length of the pendulum between the centre of suspension and the ruler, that is, to 121 inches. Therefore since V in the second case represents 1, in the 4th case 4, and in the 6th case 16, the resistance will be to the weight of the globe, in the 2d case, as 0,0030345 to 121; in the 4th, as 0,041748 to 121; in the 6th, as 0,61705 to 121.

The arc, which the point marked in the thread described in the 6th case, was of $120 - \frac{8}{9\frac{2}{3}}$, or $119\frac{5}{29}$ inches. And therefore since the radius was 121 inches, and the length of the pendulum between the point of suspension and the centre of the globe was 126 inches, the arc which the centre of the globe described was $124\frac{3}{31}$ inches. Because the greatest velocity of the oscillating body, by reason of the resistance of the air, does not fall on the lowest point of the arc described, but near the middle place of the whole arc, this velocity will be nearly the same as if the globe in its whole descent in a non-resisting medium should describe $62\frac{3}{62}$ inches, the half of that arc, and that in a cycloid, to which we have above reduced the motion of the pendulum; and therefore that velocity will be equal to that which the globe would acquire by falling perpendicularly from a height equal to the versed sine of that arc. But that versed sine in the cycloid is to that arc $62\frac{3}{62}$ as the same arc to twice the length of the pendulum 252, and therefore equal to 15,278 inches. Therefore the velocity of the pendulum is the same which a body would acquire by falling, and in its fall describing a space of 15,278 inches. Therefore with such a velocity the globe meets with a resistance which is to its weight as 0,61705 to 121, or (if we take that part only of the resistance which is in the duplicate ratio of the velocity) as 0,56752 to 121.

I found, by an hydrostatical experiment, that the weight of this wooden globe was to the weight of a globe of water of the same magnitude as 55 to 97: and therefore since 121 is to 213,4 in the same ratio, the resistance made to this globe of water, moving forwards with the above-mentioned velocity, will be to its weight as 0,56752 to 213,4, that is, as 1 to $376\frac{1}{50}$. Whence since the weight of a globe of water, in the time in which the globe with a velocity uniformly continued describes a length of 30,556 inches, will generate all that velocity in the falling globe, it is manifest that the force of resistance uniformly continued in the same time will take away a velocity, which will be less than the other in the ratio of 1 to $376\frac{1}{50}$, that is, the $\frac{1}{376\frac{1}{50}}$ part of the whole velocity. And therefore in the time that the globe, with the same velocity uniformly continued, would describe the length of its semi-diameter, or $3\frac{7}{16}$ inches, it would lose the $\frac{1}{3342}$ part of its motion.

I also counted the oscillations in which the pendulum lost $\frac{1}{4}$ part of its motion. In the following table the upper numbers denote the length of the arc described in the first descent, expressed in inches and parts of an inch; the middle numbers denote the length of the arc described in the last ascent; and in the lowest place are the numbers of the oscillations. I give an account of this experiment, as being more accurate than that in which only $\frac{1}{8}$ part of the motion was lost. I leave the calculation to such as are disposed to make it.

First descent . . .	2	4	8	16	32	64
Last ascent . . .	$1\frac{1}{2}$	3	6	12	24	48
Numb. of oscill. . .	374	272	$162\frac{1}{2}$	$83\frac{1}{3}$	$41\frac{2}{3}$	$22\frac{2}{3}$

I afterward suspended a leaden globe of 2 inches in diameter, weighing $26\frac{1}{4}$ ounces troy by the same thread, so that between the centre of the globe and the point of suspension there was an interval of $10\frac{1}{2}$ feet, and I counted the oscillations in which a given part of the motion was lost. The first of the following tables exhibits the number of oscillations in which $\frac{1}{8}$ part of the whole motion was lost; the second the number of oscillations in which there was lost $\frac{1}{4}$ part of the same.

First descent	1	2	4	8	16	32	64
Last ascent	$\frac{7}{8}$	$\frac{7}{4}$	$3\frac{1}{2}$	7	14	28	56
Numb. of oscill. . .	226	228	193	140	$90\frac{1}{2}$	53	30
First descent	1	2	4	8	16	32	64
Last ascent	$\frac{3}{4}$	$1\frac{1}{2}$	3	6	12	24	48
Numb. of oscill. . .	510	518	420	318	204	121	70

Selecting in the first table the 3d, 5th, and 7th observations, and expressing the greatest velocities in these observations particularly by the numbers 1, 4, 16 respectively, and generally by the quantity V as above, there will come out in the 3d observation $\frac{\frac{1}{2}}{193} = A + B + C$, in the 5th observation $\frac{2}{90\frac{1}{2}} = 4A + 8B + 16C$, in the 7th observation $\frac{8}{30} = 16A + 64B + 256C$. These equations reduced give $A = 0{,}001414$, $B = 0{,}000297$, $C = 0{,}000879$. And thence the resistance of the globe moving with the velocity V will be to its weight $26\frac{1}{4}$ ounces in the same ratio as $0{,}0009V + 0{,}000208V^{\frac{3}{2}} + 0{,}000659V^2$ to 121 inches, the length of the pendulum. And if we regard that part only of the resistance which is in the duplicate ratio of the velocity, it will be to the weight of the globe as $0{,}000659V^2$ to 121 inches. But this part of the resistance in the first experiment was to the weight of the wooden globe of $57\frac{7}{22}$ ounces as $0{,}002217V^2$ to 121; and thence the resistance of the wooden globe is to the resistance of the leaden one (their velocities being equal) as $57\frac{7}{22}$ into 0,002217 to $26\frac{1}{4}$ into 0,000659, that is, as $7\frac{1}{3}$ to 1. The diameters of the two globes were $6\frac{7}{8}$ and 2 inches, and the squares of these are to each other as $47\frac{1}{4}$ and 4, or $11\frac{13}{16}$ and 1, nearly. Therefore the resistances of these equally swift globes were in less than a duplicate ratio of the diameters. But we have not yet considered the resistance of the thread, which was certainly very considerable, and ought to be subducted from the resistance of the pendulums here found. I could not determine this accurately, but I found it greater than a third part of the whole resistance of the lesser pendulum; and thence I gathered that the resistances of the globes, when the resistance of the thread is subducted, are nearly in the duplicate ratio of their diameters. For the ratio of $7\frac{1}{3} - \frac{1}{3}$ to $1 - \frac{1}{3}$, or $10\frac{1}{2}$ to 1 is not very different from the duplicate ratio of the diameters $11\frac{13}{16}$ to 1.

Since the resistance of the thread is of less moment in greater globes, I tried the experiment also with a globe whose diameter was $18\frac{3}{4}$ inches. The length of the pendulum between the point of suspension and the centre of oscillation was $122\frac{1}{2}$ inches, and between the point of suspension and the knot in the thread $109\frac{1}{2}$ inches. The arc described by the knot at the first descent of the pendulum was 32 inches. The arc described by the same knot in the last ascent after five oscillations was 28 inches. The sum of the arcs, or the whole arc described in one mean oscillation, was 60 inches. The difference of the arcs 4 inches. The $\frac{1}{10}$ part of this, or the difference between the descent and ascent in one mean oscillation, is $\frac{2}{5}$ of an inch. Then as the radius $109\frac{1}{2}$ to the radius $122\frac{1}{2}$, so is the whole arc of 60 inches described by the knot in one mean oscillation to the whole arc of $67\frac{1}{8}$ inches described by the centre of the globe in one mean oscillation; and so is the difference $\frac{2}{5}$ to a new difference 0,4475. If the length of the arc described were to remain, and the length of the pendulum should be augmented in the ratio of 126 to $122\frac{1}{2}$, the time of the oscillation would be augmented, and the velocity of the pendulum would be diminished in the subduplicate of that ratio; so that the difference 0,4475 of the arcs described in the descent and subsequent ascent would remain. Then if the arc described be augmented in the ratio of $124\frac{3}{31}$ to $67\frac{1}{8}$, that difference 0,4475 would be augmented in the duplicate of that ratio, and so would become 1,5295. These things would be so upon the supposition that the resistance of the pendulum were in the duplicate ratio of the velocity. Therefore if the pendulum describe the whole arc of $124\frac{3}{31}$ inches, and its length between the point of suspension and the centre of oscillation be 126 inches, the difference of the arcs described in the descent and subsequent ascent would be 1,5295 inches. And this difference multiplied into the weight of the pendulous globe, which was 208 ounces, produces 318,136. Again; in the pendulum above-mentioned, made of a wooden globe, when its centre of oscillation, being 126 inches from the point of suspension, described the whole arc of $124\frac{3}{31}$ inches, the difference of the arcs described in the descent and ascent was $\frac{126}{121}$ into $\frac{8}{9\frac{2}{3}}$. This multiplied into the weight of the globe, which was $57\frac{7}{22}$ ounces, produces 49,396. But I multiply these differences into the weights of the globes, in order to find their resistances. For the differences arise from the resistances, and are as the resistances directly and the weights inversely. Therefore the resistances are as the numbers 318,136 and 49,396. But that part of the resistance of the lesser globe, which is in the duplicate ratio of the velocity, was to the whole resistance as 0,56752 to 0,61675, that is, as 45,453 to 49,396; whereas that part of the resistance of the greater globe is almost equal to its whole resistance; and so those parts are nearly as 318,136 and 45,453, that is, as 7 and 1. But the diameters of the globes are $18\frac{3}{4}$ and $6\frac{7}{8}$; and their squares $351\frac{9}{16}$ and $47\frac{17}{64}$ are as 7,438 and 1, that is, as the resistances of the globes 7 and 1, nearly. The difference of these ratios is scarce greater than may arise from the resistance of the thread. Therefore those parts of the resistances which are, when the globes are equal, as the squares of the velocities, are also, when the velocities are equal, as the squares of the diameters of the globes.

But the greatest of the globes I used in these experiments was not perfectly spherical, and therefore in this calculation I have, for brevity's sake, neglected some little niceties; being not very solicitous for an accurate calculus in an experiment that was not very accurate. So that I could wish that these experiments were tried again with other globes, of a larger size, more in number, and more accurately formed; since the demonstration of a vacuum depends thereon. If the globes be taken in a geometrical proportion, as suppose whose diameters are 4, 8, 16, 32 inches; one may collect from the progression observed in the experiments what would happen if the globes were still larger.

In order to compare the resistances of different fluids with each other, I made the following trials. I procured a wooden vessel 4 feet long, 1 foot broad, and 1 foot high. This vessel, being uncovered, I filled with spring water, and, having immersed pendulums therein, I made them oscillate in the water. And I found that a leaden globe weighing $166\frac{1}{6}$ ounces, and in diameter $3\frac{5}{8}$ inches, moved therein as it is set down in the following table; the length of the pendulum from the point of suspension to a certain point marked in the thread being 126 inches, and to the centre of oscillation $134\frac{3}{8}$ inches.

The arc described in the first descent, by a point marked in the thread was inches.	64	32	16	8	4	2	1	$\frac{1}{2}$	$\frac{1}{4}$
The arc described in the last ascent was inches.	48	24	12	6	3	$1\frac{1}{2}$	$\frac{3}{4}$	$\frac{3}{8}$	$\frac{3}{16}$
The difference of the arcs, proportional to the motion lost, was inches.	16	8	4	2	1	$\frac{1}{2}$	$\frac{1}{4}$	$\frac{1}{8}$	$\frac{1}{16}$
The number of the oscillations in water.			$\frac{29}{60}$	$1\frac{1}{5}$	3	7	$11\frac{1}{4}$	$12\frac{2}{3}$	$13\frac{1}{3}$
The number of the oscillations in air.	$85\frac{1}{2}$	287	535						

In the experiments of the 4th column there were equal motions lost in 535 oscillations made in the air, and $1\frac{1}{5}$ in water. The oscillations in the air were indeed a little swifter than those in the water. But if the oscillations in the water were accelerated in such a ratio that the motions of the pendulums might be equally swift in both mediums, there would be still the same number $1\frac{1}{5}$ of oscillations in the water, and by these the same quantity of motion would be lost as before; because the resistance is increased, and the square of the time diminished in the same duplicate ratio. The pendulums, therefore, being of equal velocities, there were equal motions lost in 535 oscillations in the air, and $1\frac{1}{5}$ in the water; and therefore the resistance of the pendulum in the water is to its resistance in the air as 535 to $1\frac{1}{5}$. This is the proportion of the whole resistances in the case of the 4th column.

Now let $AV + CV^2$ represent the difference of the arcs described in the descent and subsequent ascent by the globe moving in air with the greatest velocity V; and since the greatest velocity is in the case of the 4th column to the greatest velocity in the case of the 1st column as 1 to 8; and that difference of the arcs in the case of the 4th column to the difference in the case of the 1st column as $\frac{2}{535}$ to $\frac{16}{85\frac{1}{2}}$, or as $85\frac{1}{2}$ to 4280; put in these cases 1 and 8 for the velocities, and $85\frac{1}{2}$ and 4280 for the differences of the arcs, and $A + C$ will be $= 85\frac{1}{2}$, and $8A + 64C = 4280$ or $A + 8C = 535$; and then by reducing these equations, there will come out $7C = 449\frac{1}{2}$ and $C = 64\frac{3}{14}$ and $A = 21\frac{2}{7}$; and therefore the resistance, which is as $\frac{7}{11}AV + \frac{3}{4}CV^2$, will become as $13\frac{6}{11}V + 48\frac{9}{56}V^2$. Therefore in the case of the 4th column, where the velocity was 1, the whole resistance is to its part proportional to the square of the velocity as $13\frac{6}{11} + 48\frac{9}{56}$ or $61\frac{12}{17}$ to $48\frac{9}{56}$; and therefore the resistance of the pendulum in water is to that part of the resistance in air, which is proportional to the square of the velocity, and which in swift motions is the only part that deserves consideration, as $61\frac{12}{17}$ to $48\frac{9}{56}$ and 535 to $1\frac{1}{5}$ conjunctly, that is, as 571 to 1. If the whole thread of the pendulum oscillating in the water had been immersed, its resistance would have been still greater; so that the resistance of the pendulum oscillating in the water, that is, that part which is proportional to the square of the velocity, and which only needs to be considered in swift bodies, is to the resistance of the same whole pendulum, oscillating in air with the same velocity, as about 850 to 1, that is as, the density of water to the density of air, nearly.

In this calculation we ought also to have taken in that part of the resistance of the pendulum in the water which was as the square of the velocity; but I found (which will perhaps seem strange) that the resistance in the water was augmented in more than a duplicate ratio of the velocity. In searching after the cause, I thought upon this, that the vessel was too narrow for the magnitude of the pendulous globe, and by its narrowness obstructed the motion of the water as it yielded to the oscillating globe. For when I immersed a pendulous globe, whose diameter was one inch only, the resistance was augmented nearly in a duplicate ratio of the velocity. I tried this by making a pendulum of two globes, of which the lesser and lower oscillated in the water, and the greater and higher was fastened to the thread just above the water, and, by oscillating in the air, assisted the motion of the pendulum, and continued it longer. The experiments made by this contrivance proved according to the following table.

Arc descr. in first descent . .	16	8	4	2	1	$\frac{1}{2}$	$\frac{1}{4}$
Arc descr. in last ascent . .	12	6	3	$1\frac{1}{2}$	$\frac{3}{4}$	$\frac{3}{8}$	$\frac{3}{16}$
Diff. of arcs, proport. to motion lost	4	2	1	$\frac{1}{2}$	$\frac{1}{4}$	$\frac{1}{8}$	$\frac{1}{16}$
Number of oscillations . . .	$3\frac{3}{8}$	$6\frac{1}{2}$	$12\frac{1}{12}$	$21\frac{1}{5}$	34	53	$62\frac{1}{5}$

In comparing the resistances of the mediums with each other, I also caused iron pendulums to oscillate in quicksilver. The length of the iron wire was about 3 feet, and the diameter of the pendulous globe about $\frac{1}{3}$ of an inch. To the wire, just above the quicksilver, there was fixed another leaden globe of a bigness sufficient to continue the motion of the pendulum for some time. Then a vessel, that would hold about 3 pounds of quicksilver, was filled by turns with quicksilver and common water, that, by making the pendulum oscillate successively in these two different fluids, I might find the proportion of their resistances; and the resistance of the quicksilver proved to be to the resistance of water as about 13 or 14 to 1; that is, as the density of quicksilver to the density of water. When I made use of a pendulous globe something bigger, as of one whose diameter was about $\frac{1}{2}$ or $\frac{2}{3}$ of an inch, the resistance of the quicksilver proved to be to the resistance of the water as about 12 or 10 to 1. But the former experiment is more to be relied on, because in the latter the vessel was too narrow in proportion to the magnitude of the immersed globe; for the vessel ought to have been enlarged together with the globe. I intended to have repeated these experiments with larger vessels, and in melted metals, and other liquors both cold and hot; but I had not leisure to try all: and besides, from what is already described, it appears sufficiently that the resistance of bodies moving swiftly is nearly proportional to the densities of the fluids in which they move. I do not say accurately; for more tenacious fluids, of equal density, will undoubtedly resist more than those that are more liquid; as cold oil more than warm, warm oil more than rain-water, and water more than spirit of wine. But in liquors, which are sensibly fluid enough, as in air, in salt and fresh water, in spirit of wine, of turpentine, and salts, in oil cleared of its fæces by distillation and warmed, in oil of vitriol, and in mercury, and melted metals, and any other such like, that are fluid enough to retain for some time the motion impressed upon them by the agitation of the vessel, and which being poured out are easily resolved into drops, I doubt not but the rule already laid down may be accurate enough, especially if the experiments be made with larger pendulous bodies and more swiftly moved.

Lastly, since it is the opinion of some that there is a certain æthereal medium extremely rare and subtile, which freely pervades the pores of all bodies; and from such a medium, so pervading the pores of bodies, some resistance must needs arise; in order to try whether the resistance, which we experience in bodies in motion, be made upon their outward superficies only, or whether their internal parts meet with any considerable resistance upon their superficies, I thought of the following experiment. I suspended a round deal box by a thread 11 feet long, on a steel hook, by means of a ring of the same metal, so as to make a pendulum of the aforesaid length. The hook had a sharp hollow edge on its upper part, so that the upper arc of the ring pressing on the edge might move the more freely; and the thread was fastened to the lower arc of the ring. The pendulum being thus prepared, I drew it aside from the perpendicular to the distance of about 6 feet, and that in a plane perpendicular to the edge of the hook, lest the ring, while the pendulum oscillated, should slide to and fro on the edge of the hook: for the point of suspension, in which the ring touches the hook, ought to remain immovable. I therefore accurately noted the place to which the pendulum was brought, and letting it go, I marked three other places, to which it returned at the end of the 1st, 2d, and 3d oscillation. This I often repeated, that I might find those places as accurately as possible. Then I filled the box with lead and other heavy metals that were near at hand. But, first, I weighed the box when empty, and that part of the thread that went round it, and half the remaining part, extended between the hook and the suspended box; for the thread so extended always acts upon the pendulum, when drawn aside from the perpendicular, with half its weight. To this weight I added the weight of the air contained in the box. And this whole weight was about $\frac{1}{78}$ of the weight of the box when filled with the metals. Then because the box when full of the metals, by extending the thread with its weight, increased the length of the pendulum, I shortened the thread so as to make the length of the pendulum, when oscillating, the same as before. Then drawing aside the pendulum to the place first marked, and letting it go, I reckoned about 77 oscillations before the box returned to the second mark, and as many afterwards before it came to the third mark, and as many after that before it came to the fourth mark. From whence I conclude that the whole resistance of the box, when full, had not a greater proportion to the resistance of the box, when empty, than 78 to 77. For if their resistances were equal, the box, when full, by reason of its *vis insita*, which was 78 times greater than the *vis insita* of the same when empty, ought to have continued its oscillating motion so much the longer, and therefore to have returned to those marks at the end of 78 oscillations. But it returned to them at the end of 77 oscillations.

Let, therefore, A represent the resistance of the box upon its external superficies, and B the resistance of the empty box on its internal superficies; and if the resistances to the internal parts of bodies equally swift be as the matter, or the number of particles that are resisted, then 78B will be the resistance made to the internal parts of the box, when full; and therefore the whole resistance A + B of the empty box will be to the whole resistance A + 78B of the full box as 77 to 78, and, by division, A + B to 77B as 77 to 1; and thence A + B to B as 77 × 77 to 1, and, by division again, A to B as 5928 to 1. Therefore the resistance of the empty box in its internal parts will be above 5000 times less than the resistance on its external superficies. This reasoning depends upon the supposition that the greater resistance of the full box arises not from any other latent cause, but only from the action of some subtile fluid upon the included metal.

This experiment is related by memory, the paper being lost in which I had described it; so that I have been obliged to omit some fractional parts, which are slipt out of my memory; and I have no leisure to try it again. The first time I made it, the hook being weak, the full box was retarded sooner. The cause I found to be, that the hook was not strong enough to bear the weight of the box; so that, as it oscillated to and fro, the hook was bent sometimes this and sometimes that way. I therefore procured a hook of sufficient strength, so that the point of suspension might remain unmoved, and then all things happened as is above described.

SECTION VII.

Of the motion of fluids, and the resistance made to projected bodies.

PROPOSITION XXXII. THEOREM XXVI.

Suppose two similar systems of bodies consisting of an equal number of particles, and let the correspondent particles be similar and proportional, each in one system to each in the other, and have a like situation among themselves, and the same given ratio of density to each other; and let them begin to move among themselves in proportional times, and with like motions (that is, those in one system among one another, and those in the other among one another). And if the particles that are in the same system do not touch one another, except in the moments of reflexion; nor attract, nor repel each other, except with accelerative forces that are as the diameters of the correspondent particles inversely, and the squares of the velocities directly; I say, that the particles of those systems will continue to move among themselves with like motions and in proportional times.

Like bodies in like situations are said to be moved among themselves with like motions and in proportional times, when their situations at the end of those times are always found alike in respect of each other; as suppose we compare the particles in one system with the correspondent particles in the other. Hence the times will be proportional, in which similar and proportional parts of similar figures will be described by correspondent particles. Therefore if we suppose two systems of this kind, the correspondent particles, by reason of the similitude of the motions at their beginning, will continue to be moved with like motions, so long as they move without meeting one another; for if they are acted on by no forces, they will go on uniformly in right lines, by the 1st Law. But if they do agitate one another with some certain forces, and those forces are as the diameters of the correspondent particles inversely and the squares of the velocities directly, then, because the particles are in like situations, and their forces are proportional, the whole forces with which correspondent particles are agitated, and which are compounded of each of the agitating forces (by Corol. 2 of the Laws), will have like directions, and have the same effect as if they respected centres placed alike among the particles; and those whole forces will be to each other as the several forces which compose them, that is, as the diameters of the correspondent particles inversely, and the squares of the velocities directly: and therefore will cause correspondent particles to continue to describe like figures. These things will be so (by Cor. 1 and 8, Prop. IV., Book 1), if those centres are at rest; but if they are moved, yet, by reason of the similitude of the translations, their situations among the particles of the system will remain similar, so that the changes introduced into the figures described by the particles will still be similar. So that the motions of correspondent and similar particles will continue similar till their first meeting with each other; and thence will arise similar collisions, and similar reflexions; which will again beget similar motions of the particles among themselves (by what was just now shown), till they mutually fall upon one another again, and so on *ad infinitum.*

Cor. 1. Hence if any two bodies, which are similar and in like situations to the correspondent particles of the systems, begin to move amongst them in like manner and in proportional times, and their magnitudes and densities be to each other as the magnitudes and densities of the corresponding particles, these bodies will continue to be moved in like manner and in proportional times; for the case of the greater parts of both systems and of the particles is the very same.

Cor. 2. And if all the similar and similarly situated parts of both systems be at rest among themselves; and two of them, which are greater than the rest, and mutually correspondent in both systems, begin to move in lines alike posited, with any similar motion whatsoever, they will excite similar motions in the rest of the parts of the systems, and will continue to move among those parts in like manner and in proportional times; and will therefore describe spaces proportional to their diameters.

PROPOSITION XXXIII. THEOREM XXVII.

The same things being supposed, I say, that the greater parts of the systems are resisted in a ratio compounded of the duplicate ratio of their velocities, and the duplicate ratio of their diameters, and the simple ratio of the density of the parts of the systems.

For the resistance arises partly from the centripetal or centrifugal forces with which the particles of the system mutually act on each other, partly from the collisions and reflexions of the particles and the greater parts. The resistances of the first kind are to each other as the whole motive forces from which they arise, that is, as the whole accelerative forces and the quantities of matter in corresponding parts; that is (by the supposition), as the squares of the velocities directly, and the distances of the corresponding particles inversely, and the quantities of matter in the correspondent parts directly: and therefore since the distances of the particles in one system are to the correspondent distances of the particles of the other as the diameter of one particle or part in the former system to the diameter of the correspondent particle or part in the other, and since the quantities of matter are as the densities of the parts and the cubes of the diameters; the resistances are to each other as the squares of the velocities and the squares of the diameters and the densities of the parts of the systems. Q.E.D. The resistances of the latter sort are as the number of correspondent reflexions and the forces of those reflexions conjunctly; but the number of the reflexions are to each other as the velocities of the corresponding parts directly and the spaces between their reflexions inversely. And the forces of the reflexions are as the velocities and the magnitudes and the densities of the corresponding parts conjunctly; that is, as the velocities and the cubes of the diameters and the densities of the parts. And, joining all these ratios, the resistances of the corresponding parts are to each other as the squares of the velocities and the squares of the diameters and the densities of the parts conjunctly. Q.E.D.

COR. 1. Therefore if those systems are two elastic fluids, like our air, and their parts are at rest among themselves; and two similar bodies proportional in magnitude and density to the parts of the fluids, and similarly situated among those parts, be any how projected in the direction of lines similarly posited; and the accelerative forces with which the particles of the fluids mutually act upon each other are as the diameters of the bodies projected inversely and the squares of their velocities directly; those bodies will excite similar motions in the fluids in proportional times, and will describe similar spaces and proportional to their diameters.

COR. 2. Therefore in the same fluid a projected body that moves swiftly meets with a resistance that is, in the duplicate ratio of its velocity, nearly. For if the forces with which distant particles act mutually upon one another should be augmented in the duplicate ratio of the velocity, the projected body would be resisted in the same duplicate ratio accurately; and therefore in a medium, whose parts when at a distance do not act mutually with any force on one another, the resistance is in the duplicate ratio of the velocity accurately. Let there be, therefore, three mediums A, B, C, consisting of similar and equal parts regularly disposed at equal distances. Let the parts of the mediums A and B recede from each other with forces that are among themselves as T and V; and let the parts of the medium C be entirely destitute of any such forces. And if four equal bodies D, E, F, G, move in these mediums, the two first D and E in the two first A and B, and the other two F and G in the third C; and if the velocity of the body D be to the velocity of the body E, and the velocity of the body F to the velocity of the body G, in the subduplicate ratio of the force T to the force V; the resistance of the body D to the resistance of the body E, and the resistance of the body F to the resistance of the body G, will be in the duplicate ratio of the velocities; and therefore the resistance of the body D will be to the resistance of the body F as the resistance of the body E to the resistance of the body G. Let the bodies D and F be equally swift, as also the bodies E and G; and, augmenting the velocities of the bodies D and F in any ratio, and diminishing the forces of the particles of the medium B in the duplicate of the same ratio, the medium B will approach to the form and condition of the medium C at pleasure; and therefore the resistances of the equal and equally swift bodies E and G in these mediums will perpetually approach to equality, so that their difference will at last become less than any given. Therefore since the resistances of the bodies D and F are to each other as the resistances of the bodies E and G, those will also in like manner approach to the ratio of equality. Therefore the bodies D and F, when they move with very great swiftness, meet with resistances very nearly equal; and therefore since the resistance of the body F is in a duplicate ratio of the velocity, the resistance of the body D will be nearly in the same ratio.

Cor. 3. The resistance of a body moving very swift in an elastic fluid is almost the same as if the parts of the fluid were destitute of their centrifugal forces, and did not fly from each other; if so be that the elasticity of the fluid arise from the centrifugal forces of the particles, and the velocity be so great as not to allow the particles time enough to act.

Cor. 4. Therefore, since the resistances of similar and equally swift bodies, in a medium whose distant parts do not fly from each other, are as the squares of the diameters, the resistances made to bodies moving with very great and equal velocities in an elastic fluid will be as the squares of the diameters, nearly.

Cor. 5. And since similar, equal, and equally swift bodies, moving through mediums of the same density, whose particles do not fly from each other mutually, will strike against an equal quantity of matter in equal times, whether the particles of which the medium consists be more and smaller, or fewer and greater, and therefore impress on that matter an equal quantity of motion, and in return (by the 3d Law of Motion) suffer an equal re-action from the same, that is, are equally resisted; it is manifest, also, that in elastic fluids of the same density, when the bodies move with extreme swiftness, their resistances are nearly equal, whether the fluids consist of gross parts, or of parts ever so subtile. For the resistance of projectiles moving with exceedingly great celerities is not much diminished by the subtilty of the medium.

Cor. 6. All these things are so in fluids whose elastic force takes its rise from the centrifugal forces of the particles. But if that force arise from some other cause, as from the expansion of the particles after the manner of wool, or the boughs of trees, or any other cause, by which the particles are hindered from moving freely among themselves, the resistance, by reason of the lesser fluidity of the medium, will be greater than in the Corollaries above.

PROPOSITION XXXIV. THEOREM XXVIII.

If in a rare medium, consisting of equal particles freely disposed at equal distances from each other, a globe and a cylinder described on equal diameters move with equal velocities in the direction of the axis of the cylinder, the resistance of the globe will be but half so great as that of the cylinder.

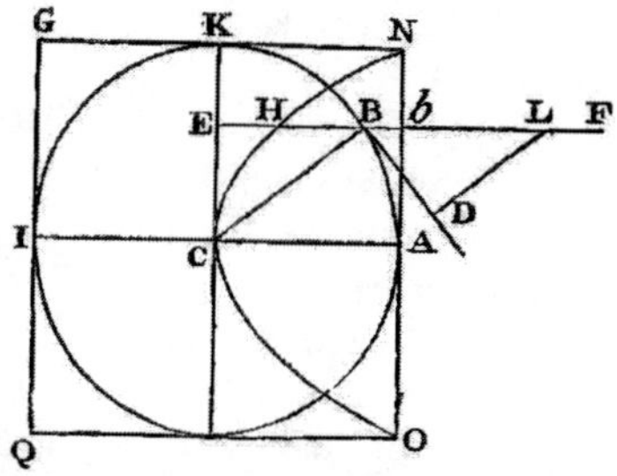

For since the action of the medium upon the body is the same (by Cor. 5 of the Laws) whether the body move in a quiescent medium, or whether the particles of the medium impinge with the same velocity upon the quiescent body, let us consider the body as if it were quiescent, and see with what force it would be impelled by the moving medium. Let, therefore, ABKI represent a spherical body described from the centre C with the semi-diameter CA, and let the particles of the medium impinge with a given velocity upon that spherical body in the directions of right lines parallel to AC; and let FB be one of those right lines. In FB take LB equal to the semi-diameter CB, and draw BD touching the sphere in B. Upon KC and BD let fall the perpendiculars BE, LD; and the force with which a particle of the medium, impinging on the globe obliquely in the direction FB, would strike the globe in B, will be to the force with which the same particle, meeting the cylinder ONGQ described about the globe with the axis ACI, would strike it perpendicularly in *b*, as LD to LB, or BE to BC. Again; the efficacy of this force to move the globe, according to the direction of its incidence FB or AC, is to the efficacy of the same to move the globe, according to the direction of its determination, that is, in the direction of the right line BC in which it impels the globe directly, as BE to BC. And, joining these ratios, the efficacy of a particle, falling upon the globe obliquely in the direction of the right line FB, to move the globe in the direction of its incidence, is to the efficacy of the same particle falling in the same line perpendicularly on the cylinder, to move it in the same direction, as BE^2 to BC^2. Therefore if in *b*E, which is perpendicular to the circular base of the cylinder NAO, and equal to the radius AC, we take *b*H equal to $\frac{BE^2}{CB}$; then *b*H will be to *b*E as the effect of the particle upon the globe to the effect of the particle upon the cylinder. And therefore the solid which is formed by all the right lines *b*H will be to the solid formed by all the right lines *b*E as the effect of all the particles upon the globe to the effect of all the particles upon the cylinder. But the former of these solids is a paraboloid whose vertex is C, its axis CA, and latus rectum CA, and the latter solid is a cylinder circumscribing the paraboloid; and it is known that a paraboloid is half its circumscribed cylinder. Therefore the whole force of the medium upon the globe is half of the entire force of the same upon the cylinder. And therefore if the particles of the medium are at rest, and the cylinder and globe move with equal velocities, the resistance of the globe will be half the resistance of the cylinder. Q.E.D.

SCHOLIUM.

By the same method other figures may be compared together as to their resistance; and those may be found which are most apt to continue their motions in resisting mediums. As if upon the circular base CEBH from the centre O, with the radius OC, and the altitude OD, one would construct a frustum CBGF of a cone, which should meet with less resistance than any other frustum constructed with the same base and altitude, and going forwards towards D in the direction of its axis: bisect the altitude OD in Q, and produce OQ to S, so that QS may be equal to QC, and S will be the vertex of the cone whose frustum is sought.

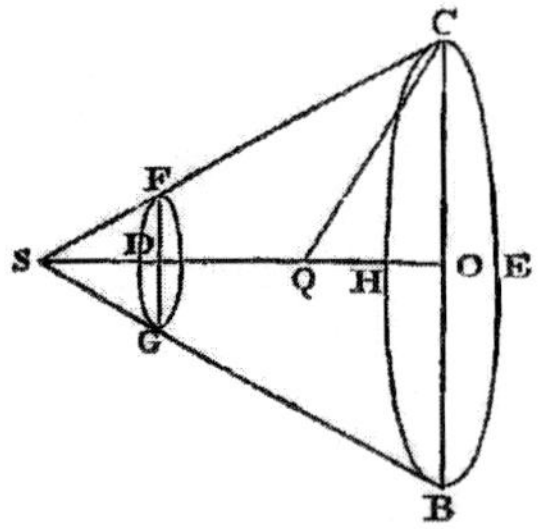

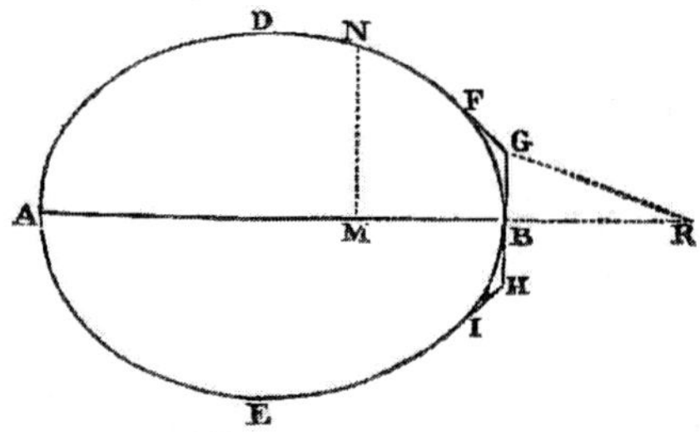

Whence, by the bye, since the angle CSB is always acute, it follows, that, if the solid ADBE be generated by the convolution of an elliptical or oval figure ADBE about its axis AB, and the generating figure be touched by three right lines FG, GH, HI, in the points F, B, and I, so that GH shall be perpendicular to the axis in the point of contact B, and FG, HI may be inclined to GH in the angles FGB, BHI of 135 degrees: the solid arising from the convolution of the figure ADFGHIE about the same axis AB will be less resisted than the former solid; if so be that both move forward in the direction of their axis AB, and that the extremity B of each go foremost. Which Proposition I conceive may be of use in the building of ships.

If the figure DNFG be such a curve, that if, from any point thereof, as N, the perpendicular NM be let fall on the axis AB, and from the given point G there be drawn the right line GR parallel to a right line touching the figure in N, and cutting the axis produced in R, MN becomes to GR as GR^3 to $4BR \times GB^2$, the solid described by the revolution of this figure about its axis AB, moving in the before-mentioned rare medium from A towards B, will be less resisted than any other circular solid whatsoever, described of the same length and breadth.

PROPOSITION XXXV. PROBLEM VII.

If a rare medium consist of very small quiescent particles of equal magnitudes, and freely disposed at equal distances from one another: to find the resistance of a globe moving uniformly forward in this medium.

CASE 1. Let a cylinder described with the same diameter and altitude be conceived to go forward with the same velocity in the direction of its axis through the same medium; and let us suppose that the particles of the medium, on which the globe or cylinder falls, fly back with as great a force of reflexion as possible. Then since the resistance of the globe (by the last Proposition) is but half the resistance of the cylinder, and since the globe is to the cylinder as 2 to 3, and since the cylinder by falling perpendicularly on the particles, and reflecting them with the utmost force, communicates to them a velocity double to its own; it follows that the cylinder, in moving forward uniformly half the length of its axis, will communicate a motion to the particles which is to the whole motion of the cylinder as the density of the medium to the density of the cylinder; and that the globe, in the time it describes one length of its diameter in moving uniformly forward, will communicate the same motion to the particles; and in the time that it describes two thirds of its diameter, will communicate a motion to the particles which is to the whole motion of the globe as the density of the medium to the density of the globe. And therefore the globe meets with a resistance, which is to the force by which its whole motion may be either taken away or generated in the time in which it describes two thirds of its diameter moving uniformly forward, as the density of the medium to the density of the globe.

CASE 2. Let us suppose that the particles of the medium incident on the globe or cylinder are not reflected; and then the cylinder falling perpendicularly on the particles will communicate its own simple velocity to them, and therefore meets a resistance but half so great as in the former case, and the globe also meets with a resistance but half so great.

CASE 3. Let us suppose the particles of the medium to fly back from the globe with a force which is neither the greatest, nor yet none at all, but with a certain mean force; then the resistance of the globe will be in the same mean ratio between the resistance in the first case and the resistance in the second. Q.E.I.

COR. 1. Hence if the globe and the particles are infinitely hard, and destitute of all elastic force, and therefore of all force of reflexion; the resistance of the globe will be to the force by which its whole motion may be destroyed or generated, in the time that the globe describes four third parts of its diameter, as the density of the medium to the density of the globe.

Cor. 2. The resistance of the globe, *cæteris paribus*, is in the duplicate ratio of the velocity.

Cor. 3. The resistance of the globe, *cæteris paribus*, is in the duplicate ratio of the diameter.

Cor. 4. The resistance of the globe is, *cæteris paribus*, as the density of the medium.

Cor. 5. The resistance of the globe is in a ratio compounded of the duplicate ratio of the velocity, and the duplicate ratio of the diameter, and the ratio of the density of the medium.

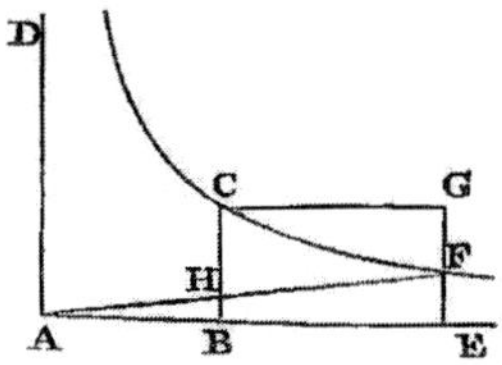

Cor. 6. The motion of the globe and its resistance may be thus expounded. Let AB be the time in which the globe may, by its resistance uniformly continued, lose its whole motion. Erect AD, BC perpendicular to AB. Let BC be that whole motion, and through the point C, the asymptotes being AD, AB, describe the hyperbola CF. Produce AB to any point E. Erect the perpendicular EF meeting the hyperbola in F. Complete the parallelogram CBEG, and draw AF meeting BC in H. Then if the globe in any time BE, with its first motion BC uniformly continued, describes in a non-resisting medium the space CBEG expounded by the area of the parallelogram, the same in a resisting medium will describe the space CBEF expounded by the area of the hyperbola; and its motion at the end of that time will be expounded by EF, the ordinate of the hyperbola, there being lost of its motion the part FG. And its resistance at the end of the same time will be expounded by the length BH, there being lost of its resistance the part CH. All these things appear by Cor. 1 and 3, Prop. V., Book II.

Cor. 7. Hence if the globe in the time T by the resistance R uniformly continued lose its whole motion M, the same globe in the time t in a resisting medium, wherein the resistance R decreases in a duplicate ratio of the velocity, will lose out of its motion M the part $\frac{tM}{T+t}$, the part $\frac{TM}{T+t}$ remaining; and will describe a space which is to the space described in the same time t, with the uniform motion M, as the logarithm of the number $\frac{T+t}{T}$ multiplied by the number 2,302585092994 is to the number $\frac{t}{T}$, because the hyperbolic area BCFE is to the rectangle BCGE in that proportion.

SCHOLIUM.

I have exhibited in this Proposition the resistance and retardation of spherical projectiles in mediums that are not continued, and shewn that this resistance is to the force by which the whole motion of the globe may be destroyed or produced in the time in which the globe can describe two thirds of its diameter, with a velocity uniformly continued, as the density of the medium to the density of the globe, if so be the globe and the particles of the medium be perfectly elastic, and are endued with the utmost force of reflexion; and that this force, where the globe and particles of the medium are infinitely hard and void of any reflecting force, is diminished one half. But in continued mediums, as water, hot oil, and quicksilver, the globe as it passes through them does not immediately strike against all the particles of the fluid that generate the resistance made to it, but presses only the particles that lie next to it, which press the particles beyond, which press other particles, and so on; and in these mediums the resistance is diminished one other half. A globe in these extremely fluid mediums meets with a resistance that is to the force by which its whole motion may be destroyed or generated in the time wherein it can describe, with that motion uniformly continued, eight third parts of its diameter, as the density of the medium to the density of the globe. This I shall endeavour to shew in what follows.

PROPOSITION XXXVI. PROBLEM VIII.

To define the motion of water running out of a cylindrical vessel through a hole made at the bottom.

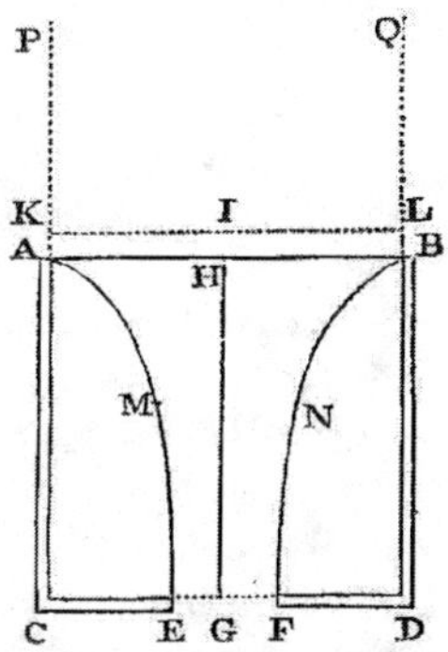

Let ACDB be a cylindrical vessel, AB the mouth of it, CD the bottom parallel to the horizon, EF a circular hole in the middle of the bottom, G the centre of the hole, and GH the axis of the cylinder perpendicular to the horizon. And suppose a cylinder of ice APQB to be of the same breadth with the cavity of the vessel, and to have the same axis, and to descend perpetually with an uniform motion, and that its parts, as soon as they touch the superficies AB, dissolve into water, and flow down by their weight into the vessel, and in their fall compose the cataract or column of water ABNFEM, passing through the hole EF, and filling up the same exactly. Let the uniform velocity of the descending ice and of the contiguous water in the circle AB be that which the water would acquire by falling through the space IH; and let IH and HG lie in the same right line; and through the point I let there be drawn the right line KL parallel to the horizon, and meeting the ice on both the sides thereof in K and L. Then the velocity of the water running out at the hole EF will be the same that it would acquire by falling from I through the space IG. Therefore, by *Galileo's* Theorems, IG will be to IH in the duplicate ratio of the velocity of the water that runs out at the hole to the velocity of the water in the circle AB, that is, in the duplicate ratio of the circle AB to the circle EF; those circles being reciprocally as the velocities of the water which in the same time and in equal quantities passes severally through each of them, and completely fills them both. We are now considering the velocity with which the water tends to the plane of the horizon. But the motion parallel to the same, by which the parts of the falling water approach to each other, is not here taken notice of; since it is neither produced by gravity, nor at all changes the motion perpendicular to the horizon which the gravity produces. We suppose, indeed, that the parts of the water cohere a little, that by their cohesion they may in falling approach to each other with motions parallel to the horizon in order to form one single cataract, and to prevent their being divided into several: but the motion parallel to the horizon arising from this cohesion does not come under our present consideration.

Case 1. Conceive now the whole cavity in the vessel, which encompasses the falling water ABNFEM, to be full of ice, so that the water may pass through the ice as through a funnel. Then if the water pass very near to the ice only, without touching it; or, which is the same thing, if by reason of the perfect smoothness of the surface of the ice, the water, though touching it, glides over it with the utmost freedom, and without the least resistance; the water will run through the hole EF with the same velocity as before, and the whole weight of the column of water ABNFEM will be all taken up as before in forcing out the water, and the bottom of the vessel will sustain the weight of the ice encompassing that column.

Let now the ice in the vessel dissolve into water; yet will the efflux of the water remain, as to its velocity, the same as before. It will not be less, because the ice now dissolved will endeavour to descend; it will not be greater, because the ice, now become water, cannot descend without hindering the descent of other water equal to its own descent. The same force ought always to generate the same velocity in the effluent water.

But the hole at the bottom of the vessel, by reason of the oblique motions of the particles of the effluent water, must be a little greater than before. For now the particles of the water do not all of them pass through the hole perpendicularly, but, flowing down on all parts from the sides of the vessel, and converging towards the hole, pass through it with oblique motions; and in tending downwards meet in a stream whose diameter is a little smaller below the hole than at the hole itself; its diameter being to the diameter of the hole as 5 to 6, or as $5\frac{1}{3}$ to $6\frac{1}{2}$, very nearly, if I took the measures of those diameters right. I procured a very thin flat plate, having a hole pierced in the middle, the diameter of the circular hole being $\frac{5}{8}$ parts of an inch. And that the stream of running waters might not be accelerated in falling, and by that acceleration become narrower, I fixed this plate not to the bottom, but to the side of the vessel, so as to make the water go out in the direction of a line parallel to the horizon. Then, when the vessel was full of water, I opened the hole to let it run out; and the diameter of the stream, measured with great accuracy at the distance of about half an inch from the hole, was $\frac{21}{40}$ of an inch. Therefore the diameter of this circular hole was to the diameter of the stream very nearly as 25 to 21. So that the water in passing through the hole converges on all sides, and, after it has run out of the vessel, becomes smaller by converging in that manner, and by becoming smaller is accelerated till it comes to the distance of half an inch from the hole, and at that distance flows in a smaller stream and with greater celerity than in the hole itself, and this in the ratio of 25×25 to 21×21, or 17 to 12, very nearly; that is, in about the subduplicate ratio of 2 to 1. Now it is certain from experiments, that the quantity of water running out in a given time through a circular hole made in the bottom of a vessel is equal to the quantity, which, flowing with the aforesaid velocity, would run out in the same time through another circular hole, whose diameter is to the diameter of the former as 21 to 25. And therefore that running water in passing through the hole itself has a velocity downwards equal to that which a heavy body would acquire in falling through half the height of the stagnant water in the vessel, nearly. But, then, after it has run out, it is still accelerated by converging, till it arrives at a distance from the hole that is nearly equal to its diameter, and acquires a velocity greater than the other in about the subduplicate ratio of 2 to 1; which velocity a heavy body would nearly acquire by falling through the whole height of the stagnant water in the vessel.

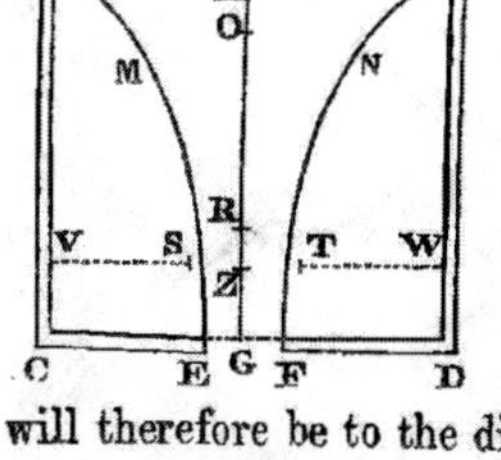

Therefore in what follows let the diameter of the stream be represented by that lesser hole which we called EF. And imagine another plane VW above the hole EF, and parallel to the plane thereof, to be placed at a distance equal to the diameter of the same hole, and to be pierced through with a greater hole ST, of such a magnitude that a stream which will exactly fill the lower hole EF may pass through it; the diameter of which hole will therefore be to the diameter of the lower hole as 25 to 21, nearly. By this means the water will run perpendicularly out at the lower hole; and the quantity of the water running out will be, according to the magnitude of this last hole, the same, very nearly, which the solution of the Problem requires. The space included between the two planes and the falling stream may be considered as the bottom of the vessel. But, to make the solution more simple and mathematical, it is better to take the lower plane alone for the bottom of the vessel, and to suppose that the water which flowed through the ice as through a funnel, and ran out of the vessel through the hole EF made in the lower plane, preserves its motion continually, and that the ice continues at rest. Therefore in what follows let ST be the diameter of a circular hole described from the centre Z, and let the stream run out of the vessel through that hole, when the water in the vessel is all fluid. And let EF be the diameter of the hole, which the stream, in falling through, exactly fills up, whether the water runs out of the vessel by that upper hole ST, or flows through the middle of the ice in the vessel, as through a funnel. And let the diameter of the upper hole ST be to the diameter of the lower EF as about 25 to 21, and let the perpendicular distance between the planes of the holes be equal to the diameter of the lesser hole EF. Then the velocity of the water downwards, in running out of the vessel through the hole ST, will be in that hole the same that a body may acquire by falling from half the height IZ; and the velocity of both the falling streams will be in the hole EF, the same which a body would acquire by falling from the whole height IG.

Case 2. If the hole EF be not in the middle of the bottom of the vessel, but in some other part thereof, the water will still run out with the same velocity as before, if the magnitude of the hole be the same. For though an heavy body takes a longer time in descending to the same depth, by an oblique line, than by a perpendicular line, yet in both cases it acquires in its descent the same velocity; as *Galileo* has demonstrated.

Case 3. The velocity of the water is the same when it runs out through a hole in the side of the vessel. For if the hole be small, so that the interval between the superficies AB and KL may vanish as to sense, and the stream of water horizontally issuing out may form a parabolic figure; from the latus rectum of this parabola may be collected, that the velocity of the effluent water is that which a body may acquire by falling the height IG or HG of the stagnant water in the vessel. For, by making an experiment, I found that if the height of the stagnant water above the hole were 20 inches, and the height of the hole above a plane parallel to the horizon were also 20 inches, a stream of water springing out from thence would fall upon the plane, at the distance of 37 inches, very nearly, from a perpendicular let fall upon that plane from the hole. For without resistance the stream would have fallen upon the plane at the distance of 40 inches, the latus rectum of the parabolic stream being 80 inches.

Case 4. If the effluent water tend upward, it will still issue forth with the same velocity. For the small stream of water springing upward, ascends with a perpendicular motion to GH or GI, the height of the stagnant water in the vessel; excepting in so far as its ascent is hindered a little by the resistance of the air; and therefore it springs out with the same velocity that it would acquire in falling from that height. Every particle of the stagnant water is equally pressed on all sides (by Prop. XIX., Book II), and, yielding to the pressure, tends always with an equal force, whether it descends through the hole in the bottom of the vessel, or gushes out in an horizontal direction through a hole in the side, or passes into a canal, and springs up from thence through a little hole made in the upper part of the canal. And it may not only be collected from reasoning, but is manifest also from the well-known experiments just mentioned, that the velocity with which the water runs out is the very same that is assigned in this Proposition.

CASE 5. The velocity of the effluent water is the same, whether the figure of the hole be circular, or square, or triangular, or any other figure equal to the circular; for the velocity of the effluent water does not depend upon the figure of the hole, but arises from its depth below the plane KL.

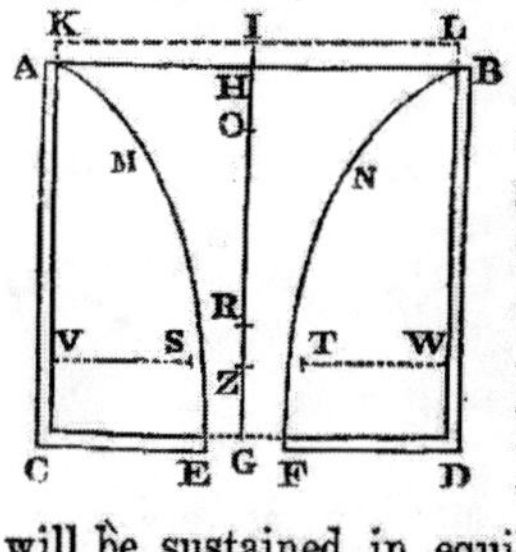

CASE 6. If the lower part of the vessel ABDC be immersed into stagnant water, and the height of the stagnant water above the bottom of the vessel be GR, the velocity with which the water that is in the vessel will run out at the hole EF into the stagnant water will be the same which the water would acquire by falling from the height IR; for the weight of all the water in the vessel that is below the superficies of the stagnant water will be sustained in equilibrio by the weight of the stagnant water, and therefore does not at all accelerate the motion of the descending water in the vessel. This case will also appear by experiments, measuring the times in which the water will run out.

COR. 1. Hence if CA the depth of the water be produced to K, so that AK may be to CK in the duplicate ratio of the area of a hole made in any part of the bottom to the area of the circle AB, the velocity of the effluent water will be equal to the velocity which the water would acquire by falling from the height KC.

COR. 2. And the force with which the whole motion of the effluent water may be generated is equal to the weight of a cylindric column of water, whose base is the hole EF, and its altitude 2GI or 2CK. For the effluent water, in the time it becomes equal to this column, may acquire, by falling by its own weight from the height GI, a velocity equal to that with which it runs out.

COR. 3. The weight of all the water in the vessel ABDC is to that part of the weight which is employed in forcing out the water as the sum of the circles AB and EF to twice the circle EF. For let IO be a mean proportional between IH and IG, and the water running out at the hole EF will, in the time that a drop falling from I would describe the altitude IG, become equal to a cylinder whose base is the circle EF and its altitude 2IG, that is, to a cylinder whose base is the circle AB, and whose altitude is 2IO. For the circle EF is to the circle AB in the subduplicate ratio of the altitude IH to the altitude IG; that is, in the simple ratio of the mean proportional IO to the altitude IG. Moreover, in the time that a drop falling from I can describe the altitude IH, the water that runs out will have become equal to a cylinder whose base is the circle AB, and its altitude 2IH; and in the time that a drop falling from I through H to G describes HG, the difference of the altitudes, the effluent water, that is, the water contained within the solid ABNFEM, will be equal to the difference of the cylinders, that is, to a cylinder whose base is AB, and its altitude 2HO. And therefore all the water contained in the vessel ABDC is to the whole falling water contained in the said solid ABNFEM as HG to 2HO, that is, as HO + OG to 2HO, or IH + IO to 2IH. But the weight of all the water in the solid ABNFEM is employed in forcing out the water; and therefore the weight of all the water in the vessel is to that part of the weight that is employed in forcing out the water as IH + IO to 2IH, and therefore as the sum of the circles EF and AB to twice the circle EF.

Cor. 4. And hence the weight of all the water in the vessel ABDC is to the other part of the weight which is sustained by the bottom of the vessel as the sum of the circles AB and EF to the difference of the same circles.

Cor. 5. And that part of the weight which the bottom of the vessel sustains is to the other part of the weight employed in forcing out the water as the difference of the circles AB and EF to twice the lesser circle EF, or as the area of the bottom to twice the hole.

Cor. 6. That part of the weight which presses upon the bottom is to the whole weight of the water perpendicularly incumbent thereon as the circle AB to the sum of the circles AB and EF, or as the circle AB to the excess of twice the circle AB above the area of the bottom. For that part of the weight which presses upon the bottom is to the weight of the whole water in the vessel as the difference of the circles AB and EF to the sum of the same circles (by Cor. 4); and the weight of the whole water in the vessel is to the weight of the whole water perpendicularly incumbent on the bottom as the circle AB to the difference of the circles AB and EF. Therefore, *ex æquo perturbatè*, that part of the weight which presses upon the bottom is to the weight of the whole water perpendicularly incumbent thereon as the circle AB to the sum of the circles AB and EF, or the excess of twice the circle AB above the bottom.

COR. 7. If in the middle of the hole EF there be placed the little circle PQ described about the centre G, and parallel to the horizon, the weight of water which that little circle sustains is greater than the weight of a third part of a cylinder of water whose base is that little circle and its height GH. For let ABNFEM be the cataract or column of falling water whose axis is GH, as above, and let all the water, whose fluidity is not requisite for the ready and quick descent of the water, be supposed to be congealed, as well round about the cataract, as above the little circle. And let PHQ be the column of water congealed above the little circle, whose vertex is H, and its altitude GH. And suppose this cataract to fall with its whole weight downwards, and not in the least to lie against or to press PHQ, but to glide freely by it without any friction, unless, perhaps, just at the very vertex of the ice, where the cataract at the beginning of its fall may tend to a concave figure. And as the congealed water AMEC, BNFD, lying round the cataract, is convex in its internal superficies AME, BNF, towards the falling cataract, so this column PHQ will be convex towards the cataract also, and will therefore be greater than a cone whose base is that little circle PQ and its altitude GH; that is, greater than a third part of a cylinder described with the same base and altitude. Now that little circle sustains the weight of this column, that is, a weight greater than the weight of the cone, or a third part of the cylinder.

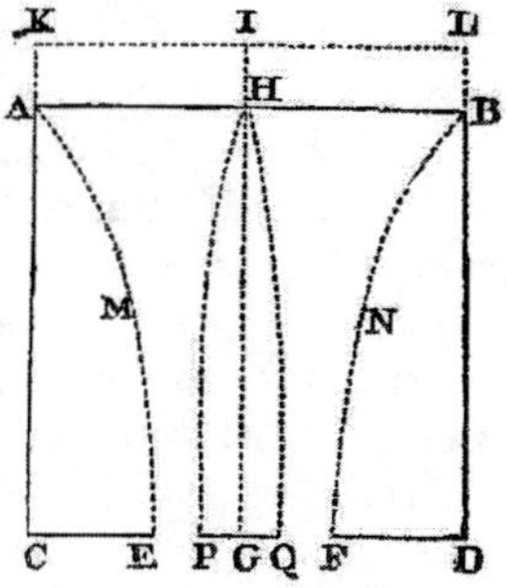

COR. 8. The weight of water which the circle PQ, when very small, sustains, seems to be less than the weight of two thirds of a cylinder of water whose base is that little circle, and its altitude HG. For, things standing as above supposed, imagine the half of a spheroid described whose base is that little circle, and its semi-axis or altitude HG. This figure will be equal to two thirds of that cylinder, and will comprehend within it the column of congealed water PHQ, the weight of which is sustained by that little circle. For though the motion of the water tends directly downwards, the external superficies of that column must yet meet the base PQ in an angle somewhat acute, because the water in its fall is perpetually accelerated, and by reason of that acceleration become narrower. Therefore, since that angle is less than a right one, this column in the lower parts thereof will lie within the hemi-spheroid. In the upper parts also it will be acute or pointed; because to make it otherwise, the horizontal motion of the water must be at the vertex infinitely more swift than its motion towards the horizon. And the less this circle PQ is, the more acute will

the vertex of this column be; and the circle being diminished *in infinitum*, the angle PHQ will be diminished *in infinitum*, and therefore the column will lie within the hemi-spheroid. Therefore that column is less than that hemi-spheroid, or than two-third parts of the cylinder whose base is that little circle, and its altitude GH. Now the little circle sustains a force of water equal to the weight of this column, the weight of the ambient water being employed in causing its efflux out at the hole.

COR. 9. The weight of water which the little circle PQ sustains, when it is very small, is very nearly equal to the weight of a cylinder of water whose base is that little circle, and its altitude $\frac{1}{2}$GH; for this weight is an arithmetical mean between the weights of the cone and the hemi-spheroid above mentioned. But if that little circle be not very small, but on the contrary increased till it be equal to the hole EF, it will sustain the weight of all the water lying perpendicularly above it, that is, the weight of a cylinder of water whose base is that little circle, and its altitude GH.

COR. 10. And (as far as I can judge) the weight which this little circle sustains is always to the weight of a cylinder of water whose base is that little circle, and its altitude $\frac{1}{2}$GH, as EF^2 to $EF^2 - \frac{1}{2}PQ^2$, or as the circle EF to the excess of this circle above half the little circle PQ, very nearly.

LEMMA IV.

If a cylinder move uniformly forward in the direction of its length, the resistance made thereto is not at all changed by augmenting or diminishing that length; and is therefore the same with the resistance of a circle, described with the same diameter, and moving forward with the same velocity in the direction of a right line perpendicular to its plane.

For the sides are not at all opposed to the motion; and a cylinder becomes a circle when its length is diminished *in infinitum*.

PROPOSITION XXXVII. THEOREM XXIX.

If a cylinder move uninformly forward in a compressed, infinite, and non-elastic fluid, in the direction of its length, the resistance arising from the magnitude of its transverse section is to the force by which its whole motion may be destroyed or generated, in the time that it moves four times its length, as the density of the medium to the density of the cylinder, nearly.

For let the vessel ABDC touch the surface of stagnant water with its bottom CD, and let the water run out of this vessel into the stagnant water through the cylindric canal EFTS perpendicular to the horizon; and let the little circle PQ be placed parallel to the horizon any where in the

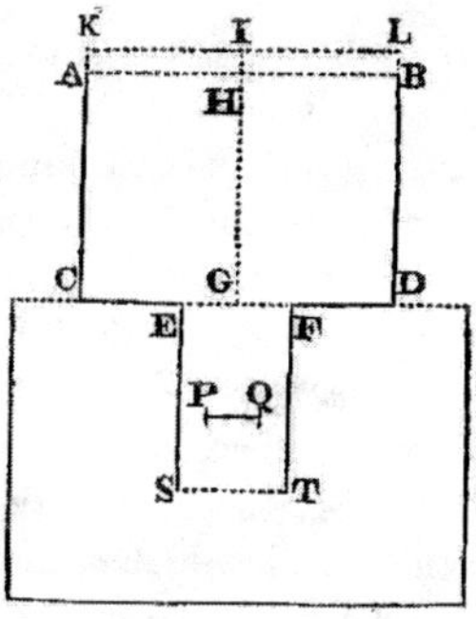

middle of the canal; and produce CA to K, so that AK may be to CK in the duplicate of the ratio, which the excess of the orifice of the canal EF above the little circle PQ bears to the circle AB. Then it is manifest (by Case 5, Case 6, and Cor. 1, Prop. XXXVI) that the velocity of the water passing through the annular space between the little circle and the sides of the vessel will be the very same which the water would acquire by falling, and in its fall describing the altitude KC or IG.

And (by Cor. 10, Prop. XXXVI) if the breadth of the vessel be infinite, so that the lineola HI may vanish, and the altitudes IG, HG become equal; the force of the water that flows down and presses upon the circle will be to the weight of a cylinder whose base is that little circle, and the altitude $\frac{1}{2}$IG, as EF^2 to $EF^2 - \frac{1}{2}PQ^2$, very nearly. For the force of the water flowing downward uniformly through the whole canal will be the same upon the little circle PQ in whatsoever part of the canal it be placed.

Let now the orifices of the canal EF, ST be closed, and let the little circle ascend in the fluid compressed on every side, and by its ascent let it oblige the water that lies above it to descend through the annular space between the little circle and the sides of the canal. Then will the velocity of the ascending little circle be to the velocity of the descending water as the difference of the circles EF and PQ is to the circle PQ; and the velocity of the ascending little circle will be to the sum of the velocities, that is, to the relative velocity of the descending water with which it passes by the little circle in its ascent, as the difference of the circles EF and PQ to the circle EF, or as $EF^2 - PQ^2$ to EF^2. Let that relative velocity be equal to the velocity with which it was shewn above that the water would pass through the annular space, if the circle were to remain unmoved, that is, to the velocity which the water would acquire by falling, and in its fall describing the altitude IG; and the force of the water upon the ascending circle will be the same as before (by Cor. 5, of the Laws of Motion); that is, the resistance of the ascending little circle will be to the weight of a cylinder of water whose base is that little circle, and its altitude $\frac{1}{2}$IG, as EF^2 to $EF^2 - \frac{1}{2}PQ^2$, nearly. But the velocity of the little circle will be to the velocity which the water acquires by falling, and in its fall describing the altitude IG, as $EF^2 - PQ^2$ to EF^2.

Let the breadth of the canal be increased *in infinitum;* and the ratios between $EF^2 - PQ^2$ and EF^2, and between EF^2 and $EF^2 - \frac{1}{2}PQ^2$, will become at last ratios of equality. And therefore the velocity of the little circle will now be the same which the water would acquire in falling, and in its fall describing the altitude IG; and the resistance will become equal to the weight of a cylinder whose base is that little circle, and its altitude half the altitude IG, from which the cylinder must fall to acquire the velocity of the ascending circle; and with this velocity the cylinder in the time of its fall will describe four times its length. But the resistance of the cylinder moving forward with this velocity in the direction of its length is the same with the resistance of the little circle (by Lem. IV), and is therefore nearly equal to the force by which its motion may be generated while it describes four times its length.

If the length of the cylinder be augmented or diminished, its motion, and the time in which it describes four times its length, will be augmented or diminished in the same ratio; and therefore the force by which the motion, so increased or diminished, may be destroyed or generated, will continue the same; because the time is increased or diminished in the same proportion; and therefore that force remains still equal to the resistance of the cylinder, because (by Lem. IV) that resistance will also remain the same.

If the density of the cylinder be augmented or diminished, its motion, and the force by which its motion may be generated or destroyed in the same time, will be augmented or diminished in the same ratio. Therefore the resistance of any cylinder whatsoever will be to the force by which its whole motion may be generated or destroyed, in the time during which it moves four times its length, as the density of the medium to the density of the cylinder, nearly. Q.E.D.

A fluid must be compressed to become continued; it must be continued and non-elastic, that all the pressure arising from its compression may be propagated in an instant; and so, acting equally upon all parts of the body moved, may produce no change of the resistance. The pressure arising from the motion of the body is spent in generating a motion in the parts of the fluid, and this creates the resistance. But the pressure arising from the compression of the fluid, be it ever so forcible, if it be propagated in an instant, generates no motion in the parts of a continued fluid, produces no change at all of motion therein; and therefore neither augments nor lessens the resistance. This is certain, that the action of the fluid arising from the compression cannot be stronger on the hinder parts of the body moved than on its fore parts, and therefore cannot lessen the resistance described in this proposition. And if its propagation be infinitely swifter than the motion of the body pressed, it will not be stronger on the fore parts than on the hinder parts. But that action will be infinitely swifter, and propagated in an instant, if the fluid be continued and non-elastic.

Cor. 1. The resistances, made to cylinders going uniformly forward in the direction of their lengths through continued infinite mediums, are in a ratio compounded of the duplicate ratio of the velocities and the duplicate ratio of the diameters, and the ratio of the density of the mediums.

COR. 2. If the breadth of the canal be not infinitely increased but the cylinder go forward in the direction of its length through an included quiescent medium, its axis all the while coinciding with the axis of the canal, its resistance will be to the force by which its whole motion, in the time in which it describes four times its length, may be generated or destroyed, in a ratio compounded of the ratio of EF^2 to $EF^2 - \frac{1}{2}PQ^2$ once, and the ratio of EF^2 to $EF^2 - PQ^2$ twice, and the ratio of the density of the medium to the density of the cylinder.

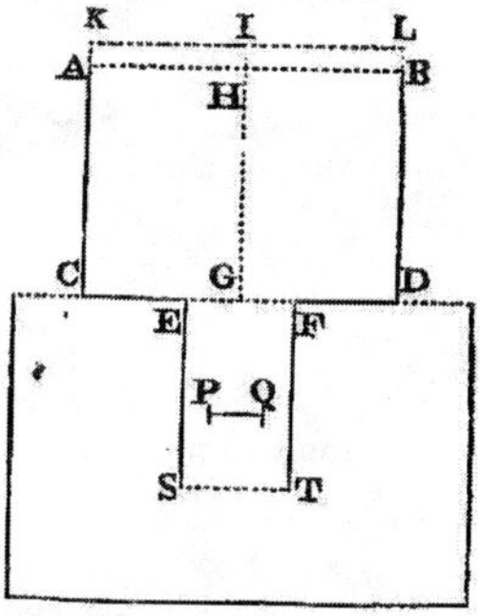

COR. 3. The same thing supposed, and that a length L is to the quadruple of the length of the cylinder in a ratio compounded of the ratio $EF^2 - \frac{1}{2}PQ^2$ to EF^2 once, and the ratio of $EF^2 - PQ^2$ to EF^2 twice; the resistance of the cylinder will be to the force by which its whole motion, in the time during which it describes the length L, may be destroyed or generated, as the density of the medium to the density of the cylinder.

SCHOLIUM.

In this proposition we have investigated that resistance alone which arises from the magnitude of the transverse section of the cylinder, neglecting that part of the same which may arise from the obliquity of the motions. For as, in Case 1, of Prop. XXXVI., the obliquity of the motions with which the parts of the water in the vessel converged on every side to the hole EF hindered the efflux of the water through the hole, so, in this Proposition, the obliquity of the motions, with which the parts of the water, pressed by the antecedent extremity of the cylinder, yield to the pressure, and diverge on all sides, retards their passage through the places that lie round that antecedent extremity, toward the hinder parts of the cylinder, and causes the fluid to be moved to a greater distance; which increases the resistance, and that in the same ratio almost in which it diminished the efflux of the water out of the vessel, that is, in the duplicate ratio of 25 to 21, nearly. And as, in Case 1, of that Proposition, we made the parts of the water pass through the hole EF perpendicularly and in the greatest plenty, by supposing all the water in the vessel lying round the cataract to be frozen, and that part of the water whose motion was oblique and useless to remain without motion, so in this Proposition, that the obliquity of the motions may be taken away, and the parts of the water may give the freest passage to the cylinder, by yielding to it with the most direct and quick motion possible, so that only so much resistance may remain as arises from the magnitude of the transverse section, and which is incapable of diminution, unless by diminishing the diameter of the cylinder; we must conceive those parts of the fluid whose motions are oblique and useless, and produce resistance, to be at rest among themselves at both extremities of the cylinder, and there to cohere, and be joined to the cylinder. Let ABCD be a rectangle, and let AE and BE be two parabolic arcs, described with the axis AB, and with a latus rectum that is to the space HG, which must be described by the cylinder in falling, in order to acquire the velocity with which it moves, as HG to $\frac{1}{2}$AB. Let CF and DF be two other parabolic arcs described with the axis CD, and a latus rectum quadruple of the former; and by the convolution of the figure about the axis EF let there be generated a solid, whose middle part ABDC is the cylinder we are here speaking of, and whose extreme parts ABE and CDF contain the parts of the fluid at rest among themselves, and concreted into two hard bodies, adhering to the cylinder at each end like a head and tail. Then if this solid EACFDB move in the direction of the length of its axis FE toward the parts beyond E, the resistance will be the same which we have here determined in this Proposition, nearly; that is, it will have the same ratio to the force with which the whole motion of the cylinder may be destroyed or generated, in the time that it is describing the length 4AC with that motion uniformly continued, as the density of the fluid has to the density of the cylinder, nearly. And (by Cor. 7, Prop. XXXVI) the resistance must be to this force in the ratio of 2 to 3, at the least.

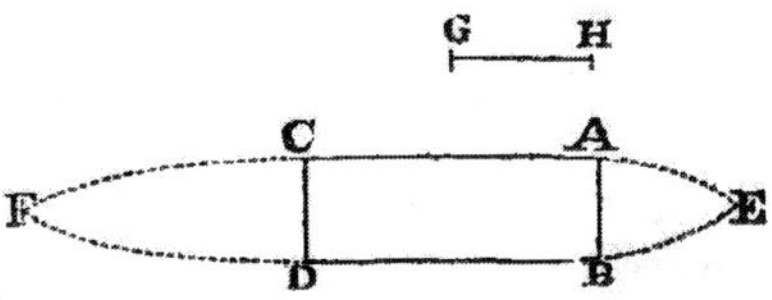

LEMMA V.

If a cylinder, a sphere, and a spheroid, of equal breadths be placed successively in the middle of a cylindric canal, so that their axes may coincide with the axis of the canal, these bodies will equally hinder the passage of the water through the canal.

For the spaces lying between the sides of the canal, and the cylinder, sphere, and spheroid, through which the water passes, are equal; and the water will pass equally through equal spaces.

This is true, upon the supposition that all the water above the cylinder, sphere, or spheroid, whose fluidity is not necessary to make the passage of the water the quickest possible, is congealed, as was explained above in Cor. 7, Prop. XXXVI.

LEMMA VI.

The same supposition remaining, the fore-mentioned bodies are equally acted on by the water flowing through the canal.

This appears by Lem. V and the third Law. For the water and the bodies act upon each other mutually and equally.

LEMMA VII.

If the water be at rest in the canal, and these bodies move with equal velocity and the contrary way through the canal, their resistances will be equal among themselves.

This appears from the last Lemma, for the relative motions remain the same among themselves.

SCHOLIUM.

The case is the same of all convex and round bodies, whose axes coincide with the axis of the canal. Some difference may arise from a greater or less friction; but in these *Lemmata* we suppose the bodies to be perfectly smooth, and the medium to be void of all tenacity and friction; and that those parts of the fluid which by their oblique and superfluous motions may disturb, hinder, and retard the flux of the water through the canal, are at rest among themselves; being fixed like water by frost, and adhering to the fore and hinder parts of the bodies in the manner explained in the Scholium of the last Proposition; for in what follows we consider the very least resistance that round bodies described with the greatest given transverse sections can possibly meet with.

Bodies swimming upon fluids, when they move straight forward, cause the fluid to ascend at their fore parts and subside at their hinder parts, especially if they are of an obtuse figure; and thence they meet with a little more resistance than if they were acute at the head and tail. And bodies moving in elastic fluids, if they are obtuse behind and before, condense the fluid a little more at their fore parts, and relax the same at their hinder parts; and therefore meet also with a little more resistance than if they were acute at the head and tail. But in these Lemmas and Propositions we are not treating of elastic but non-elastic fluids; not of bodies floating on the surface of the fluid, but deeply immersed therein. And when the resistance of bodies in non-elastic fluids is once known, we may then augment this resistance a little in elastic fluids, as our air; and in the surfaces of stagnating fluids, as lakes and seas.

PROPOSITION XXXVIII. THEOREM XXX.

If a globe move uniformly forward in a compressed, infinite, and non-elastic fluid, its resistance is to the force by which its whole motion may be destroyed or generated, in the time that it describes eight third parts of its diameter, as the density of the fluid to the density of the globe, very nearly.

For the globe is to its circumscribed cylinder as two to three; and therefore the force which can destroy all the motion of the cylinder, while the same cylinder is describing the length of four of its diameters, will destroy all the motion of the globe, while the globe is describing two thirds of this length, that is, eight third parts of its own diameter. Now the resistance of the cylinder is to this force very nearly as the density of the fluid to the density of the cylinder or globe (by Prop. XXXVII), and the resistance of the globe is equal to the resistance of the cylinder (by Lem. V, VI, and VII). Q.E.D.

Cor. 1. The resistances of globes in infinite compressed mediums are in a ratio compounded of the duplicate ratio of the velocity, and the duplicate ratio of the diameter, and the ratio of the density of the mediums.

Cor. 2. The greatest velocity, with which a globe can descend by its comparative weight through a resisting fluid, is the same which it may acquire by falling with the same weight, and without any resistance, and in its fall describing a space that is, to four third parts of its diameter as the density of the globe to the density of the fluid. For the globe in the time of its fall, moving with the velocity acquired in falling, will describe a space that will be to eight third parts of its diameter as the density of the globe to the density of the fluid; and the force of its weight which generates this motion will be to the force that can generate the same motion, in the time that the globe describes eight third parts of its diameter, with the same velocity as the density of the fluid to the density of the globe; and therefore (by this Proposition) the force of weight will be equal to the force of resistance, and therefore cannot accelerate the globe.

Cor. 3. If there be given both the density of the globe and its velocity at the beginning of the motion, and the density of the compressed quiescent fluid in which the globe moves, there is given at any time both the velocity of the globe and its resistance, and the space described by it (by Cor. 7, Prop. XXXV).

Cor. 4. A globe moving in a compressed quiescent fluid of the same density with itself will lose half its motion before it can describe the length of two of its diameters (by the same Cor. 7).

PROPOSITION XXXIX. THEOREM XXXI.

If a globe move uniformly forward through a fluid inclosed and compressed in a cylindric canal, its resistance is to the force by which its whole motion may be generated or destroyed, in the time in which it describes eight third parts of its diameter, in a ratio compounded of the ratio of the orifice of the canal to the excess of that orifice above half the greatest circle of the globe; and the duplicate ratio of the orifice of the canal to the excess of that orifice above the greatest circle of the globe; and the ratio of the density of the fluid to the density of the globe, nearly.

This appears by Cor. 2, Prop. XXXVII, and the demonstration proceeds in the same manner as in the foregoing Proposition.

SCHOLIUM.

In the last two Propositions we suppose (as was done before in Lem. V) that all the water which precedes the globe, and whose fluidity increases the resistance of the same, is congealed. Now if that water becomes fluid, it will somewhat increase the resistance. But in these Propositions that increase is so small, that it may be neglected, because the convex superficies of the globe produces the very same effect almost as the congelation of the water.

PROPOSITION XL. PROBLEM IX.

To find by phenomena the resistance of a globe moving through a perfectly fluid compressed medium.

Let A be the weight of the globe *in vacuo*, B its weight in the resisting medium, D the diameter of the globe, F a space which is to $\frac{4}{3}$D as the density of the globe to the density of the medium, that is, as A to A — B, G the time in which the globe falling with the weight B without resistance describes the space F, and H the velocity which the body acquires by that fall. Then H will be the greatest velocity with which the globe can possibly descend with the weight B in the resisting medium, by Cor. 2, Prop. XXXVIII; and the resistance which the globe meets with, when descending with that velocity, will be equal to its weight B; and the resistance it meets with in any other velocity will be to the weight B in the duplicate ratio of that velocity to the greatest velocity H, by Cor. 1, Prop. XXXVIII.

This is the resistance that arises from the inactivity of the matter of the fluid. That resistance which arises from the elasticity, tenacity, and friction of its parts, may be thus investigated.

Let the globe be let fall so that it may descend in the fluid by the weight B; and let P be the time of falling, and let that time be expressed in seconds, if the time G be given in seconds. Find the absolute number N agreeing to the logarithm $0,4342944819 \frac{2P}{G}$, and let L be the logarithm of the number $\frac{N+1}{N}$; and the velocity acquired in falling will be

$\frac{N-1}{N+1}$ H, and the height described will be $\frac{2PF}{G}$ — 1,3862943611F + 4,605170186LF. If the fluid be of a sufficient depth, we may neglect the term 4,605170186LF; and $\frac{2PF}{G}$ — 1,3862943611F will be the altitude described, nearly. These things appear by Prop. IX, Book II, and its Corollaries, and are true upon this supposition, that the globe meets with no other resistance but that which arises from the inactivity of matter. Now if it really meet with any resistance of another kind, the descent will be slower, and from the quantity of that retardation will be known the quantity of this new resistance.

That the velocity and descent of a body falling in a fluid might more easily be known, I have composed the following table; the first column of which denotes the times of descent; the second shews the velocities acquired in falling, the greatest velocity being 100000000; the third exhibits the spaces described by falling in those times, 2F being the space which the body describes in the time G with the greatest velocity; and the fourth gives the spaces described with the greatest velocity in the same times. The numbers in the fourth column are $\frac{2P}{G}$, and by subducting the number 1,3862944 — 4,6051702L, are found the numbers in the third column; and these numbers must be multiplied by the space F to obtain the spaces described in falling. A fifth column is added to all these, containing the spaces described in the same times by a body falling *in vacuo* with the force of B its comparative weight.

The Times P.	Velocities of the body falling in the fluid.	The spaces described in falling in the fluid.	The spaces described with the greatest motion.	The spaces described by falling *in vacuo*.
0,001G	$99999\frac{29}{30}$	0,000001F	0,002F	0,000001F
0,01G	999967	0,0001F	0,02F	0,0001F
0,1G	9966799	0,0099834F	0,2F	0,01F
0,2G	19737532	0,0397361F	0,4F	0,04F
0,3G	29131261	0,0886815F	0,6F	0,09F
0,4G	37994896	0,1559070F	0,8F	0,16F
0,5G	46211716	0,2402290F	1,0F	0,25F
0,6G	53704957	0,3402706F	1,2F	0,36F
0,7G	60436778	0,4545405F	1,4F	0,49F
0,8G	66403677	0,5815071F	1,6F	0,64F
0,9G	71629787	0,7196609F	1,8F	0,81F
1G	76159416	0,8675617F	2F	1F
2G	96402758	2,6500055F	4F	4F
3G	99505475	4,6186570F	6F	9F
4G	99932930	6,6143765F	8F	16F
5G	99990920	8,6137964F	10F	25F
6G	99998771	10,6137179F	12F	36F
7G	99999834	12,6137073F	14F	49F
8G	99999980	14,6137059F	16F	64F
9G	99999997	16,6137057F	18F	81F
10G	$99999999\frac{3}{5}$	18,6137056F	20F	100F

SCHOLIUM.

In order to investigate the resistances of fluids from experiments, I procured a square wooden vessel, whose length and breadth on the inside was 9 inches *English* measure, and its depth 9 feet $\frac{1}{2}$; this I filled with rain-water: and having provided globes made up of wax, and lead included therein, I noted the times of the descents of these globes, the height through which they descended being 112 inches. A solid cubic foot of *English* measure contains 76 pounds *troy* weight of rain water; and a solid inch contains $\frac{19}{36}$ ounces *troy* weight, or $253\frac{1}{3}$ grains; and a globe of water of one inch in diameter contains 132,645 grains in air, or 132,8 grains *in vacuo;* and any other globe will be as the excess of its weight *in vacuo* above its weight in water.

EXPER. 1. A globe whose weight was $156\frac{1}{4}$ grains in air, and 77 grains in water, described the whole height of 112 inches in 4 seconds. And, upon repeating the experiment, the globe spent again the very same time of 4 seconds in falling.

The weight of this globe *in vacuo* is $156\frac{13}{38}$ grains; and the excess of this weight above the weight of the globe in water is $79\frac{13}{38}$ grains. Hence the diameter of the globe appears to be 0,84224 parts of an inch. Then it will be, as that excess to the weight of the globe *in vacuo*, so is the density of the water to the density of the globe; and so is $\frac{8}{3}$ parts of the diameter of the globe (viz. 2,24597 inches) to the space 2F, which will be therefore 4,4256 inches. Now a globe falling *in vacuo* with its whole weight of $156\frac{13}{38}$ grains in one second of time will describe $193\frac{1}{3}$ inches; and falling in water in the same time with the weight of 77 grains without resistance, will describe 95,219 inches; and in the time G, which is to one second of time in the subduplicate ratio of the space F, or of 2,2128 inches to 95,219 inches, will describe 2,2128 inches, and will acquire the greatest velocity H with which it is capable of descending in water. Therefore the time G is 0″,15244. And in this time G, with that greatest velocity H, the globe will describe the space 2F, which is 4,4256 inches; and therefore in 4 seconds will describe a space of 116,1245 inches. Subduct the space 1,3862944F, or 3,0676 inches, and there will remain a space of 113,0569 inches, which the globe falling through water in a very wide vessel will describe in 4 seconds. But this space, by reason of the narrowness of the wooden vessel before mentioned, ought to be diminished in a ratio compounded of the subduplicate ratio of the orifice of the vessel to the excess of this orifice above half a great circle of the globe, and of the simple ratio of the same orifice to its excess above a great circle of the globe, that is, in a ratio of 1 to 0,9914. This done, we have a space of 112,08 inches, which a globe falling through the water in this wooden vessel in 4 seconds of time ought nearly to describe by this theory; but it described 112 inches by the experiment.

Exper. 2. Three equal globes, whose weights were severally $76\frac{1}{3}$ grains in air, and $5\frac{1}{16}$ grains in water, were let fall successively; and every one fell through the water in 15 seconds of time, describing in its fall a height of 112 inches.

By computation, the weight of each globe *in vacuo* is $76\frac{5}{12}$ grains; the excess of this weight above the weight in water is 71 grains $\frac{17}{48}$; the diameter of the globe 0,81296 of an inch; $\frac{8}{3}$ parts of this diameter 2,16789 inches; the space 2F is 2,3217 inches; the space which a globe of $5\frac{1}{16}$ grains in weight would describe in one second without resistance, 12,808 inches, and the time G0″,301056. Therefore the globe, with the greatest velocity it is capable of receiving from a weight of $5\frac{1}{16}$ grains in its descent through water, will describe in the time 0″,301056 the space of 2,3217 inches; and in 15 seconds the space 115,678 inches. Subduct the space 1,3862944F, or 1,609 inches, and there remains the space 114,069 inches; which therefore the falling globe ought to describe in the same time, if the vessel were very wide. But because our vessel was narrow, the space ought to be diminished by about 0,895 of an inch. And so the space will remain 113,174 inches, which a globe falling in this vessel ought nearly to describe in 15 seconds, by the theory. But by the experiment it described 112 inches. The difference is not sensible.

Exper. 3. Three equal globes, whose weights were severally 121 grains in air, and 1 grain in water, were successively let fall; and they fell through the water in the times 46″, 47″, and 50″, describing a height of 112 inches.

By the theory, these globes ought to have fallen in about 40″. Now whether their falling more slowly were occasioned from hence, that in slow motions the resistance arising from the force of inactivity does really bear a less proportion to the resistance arising from other causes; or whether it is to be attributed to little bubbles that might chance to stick to the globes, or to the rarefaction of the wax by the warmth of the weather, or of the hand that let them fall; or, lastly, whether it proceeded from some insensible errors in weighing the globes in the water, I am not certain. Therefore the weight of the globe in water should be of several grains, that the experiment may be certain, and to be depended on.

Exper. 4. I began the foregoing experiments to investigate the resistances of fluids, before I was acquainted with the theory laid down in the Propositions immediately preceding. Afterward, in order to examine the theory after it was discovered, I procured a wooden vessel, whose breadth on the inside was $8\frac{2}{3}$ inches, and its depth 15 feet and $\frac{1}{3}$. Then I made four globes of wax, with lead included, each of which weighed $139\frac{1}{4}$ grains in air, and $7\frac{1}{8}$ grains in water. These I let fall, measuring the times of their falling in the water with a pendulum oscillating to half seconds. The globes were cold, and had remained so some time, both when they were weighed and when they were let fall; because warmth rarefies the wax, and by rarefying it diminishes the weight of the globe in the water; and wax, when rarefied, is not instantly reduced by cold to its former density. Before they were let fall, they were totally immersed under water, lest, by the weight of any part of them that might chance to be above the water, their descent should be accelerated in its beginning. Then, when after their immersion they were perfectly at rest, they were let go with the greatest care, that they might not receive any impulse from the hand that let them down. And they fell successively in the times of $47\frac{1}{2}$, $48\frac{1}{2}$, 50, and 51 oscillations, describing a height of 15 feet and 2 inches. But the weather was now a little colder than when the globes were weighed, and therefore I repeated the experiment another day; and then the globes fell in the times of 49, $49\frac{1}{2}$, 50, and 53; and at a third trial in the times of $49\frac{1}{2}$, 50, 51, and 53 oscillations. And by making the experiment several times over, I found that the globes fell mostly in the times of $49\frac{1}{2}$ and 50 oscillations. When they fell slower, I suspect them to have been retarded by striking against the sides of the vessel.

Now, computing from the theory, the weight of the globe *in vacuo* is $139\frac{2}{5}$ grains; the excess of this weight above the weight of the globe in water $132\frac{1}{40}$ grains; the diameter of the globe 0,99868 of an inch; $\frac{8}{3}$ parts of the diameter 2,66315 inches; the space 2F 2,8066 inches; the space which a globe weighing $7\frac{1}{8}$ grains falling without resistance describes in a second of time 9,88164 inches; and the time G0″,376843. Therefore the globe with the greatest velocity with which it is capable of descending through the water by the force of a weight of $7\frac{1}{8}$ grains, will in the time 0″,376843 describe a space of 2,8066 inches, and in one second of time a space of 7,44766 inches, and in the time 25″, or in 50 oscillations, the space 186,1915 inches. Subduct the space 1,386294F, or 1,9454 inches, and there will remain the space 184,2461 inches which the globe will describe in that time in a very wide vessel. Because our vessel was narrow, let this space be diminished in a ratio compounded of the subduplicate ratio of the orifice of the vessel to the excess of this orifice above half a great circle of the globe, and of the simple ratio of the same orifice to its excess above a great circle of the globe; and we shall have the space of 181,86 inches, which the globe ought by the theory to describe in this vessel in the time of 50 oscillations, nearly. But it described the space of 182 inches, by experiment, in $49\frac{1}{2}$ or 50 oscillations.

EXPER. 5. Four globes weighing $154\frac{3}{8}$ grains in air, and $21\frac{1}{2}$ grains in water, being let fall several times, fell in the times of $28\frac{1}{2}$, 29, $29\frac{1}{2}$, and 30, and sometimes of 31, 32, and 33 oscillations, describing a height of 15 feet and 2 inches.

They ought by the theory to have fallen in the time of 29 oscillations, nearly.

Exper. 6. Five globes, weighing $212\frac{3}{8}$ grains in air, and $79\frac{1}{2}$ in water, being several times let fall, fell in the times of 15, $15\frac{1}{2}$, 16, 17, and 18 oscillations, describing a height of 15 feet and 2 inches.

By the theory they ought to have fallen in the time of 15 oscillations, nearly.

Exper. 7. Four globes, weighing $293\frac{3}{8}$ grains in air, and $35\frac{7}{8}$ grains in water, being let fall several times, fell in the times of $29\frac{1}{2}$ 30, $30\frac{1}{2}$ 31, 32, and 33 oscillations, describing a height of 15 feet and 1 inch and $\frac{1}{2}$.

By the theory they ought to have fallen in the time of 28 oscillations, nearly.

In searching for the cause that occasioned these globes of the same weight and magnitude to fall, some swifter and some slower, I hit upon this; that the globes, when they were first let go and began to fall, oscillated about their centres; that side which chanced to be the heavier descending first, and producing an oscillating motion. Now by oscillating thus, the globe communicates a greater motion to the water than if it descended without any oscillations; and by this communication loses part of its own motion with which it should descend; and therefore as this oscillation is greater or less, it will be more or less retarded. Besides, the globe always recedes from that side of itself which is descending in the oscillation, and by so receding comes nearer to the sides of the vessel, so as even to strike against them sometimes. And the heavier the globes are, the stronger this oscillation is; and the greater they are, the more is the water agitated by it. Therefore to diminish this oscillation of the globes, I made new ones of lead and wax, sticking the lead in one side of the globe very near its surface; and I let fall the globe in such a manner, that, as near as possible, the heavier side might be lowest at the beginning of the descent. By this means the oscillations became much less than before, and the times in which the globes fell were not so unequal: as in the following experiments.

Exper. 8. Four globes weighing 139 grains in air, and $6\frac{1}{2}$ in water, were let fall several times, and fell mostly in the time of 51 oscillations, never in more than 52, or in fewer than 50, describing a height of 182 inches.

By the theory they ought to fall in about the time of 52 oscillations.

Exper. 9. Four globes weighing $273\frac{1}{4}$ grains in air, and $140\frac{3}{4}$ in water, being several times let fall, fell in never fewer than 12, and never more than 13 oscillations, describing a height of 182 inches.

These globes by the theory ought to have fallen in the time of $11\frac{1}{3}$ oscillations, nearly.

Exper. 10. Four globes, weighing 384 grains in air, and $119\frac{1}{2}$ in water, being let fall several times, fell in the times of $17\frac{3}{4}$ 18, $18\frac{1}{2}$, and 19 oscillations, describing a height of $181\frac{1}{2}$ inches. And when they fell in the time of 19 oscillations, I sometimes heard them hit against the sides of the vessel before they reached the bottom.

By the theory they ought to have fallen in the time of $15\frac{5}{9}$ oscillations, nearly.

EXPER. 11. Three equal globes, weighing 48 grains in the air, and $3\frac{29}{32}$ in water, being several times let fall, fell in the times of $43\frac{1}{2}$, 44, $44\frac{1}{2}$, 45, and 46 oscillations, and mostly in 44 and 45, describing a height of $182\frac{1}{2}$ inches, nearly.

By the theory they ought to have fallen in the time of 46 oscillations and $\frac{5}{9}$, nearly.

EXPER. 12. Three equal globes, weighing 141 grains in air, and $4\frac{3}{8}$ in water, being let fall several times, fell in the times of 61, 62, 63, 64, and 65 oscillations, describing a space of 182 inches.

And by the theory they ought to have fallen in $64\frac{1}{2}$ oscillations, nearly.

From these experiments it is manifest, that when the globes fell slowly, as in the second, fourth, fifth, eighth, eleventh, and twelfth experiments, the times of falling are rightly exhibited by the theory; but when the globes fell more swiftly, as in the sixth, ninth, and tenth experiments, the resistance was somewhat greater than in the duplicate ratio of the velocity. For the globes in falling oscillate a little; and this oscillation, in those globes that are light and fall slowly, soon ceases by the weakness of the motion; but in greater and heavier globes, the motion being strong, it continues longer, and is not to be checked by the ambient water till after several oscillations. Besides, the more swiftly the globes move, the less are they pressed by the fluid at their hinder parts; and if the velocity be perpetually increased, they will at last leave an empty space behind them, unless the compression of the fluid be increased at the same time. For the compression of the fluid ought to be increased (by Prop. XXXII and XXXIII) in the duplicate ratio of the velocity, in order to preserve the resistance in the same duplicate ratio. But because this is not done, the globes that move swiftly are not so much pressed at their hinder parts as the others; and by the defect of this pressure it comes to pass that their resistance is a little greater than in a duplicate ratio of their velocity.

So that the theory agrees with the phænomena of bodies falling in water. It remains that we examine the phænomena of bodies falling in air.

EXPER. 13. From the top of St. *Paul's* Church in *London*, in *June* 1710, there were let fall together two glass globes, one full of quicksilver, the other of air; and in their fall they described a height of 220 *English* feet. A wooden table was suspended upon iron hinges on one side, and the other side of the same was supported by a wooden pin. The two globes lying upon this table were let fall together by pulling out the pin by means of an iron wire reaching from thence quite down to the ground; so that, the pin being removed, the table, which had then no support but the iron hinges, fell downward, and turning round upon the hinges, gave leave to the globes to drop off from it. At the same instant, with the same pull of the iron wire that took out the pin, a pendulum oscillating to seconds was let go, and began to oscillate. The diameters and weights of the globes, and their times of falling, are exhibited in the following table.

The globes filled with mercury.			*The globes full of air.*		
Weights.	Diameters.	Times in falling.	Weights	Diameters.	Times in falling.
908 grains	0,8 of an inch	4″	510 grains	5,1 inches	8″½
983	0,8	4 —	642	5,2	8
866	0,8	4	599	5,1	8
747	0,75	4 +	515	5,0	8¼
808	0,75	4	483	5,0	8½
784	0,75	4 +	641	5,2	8

But the times observed must be corrected; for the globes of mercury (by *Galileo's* theory), in 4 seconds of time, will describe 257 *English* feet, and 220 feet in only 3″42‴. So that the wooden table, when the pin was taken out, did not turn upon its hinges so quickly as it ought to have done; and the slowness of that revolution hindered the descent of the globes at the beginning. For the globes lay about the middle of the table, and indeed were rather nearer to the axis upon which it turned than to the pin. And hence the times of falling were prolonged about 18‴; and therefore ought to be corrected by subducting that excess, especially in the larger globes, which, by reason of the largeness of their diameters, lay longer upon the revolving table than the others. This being done, the times in which the six larger globes fell will come forth 8″ 12‴, 7″ 42‴, 7″ 42‴, 7″ 57‴, 8″ 12‴ and 7″ 42‴.

Therefore the fifth in order among the globes that were full of air being 5 inches in diameter, and 483 grains in weight, fell in 8″ 12‴, describing a space of 220 feet. The weight of a bulk of water equal to this globe is 16600 grains; and the weight of an equal bulk of air is $\frac{16600}{860}$ grains, or $19\frac{3}{10}$ grains; and therefore the weight of the globe *in vacuo* is $502\frac{3}{10}$ grains; and this weight is to the weight of a bulk of air equal to the globe as $502\frac{3}{10}$ to $19\frac{3}{10}$; and so is 2F to $\frac{8}{3}$ of the diameter of the globe, that is, to $13\frac{1}{3}$ inches. Whence 2F becomes 28 feet 11 inches. A globe, falling *in vacuo* with its whole weight of $502\frac{3}{10}$ grains, will in one second of time describe $193\frac{1}{3}$ inches as above; and with the weight of 483 grains will describe 185,905 inches; and with that weight 483 grains *in vacuo* will describe the space F, or 14 feet $5\frac{1}{2}$ inches, in the time of 57‴ 58⁗, and acquire the greatest velocity it is capable of descending with in the air. With this velocity the globe in 8″ 12‴ of time will describe 245 feet and $5\frac{1}{3}$ inches. Subduct 1,3863F, or 20 feet and $\frac{1}{2}$ an inch, and there remain 225 feet 5 inches. This space, therefore, the falling globe ought by the theory to describe in 8″ 12‴. But by the experiment it described a space of 220 feet. The difference is insensible.

By like calculations applied to the other globes full of air, I composed the following table.

The weights of the globes.	The diameters.	The times of falling from a height of 220 feet.		The spaces which they would describe by the theory.		The excesses.	
510 grains	5,1 inches	8″	12‴	226 feet	11 inch.	6 feet	11 inch.
642	5,2	7	42	230	9	10	9
599	5,1	7	42	227	10	7	0
515	5	7	57	224	5	4	5
483	5	8	12	225	5	5	5
641	5,2	7	42	230	7	10	7

EXPER. 14. *Anno* 1719, in the month of *July*, Dr. *Desaguliers* made some experiments of this kind again, by forming hogs' bladders into spherical orbs; which was done by means of a concave wooden sphere, which the bladders, being wetted well first, were put into. After that being blown full of air, they were obliged to fill up the spherical cavity that contained them; and then, when dry, were taken out. These were let fall from the lantern on the top of the cupola of the same church, namely, from a height of 272 feet; and at the same moment of time there was let fall a leaden globe, whose weight was about 2 pounds *troy* weight. And in the mean time some persons standing in the upper part of the church where the globes were let fall observed the whole times of falling; and others standing on the ground observed the differences of the times between the fall of the leaden weight and the fall of the bladder. The times were measured by pendulums oscillating to half seconds. And one of those that stood upon the ground had a machine vibrating four times in one second; and another had another machine accurately made with a pendulum vibrating four times in a second also. One of those also who stood at the top of the church had a like machine; and these instruments were so contrived, that their motions could be stopped or renewed at pleasure. Now the leaden globe fell in about four seconds and $\frac{1}{4}$ of time; and from the addition of this time to the difference of time above spoken of, was collected the whole time in which the bladder was falling. The times which the five bladders spent in falling, after the leaden globe had reached the ground, were, the first time, $14\frac{3}{4}$″, $12\frac{3}{4}$″, $14\frac{5}{8}$″, $17\frac{3}{4}$″, and $16\frac{7}{8}$″; and the second time, $14\frac{1}{2}$″, $14\frac{1}{4}$″, 14″, 19″, and $16\frac{3}{4}$″. Add to these $4\frac{1}{4}$″, the time in which the leaden globe was falling, and the whole times in which the five bladders fell were, the first time, 19″, 17″, $18\frac{7}{8}$″, 22″, and $21\frac{1}{8}$″; and the second time, $18\frac{3}{4}$″, $18\frac{1}{2}$″, $18\frac{1}{4}$″, $23\frac{1}{4}$″, and 21″. The times observed at the top of the church were, the first time, $19\frac{3}{8}$″, $17\frac{1}{4}$″, $18\frac{3}{4}$″, $22\frac{1}{8}$″, and $21\frac{5}{8}$″; and the second time, 19″, $18\frac{5}{8}$″, $18\frac{3}{8}$″, 24″. and $21\frac{1}{4}$″. But the bladders did not always fall directly down, but sometimes fluttered a little in the air, and waved to and fro, as

they were descending. And by these motions the times of their falling were prolonged, and increased by half a second sometimes, and sometimes by a whole second. The second and fourth bladder fell most directly the first time, and the first and third the second time. The fifth bladder was wrinkled, and by its wrinkles was a little retarded. I found their diameters by their circumferences measured with a very fine thread wound about them twice. In the following table I have compared the experiments with the theory; making the density of air to be to the density of rain-water as 1 to 860, and computing the spaces which by the theory the globes ought to describe in falling.

The weights of the bladders.	The diameters.	The times of falling from a height of 272 feet.	The spaces which by the theory ought to have been described in those times.	The difference between the theory and the experiments.
128 grains	5,28 inches	19″	271 feet 11 in.	— 0 ft. 1 in.
156	5,19	17	272 0½	+ 0 0½
137½	5,3	18	272 7	+ 0 7
97½	5,26	22	277 4	+ 5 4
99⅛	5	21⅛	282 0	+ 10 0

Our theory, therefore, exhibits rightly, within a very little, all the resistance that globes moving either in air or in water meet with; which appears to be proportional to the densities of the fluids in globes of equal velocities and magnitudes.

In the Scholium subjoined to the sixth Section, we shewed, by experiments of pendulums, that the resistances of equal and equally swift globes moving in air, water, and quicksilver, are as the densities of the fluids. We here prove the same more accurately by experiments of bodies falling in air and water. For pendulums at each oscillation excite a motion in the fluid always contrary to the motion of the pendulum in its return; and the resistance arising from this motion, as also the resistance of the thread by which the pendulum is suspended, makes the whole resistance of a pendulum greater than the resistance deduced from the experiments of falling bodies. For by the experiments of pendulums described in that Scholium, a globe of the same density as water in describing the length of its semidiameter in air would lose the $\frac{1}{3342}$ part of its motion. But by the theory delivered in this seventh Section, and confirmed by experiments of falling bodies, the same globe in describing the same length would lose only a part of its motion equal to $\frac{1}{4586}$, supposing the density of water to be to the density of air as 860 to 1. Therefore the resistances were found greater by the experiments of pendulums (for the reasons just mentioned) than by the experiments of falling globes; and that in the ratio of about 4 to 3. But yet since the resistances of pendulums oscillating in air, water, and quicksilver, are alike increased by like causes, the proportion of the resistances in these mediums will be rightly enough exhibited by the experiments of pendulums, as well as by the experiments of falling bodies. And from all this it may be concluded, that the resistances of bodies, moving in any fluids whatsoever, though of the most extreme fluidity, are, *cæteris paribus*, as the densities of the fluids.

These things being thus established, we may now determine what part of its motion any globe projected in any fluid whatsoever would nearly lose in a given time. Let D be the diameter of the globe, and V its velocity at the beginning of its motion, and T the time in which a globe with the velocity V can describe *in vacuo* a space that is, to the space $\frac{8}{3}$D as the density of the globe to the density of the fluid; and the globe projected in that fluid will, in any other time t lose the part $\frac{tV}{T + t}$, the part $\frac{TV}{T + t}$ remaining; and will describe a space, which will be to that described in the same time *in vacuo* with the uniform velocity V, as the logarithm of the number $\frac{T + t}{T}$ multiplied by the number 2,302585093 is to the number $\frac{t}{T}$, by Cor. 7, Prop. XXXV. In slow motions the resistance may be a little less, because the figure of a globe is more adapted to motion than the figure of a cylinder described with the same diameter. In swift motions the resistance may be a little greater, because the elasticity and compression of the fluid do not increase in the duplicate ratio of the velocity. But these little niceties I take no notice of.

And though air, water, quicksilver, and the like fluids, by the division of their parts *in infinitum*, should be subtilized, and become mediums infinitely fluid, nevertheless, the resistance they would make to projected globes would be the same. For the resistance considered in the preceding Propositions arises from the inactivity of the matter; and the inactivity of matter is essential to bodies, and always proportional to the quantity of matter. By the division of the parts of the fluid the resistance arising from the tenacity and friction of the parts may be indeed diminished; but the quantity of matter will not be at all diminished by this division; and if the quantity of matter be the same, its force of inactivity will be the same; and therefore the resistance here spoken of will be the same, as being always proportional to that force. To diminish this resistance, the quantity of matter in the spaces through which the bodies move must be diminished; and therefore the celestial spaces, through which the globes of the planets and comets are perpetually passing towards all parts, with the utmost freedom, and without the least sensible diminution of their motion, must be utterly void of any corporeal fluid, excepting, perhaps, some extremely rare vapours and the rays of light.

Projectiles excite a motion in fluids as they pass through them, and this motion arises from the excess of the pressure of the fluid at the fore parts of the projectile above the pressure of the same at the hinder parts; and cannot be less in mediums infinitely fluid than it is in air, water, and quicksilver, in proportion to the density of matter in each. Now this excess of pressure does, in proportion to its quantity, not only excite a motion in the fluid, but also acts upon the projectile so as to retard its motion; and therefore the resistance in every fluid is as the motion excited by the projectile in the fluid; and cannot be less in the most subtile æther in proportion to the density of that æther, than it is in air, water, and quicksilver, in proportion to the densities of those fluids.

SECTION VIII.

Of motion propagated through fluids.

PROPOSITION XLI. THEOREM XXXII.

A pressure is not propagated through a fluid in rectilinear directions unless where the particles of the fluid lie in a right line.

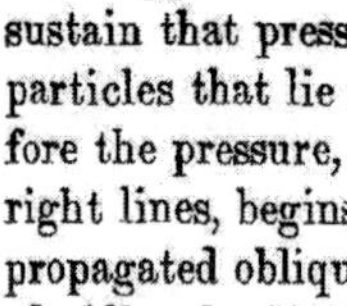

If the particles a, b, c, d, e, lie in a right line, the pressure may be indeed directly propagated from a to e; but then the particle e will urge the obliquely posited particles f and g obliquely, and those particles f and g will not sustain this pressure, unless they be supported by the particles h and k lying beyond them; but the particles that support them are also pressed by them; and those particles cannot sustain that pressure, without being supported by, and pressing upon, those particles that lie still farther, as l and m, and so on *in infinitum.* Therefore the pressure, as soon as it is propagated to particles that lie out of right lines, begins to deflect towards one hand and the other, and will be propagated obliquely *in infinitum;* and after it has begun to be propagated obliquely, if it reaches more distant particles lying out of the right line, it will deflect again on each hand; and this it will do as often as it lights on particles that do not lie exactly in a right line. Q.E.D.

Cor. If any part of a pressure, propagated through a fluid from a given point, be intercepted by any obstacle, the remaining part, which is not intercepted, will deflect into the spaces behind the obstacle. This may be demonstrated also after the following manner. Let a pressure be propagated from the point A towards any part, and, if it be possible, in rectilinear

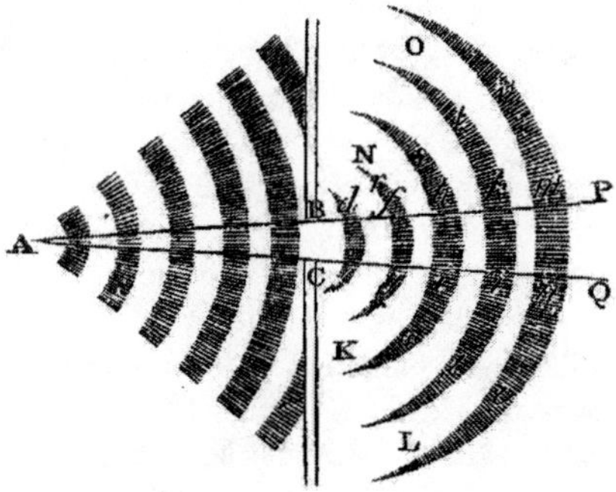

directions; and the obstacle NBCK being perforated in BC, let all the pressure be intercepted but the coniform part APQ passing through the circular hole BC. Let the cone APQ be divided into frustums by the transverse planes, *de*, *fg*, *hi*. Then while the cone ABC, propagating the pressure, urges the conic frustum *degf* beyond it on the superficies *de*, and this frustum urges the next frustum *fgih* on the superficies *fg*, and that frustum urges a third frustum, and so *in infinitum;* it is manifest (by the third Law) that the first frustum *defg* is, by the re-action of the second frustum *fghi*, as much urged and pressed on the superficies *fg*, as it urges and presses that second frustum. Therefore the frustum *degf* is compressed on both sides, that is, between the cone A*de* and the frustum *fhig;* and therefore (by Case 6, Prop. XIX) cannot preserve its figure, unless it be compressed with the same force on all sides. Therefore with the same force with which it is pressed on the superficies *de*, *fg*, it will endeavour to break forth at the sides *df*, *eg;* and there (being not in the least tenacious or hard, but perfectly fluid) it will run out, expanding itself, unless there be an ambient fluid opposing that endeavour. Therefore, by the effort it makes to run out, it will press the ambient fluid, at its sides *df*, *eg*, with the same force that it does the frustum *fghi;* and therefore, the pressure will be propagated as much from the sides *df*, *eg*, into the spaces NO, KL this way and that way, as it is propagated from the superficies *fg* towards PQ. Q.E.D.

PROPOSITION XLII. THEOREM XXXIII.

All motion propagated through a fluid diverges from a rectilinear progress into the unmoved spaces.

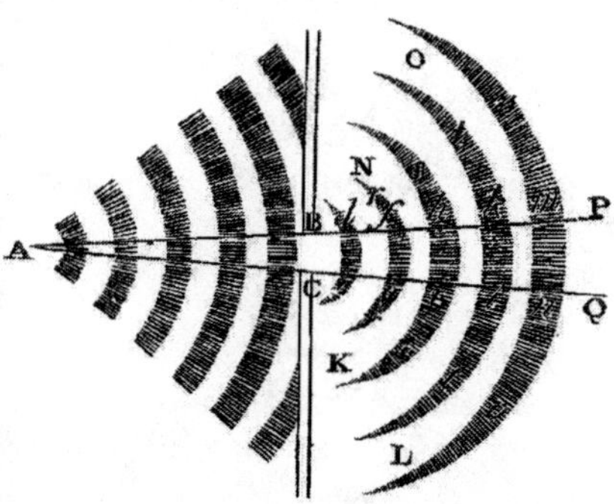

Case 1. Let a motion be propagated from the point A through the hole BC, and, if it be possible, let it proceed in the conic space BCQP according to right lines diverging from the point A. And let us first suppose this motion to be that of waves in the surface of standing water; and let *de*, *fg*, *hi*, *kl*, &c., be the tops of the several waves, divided from each other by as many intermediate valleys or hollows. Then, because the water in the

ridges of the waves is higher than in the unmoved parts of the fluid KL, NO, it will run down from off the tops of those ridges, *e*, *g*, *i*, *l*, &c., *d*, *f*, *h*, *k*, &c., this way and that way towards KL and NO; and because the water is more depressed in the hollows of the waves than in the unmoved parts of the fluid KL, NO, it will run down into those hollows out of those unmoved parts. By the first deflux the ridges of the waves will dilate themselves this way and that way, and be propagated towards KL and NO. And because the motion of the waves from A towards PQ is carried on by a continual deflux from the ridges of the waves into the hollows next to them, and therefore cannot be swifter than in proportion to the celerity of the descent; and the descent of the water on each side towards KL and NO must be performed with the same velocity; it follows that the dilatation of the waves on each side towards KL and NO will be propagated with the same velocity as the waves themselves go forward with directly from A to PQ. And therefore the whole space this way and that way towards KL and NO will be filled by the dilated waves *rfgr*, *shis*, *tklt*, *vmnv*, &c. Q.E.D. That these things are so, any one may find by making the experiment in still water.

Case 2. Let us suppose that *de*, *fg*, *hi*, *kl*, *mn*, represent pulses successively propagated from the point A through an elastic medium. Conceive the pulses to be propagated by successive condensations and rarefactions of the medium, so that the densest part of every pulse may occupy a spherical superficies described about the centre A, and that equal intervals intervene between the successive pulses. Let the lines *de*, *fg*, *hi*, *kl*, &c., represent the densest parts of the pulses, propagated through the hole BC; and because the medium is denser there than in the spaces on either side towards KL and NO, it will dilate itself as well towards those spaces KL, NO, on each hand, as towards the rare intervals between the pulses; and thence the medium, becoming always more rare next the intervals, and more dense next the pulses, will partake of their motion. And because the progressive motion of the pulses arises from the perpetual relaxation of the denser parts towards the antecedent rare intervals; and since the pulses will relax themselves on each hand towards the quiescent parts of the medium KL, NO, with very near the same celerity; therefore the pulses will dilate themselves on all sides into the unmoved parts KL, NO, with almost the same celerity with which they are propagated directly from the centre A; and therefore will fill up the whole space KLON. Q.E.D. And we find the same by experience also in sounds which are heard through a mountain interposed; and, if they come into a chamber through the window, dilate themselves into all the parts of the room, and are heard in every corner; and not as reflected from the opposite walls, but directly propagated from the window, as far as our sense can judge.

Case 3. Let us suppose, lastly, that a motion of any kind is propagated from A through the hole BC. Then since the cause of this propagation is that the parts of the medium that are near the centre A disturb and agitate those which lie farther from it; and since the parts which are urged are fluid, and therefore recede every way towards those spaces where they are less pressed, they will by consequence recede towards all the parts of the quiescent medium; as well to the parts on each hand, as KL and NO, as to those right before, as PQ; and by this means all the motion, as soon as it has passed through the hole BC, will begin to dilate itself, and from thence, as from its principle and centre, will be propagated directly every way. Q.E.D.

PROPOSITION XLIII. THEOREM XXXIV.

Every tremulous body in an elastic medium propagates the motion of the pulses on every side right forward; but in a non-elastic medium excites a circular motion.

CASE. 1. The parts of the tremulous body, alternately going and returning, do in going urge and drive before them those parts of the medium that lie nearest, and by that impulse compress and condense them; and in returning suffer those compressed parts to recede again, and expand themselves. Therefore the parts of the medium that lie nearest to the tremulous body move to and fro by turns, in like manner as the parts of the tremulous body itself do; and for the same cause that the parts of this body agitate these parts of the medium, these parts, being agitated by like tremors, will in their turn agitate others next to themselves; and these others, agitated in like manner, will agitate those that lie beyond them, and so on *in infinitum.* And in the same manner as the first parts of the medium were condensed in going, and relaxed in returning, so will the other parts be condensed every time they go, and expand themselves every time they return. And therefore they will not be all going and all returning at the same instant (for in that case they would always preserve determined distances from each other, and there could be no alternate condensation and rarefaction); but since, in the places where they are condensed, they approach to, and, in the places where they are rarefied, recede from each other, therefore some of them will be going while others are returning; and so on *in infinitum.* The parts so going, and in their going condensed, are pulses, by reason of the progressive motion with which they strike obstacles in their way; and therefore the successive pulses produced by a tremulous body will be propagated in rectilinear directions; and that at nearly equal distances from each other, because of the equal intervals of time in which the body, by its several tremors produces the several pulses. And though the parts of the tremulous body go and return in some certain and determinate direction, yet the pulses propagated from thence through the medium will dilate themselves towards the sides, by the foregoing Proposition; and will be propagated on all sides from that tremulous body, as from a common centre, in superficies nearly spherical and concentrical. An example of this we have in waves excited by shaking a finger in water, which proceed not only forward and backward agreeably to the motion of the finger, but spread themselves in the manner of concentrical circles all round the finger, and are propagated on every side. For the gravity of the water supplies the place of elastic force.

Case 2. If the medium be not elastic, then, because its parts cannot be condensed by the pressure arising from the vibrating parts of the tremulous body, the motion will be propagated in an instant towards the parts where the medium yields most easily, that is, to the parts which the tremulous body would otherwise leave vacuous behind it. The case is the same with that of a body projected in any medium whatever. A medium yielding to projectiles does not recede *in infinitum*, but with a circular motion comes round to the spaces which the body leaves behind it. Therefore as often as a tremulous body tends to any part, the medium yielding to it comes round in a circle to the parts which the body leaves; and as often as the body returns to the first place, the medium will be driven from the place it came round to, and return to its original place. And though the tremulous body be not firm and hard, but every way flexible, yet if it continue of a given magnitude, since it cannot impel the medium by its tremors any where without yielding to it somewhere else, the medium receding from the parts of the body where it is pressed will always come round in a circle to the parts that yield to it. Q.E.D.

Cor. It is a mistake, therefore, to think, as some have done, that the agitation of the parts of flame conduces to the propagation of a pressure in rectilinear directions through an ambient medium. A pressure of that kind must be derived not from the agitation only of the parts of flame, but from the dilatation of the whole.

PROPOSITION XLIV. THEOREM XXXV.

If water ascend and descend alternately in the erected legs KL, MN, of a canal or pipe; and a pendulum be constructed whose length between the point of suspension and the centre of oscillation is equal to half the length of the water in the canal; I say, that the water will ascend and descend in the same times in which the pendulum oscillates.

I measure the length of the water along the axes of the canal and its legs, and make it equal to the sum of those axes; and take no notice of the resistance of the water arising from its attrition by the sides of the canal. Let, therefore, AB, CD, represent the mean height of the water in both legs; and when the water in the leg KL ascends to the height EF, the water will descend in the leg MN to the height GH. Let P be a pendulous body, VP the thread, V the point of suspension, RPQS the cycloid which

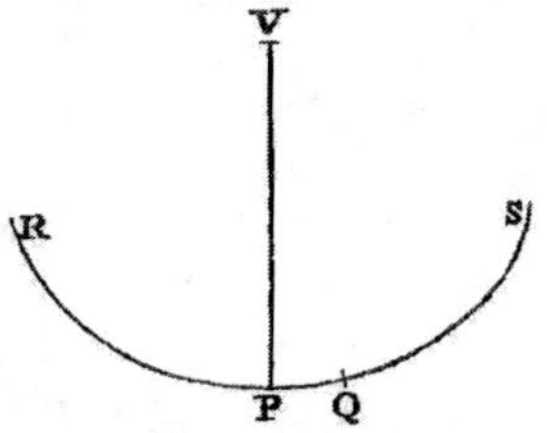

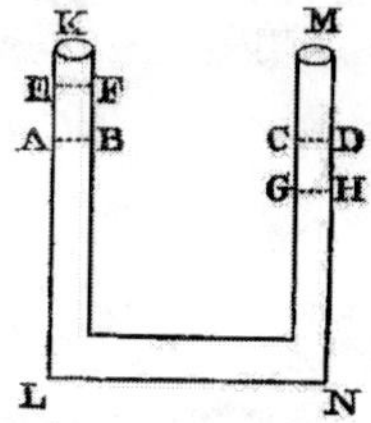

the pendulum describes, P its lowest point, PQ an arc equal to the height AE. The force with which the motion of the water is accelerated and retarded alternately is the excess of the weight of the water in one leg above the weight in the other; and, therefore, when the water in the leg KL ascends to EF, and in the other leg descends to GH, that force is double the weight of the water EABF, and therefore is to the weight of the whole water as AE or PQ to VP or PR. The force also with which the body P is accelerated or retarded in any place, as Q, of a cycloid, is (by Cor. Prop. LI) to its whole weight as its distance PQ from the lowest place P to the length PR of the cycloid. Therefore the motive forces of the water and pendulum, describing the equal spaces AE, PQ, are as the weights to be moved; and therefore if the water and pendulum are quiescent at first, those forces will move them in equal times, and will cause them to go and return together with a reciprocal motion. Q.E.D.

COR. 1. Therefore the reciprocations of the water in ascending and descending are all performed in equal times, whether the motion be more or less intense or remiss.

COR. 2. If the length of the whole water in the canal be of $6\frac{1}{9}$ feet of *French* measure, the water will descend in one second of time, and will ascend in another second, and so on by turns *in infinitum;* for a pendulum of $3\frac{1}{18}$ such feet in length will oscillate in one second of time.

COR. 3. But if the length of the water be increased or diminished, the time of the reciprocation will be increased or diminished in the subduplicate ratio of the length.

PROPOSITION XLV. THEOREM XXXVI.

The velocity of waves is in the subduplicate ratio of the breadths.

This follows from the construction of the following Proposition.

PROPOSITION XLVI. PROBLEM X.

To find the velocity of waves.

Let a pendulum be constructed, whose length between the point of suspension and the centre of oscillation is equal to the breadth of the waves;

and in the time that the pendulum will perform one single oscillation the waves will advance forward nearly a space equal to their breadth.

That which I call the breadth of the waves is the transverse measure lying between the deepest part of the hollows, or the tops of the ridges. Let

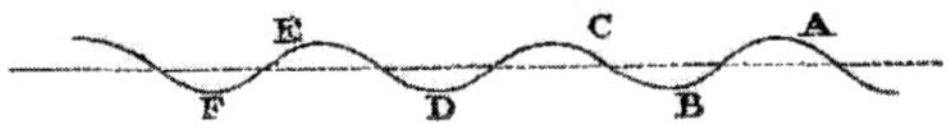

ABCDEF represent the surface of stagnant water ascending and descending in successive waves; and let A, C, E, &c., be the tops of the waves; and let B, D, F, &c., be the intermediate hollows. Because the motion of the waves is carried on by the successive ascent and descent of the water, so that the parts thereof, as A, C, E, &c., which are highest at one time become lowest immediately after; and because the motive force, by which the highest parts descend and the lowest ascend, is the weight of the elevated water, that alternate ascent and descent will be analogous to the reciprocal motion of the water in the canal, and observe the same laws as to the times of its ascent and descent; and therefore (by Prop. XLIV) if the distances between the highest places of the waves A, C, E, and the lowest B, D, F, be equal to twice the length of any pendulum, the highest parts A, C, E, will become the lowest in the time of one oscillation, and in the time of another oscillation will ascend again. Therefore between the passage of each wave, the time of two oscillations will intervene; that is, the wave will describe its breadth in the time that pendulum will oscillate twice; but a pendulum of four times that length, and which therefore is equal to the breadth of the waves, will just oscillate once in that time. Q.E.I.

Cor. 1. Therefore waves, whose breadth is equal to $3\frac{1}{18}$ *French* feet, will advance through a space equal to their breadth in one second of time; and therefore in one minute will go over a space of $183\frac{1}{3}$ feet; and in an hour a space of 11000 feet, nearly.

Cor. 2. And the velocity of greater or less waves will be augmented or diminished in the subduplicate ratio of their breadth.

These things are true upon the supposition that the parts of water ascend or descend in a right line; but, in truth, that ascent and descent is rather performed in a circle; and therefore I propose the time defined by this Proposition as only near the truth.

PROPOSITION XLVII. THEOREM XXXVII.

If pulses are propagated through a fluid, the several particles of the fluid, going and returning with the shortest reciprocal motion, are always accelerated or retarded according to the law of the oscillating pendulum.

Let AB, BC, CD, &c., represent equal distances of successive pulses; ABC the line of direction of the motion of the successive pulses propagated from A to B; E, F, G three physical points of the quiescent medium situate in the right line AC at equal distances from each other; Ee, Ff, Gg, equal spaces of extreme shortness, through which those points go and return with a reciprocal motion in each vibration; ϵ, ϕ, γ, any intermediate places of the same points; EF, FG physical lineolæ, or linear parts of the medium lying between those points, and successively transferred into the places $\epsilon\phi$, $\phi\gamma$, and ef, fg. Let there be drawn the right line PS equal to the right line Ee. Bisect the same in O, and from the centre O, with the interval OP, describe the circle SIPi. Let the whole time of one vibration; with its proportional parts, be expounded by the whole circumference of this circle and its parts, in such sort, that, when any time PH or PHSh is completed, if there be let fall to PS the perpendicular HL or hl, and there be taken Eϵ equal to PL or Pl, the physical point E may be found in ϵ. A point, as E, moving according to this law with a reciprocal motion, in its going from E through ϵ to e, and returning again through ϵ to E, will perform its several vibrations with the same degrees of acceleration and retardation with those of an oscillating pendulum. We are now to prove that the several physical points of the medium will be agitated with such a kind of motion. Let us suppose, then, that a medium hath such a motion excited in it from any cause whatsoever, and consider what will follow from thence.

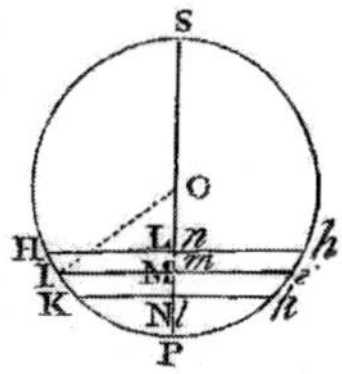

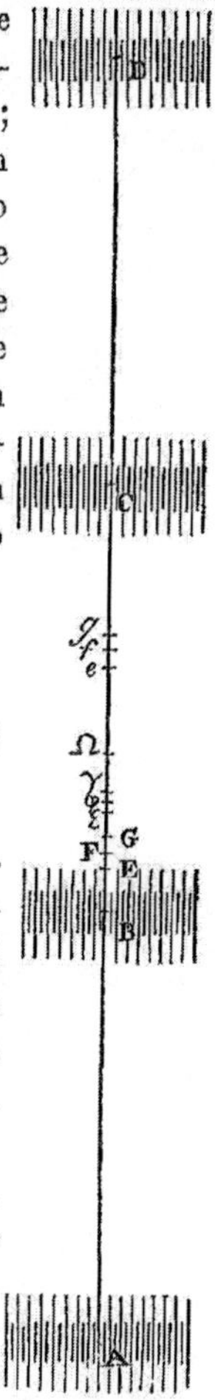

In the circumference PHSh let there be taken the equal arcs, HI, IK, or hi, ik, having the same ratio to the whole circumference as the equal right lines EF, FG have to BC, the whole interval of the pulses. Let fall the perpendiculars IM, KN, or im, kn; then because the points E, F, G are successively agitated with like motions, and perform their entire vibrations composed of their going and return, while the pulse is transferred from B to C; if PH or PHSh be the time elapsed since the beginning of the motion of the point E, then will PI or PHSi be the time elapsed since the beginning of the motion of the point F, and PK or PHSk the time elapsed since the beginning of the motion of the point G; and therefore Eϵ, Fϕ, Gγ, will be respectively equal to PL, PM, PN, while the points are going, and to Pl, Pm, Pn, when the points are returning. Therefore $\epsilon\gamma$ or EG + Gγ — Eϵ will, when the points are going, be equal to EG — LN, and in their return equal to EG + ln. But $\varepsilon\gamma$ is the breadth or expansion of the part EG of the medium in the place $\varepsilon\gamma$; and therefore the expansion of that part in its going is to its mean expansion as EG — LN to EG; and in its return, as EG + ln or EG + LN to EG. Therefore since LN is to KH as IM to the radius OP, and KH to EG as the circumference PHShP to BC; that is, if we put V for the radius of a circle whose circumference is equal to BC the interval of the pulses, as OP to V; and, *ex æquo*, LN to EG as IM to V; the expansion of the part EG, or of the physical point F in the place $\varepsilon\gamma$, to the mean expansion of the same part in its first place EG, will be as V — IM to V in going, and as V + im to V in its return. Hence the elastic force of the point F in the place $\varepsilon\gamma$ to its mean elastic force in the place EG is as $\frac{1}{\text{V}-\text{IM}}$ to $\frac{1}{\text{V}}$ in its going, and as $\frac{1}{\text{V}+im}$ to $\frac{1}{\text{V}}$ in its return. And by the same reasoning the elastic forces of the physical points E and G in going are as $\frac{1}{\text{V}-\text{HL}}$ and $\frac{1}{\text{V}-\text{KN}}$ to $\frac{1}{\text{V}}$; and the difference of the forces to the mean elastic force of the medium as $\frac{\text{HL}-\text{KN}}{\text{VV}-\text{V}\times\text{HL}-\text{V}\times\text{KN}+\text{HL}\times\text{KN}}$ to $\frac{1}{\text{V}}$; that is, as $\frac{\text{HL}-\text{KN}}{\text{VV}}$ to $\frac{1}{\text{V}}$, or as HL — KN to V; if we suppose (by reason of the very short extent of the vibrations) HL and KN to be indefinitely less than the quantity V. Therefore since the quantity V is given, the difference of the forces is as HL — KN; that is (because HL — KN is proportional to HK, and OM to OI or OP; and because HK and OP are given) as OM; that is, if Ff be bisected in Ω, as $\Omega\phi$. And for the same reason the difference of the elastic forces of the physical points ε and γ, in the return of the physical lineola $\varepsilon\gamma$, is as $\Omega\phi$. But that difference (that is, the excess of the elastic force of the point ε above the elastic force of the point γ) is the very force by which the intervening physical lineola $\varepsilon\gamma$ of the medium is accelerated in going, and retarded in returning; and therefore the accelerative force of the physical lineola $\varepsilon\gamma$ is as its distance from Ω, the middle place of the vibration. Therefore (by Prop. XXXVIII, Book I) the time is rightly expounded by the arc PI; and the linear part of the medium $\varepsilon\gamma$ is moved according to the law abovementioned, that is, according to the law of a pendulum oscillating; and the case is the same of all the linear parts of which the whole medium is compounded. Q.E.D.

Cor. Hence it appears that the number of the pulses propagated is the same with the number of the vibrations of the tremulous body, and is not multiplied in their progress. For the physical lineola $\varepsilon\gamma$ as soon as it returns to its first place is at rest; neither will it move again, unless it receives a new motion either from the impulse of the tremulous body, or of the pulses propagated from that body. As soon, therefore, as the pulses cease to be propagated from the tremulous body, it will return to a state of rest, and move no more.

PROPOSITION XLVIII. THEOREM XXXVIII.

The velocities of pulses propagated in an elastic fluid are in a ratio compounded of the subduplicate ratio of the elastic force directly, and the subduplicate ratio of the density inversely; supposing the elastic force of the fluid to be proportional to its condensation.

CASE 1. If the mediums be homogeneous, and the distances of the pulses in those mediums be equal amongst themselves, but the motion in one medium is more intense than in the other, the contractions and dilatations of the correspondent parts will be as those motions; not that this proportion is perfectly accurate. However, if the contractions and dilatations are not exceedingly intense, the error will not be sensible; and therefore this proportion may be considered as physically exact. Now the motive elastic forces are as the contractions and dilatations; and the velocities generated in the same time in equal parts are as the forces. Therefore equal and corresponding parts of corresponding pulses will go and return together, through spaces proportional to their contractions and dilatations, with velocities that are as those spaces; and therefore the pulses, which in the time of one going and returning advance forward a space equal to their breadth, and are always succeeding into the places of the pulses that immediately go before them, will, by reason of the equality of the distances, go forward in both mediums with equal velocity.

CASE 2. If the distances of the pulses or their lengths are greater in one medium than in another, let us suppose that the correspondent parts describe spaces, in going and returning, each time proportional to the breadths of the pulses; then will their contractions and dilatations be equal; and therefore if the mediums are homogeneous, the motive elastic forces, which agitate them with a reciprocal motion, will be equal also. Now the matter to be moved by these forces is as the breadth of the pulses; and the space through which they move every time they go and return is in the same ratio. And, moreover, the time of one going and returning is in a ratio compounded of the subduplicate ratio of the matter, and the subduplicate ratio of the space; and therefore is as the space. But the pulses advance a space equal to their breadths in the times of going once and returning once; that is, they go over spaces proportional to the times, and therefore are equally swift.

CASE 3. And therefore in mediums of equal density and elastic force, all the pulses are equally swift. Now if the density or the elastic force of the medium were augmented, then, because the motive force is increased in the ratio of the elastic force, and the matter to be moved is increased in the ratio of the density, the time which is necessary for producing the same motion as before will be increased in the subduplicate ratio of the density, and will be diminished in the subduplicate ratio of the elastic force. And therefore the velocity of the pulses will be in a ratio compounded of the subduplicate ratio of the density of the medium inversely, and the subduplicate ratio of the elastic force directly. Q.E.D.

This Proposition will be made more clear from the construction of the following Problem.

PROPOSITION XLIX. PROBLEM XI.

The density and elastic force of a medium being given, to find the velocity of the pulses.

Suppose the medium to be pressed by an incumbent weight after the manner of our air; and let A be the height of a homogeneous medium, whose weight is equal to the incumbent weight, and whose density is the same with the density of the compressed medium in which the pulses are propagated. Suppose a pendulum to be constructed whose length between the point of suspension and the centre of oscillation is A: and in the time in which that pendulum will perform one entire oscillation composed of its going and returning, the pulse will be propagated right onwards through a space equal to the circumference of a circle described with the radius A.

For, letting those things stand which were constructed in Prop. XLVII, if any physical line, as EF, describing the space PS in each vibration, be acted on in the extremities P and S of every going and return that it makes by an elastic force that is equal to its weight, it will perform its several vibrations in the time in which the same might oscillate in a cycloid whose whole perimeter is equal to the length PS; and that because equal forces will impel equal corpuscles through equal spaces in the same or equal times. Therefore since the times of the oscillations are in the subduplicate ratio of the lengths of the pendulums, and the length of the pendulum is equal to half the arc of the whole cycloid, the time of one vibration would be to the time of the oscillation of a pendulum whose length is A in the subduplicate ratio of the length $\frac{1}{2}$PS or PO to the length A. But the elastic force with which the physical lineola EG is urged, when it is found in its extreme places P, S, was (in the demonstration of Prop. XLVII) to its whole elastic force as HL — KN to V, that is (since the point K now falls upon P), as HK to V: and all that force, or which is the same thing, the incumbent weight by which the lineola EG is compressed, is to the weight of the lineola as the altitude A of the incumbent weight to EG the length of the lineola; and therefore, *ex æquo*, the force with which the lineola EG is urged in the places P and S is to the weight of that lineola as HK × A to V × EG; or as PO × A to VV; because HK was to EG as PO to V. Therefore since the times in which equal bodies are impelled through equal spaces are reciprocally in the subduplicate ratio of the forces, the time of one vibration, produced by the action of that elastic force, will be to the time of a vibration, produced by the impulse of the weight in a subduplicate ratio of VV to PO × A, and therefore to the time of the oscillation of a pendulum whose length is A in the subduplicate ratio of VV to PO × A, and the subduplicate ratio of PO to A conjunctly; that is, in the entire ratio of V to A. But in the time of one vibration composed of the going and returning of the pendulum, the pulse will be propagated right onward through a space equal to its breadth BC. Therefore the time in which a pulse runs over the space BC is to the time of one oscillation composed of the going and returning of the pendulum as V to A, that is, as BC to the circumference of a circle whose radius is A. But the time in which the pulse will run over the space BC is to the time in which it will run over a length equal to that circumference in the same ratio; and therefore in the time of such an oscillation the pulse will run over a length equal to that circumference. Q.E.D.

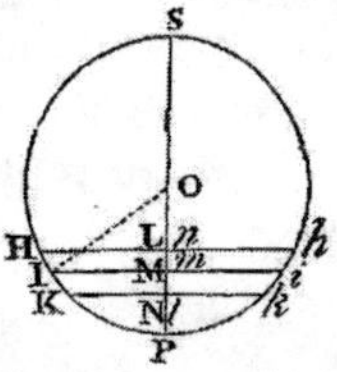

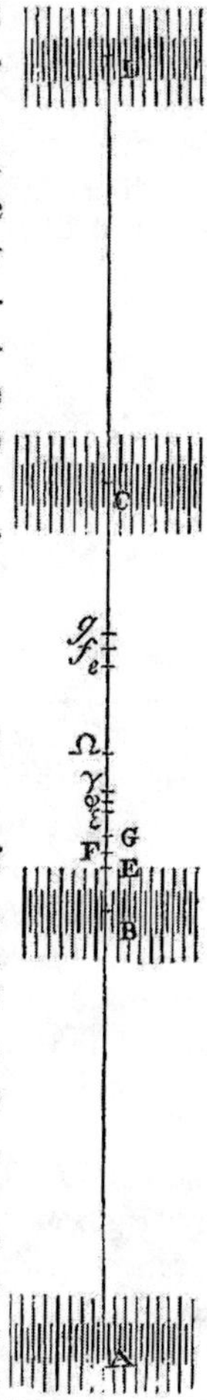

COR. 1. The velocity of the pulses is equal to that which heavy bodies acquire by falling with an equally accelerated motion, and in their fall describing half the altitude A. For the pulse will, in the time of this fall, supposing it to move with the velocity acquired by that fall, run over a space that will be equal to the whole altitude A; and therefore in the time of one oscillation composed of one going and return, will go over a space equal to the circumference of a circle described with the radius A; for the time of the fall is to the time of oscillation as the radius of a circle to its circumference.

COR. 2. Therefore since that altitude A is as the elastic force of the fluid directly, and the density of the same inversely, the velocity of the pulses will be in a ratio compounded of the subduplicate ratio of the density inversely, and the subduplicate ratio of the elastic force directly.

PROPOSITION L. PROBLEM XII.

To find the distances of the pulses.

Let the number of the vibrations of the body, by whose tremor the pulses are produced, be found to any given time. By that number divide the space which a pulse can go over in the same time, and the part found will be the breadth of one pulse. Q.E.I.

SCHOLIUM.

The last Propositions respect the motions of light and sounds; for since light is propagated in right lines, it is certain that it cannot consist in action alone (by Prop. XLI and XLII). As to sounds, since they arise from tremulous bodies, they can be nothing else but pulses of the air propagated through it (by Prop. XLIII); and this is confirmed by the tremors which sounds, if they be loud and deep, excite in the bodies near them, as we experience in the sound of drums; for quick and short tremors are less easily excited. But it is well known that any sounds, falling upon strings in unison with the sonorous bodies, excite tremors in those strings. This is also confirmed from the velocity of sounds; for since the specific gravities of rain-water and quicksilver are to one another as about 1 to $13\frac{2}{3}$, and when the mercury in the barometer is at the height of 30 inches of our measure, the specific gravities of the air and of rain-water are to one another as about 1 to 870, therefore the specific gravity of air and quicksilver are to each other as 1 to 11890. Therefore when the height of the quicksilver is at 30 inches, a height of uniform air, whose weight would be sufficient to compress our air to the density we find it to be of, must be equal to 356700 inches, or 29725 feet of our measure; and this is that very height of the medium, which I have called A in the construction of the foregoing Proposition. A circle whose radius is 29725 feet is 186768 feet in circumference. And since a pendulum $39\frac{1}{5}$ inches in length completes one oscillation, composed of its going and return, in two seconds of time, as is commonly known, it follows that a pendulum 29725 feet, or 356700 inches in length will perform a like oscillation in $190\frac{3}{4}$ seconds. Therefore in that time a sound will go right onwards 186768 feet, and therefore in one second 979 feet.

But in this computation we have made no allowance for the crassitude of the solid particles of the air, by which the sound is propagated instantaneously. Because the weight of air is to the weight of water as 1 to 870, and because salts are almost twice as dense as water; if the particles of air are supposed to be of near the same density as those of water or salt, and the rarity of the air arises from the intervals of the particles; the diameter of one particle of air will be to the interval between the centres of the particles as 1 to about 9 or 10, and to the interval between the particles themselves as 1 to 8 or 9. Therefore to 979 feet, which, according to the above calculation, a sound will advance forward in one second of time, we may add $\frac{979}{9}$, or about 109 feet, to compensate for the crassitude of the particles of the air: and then a sound will go forward about 1088 feet in one second of time.

Moreover, the vapours floating in the air being of another spring, and a different tone, will hardly, if at all, partake of the motion of the true air in which the sounds are propagated. Now if these vapours remain unmoved, that motion will be propagated the swifter through the true air alone, and that in the subduplicate ratio of the defect of the matter. So if the atmosphere consist of ten parts of true air and one part of vapours, the motion of sounds will be swifter in the subduplicate ratio of 11 to 10, or very nearly in the entire ratio of 21 to 20, than if it were propagated through eleven parts of true air: and therefore the motion of sounds above discovered must be increased in that ratio. By this means the sound will pass through 1142 feet in one second of time.

These things will be found true in spring and autumn, when the air is rarefied by the gentle warmth of those seasons, and by that means its elastic force becomes somewhat more intense. But in winter, when the air is condensed by the cold, and its elastic force is somewhat remitted, the motion of sounds will be slower in a subduplicate ratio of the density; and, on the other hand, swifter in the summer.

Now by experiments it actually appears that sounds do really advance in one second of time about 1142 feet of *English* measure, or 1070 feet of *French* measure.

The velocity of sounds being known, the intervals of the pulses are known also. For M. *Sauveur*, by some experiments that he made, found that an open pipe about five *Paris* feet in length gives a sound of the same tone with a viol-string that vibrates a hundred times in one second. Therefore there are near 100 pulses in a space of 1070 *Paris* feet, which a sound runs over in a second of time; and therefore one pulse fills up a space of about $10\frac{7}{10}$ *Paris* feet, that is, about twice the length of the pipe. From whence it is probable that the breadths of the pulses, in all sounds made in open pipes, are equal to twice the length of the pipes.

Moreover, from the Corollary of Prop. XLVII appears the reason why the sounds immediately cease with the motion of the sonorous body, and why they are heard no longer when we are at a great distance from the sonorous bodies than when we are very near them. And besides, from the foregoing principles, it plainly appears how it comes to pass that sounds are so mightily increased in speaking-trumpets; for all reciprocal motion uses to be increased by the generating cause at each return. And in tubes hindering the dilatation of the sounds, the motion decays more slowly, and recurs more forcibly; and therefore is the more increased by the new motion impressed at each return. And these are the principal phænomena of sounds.

SECTION IX.

Of the circular motion of fluids.

HYPOTHESIS.

The resistance arising from the want of lubricity in the parts of a fluid, is, cæteris paribus, *proportional to the velocity with which the parts of the fluid are separated from each other.*

PROPOSITION LI. THEOREM XXXIX.

If a solid cylinder infinitely long, in an uniform and infinite fluid, revolve with an uniform motion about an axis given in position, and the fluid be forced round by only this impulse of the cylinder, and every part of the fluid persevere uniformly in its motion; I say, that the periodic times of the parts of the fluid are as their distances from the axis of the cylinder.

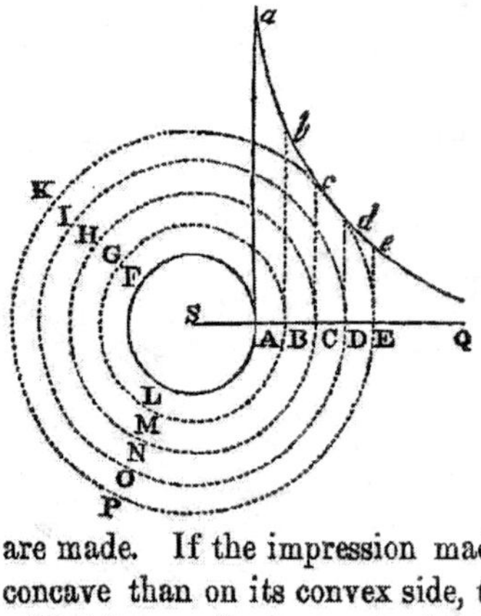

Let AFL be a cylinder turning uniformly about the axis S, and let the concentric circles BGM, CHN, DIO, EKP, &c., divide the fluid into innumerable concentric cylindric solid orbs of the same thickness. Then, because the fluid is homogeneous, the impressions which the contiguous orbs make upon each other mutually will be (by the Hypothesis) as their translations from each other, and as the contiguous superficies upon which the impressions are made. If the impression made upon any orb be greater or less on its concave than on its convex side, the stronger impression will prevail, and will either accelerate or retard the motion of the orb, according as it agrees with, or is contrary to, the motion of the same. Therefore, that every orb may persevere uniformly in its motion, the impressions made on both sides must be equal and their directions contrary. Therefore since the impressions are as the contiguous superficies, and as their translations from one another, the translations will be inversely as the superficies, that is, inversely as the distances of the superficies from the axis. But the differences of the angular motions about the axis are as those translations applied to the distances, or as the translations directly and the distances inversely; that is, joining these ratios together, as the squares of the distances inversely. Therefore if there be erected the lines A*a*, B*b*, C*c*, D*d*, E*e*, &c., perpendicular to the several parts of the infinite right line SABCDEQ, and reciprocally proportional to the squares of SA, SB, SC, SD, SE, &c., and through the extremities of those perpendiculars there be supposed to pass an hyperbolic curve, the sums of the differences, that is, the whole angular motions, will be as the correspondent sums of the lines A*a*, B*b*, C*c*, D*d*, E*e*, that is (if to constitute a medium uniformly fluid the number of the orbs be increased and their breadth diminished *in infinitum*), as the hyperbolic areas A*a*Q, B*b*Q, C*c*Q, D*d*Q, E*e*Q, &c., analogous to the sums; and the times, reciprocally proportional to the angular motions, will be also reciprocally proportional to those areas. Therefore the periodic time of any particle as D, is reciprocally as the area D*d*Q, that is (as appears from the known methods of quadratures of curves), directly as the distance SD. Q.E.D.

COR. 1. Hence the angular motions of the particles of the fluid are reciprocally as their distances from the axis of the cylinder, and the absolute velocities are equal.

COR. 2. If a fluid be contained in a cylindric vessel of an infinite length, and contain another cylinder within, and both the cylinders revolve about one common axis, and the times of their revolutions be as their semidiameters, and every part of the fluid perseveres in its motion, the periodic times of the several parts will be as the distances from the axis of the cylinders.

COR. 3. If there be added or taken away any common quantity of angular motion from the cylinder and fluid moving in this manner; yet because this new motion will not alter the mutual attrition of the parts of the fluid, the motion of the parts among themselves will not be changed; for the translations of the parts from one another depend upon the attrition. Any part will persevere in that motion, which, by the attrition made on both sides with contrary directions, is no more accelerated than it is retarded.

COR. 4. Therefore if there be taken away from this whole system of the cylinders and the fluid all the angular motion of the outward cylinder, we shall have the motion of the fluid in a quiescent cylinder.

COR. 5. Therefore if the fluid and outward cylinder are at rest, and the inward cylinder revolve uniformly, there will be communicated a circular motion to the fluid, which will be propagated by degrees through the whole fluid; and will go on continually increasing, till such time as the several parts of the fluid acquire the motion determined in Cor. 4.

COR. 6. And because the fluid endeavours to propagate its motion still farther, its impulse will carry the outmost cylinder also about with it, unless the cylinder be violently detained; and accelerate its motion till the periodic times of both cylinders become equal among themselves. But if the outward cylinder be violently detained, it will make an effort to retard the motion of the fluid; and unless the inward cylinder preserve that motion by means of some external force impressed thereon, it will make it cease by degrees.

All these things will be found true by making the experiment in deep standing water.

PROPOSITION LII. THEOREM XL.

If a solid sphere, in an uniform and infinite fluid, revolves about an axis given in position with an uniform motion, and the fluid be forced round by only this impulse of the sphere; and every part of the fluid perseveres uniformly in its motion; I say, that the periodic times of the parts of the fluid are as the squares of their distances from the centre of the sphere.

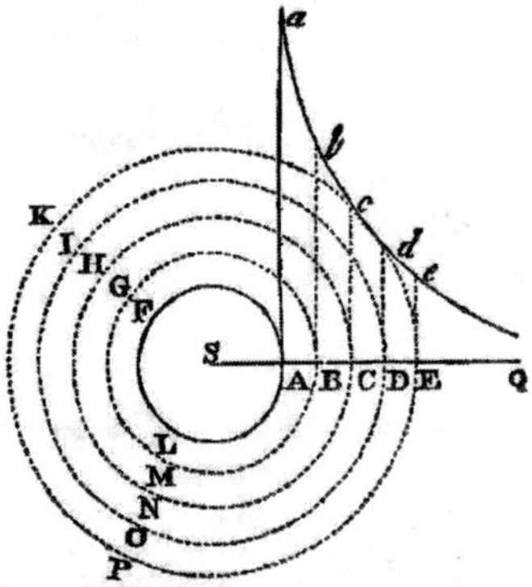

Case 1. Let AFL be a sphere turning uniformly about the axis S, and let the concentric circles BGM, CHN, DIO, EKP, &c., divide the fluid into innumerable concentric orbs of the same thickness. Suppose those orbs to be solid; and, because the fluid is homogeneous, the impressions which the contiguous orbs make one upon another will be (by the supposition) as their translations from one another, and the contiguous superficies upon which the impressions are made. If the impression upon any orb be greater or less upon its concave than upon its convex side, the more forcible impression will prevail, and will either accelerate or retard the velocity of the orb, according as it is directed with a conspiring or contrary motion to that of the orb. Therefore that every orb may persevere uniformly in its motion, it is necessary that the impressions made upon both sides of the orb should be equal, and have contrary directions. Therefore since the impressions are as the contiguous superficies, and as their translations from one another, the translations will be inversely as the superficies, that is, inversely as the squares of the distances of the superficies from the centre. But the differences of the angular motions about the axis are as those translations applied to the distances, or as the translations directly and the distances inversely; that is, by compounding those ratios, as the cubes of the distances inversely. Therefore if upon the several parts of the infinite right line SABCDEQ

there be erected the perpendiculars A*a*, B*b*, C*c*, D*d*, E*e*, &c., reciprocally proportional to the cubes of SA, SB, SC, SD, SE, &c., the sums of the differences, that is, the whole angular motions will be as the corresponding sums of the lines A*a*, B*b*, C*c*, D*d*, E*e*, &c., that is (if to constitute an uniformly fluid medium the number of the orbs be increased and their thickness diminished *in infinitum*), as the hyperbolic areas A*a*Q, B*b*Q, C*c*Q, D*d*Q, E*e*Q, &c., analogous to the sums; and the periodic times being reciprocally proportional to the angular motions, will be also reciprocally proportional to those areas. Therefore the periodic time of any orb DIO is reciprocally as the area D*d*Q, that is (by the known methods of quadratures), directly as the square of the distance SD. Which was first to be demonstrated.

CASE 2. From the centre of the sphere let there be drawn a great number of indefinite right lines, making given angles with the axis, exceeding one another by equal differences; and, by these lines revolving about the axis, conceive the orbs to be cut into innumerable annuli; then will every annulus have four annuli contiguous to it, that is, one on its inside, one on its outside, and two on each hand. Now each of these annuli cannot be impelled equally and with contrary directions by the attrition of the interior and exterior annuli, unless the motion be communicated according to the law which we demonstrated in Case 1. This appears from that demonstration. And therefore any series of annuli, taken in any right line extending itself *in infinitum* from the globe, will move according to the law of Case 1, except we should imagine it hindered by the attrition of the annuli on each side of it. But now in a motion, according to this law, no such is, and therefore cannot be, any obstacle to the motions persevering according to that law. If annuli at equal distances from the centre revolve either more swiftly or more slowly near the poles than near the ecliptic, they will be accelerated if slow, and retarded if swift, by their mutual attrition; and so the periodic times will continually approach to equality, according to the law of Case 1. Therefore this attrition will not at all hinder the motion from going on according to the law of Case 1, and therefore that law will take place; that is, the periodic times of the several annuli will be as the squares of their distances from the centre of the globe. Which was to be demonstrated in the second place.

CASE 3. Let now every annulus be divided by transverse sections into innumerable particles constituting a substance absolutely and uniformly fluid; and because these sections do not at all respect the law of circular motion, but only serve to produce a fluid substance, the law of circular motion will continue the same as before. All the very small annuli will either not at all change their asperity and force of mutual attrition upon account of these sections, or else they will change the same equally. Therefore the proportion of the causes remaining the same, the proportion of the effects

will remain the same also; that is, the proportion of the motions and the periodic times. Q.E.D. But now as the circular motion, and the centrifugal force thence arising, is greater at the ecliptic than at the poles, there must be some cause operating to retain the several particles in their circles; otherwise the matter that is at the ecliptic will always recede from the centre, and come round about to the poles by the outside of the vortex, and from thence return by the axis to the ecliptic with a perpetual circulation.

Cor. 1. Hence the angular motions of the parts of the fluid about the axis of the globe are reciprocally as the squares of the distances from the centre of the globe, and the absolute velocities are reciprocally as the same squares applied to the distances from the axis.

Cor. 2. If a globe revolve with a uniform motion about an axis of a given position in a similar and infinite quiescent fluid with an uniform motion, it will communicate a whirling motion to the fluid like that of a vortex, and that motion will by degrees be propagated onward *in infinitum;* and this motion will be increased continually in every part of the fluid, till the periodical times of the several parts become as the squares of the distances from the centre of the globe.

Cor. 3. Because the inward parts of the vortex are by reason of their greater velocity continually pressing upon and driving forward the external parts, and by that action are perpetually communicating motion to them, and at the same time those exterior parts communicate the same quantity of motion to those that lie still beyond them, and by this action preserve the quantity of their motion continually unchanged, it is plain that the motion is perpetually transferred from the centre to the circumference of the vortex, till it is quite swallowed up and lost in the boundless extent of that circumference. The matter between any two spherical superficies concentrical to the vortex will never be accelerated; because that matter will be always transferring the motion it receives from the matter nearer the centre to that matter which lies nearer the circumference.

Cor. 4. Therefore, in order to continue a vortex in the same state of motion, some active principle is required from which the globe may receive continually the same quantity of motion which it is always communicating to the matter of the vortex. Without such a principle it will undoubtedly come to pass that the globe and the inward parts of the vortex, being always propagating their motion to the outward parts, and not receiving any new motion, will gradually move slower and slower, and at last be carried round no longer.

Cor. 5. If another globe should be swimming in the same vortex at a certain distance from its centre, and in the mean time by some force revolve constantly about an axis of a given inclination, the motion of this globe will drive the fluid round after the manner of a vortex; and at first this

new and small vortex will revolve with its globe about the centre of the other; and in the mean time its motion will creep on farther and farther, and by degrees be propagated *in infinitum*, after the manner of the first vortex. And for the same reason that the globe of the new vortex was carried about before by the motion of the other vortex, the globe of this other will be carried about by the motion of this new vortex, so that the two globes will revolve about some intermediate point, and by reason of that circular motion mutually fly from each other, unless some force restrains them. Afterward, if the constantly impressed forces, by which the globes persevere in their motions, should cease, and every thing be left to act according to the laws of mechanics, the motion of the globes will languish by degrees (for the reason assigned in Cor. 3 and 4), and the vortices at last will quite stand still.

COR. 6. If several globes in given places should constantly revolve with determined velocities about axes given in position, there would arise from them as many vortices going on *in infinitum*. For upon the same account that any one globe propagates its motion *in infinitum*, each globe apart will propagate its own motion *in infinitum* also; so that every part of the infinite fluid will be agitated with a motion resulting from the actions of all the globes. Therefore the vortices will not be confined by any certain limits, but by degrees run mutually into each other; and by the mutual actions of the vortices on each other, the globes will be perpetually moved from their places, as was shewn in the last Corollary; neither can they possibly keep any certain position among themselves, unless some force restrains them. But if those forces, which are constantly impressed upon the globes to continue these motions, should cease, the matter (for the reason assigned in Cor. 3 and 4) will gradually stop, and cease to move in vortices.

COR. 7. If a similar fluid be inclosed in a spherical vessel, and, by the uniform rotation of a globe in its centre, is driven round in a vortex; and the globe and vessel revolve the same way about the same axis, and their periodical times be as the squares of the semi-diameters; the parts of the fluid will not go on in their motions without acceleration or retardation, till their periodical times are as the squares of their distances from the centre of the vortex. No constitution of a vortex can be permanent but this.

COR. 8. If the vessel, the inclosed fluid, and the globe, retain this motion, and revolve besides with a common angular motion about any given axis, because the mutual attrition of the parts of the fluid is not changed by this motion, the motions of the parts among each other will not be changed; for the translations of the parts among themselves depend upon this attrition. Any part will persevere in that motion in which its attri-

tion on one side retards it just as much as its attrition on the other side accelerates it.

Cor. 9. Therefore if the vessel be quiescent, and the motion of the globe be given, the motion of the fluid will be given. For conceive a plane to pass through the axis of the globe, and to revolve with a contrary motion; and suppose the sum of the time of this revolution and of the revolution of the globe to be to the time of the revolution of the globe as the square of the semi-diameter of the vessel to the square of the semi-diameter of the globe; and the periodic times of the parts of the fluid in respect of this plane will be as the squares of their distances from the centre of the globe.

Cor. 10. Therefore if the vessel move about the same axis with the globe, or with a given velocity about a different one, the motion of the fluid will be given. For if from the whole system we take away the angular motion of the vessel, all the motions will remain the same among themselves as before, by Cor. 8, and those motions will be given by Cor. 9.

Cor. 11. If the vessel and the fluid are quiescent, and the globe revolves with an uniform motion, that motion will be propagated by degrees through the whole fluid to the vessel, and the vessel will be carried round by it, unless violently detained; and the fluid and the vessel will be continually accelerated till their periodic times become equal to the periodic times of the globe. If the vessel be either withheld by some force, or revolve with any constant and uniform motion, the medium will come by little and little to the state of motion defined in Cor. 8, 9, 10, nor will it ever persevere in any other state. But if then the forces, by which the globe and vessel revolve with certain motions, should cease, and the whole system be left to act according to the mechanical laws, the vessel and globe, by means of the intervening fluid, will act upon each other, and will continue to propagate their motions through the fluid to each other, till their periodic times become equal among themselves, and the whole system revolves together like one solid body.

SCHOLIUM.

In all these reasonings I suppose the fluid to consist of matter of uniform density and fluidity; I mean, that the fluid is such, that a globe placed any where therein may propagate with the same motion of its own, at distances from itself continually equal, similar and equal motions in the fluid in the same interval of time. The matter by its circular motion endeavours to recede from the axis of the vortex, and therefore presses all the matter that lies beyond. This pressure makes the attrition greater, and the separation of the parts more difficult; and by consequence diminishes the fluidity of the matter. Again; if the parts of the fluid are in any one place denser or larger than in the others, the fluidity will be less in that place, because there are fewer superficies where the parts can be separated

from each other. In these cases I suppose the defect of the fluidity to be supplied by the smoothness or softness of the parts, or some other condition; otherwise the matter where it is less fluid will cohere more, and be more sluggish, and therefore will receive the motion more slowly, and propagate it farther than agrees with the ratio above assigned. If the vessel be not spherical, the particles will move in lines not circular, but answering to the figure of the vessel; and the periodic times will be nearly as the squares of the mean distances from the centre. In the parts between the centre and the circumference the motions will be slower where the spaces are wide, and swifter where narrow; but yet the particles will not tend to the circumference at all the more for their greater swiftness; for they then describe arcs of less curvity, and the conatus of receding from the centre is as much diminished by the diminution of this curvature as it is augmented by the increase of the velocity. As they go out of narrow into wide spaces, they recede a little farther from the centre, but in doing so are retarded; and when they come out of wide into narrow spaces, they are again accelerated; and so each particle is retarded and accelerated by turns for ever. These things will come to pass in a rigid vessel; for the state of vortices in an infinite fluid is known by Cor. 6 of this Proposition.

I have endeavoured in this Proposition to investigate the properties of vortices, that I might find whether the celestial phænomena can be explained by them; for the phænomenon is this, that the periodic times of the planets revolving about Jupiter are in the sesquiplicate ratio of their distances from Jupiter's centre; and the same rule obtains also among the planets that revolve about the sun. And these rules obtain also with the greatest accuracy, as far as has been yet discovered by astronomical obsertion. Therefore if those planets are carried round in vortices revolving about Jupiter and the sun, the vortices must revolve according to that law. But here we found the periodic times of the parts of the vortex to be in the duplicate ratio of the distances from the centre of motion; and this ratio cannot be diminished and reduced to the sesquiplicate, unless either the matter of the vortex be more fluid the farther it is from the centre, or the resistance arising from the want of lubricity in the parts of the fluid should, as the velocity with which the parts of the fluid are separated goes on increasing, be augmented with it in a greater ratio than that in which the velocity increases. But neither of these suppositions seem reasonable. The more gross and less fluid parts will tend to the circumference, unless they are heavy towards the centre. And though, for the sake of demonstration, I proposed, at the beginning of this Section, an Hypothesis that the resistance is proportional to the velocity, nevertheless, it is in truth probable that the resistance is in a less ratio than that of the velocity; which granted, the periodic times of the parts of the vortex will be in a greater than the duplicate ratio of the distances from its centre. If, as some think, the vortices move more swiftly near the centre, then slower to a certain limit, then again swifter near the circumference, certainly neither the sesquiplicate, nor any other certain and determinate ratio, can obtain in them. Let philosophers then see how that phænomenon of the sesquiplicate ratio can be accounted for by vortices.

PROPOSITION LIII. THEOREM XLI.

Bodies carried about in a vortex, and returning in the same orb, are of the same density with the vortex, and are moved according to the same law with the parts of the vortex, as to velocity and direction of motion.

For if any small part of the vortex, whose particles or physical points preserve a given situation among each other, be supposed to be congealed, this particle will move according to the same law as before, since no change is made either in its density, *vis insita*, or figure. And again; if a congealed or solid part of the vortex be of the same density with the rest of the vortex, and be resolved into a fluid, this will move according to the same law as before, except in so far as its particles, now become fluid, may be moved among themselves. Neglect, therefore, the motion of the particles among themselves as not at all concerning the progressive motion of the whole, and the motion of the whole will be the same as before. But this motion will be the same with the motion of other parts of the vortex at equal distances from the centre; because the solid, now resolved into a fluid, is become perfectly like to the other parts of the vortex. Therefore a solid, if it be of the same density with the matter of the vortex, will move with the same motion as the parts thereof, being relatively at rest in the matter that surrounds it. If it be more dense, it will endeavour more than before to recede from the centre; and therefore overcoming that force of the vortex, by which, being, as it were, kept in equilibrio, it was retained in its orbit, it will recede from the centre, and in its revolution describe a spiral, returning no longer into the same orbit. And, by the same argument, if it be more rare, it will approach to the centre. Therefore it can never continually go round in the same orbit, unless it be of the same density with the fluid. But we have shewn in that case that it would revolve according to the same law with those parts of the fluid that are at the same or equal distances from the centre of the vortex.

Cor. 1. Therefore a solid revolving in a vortex, and continually going round in the same orbit, is relatively quiescent in the fluid that carries it.

Cor. 2. And if the vortex be of an uniform density, the same body may revolve at any distance from the centre of the vortex.

SCHOLIUM.

Hence it is manifest that the planets are not carried round in corporeal vortices; for, according to the *Copernican* hypothesis, the planets going round the sun revolve in ellipses, having the sun in their common focus; and by radii drawn to the sun describe areas proportional to the times. But now the parts of a vortex can never revolve with such a motion. Let AD, BE, CF, represent three orbits described about the sun S, of which let the utmost circle CF be concentric to the sun; and let the aphelia of the two innermost be A, B; and their perihelia D, E. Therefore a body revolving in the orb CF, describing, by a radius drawn to the sun, areas proportional to the times, will move with an uniform motion. And, according to the laws of astronomy, the body revolving in the orb BE will move slower in its aphelion B, and swifter in its perihelion E; whereas, according to the laws of mechanics, the matter of the vortex ought to move more swiftly in the narrow space between A and C than in the wide space between D and F; that is, more swiftly in the aphelion than in the perihelion. Now these two conclusions contradict each other. So at the beginning of the sign of Virgo, where the aphelion of Mars is at present, the distance between the orbits of Mars and Venus is to the distance between the same orbits, at the beginning of the sign of Pisces, as about 3 to 2; and therefore the matter of the vortex between those orbits ought to be swifter at the beginning of Pisces than at the beginning of Virgo in the ratio of 3 to 2; for the narrower the space is through which the same quantity of matter passes in the same time of one revolution, the greater will be the velocity with which it passes through it. Therefore if the earth being relatively at rest in this celestial matter should be carried round by it, and revolve together with it about the sun, the velocity of the earth at the beginning of Pisces would be to its velocity at the beginning of Virgo in a sesquialteral ratio. Therefore the sun's apparent diurnal motion at the beginning of Virgo ought to be above 70 minutes, and at the beginning of Pisces less than 48 minutes; whereas, on the contrary, that apparent motion of the sun is really greater at the beginning of Pisces than at the beginning of Virgo, as experience testifies; and therefore the earth is swifter at the beginning of Virgo than at the beginning of Pisces; so that the hypothesis of vortices is utterly irreconcileable with astronomical phænomena, and rather serves to perplex than explain the heavenly motions. How these motions are performed in free spaces without vortices, may be understood by the first Book; and I shall now more fully treat of it in the following Book.

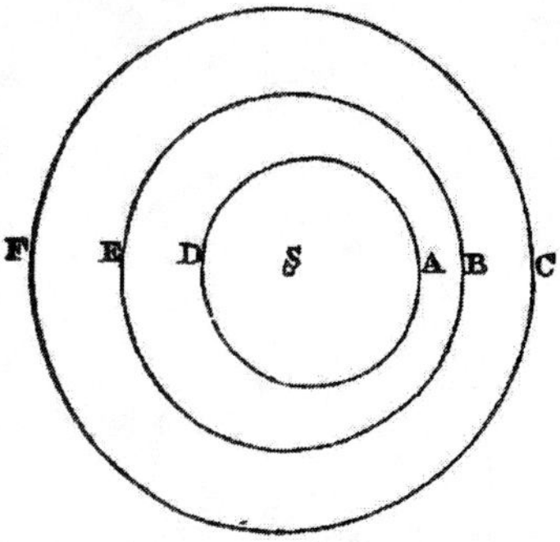

BOOK III.

BOOK III.

In the preceding Books I have laid down the principles of philosophy; principles not philosophical, but mathematical; such, to wit, as we may build our reasonings upon in philosophical inquiries. These principles are the laws and conditions of certain motions, and powers or forces, which chiefly have respect to philosophy; but, lest they should have appeared of themselves dry and barren, I have illustrated them here and there with some philosophical scholiums, giving an account of such things as are of more general nature, and which philosophy seems chiefly to be founded on; such as the density and the resistance of bodies, spaces void of all bodies, and the motion of light and sounds. It remains that, from the same principles, I now demonstrate the frame of the System of the World. Upon this subject I had, indeed, composed the third Book in a popular method, that it might be read by many; but afterward, considering that such as had not sufficiently entered into the principles could not easily discern the strength of the consequences, nor lay aside the prejudices to which they had been many years accustomed, therefore, to prevent the disputes which might be raised upon such accounts, I chose to reduce the substance of this Book into the form of Propositions (in the mathematical way), which should be read by those only who had first made themselves masters of the principles established in the preceding Books: not that I would advise any one to the previous study of every Proposition of those Books; for they abound with such as might cost too much time, even to readers of good mathematical learning. It is enough if one carefully reads the Definitions, the Laws of Motion, and the first three Sections of the first Book. He may then pass on to this Book, and consult such of the remaining Propositions of the first two Books, as the references in this, and his occasions, shall require.

RULES OF REASONING IN PHILOSOPHY.

RULE I.

We are to admit no more causes of natural things than such as are both true and sufficient to explain their appearances.

To this purpose the philosophers say that Nature does nothing in vain, and more is in vain when less will serve; for Nature is pleased with simplicity, and affects not the pomp of superfluous causes.

RULE II.

Therefore to the same natural effects we must, as far as possible, assign the same causes.

As to respiration in a man and in a beast; the descent of stones in *Europe* and in *America;* the light of our culinary fire and of the sun; the reflection of light in the earth, and in the planets.

RULE III.

The qualities of bodies, which admit neither intension nor remission of degrees, and which are found to belong to all bodies within the reach of our experiments, are to be esteemed the universal qualities of all bodies whatsoever.

For since the qualities of bodies are only known to us by experiments, we are to hold for universal all such as universally agree with experiments; and such as are not liable to diminution can never be quite taken away. We are certainly not to relinquish the evidence of experiments for the sake of dreams and vain fictions of our own devising; nor are we to recede from the analogy of Nature, which uses to be simple, and always consonant to itself. We no other way know the extension of bodies than by our senses, nor do these reach it in all bodies; but because we perceive extension in all that are sensible, therefore we ascribe it universally to all others also. That abundance of bodies are hard, we learn by experience; and because the hardness of the whole arises from the hardness of the parts, we therefore justly infer the hardness of the undivided particles not only of the bodies we feel but of all others. That all bodies are impenetrable, we gather not from reason, but from sensation. The bodies which we handle we find impenetrable, and thence conclude impenetrability to be an universal property of all bodies whatsoever. That all bodies are moveable, and endowed with certain powers (which we call the *vires inertiæ*) of persevering in their motion, or in their rest, we only infer from the like properties observed in the bodies which we have seen. The extension, hardness, impenetrability, mobility, and *vis inertiæ* of the whole, result from the extension, hardness, impenetrability, mobility, and *vires inertiæ* of the parts; and thence we conclude the least particles of all bodies to be also all extended, and hard, and impenetrable, and moveable, and endowed with their proper *vires inertiæ*. And this is the foundation of all philosophy. Moreover, that the divided but contiguous particles of bodies may be separated from one another, is matter of observation; and, in the particles that remain undivided, our minds are able to distinguish yet lesser parts, as is mathematically demonstrated. But whether the parts so distinguished, and not yet divided, may, by the powers of Nature, be actually divided and separated from one another, we cannot certainly determine. Yet, had we the proof of but one experiment that any undivided particle, in breaking a hard and solid body, suffered a division, we might by virtue of this rule conclude that the undivided as well as the divided particles may be divided and actually separated to infinity.

Lastly, if it universally appears, by experiments and astronomical observations, that all bodies about the earth gravitate towards the earth, and that in proportion to the quantity of matter which they severally contain; that the moon likewise, according to the quantity of its matter, gravitates towards the earth; that, on the other hand, our sea gravitates towards the moon; and all the planets mutually one towards another; and the comets in like manner towards the sun; we must, in consequence of this rule, universally allow that all bodies whatsoever are endowed with a principle of mutual gravitation. For the argument from the appearances concludes with more force for the universal gravitation of all bodies than for their impenetrability; of which, among those in the celestial regions, we have no experiments, nor any manner of observation. Not that I affirm gravity to be essential to bodies: by their *vis insita* I mean nothing but their *vis inertiæ*. This is immutable. Their gravity is diminished as they recede from the earth.

RULE IV.

In experimental philosophy we are to look upon propositions collected by general induction from phænomena as accurately or very nearly true, notwithstanding any contrary hypotheses that may be imagined, till such time as other phænomena occur, by which they may either be made more accurate, or liable to exceptions.

This rule we must follow, that the argument of induction may not be evaded by hypotheses.

PHÆNOMENA, OR APPEARANCES.

PHÆNOMENON I.

That the circumjovial planets, by radii drawn to Jupiter's centre, describe areas proportional to the times of description; and that their periodic times, the fixed stars being at rest, are in the sesquiplicate proportion of their distances from its centre.

This we know from astronomical observations. For the orbits of these planets differ but insensibly from circles concentric to Jupiter; and their motions in those circles are found to be uniform. And all astronomers agree that their periodic times are in the sesquiplicate proportion of the semi-diameters of their orbits; and so it manifestly appears from the following table.

The periodic times of the satellites of Jupiter.

$1^d. 18^h. 27'. 34''.$ $3^d. 13^h. 13' 42''.$ $7^d. 3^h. 42' 36''.$ $16^d. 16^h. 32' 9''.$

The distances of the satellites from Jupiter's centre.

From the observations of	1	2	3	4	
Borelli	$5\frac{2}{3}$	$8\frac{2}{3}$	14	$24\frac{2}{3}$	*semi-diameter of Jupiter.*
Townly *by the Microm.* . . .	5,52	8,78	13,47	24,72	*semi-diameter of Jupiter.*
Cassini *by the Telescope* . . .	5	8	13	23	*semi-diameter of Jupiter.*
Cassini *by the eclip. of the satel.* .	$5\frac{2}{3}$	9	$14\frac{23}{60}$	$25\frac{3}{10}$	*semi-diameter of Jupiter.*
From the periodic times	5,667	9,017	14,384	25,299	

Mr. *Pound* has determined, by the help of excellent micrometers, the diameters of Jupiter and the elongation of its satellites after the following manner. The greatest heliocentric elongation of the fourth satellite from Jupiter's centre was taken with a micrometer in a 15 feet telescope, and at the mean distance of Jupiter from the earth was found about $8' 16''$. The elongation of the third satellite was taken with a micrometer in a telescope of 123 feet, and at the same distance of Jupiter from the earth was found $4' 42''$. The greatest elongations of the other satellites, at the same distance of Jupiter from the earth, are found from the periodic times to be $2' 56'' 47'''$, and $1' 51'' 6'''$.

The diameter of Jupiter taken with the micrometer in a 123 feet telescope several times, and reduced to Jupiter's mean distance from the earth, proved always less than $40''$, never less than $38''$, generally $39''$. This diameter in shorter telescopes is $40''$, or $41''$; for Jupiter's light is a little dilated by the unequal refrangibility of the rays, and this dilatation bears a less ratio to the diameter of Jupiter in the longer and more perfect telescopes than in those which are shorter and less perfect. The times in which two satellites, the first and the third, passed over Jupiter's body, were observed, from the beginning of the ingress to the beginning of the egress, and from the complete ingress to the complete egress, with the long telescope. And from the transit of the first satellite, the diameter of Jupiter at its mean distance from the earth came forth $37\frac{1}{8}''$, and from the transit of the third $37\frac{3}{8}''$. There was observed also the time in which the shadow of the first satellite passed over Jupiter's body, and thence the diameter of Jupiter at its mean distance from the earth came out about $37''$. Let us suppose its diameter to be $37\frac{1}{4}''$ very nearly, and then the greatest elongations of the first, second, third, and fourth satellite will be respectively equal to 5,965, 9,494, 15,141, and 26,63 semi-diameters of Jupiter.

PHÆNOMENON II.

That the circumsaturnal planets, by radii drawn to Saturn's centre, describe areas proportional to the times of description; and that their periodic times, the fixed stars being at rest, are in the sesquiplicate proportion of their distances from its centre.

For, as *Cassini* from his own observations has determined, their distances from Saturn's centre and their periodic times are as follow.

The periodic times of the satellites of Saturn.

$1^d. 21^h. 18' 27''$. $2^d. 17^h. 41' 22''$. $4^d. 12^h. 25' 12''$. $15^d. 22^h. 41' 14''$. $79^d. 7^h. 48' 00''$.

The distances of the satellites from Saturn's centre, in semi-diameters of its ring.

From observations	$1\frac{19}{20}$.	$2\frac{1}{2}$.	$3\frac{1}{2}$.	8.	24.
From the periodic times . . .	1,93.	2,47.	3,45.	8.	23,35.

The greatest elongation of the fourth satellite from Saturn's centre is commonly determined from the observations to be eight of those semi-diameters very nearly. But the greatest elongation of this satellite from Saturn's centre, when taken with an excellent micrometer in Mr. *Huygens'* telescope of 123 feet, appeared to be eight semi-diameters and $\frac{7}{10}$ of a semi-diameter. And from this observation and the periodic times the distances of the satellites from Saturn's centre in semi-diameters of the ring are 2,1. 2,69. 3,75. 8,7. and 25,35. The diameter of Saturn observed in the same telescope was found to be to the diameter of the ring as 3 to 7; and the diameter of the ring, *May* 28–29, 1719, was found to be $43''$; and thence the diameter of the ring when Saturn is at its mean distance from the earth is $42''$, and the diameter of Saturn $18''$. These things appear so in very long and excellent telescopes, because in such telescopes the apparent magnitudes of the heavenly bodies bear a greater proportion to the dilatation of light in the extremities of those bodies than in shorter telescopes.

If we, then, reject all the spurious light, the diameter of Saturn will not amount to more than 16″.

PHÆNOMENON III.

That the five primary planets, Mercury, Venus, Mars, Jupiter, and Saturn, with their several orbits, encompass the sun.

That Mercury and Venus revolve about the sun, is evident from their moon-like appearances. When they shine out with a full face, they are, in respect of us, beyond or above the sun; when they appear half full, they are about the same height on one side or other of the sun; when horned, they are below or between us and the sun; and they are sometimes, *when directly under*, seen like spots traversing the sun's disk. That Mars surrounds the sun, is as plain from its full face when near its conjunction with the sun, and from the gibbous figure which it shews in its quadratures. And the same thing is demonstrable of Jupiter and Saturn, from their appearing full in all situations; for the shadows of their satellites that appear sometimes upon their disks make it plain that the light they shine with is not their own, but borrowed from the sun.

PHÆNOMENON IV.

That the fixed stars being at rest, the periodic times of the five primary planets, and (whether of the sun about the earth, or) of the earth about the sun, are in the sesquiplicate proportion of their mean distances from the sun.

This proportion, first observed by *Kepler*, is now received by all astronomers; for the periodic times are the same, and the dimensions of the orbits are the same, whether the sun revolves about the earth, or the earth about the sun. And as to the measures of the periodic times, all astronomers are agreed about them. But for the dimensions of the orbits, *Kepler* and *Bullialdus*, above all others, have determined them from observations with the greatest accuracy; and the mean distances corresponding to the periodic times differ but insensibly from those which they have assigned, and for the most part fall in between them; as we may see from the following table.

The periodic times with respect to the fixed stars, of the planets and earth revolving about the sun, in days and decimal parts of a day.

♄	♃	♂	♁	♀	☿
10759,275.	4332,514.	686,9785.	365,2565.	224,6176.	87,9692.

The mean distances of the planets and of the earth from the sun.

	♄	♃	♂
According to *Kepler*	951000.	519650.	152350.
" to *Bullialdus*	954198.	522520.	152350.
" to the periodic times	954006.	520096.	152369.

	♂	♀	☿
According to *Kepler*	100000.	72400.	38806.
" to *Bullialdus*	100000.	72398.	38585.
" to the periodic times	100000.	72333.	38710.

As to Mercury and Venus, there can be no doubt about their distances from the sun; for they are determined by the elongations of those planets from the sun; and for the distances of the superior planets, all dispute is cut off by the eclipses of the satellites of Jupiter. For by those eclipses the position of the shadow which Jupiter projects is determined; whence we have the heliocentric longitude of Jupiter. And from its heliocentric and geocentric longitudes compared together, we determine its distance.

PHÆNOMENON V.

Then the primary planets, by radii drawn to the earth, describe areas no wise proportional to the times; but that the areas which they describe by radii drawn to the sun are proportional to the times of description.

For to the earth they appear sometimes direct, sometimes stationary, nay, and sometimes retrograde. But from the sun they are always seen direct, and to proceed with a motion nearly uniform, that is to say, a little swifter in the perihelion and a little slower in the aphelion distances, so as to maintain an equality in the description of the areas. This a noted proposition among astronomers, and particularly demonstrable in Jupiter, from the eclipses of his satellites; by the help of which eclipses, as we have said, the heliocentric longitudes of that planet, and its distances from the sun, are determined.

PHÆNOMENON VI.

That the moon, by a radius drawn to the earth's centre, describes an area proportional to the time of description.

This we gather from the apparent motion of the moon, compared with its apparent diameter. It is true that the motion of the moon is a little disturbed by the action of the sun: but in laying down these Phænomena, I neglect those small and inconsiderable errors.

PROPOSITIONS.

PROPOSITION I. THEOREM I.

That the forces by which the circumjovial planets are continually drawn off from rectilinear motions, and retained in their proper orbits, tend to Jupiter's centre; and are reciprocally as the squares of the distances of the places of those planets *from that centre.*

The former part of this Proposition appears from Phæn. I, and Prop. II or III, Book I; the latter from Phæn. I, and Cor. 6, Prop. IV, of the same Book.

The same thing we are to understand of the planets which encompass Saturn, by Phæn. II.

PROPOSITION II. THEOREM II.

That the forces by which the primary planets are continually drawn off from rectilinear motions, and retained in their proper orbits, tend to the sun; and are reciprocally as the squares of the distances of the places of those planets *from the sun's centre.*

The former part of the Proposition is manifest from Phæn. V, and Prop. II, Book I; the latter from Phæn. IV, and *Cor.* 6, Prop. IV, of the same Book. But this part of the Proposition is, with great accuracy, demonstrable from the quiescence of the aphelion points; for a very small aberration from the *reciprocal* duplicate proportion would (by Cor. 1, Prop. XLV, Book I) produce a motion of the apsides sensible enough in every single revolution, and in many of them enormously great.

PROPOSITION III. THEOREM III.

That the force by which the moon is retained in its orbit tends to the earth; and is reciprocally as the square of the distance of its place from the earth's centre.

The former part of the Proposition is evident from Phæn. VI, and Prop. II or III, Book I; the latter from the very slow motion of the moon's apogee; which in every single revolution amounting but to 3° 3′ *in consequentia*, may be neglected. For (by Cor. 1, Prop. XLV, Book I) it appears, that, if the distance of the moon from the earth's centre is to the semi-diameter of the earth as D to 1, the force, from which such a motion will result, is reciprocally as $D^{2\frac{4}{243}}$, i. e., reciprocally as the power of D, whose exponent is $2\frac{4}{243}$; that is to say, in the proportion of the distance something greater than reciprocally duplicate, but which comes $59\frac{3}{4}$ times nearer to the duplicate than to the triplicate proportion. But in regard that this motion is owing to the action of the sun (as we shall afterwards shew), it is here to be neglected. The action of the sun, attracting the moon from the earth, is nearly as the moon's distance from the earth; and therefore (by what we have shewed in Cor. 2, Prop. XLV, Book I) is to the centripetal force of the moon as 2 to 357,45, or nearly so; that is, as 1 to $178\frac{29}{40}$. And if we neglect so inconsiderable a force of the sun, the remaining force, by which the moon is retained in its orb, will be reciprocally as D^2. This will yet more fully appear from comparing this force with the force of gravity, as is done in the next Proposition.

COR. If we augment the mean centripetal force by which the moon is retained in its orb, first in the proportion of $177\frac{29}{40}$ to $178\frac{29}{40}$, and then in the duplicate proportion of the semi-diameter of the earth to the mean distance of the centres of the moon and earth, we shall have the centripetal force of the moon at the surface of the earth; supposing this force, in descending to the earth's surface, continually to increase in the reciprocal duplicate proportion of the height.

PROPOSITION IV. THEOREM IV.

That the moon gravitates towards the earth, and by the force of gravity is continually drawn off from a rectilinear motion, and retained in its orbit.

The mean distance of the moon from the earth in the syzygies in semi-diameters of the earth, is, according to *Ptolemy* and most astronomers, 59; according to *Vendelin* and *Huygens*, 60; to *Copernicus*, $60\frac{1}{3}$; to *Street*, $60\frac{2}{5}$; and to *Tycho*, $56\frac{1}{2}$. But *Tycho*, and all that follow his tables of refraction, making the refractions of the sun and moon (altogether against the nature of light) to exceed the refractions of the fixed stars, and that by four or five minutes *near the horizon*, did thereby increase the moon's *horizontal* parallax by a like number of minutes, that is, by a twelfth or fifteenth part of the whole parallax. Correct this error, and the distance will become about $60\frac{1}{2}$ semi-diameters of the earth, near to what others have assigned. Let us assume the mean distance of 60 diameters in the syzygies; and suppose one revolution of the moon, in respect of the fixed stars, to be completed in $27^d. 7^h. 43'$, as astronomers have determined; and the circumference of the earth to amount to 123249600 *Paris* feet, as the French have found by mensuration. And now if we imagine the moon, deprived of all motion, to be let go, so as to descend towards the earth with the impulse of all that force by which (by Cor. Prop. III) it is retained in its orb, it will in the space of one minute of time, describe in its fall $15\frac{1}{12}$ *Paris* feet. This we gather by a calculus, founded either upon Prop. XXXVI, Book I, or (which comes to the same thing) upon Cor. 9, Prop. IV, of the same Book. For the versed sine of that arc, which the moon, in the space of one minute of time, would by its mean motion describe at the distance of 60 semi-diameters of the earth, is nearly $15\frac{1}{12}$ *Paris* feet, or more accurately 15 feet, 1 inch, and 1 line $\frac{4}{9}$. Wherefore, since that force, in approaching to the earth, increases in the reciprocal duplicate proportion of the distance, and, upon that account, at the surface of the earth, is 60×60 times greater than at the moon, a body in our regions, falling with that force, ought in the space of one minute of time, to describe $60 \times 60 \times 15\frac{1}{12}$ *Paris* feet; and, in the space of one second of time, to describe $15\frac{1}{12}$ of those feet; or more accurately 15 feet, 1 inch, and 1 line $\frac{4}{9}$. And with this very force we actually find that bodies here upon earth do really descend; for a pendulum oscillating seconds in the latitude of *Paris* will be 3 *Paris* feet, and 8 lines $\frac{1}{2}$ in length, as Mr. *Huygens* has observed. And the space which a heavy body describes by falling in one second of time is to half the length of this pendulum in the duplicate ratio of the circumference of a circle to its diameter (as Mr. *Huygens* has also shewn), and is therefore 15 *Paris* feet, 1 inch, 1 line $\frac{7}{9}$. And therefore the force by which the moon is retained in its orbit becomes, at the very surface of the earth, equal to the force of gravity which we observe in heavy bodies there. And therefore (by Rule I and II) the force by which the moon is retained in its orbit is that very same force which we commonly call gravity; for, were gravity another force different from that, then bodies descending to the earth with the joint impulse of both forces would fall with a double velocity, and in the space of one second of time would describe $30\frac{1}{6}$ *Paris* feet; altogether against experience.

This calculus is founded on the hypothesis of the earth's standing still; for if both earth and moon move about the sun, and at the same time about their common centre of gravity, the distance of the centres of the moon and earth from one another will be $60\frac{1}{2}$ semi-diameters of the earth; as may be found by a computation from Prop. LX, Book I.

SCHOLIUM.

The demonstration of this Proposition may be more diffusely explained after the following manner. Suppose several moons to revolve about the earth, as in the system of Jupiter or Saturn; the periodic times of these moons (by the argument of induction) would observe the same law which *Kepler* found to obtain among the planets; and therefore their centripetal forces would be reciprocally as the squares of the distances from the centre of the earth, by Prop. I, of this Book. Now if the lowest of these were very small, and were so near the earth as almost to touch the tops of the highest mountains, the centripetal force thereof, retaining it in its orb, would be very nearly equal to the weights of any *terrestrial* bodies that should be found upon the tops of those mountains, as may be known by the foregoing computation. Therefore if the same little moon should be deserted by its centrifugal force that carries it through its orb, and so be disabled from going onward therein, it would descend to the earth; and that with the same velocity as heavy bodies do actually fall with upon the tops of those very mountains; because of the equality of the forces that oblige them both to descend. And if the force by which that lowest moon would descend were different from gravity, and if that moon were to gravitate towards the earth, as we find terrestrial bodies do upon the tops of mountains, it would then descend with twice the velocity, as being impelled by both these forces conspiring together. Therefore since both these forces, that is, the gravity of heavy bodies, and the centripetal forces of the moons, respect the centre of the earth, and are similar and equal between themselves, they will (by Rule I and II) have one and the same cause. And therefore the force which retains the moon in its orbit is that very force which we commonly call gravity; because otherwise this little moon at the top of a mountain must either be without gravity, or fall twice as swiftly as heavy bodies are wont to do.

PROPOSITION V. THEOREM V.

That the circumjovial planets gravitate towards Jupiter; the circumsaturnal towards Saturn; the circumsolar towards the sun; and by the forces of their gravity are drawn off from rectilinear motions, and retained in curvilinear orbits.

For the revolutions of the circumjovial planets about Jupiter, of the circumsaturnal about Saturn, and of Mercury and Venus, and the other circumsolar planets, about the sun, are appearances of the same sort with the revolution of the moon about the earth; and therefore, by Rule II, must be owing to the same sort of causes; especially since it has been demonstrated, that the forces upon which those revolutions depend tend to the centres of Jupiter, of Saturn, and of the sun; and that those forces, in receding from Jupiter, from Saturn, and from the sun, decrease in the same proportion, and according to the same law, as the force of gravity does in receding from the earth.

Cor. 1. There is, therefore, a power of gravity tending to all the planets; for, doubtless, Venus, Mercury, and the rest, are bodies of the same sort with Jupiter and Saturn. And since all attraction (by Law III) is mutual, Jupiter will therefore gravitate towards all his own satellites, Saturn towards his, the earth towards the moon, and the sun towards all the primary planets.

Cor. 2. The force of gravity which tends to any one planet is reciprocally as the square of the distance of places from that planet's centre.

Cor. 3. All the planets do mutually gravitate towards one another, by Cor. 1 and 2. And hence it is that Jupiter and Saturn, when near their conjunction, by their mutual attractions sensibly disturb each other's motions. So the sun disturbs the motions of the moon; and both sun and moon disturb our sea, as we shall hereafter explain.

SCHOLIUM.

The force which retains the celestial bodies in their orbits has been hitherto called centripetal force; but it being now made plain that it can be no other than a gravitating force, we shall hereafter call it gravity. For the cause of that centripetal force which retains the moon in its orbit will extend itself to all the planets, by Rule I, II, and IV.

PROPOSITION VI. THEOREM VI.

That all bodies gravitate towards every planet; and that the weights of bodies towards any the same *planet, at equal distances from the centre of the planet, are proportional to the quantities of matter which they severally contain.*

It has been, now of a long time, observed by others, that all sorts of heavy bodies (allowance being made for the inequality of retardation which they suffer from a small power of resistance in the air) descend to the earth *from equal heights* in equal times; and that equality of times we may distinguish to a great accuracy, by the help of pendulums. I tried the thing in gold, silver, lead, glass, sand, common salt, wood, water, and wheat. I provided two wooden boxes, round and equal: I filled the one with wood, and suspended an equal weight of gold (as exactly as I could) in the centre of oscillation of the other. The boxes hanging by equal threads of 11 feet made a couple of pendulums perfectly equal in weight and figure, and equally receiving the resistance of the air. And, placing the one by the other, I observed them to play together forward and backward, for a long time, with equal vibrations. And therefore the quantity of matter in the gold (by Cor. 1 and 6, Prop. XXIV, Book II) was to the quantity of matter in the wood as the action of the motive force (or *vis motrix*) upon all the gold to the action of the same upon all the wood; that is, as the weight of the one to the weight of the other: and the like happened in the other bodies. By these experiments, in bodies of the same weight, I could manifestly have discovered a difference of matter less than the thousandth part of the whole, had any such been. But, without all doubt, the nature of gravity towards the planets is the same as towards the earth. For, should we imagine our terrestrial bodies removed to the orb of the moon, and there, together with the moon, deprived of all motion, to be let go, so as to fall together towards the earth, it is certain, from what we have demonstrated before, that, in equal times, they would describe equal spaces with the moon, and of consequence are to the moon, in quantity of matter, as their weights to its weight. Moreover, since the satellites of Jupiter perform their revolutions in times which observe the sesquiplicate proportion of their distances from Jupiter's centre, their accelerative gravities towards Jupiter will be reciprocally as the squares of their distances from Jupiter's centre; that is, equal, at equal distances. And, therefore, these satellites, if supposed to fall *towards Jupiter* from equal heights, would describe equal spaces in equal times, in like manner as heavy bodies do on our earth. And, by the same argument, if the circumsolar planets were supposed to be let fall at equal distances from the sun, they would, in their descent towards the sun, describe equal spaces in equal times. But forces which equally accelerate unequal bodies must be as those bodies: that is to say, the weights of the planets *towards the sun* must be as their quantities of matter. Further, that the weights of Jupiter and of his satellites towards the sun are proportional to the several quantities of their matter, appears from the exceedingly regular motions of the satellites (by Cor. 3, Prop. LXV, Book I). For if some of those bodies were more strongly attracted to the sun in proportion to their quantity of matter than others, the motions of the satellites would be disturbed by that inequality of attraction (by Cor. 2, Prop. LXV, Book I). If, at equal distances from the sun, any satellite, in proportion to the quantity of its matter, did gravitate towards the sun with a force greater than Jupiter in proportion to his, according to any given proportion, suppose of *d* to *e*; then the distance between the centres of the sun and of the satellite's orbit would be always greater than the distance between the centres of the sun and of Jupiter nearly in the subduplicate of that proportion: as by some computations I have found. And if the satellite did gravitate towards the sun with a force, lesser in the proportion of *e* to *d*, the distance of the centre of the satellite's orb from the sun would be less than the distance of the centre of Jupiter from the sun in the subduplicate of the same proportion. Therefore if, at equal distances from the sun, the accelerative gravity of any satellite towards the sun were greater or less than the accelerative gravity of Jupiter towards the sun but by one $\frac{1}{1000}$ part of the whole gravity, the distance of the centre of the satellite's orbit from the sun would be greater or less than the distance of Jupiter from the sun by one $\frac{1}{5000}$ part of the whole distance; that is, by a fifth part of the distance of the utmost satellite from the centre of Jupiter; an eccentricity of the orbit which would be very sensible. But the orbits of the satellites are concentric to Jupiter, and therefore the accelerative gravities of Jupiter, and of all its satellites towards the sun, are equal among themselves. And by the same argument, the weights of Saturn and of his satellites towards the sun, at equal distances from the sun, are as their several quantities of matter; and the weights of the moon and of the earth towards the sun are either none, or accurately proportional to the masses of matter which they contain. But some they are, by Cor. 1 and 3, Prop. V.

But further; the weights of all the parts of every planet towards any other planet are one to another as the matter in the several parts; for if some parts did gravitate more, others less, than for the quantity of their matter, then the whole planet, according to the sort of parts with which it most abounds, would gravitate more or less than in proportion to the quantity of matter in the whole. Nor is it of any moment whether these parts are external or internal; for if, for example, we should imagine the terrestrial bodies with us to be raised up to the orb of the moon, to be there compared with its body; if the weights of such bodies were to the weights of the external parts of the moon as the quantities of matter in the one and in the other respectively; but to the weights of the internal parts in a greater or less proportion, then likewise the weights of those bodies would be to the weight of the whole moon in a greater or less proportion; against what we have shewed above.

Cor. 1. Hence the weights of bodies do not depend upon their forms and textures; for if the weights could be altered with the forms, they would be greater or less, according to the variety of forms, in equal matter; altogether against experience.

Cor. 2. Universally, all bodies about the earth gravitate towards the earth; and the weights of all, at equal distances from the earth's centre, are as the quantities of matter which they severally contain. This is the quality of all bodies within the reach of our experiments; and therefore (by Rule III) to be affirmed of all bodies whatsoever. If the *æther*, or any other body, were either altogether void of gravity, or were to gravitate less in proportion to its quantity of matter, then, because (according to *Aristotle, Des Cartes,* and others) there is no difference betwixt that and other bodies but in *mere* form of matter, by a successive change from form to form, it might be changed at last into a body of the same condition with those which gravitate most in proportion to their quantity of matter; and, on the other hand, the heaviest bodies, acquiring the first form of that body, might by degrees quite lose their gravity. And therefore the weights would depend upon the forms of bodies, and with those forms might be changed: contrary to what was proved in the preceding Corollary.

Cor. 3. All spaces are not equally full; for if all spaces were equally full, then the specific gravity of the fluid which fills the region of the air, on account of the extreme density of the matter, would fall nothing short of the specific gravity of quicksilver, or gold, or any other the most dense body; and, therefore, neither gold, nor any other body, could descend in air; for bodies do not descend in fluids, unless they are specifically heavier than the fluids. And if the quantity of matter in a given space can, by any rarefaction, be diminished, what should hinder a diminution to infinity?

Cor. 4. If all the solid particles of all bodies are of the same density, nor can be rarefied without pores, a void, space, or vacuum must be granted.

By bodies of the same density, I mean those whose *vires inertiæ* are in the proportion of their bulks.

Cor. 5. The power of gravity is of a different nature from the power of magnetism; for the magnetic attraction is not as the matter attracted. Some bodies are attracted more by the magnet; others less; most bodies not at all. The power of magnetism in one and the same body may be increased and diminished; and is sometimes far stronger, for the quantity of matter, than the power of gravity; and in receding from the magnet decreases not in the duplicate but almost in the triplicate proportion of the distance, as nearly as I could judge from some rude observations.

PROPOSITION VII. THEOREM VII.

That there is a power of gravity tending to all bodies, proportional to the several quantities of matter which they contain.

That all the planets mutually gravitate one towards another, we have proved before; as well as that the force of gravity towards every one of them, considered apart, is reciprocally as the square of the distance of places from the centre of the planet. And thence (by Prop. LXIX, Book I, and its Corollaries) it follows, that the gravity tending towards all the planets is proportional to the matter which they contain.

Moreover, since all the parts of any planet A gravitate towards any other planet B; and the gravity of every part is to the gravity of the whole as the matter of the part to the matter of the whole; and (by Law III) to every action corresponds an equal re-action; therefore the planet B will, on the other hand, gravitate towards all the parts of the planet A; and its gravity towards any one part will be to the gravity towards the whole as the matter of the part to the matter of the whole. Q.E.D.

Cor. 1. Therefore the force of gravity towards any whole planet arises from, and is compounded of, the forces of gravity towards all its parts. Magnetic and electric attractions afford us examples of this; for all attraction towards the whole arises from the attractions towards the several parts. The thing may be easily understood in gravity, if we consider a greater planet, as formed of a number of lesser planets, meeting together in one globe; for *hence it would appear that* the force of the whole must arise from the forces of the component parts. If it is objected, that, according to this law, all bodies with us must mutually gravitate one towards another, whereas no such gravitation any where appears, I answer, that since the gravitation towards these bodies is to the gravitation towards the whole earth as these bodies are to the whole earth, the gravitation towards them must be far less than to fall under the observation of our senses.

Cor. 2. The force of gravity towards the several equal particles of any body is reciprocally as the square of the distance of places from the particles; as appears from Cor. 3, Prop. LXXIV, Book I.

PROPOSITION VIII. THEOREM VIII.

In two spheres mutually gravitating each towards the other, if the matter in places on all sides round about and equi-distant from the centres is similar, the weight of either sphere towards the other will be reciprocally as the square of the distance between their centres.

After I had found that the force of gravity towards a whole planet did arise from and was compounded of the forces of gravity towards all its parts, and towards every one part was in the reciprocal proportion of the squares of the distances from the part, I was yet in doubt whether that reciprocal duplicate proportion did accurately hold, or but nearly so, in the total force compounded of so many partial ones; for it might be that the proportion which accurately enough took place in greater distances should be wide of the truth near the surface of the planet, where the distances of the particles are unequal, and their situation dissimilar. But by the help of Prop. LXXV and LXXVI, Book I, and their Corollaries, I was at last satisfied of the truth of the Proposition, as it now lies before us.

Cor. 1. Hence we may find and compare together the weights of bodies towards different planets; for the weights of bodies revolving in circles about planets are (by Cor. 2, Prop. IV, Book I) as the diameters of the circles directly, and the squares of their periodic times reciprocally; and their weights at the surfaces of the planets, or at any other distances from their centres, are (by this Prop.) greater or less in the reciprocal duplicate proportion of the distances. Thus from the periodic times of Venus, revolving about the sun, in $224^{d}.\ 16\frac{3}{4}^{h}$, of the utmost circumjovial satellite revolving about Jupiter, in $16^{d}.\ 16\frac{8}{15}^{h}$.; of the Huygenian satellite about Saturn in $15^{d}.\ 22\frac{2}{3}^{h}$.; and of the moon about the earth in $27^{d}.\ 7^{h}.\ 43'$; compared with the mean distance of Venus from the sun, and with the greatest heliocentric elongations of the outmost circumjovial satellite from Jupiter's centre, $8'\ 16''$; of the Huygenian satellite from the centre of Saturn, $3'\ 4''$; and of the moon from the earth, $10'\ 33''$: by computation I found that the weight of equal bodies, at equal distances from the centres of the sun, of Jupiter, of Saturn, and of the earth, towards the sun, Jupiter, Saturn, and the earth, were one to another, as 1, $\frac{1}{1067}$, $\frac{1}{3021}$, and $\frac{1}{169282}$ respectively. Then because as the distances are increased or diminished, the weights are diminished or increased in a duplicate ratio, the weights of equal bodies towards the sun, Jupiter, Saturn, and the earth, at the distances 10000, 997, 791, and 109 from their centres, that is, at their very superficies, will be as 10000, 943, 529, and 435 respectively. How much the weights of bodies are at the superficies of the moon, will be shewn hereafter.

Cor. 2. Hence likewise we discover the quantity of matter in the several planets; for their quantities of matter are as the forces of gravity at equal distances from their centres; that is, in the sun, Jupiter, Saturn, and the earth, as 1, $\frac{1}{1067}$, $\frac{1}{3021}$, and $\frac{1}{169282}$ respectively. If the parallax of the sun be taken greater or less than 10″ 30‴, the quantity of matter in the earth must be augmented or diminished in the triplicate of that proportion.

COR. 3. Hence also we find the densities of the planets; for (by Prop. LXXII, Book I) the weights of equal and similar bodies towards similar spheres are, at the surfaces of those spheres, as the diameters of the spheres; and therefore the densities of dissimilar spheres are as those weights applied to the diameters of the spheres. But the true diameters of the Sun, Jupiter, Saturn, and the earth, were one to another as 10000, 997, 791, and 109; and the weights towards the same as 10000, 943, 529, and 435 respectively; and therefore their densities are as 100, 94$\frac{1}{2}$, 67, and 400. The density of the earth, which comes out by this computation, does not depend upon the parallax of the sun, but is determined by the parallax of the moon, and therefore is here truly defined. The sun, therefore, is a little denser than Jupiter, and Jupiter than Saturn, and the earth four times denser than the sun; for the sun, by its great heat, is kept in a sort of a rarefied state. The moon is denser than the earth, as shall appear afterward.

COR. 4. The smaller the planets are, they are, *cæteris paribus*, of so much the greater density; for so the powers of gravity on their several surfaces come nearer to equality. They are likewise, *cæteris paribus*, of the greater density, as they are nearer to the sun. So Jupiter is more dense than Saturn, and the earth than Jupiter; for the planets were to be placed at different distances from the sun, that, according to their degrees of density, they might enjoy a greater or less proportion to the sun's heat. Our water, if it were removed as far as the orb of Saturn, would be converted into ice, and in the orb of Mercury would quickly fly away in vapour; for the light of the sun, to which its heat is proportional, is seven times denser in the orb of Mercury than with us: and by the thermometer I have found that a sevenfold heat of our summer sun will make water boil. Nor are we to doubt that the matter of Mercury is adapted to its heat, and is therefore more dense than the matter of our earth; since, in a denser matter, the operations of Nature require a stronger heat.

PROPOSITION IX. THEOREM IX.

That the force of gravity, considered downward from the surface of the planets, decreases nearly in the proportion of the distances from their centres.

If the matter of the planet were of an uniform density, this Proposition would be accurately true (by Prop. LXXIII, Book I). The error, therefore, can be no greater than what may arise from the inequality of the density.

PROPOSITION X. THEOREM X.

That the motions of the planets in the heavens may subsist an exceedingly long time.

In the Scholium of Prop. XL, Book II, I have shewed that a globe of water frozen into ice, and moving freely in our air, in the time that it would describe the length of its semi-diameter, would lose by the resistance of the air $\frac{1}{4586}$ part of its motion; and the same proportion holds nearly in all globes, how great soever, and moved with whatever velocity. But that our globe of earth is of greater density than it would be if the whole consisted of water only, I thus make out. If the whole consisted of water only, whatever was of less density than water, because of its less specific gravity, would emerge and float above. And upon this account, if a globe of terrestrial matter, covered on all sides with water, was less dense than water, it would emerge somewhere; and, the subsiding water falling back, would be gathered to the opposite side. And such is the condition of our earth, which in a great measure is covered with seas. The earth, if it was not for its greater density, would emerge from the seas, and, according to its degree of levity, would be raised more or less above their surface, the water of the seas flowing backward to the opposite side. By the same argument, the spots of the sun, which float upon the lucid matter thereof, are lighter than that matter; and, however the planets have been formed while they were yet in fluid masses, all the heavier matter subsided to the centre. Since, therefore, the common matter of our earth on the surface thereof is about twice as heavy as water, and a little lower, in mines, is found about three, or four, or even five times more heavy, it is probable that the quantity of the whole matter of the earth may be five or six times greater than if it consisted all of water; especially since I have before shewed that the earth is about four times more dense than Jupiter. If, therefore, Jupiter is a little more dense than water, in the space of thirty days, in which that planet describes the length of 459 of its semi-diameters, it would, in a medium of the same density with our air, lose almost a tenth part of its motion. But since the resistance of mediums decreases in proportion to their weight or density, so that water, which is $13\frac{2}{5}$ times lighter than quicksilver, resists less in that proportion; and air, which is 860 times lighter than water, resists less in the same proportion; therefore in the heavens, where the weight of the medium in which the planets move is immensely diminished, the resistance will almost vanish.

It is shewn in the Scholium of Prop. XXII, Book II, that at the height of 200 miles above the earth the air is more rare than it is at the superficies of the earth in the ratio of 30 to 0,0000000000003998, or as

75000000000000 to 1 nearly. And hence the planet Jupiter, revolving in a medium of the same density with that superior air, would not lose by the resistance of the medium the 1000000th part of its motion in 1000000 years. In the spaces near the earth the resistance is produced only by the air, exhalations, and vapours. When these are carefully exhausted by the air-pump from under the receiver, heavy bodies fall within the receiver with perfect freedom, and without the least sensible resistance: gold itself, and the lightest down, let fall together, will descend with equal velocity; and though they fall through a space of four, six, and eight feet, they will come to the bottom at the same time; as appears from experiments. And therefore the celestial regions being perfectly void of air and exhalations, the planets and comets meeting no sensible resistance in those spaces will continue their motions through them for an immense tract of time.

HYPOTHESIS I.

That the centre of the system of the world is immovable.

This is acknowledged by all, while some contend that the earth, others that the sun, is fixed in that centre. Let us see what may from hence follow.

PROPOSITION XI. THEOREM XI.

That the common centre of gravity of the earth, the sun, and all the planets, is immovable.

For (by Cor. 4 of the Laws) that centre either is at rest, or moves uniformly forward in a right line; but if that centre moved, the centre of the world would move also, against the Hypothesis.

PROPOSITION XII. THEOREM XII.

That the sun is agitated by a perpetual motion, but never recedes far from the common centre of gravity of all the planets.

For since (by Cor. 2, Prop. VIII) the quantity of matter in the sun is to the quantity of matter in Jupiter as 1067 to 1; and the distance of Jupiter from the sun is to the semi-diameter of the sun in a proportion but a small matter greater, the common centre of gravity of Jupiter and the sun will fall upon a point a little without the surface of the sun. By the same argument, since the quantity of matter in the sun is to the quantity of matter in Saturn as 3021 to 1, and the distance of Saturn from the sun is to the semi-diameter of the sun in a proportion but a small matter less, the common centre of gravity of Saturn and the sun will fall upon a point a little within the surface of the sun. And, pursuing the principles of this computation, we should find that though the earth and all the planets were placed on one side of the sun, the distance of the common centre of gravity of all from the centre of the sun would scarcely amount to one diameter of the sun. In other cases, the distances of those centres are always less; and therefore, since that centre of gravity is in perpetual rest, the sun, according to the various positions of the planets, must perpetually be moved every way, but will never recede far from that centre.

Cor. Hence the common centre of gravity of the earth, the sun, and all the planets, is to be esteemed the centre of the world; for since the earth, the sun, and all the planets, mutually gravitate one towards another, and are therefore, according to their powers of gravity, in perpetual agitation, as the Laws of Motion require, it is plain that their moveable centres cannot be taken for the immovable centre of the world. If that body were to be placed in the centre, towards which other bodies gravitate most (according to common opinion), that privilege ought to be allowed to the sun; but since the sun itself is moved, a fixed point is to be chosen from which the centre of the sun recedes least, and from which it would recede yet less if the body of the sun were denser and greater, and therefore less apt to be moved.

PROPOSITION XIII. THEOREM XIII.

The planets move in ellipses which have their common focus in the centre of the sun; and, by radii drawn to that centre, they describe areas proportional to the times of description.

We have discoursed above of these motions from the Phænomena. Now that we know the principles on which they depend, from those principles we deduce the motions of the heavens *à priori*. Because the weights of the planets towards the sun are reciprocally as the squares of their distances from the sun's centre, if the sun was at rest, and the other planets did not mutually act one upon another, their orbits would be ellipses, having the sun in their common focus; and they would describe areas proportional to the times *of description*, by Prop. I and XI, and Cor. 1, Prop. XIII, Book I. But the mutual actions of the planets one upon another are so very small, that they may be neglected; and by Prop. LXVI, Book I, they less disturb the motions of the planets around the sun in motion than if those motions were performed about the sun at rest.

It is true, that the action of Jupiter upon Saturn is not to be neglected; for the force of gravity towards Jupiter is to the force of gravity towards the sun (at equal distances, Cor. 2, Prop. VIII) as 1 to 1067; and therefore in the conjunction of Jupiter and Saturn, because the distance of Saturn from Jupiter is to the distance of Saturn from the sun almost as 4 to 9, the gravity of Saturn towards Jupiter will be to the gravity of Saturn towards the sun as 81 to 16 $\times$ 1067; or, as 1 to about 211. And hence arises a perturbation of the orb of Saturn in every conjunction of this planet with Jupiter, so sensible, that astronomers are puzzled with it. As the planet is differently situated in these conjunctions, its eccentricity is sometimes augmented, sometimes diminished; its aphelion is sometimes carried forward, sometimes backward, and its mean motion is by turns accelerated and retarded; yet the whole error in its motion about the sun, though arising from so great a force, may be almost avoided (except in the mean motion) by placing the lower focus of its orbit in the common centre of gravity of Jupiter and the sun (according to Prop. LXVII, Book I), and therefore that error, when it is greatest, scarcely exceeds two minutes; and the greatest error in the mean motion scarcely exceeds two minutes yearly. But in the conjunction of Jupiter and Saturn, the accelerative forces of gravity of the sun towards Saturn, of Jupiter towards Saturn, and of Jupiter towards the sun, are almost as 16, 81, and $\frac{16 \times 81 \times 3021}{25}$, or 156609; and therefore the difference of the forces of gravity of the sun towards Saturn, and of Jupiter towards Saturn, is to the force of gravity of Jupiter towards the sun as 65 to 156609, or as 1 to 2409. But the greatest power of Saturn to disturb the motion of Jupiter is proportional to this difference; and therefore the perturbation of the orbit of Jupiter is much less than that of Saturn's. The perturbations of the other orbits are yet far less, except that the orbit of the earth is sensibly disturbed by the moon. The common centre of gravity of the earth and moon moves in an ellipsis about the sun in the focus thereof, and, by a radius drawn to the sun, describes areas proportional to the times of description. But the earth in the mean time by a menstrual motion is revolved about this common centre.

PROPOSITION XIV. THEOREM XIV.

The aphelions and nodes of the orbits of the planets are fixed.

The aphelions are immovable by Prop. XI, Book I; and so are the planes of the orbits, by Prop. I of the same Book. And if the planes are fixed, the nodes must be so too. It is true, that some inequalities may arise from the mutual actions of the planets and comets in their revolutions; but these will be so small, that they may be here passed by.

Cor. 1. The fixed stars are immovable, seeing they keep the same position to the aphelions and nodes of the planets.

Cor. 2. And since these stars are liable to no sensible parallax from the annual motion of the earth, they can have no force, because of their immense distance, to produce any sensible effect in our system. Not to mention that the fixed stars, every where promiscuously dispersed in the heavens, by their contrary attractions destroy their mutual actions, by Prop. LXX, Book I.

SCHOLIUM.

Since the planets near the sun (viz. Mercury, Venus, the Earth, and

Mars) are so small that they can act with but little force upon each other, therefore their aphelions and nodes must be fixed, excepting in so far as they are disturbed by the actions of Jupiter and Saturn, and other higher bodies. And hence we may find, by the theory of gravity, that their aphelions move a little *in consequentia,* in respect of the fixed stars, and that in the sesquiplicate proportion of their several distances from the sun. So that if the aphelion of Mars, in the space of a hundred years, is carried 33′ 20″ *in consequentia,* in respect of the fixed stars, the aphelions of the Earth, of Venus, and of Mercury, will in a hundred years be carried forwards 17′ 40″, 10′ 53″, and 4′ 16″, respectively. But these motions are so inconsiderable, that we have neglected them in this Proposition.

PROPOSITION XV. PROBLEM I.

To find the principal diameters of the orbits of the planets.

They are to be taken in the sub-sesquiplicate proportion of the periodic times, by Prop. XV, Book I, and then to be severally augmented in the proportion of the sum of the masses of matter in the sun and each planet to the first of two mean proportionals betwixt that sum and the quantity of matter in the sun, by Prop. LX, Book I.

PROPOSITION XVI. PROBLEM II.

To find the eccentricities and aphelions of the planets.

This Problem is resolved by Prop. XVIII, Book I.

PROPOSITION XVII. THEOREM XV.

That the diurnal motions of the planets are uniform, and that the libration of the moon arises from its diurnal motion.

The Proposition is proved from the first Law of Motion, and Cor. 22, Prop. LXVI, Book I. Jupiter, with respect to the fixed stars, revolves in $9^h.\ 56'$; Mars in $24^h.\ 39'$; Venus in about $23^h.$; the Earth in $23^h.\ 56'$; the Sun in $25\frac{1}{2}$ days, and the moon in 27 days, 7 hours, 43′. These things appear by the Phænomena. The spots in the sun's body return to the same situation on the sun's disk, with respect to the earth, in $27\frac{1}{2}$ days; and therefore with respect to the fixed stars the sun revolves in about $25\frac{1}{2}$ days. But because the lunar day, arising from its uniform revolution about its axis, is menstrual, *that is, equal to the time of its periodic revolution in its orb,* therefore the same face of the moon will be always nearly turned to the upper focus of its orb; but, as the situation of that focus requires, will deviate a little to one side and to the other from the earth in the lower focus; and this is the libration in longitude; for the libration in latitude arises from the moon's latitude, and the inclination of its axis to the plane of the ecliptic. This theory of the libration of the moon, Mr. *N. Mercator,* in his Astronomy, published at the beginning of the year 1676, explained more fully out of the letters I sent him. The utmost satellite of Saturn seems to revolve about its axis with a motion like this of the moon, respecting Saturn continually with the same face; for in its revolution round Saturn, as often as it comes to the eastern part of its orbit, it is scarcely visible, and generally quite disappears; which is like to be occasioned by some spots in that part of its body, which is then turned towards the earth, as M. *Cassini* has observed. So also the utmost satellite of Jupiter seems to revolve about its axis with a like motion, because in that part of its body which is turned from Jupiter it has a spot, which always appears as if it were in Jupiter's own body, whenever the satellite passes between Jupiter and our eye.

PROPOSITION XVIII. THEOREM XVI.

That the axes of the planets are less than the diameters drawn perpendicular to the axes.

The equal gravitation of the parts on all sides would give a spherical figure to the planets, if it was not for their diurnal revolution in a circle. By that circular motion it comes to pass that the parts receding from the axis endeavour to ascend about the equator; and therefore if the matter is in a fluid state, by its ascent towards the equator it will enlarge the diameters there, and by its descent towards the poles it will shorten the axis. So the diameter of Jupiter (by the concurring observations of astronomers) is found shorter betwixt pole and pole than from east to west. And, by the same argument, if our earth was not higher about the equator than at the poles, the seas would subside about the poles, and, rising towards the equator, would lay all things there under water.

PROPOSITION XIX. PROBLEM III.

To find the proportion of the axis of a planet to the diameters perpendicular thereto.

Our countryman, Mr. *Norwood*, measuring a distance of 905751 feet of *London* measure between *London* and *York*, in 1635, and observing the difference of latitudes to be 2° 28′, determined the measure of one degree to be 367196 feet of *London* measure, that is 57300 *Paris* toises. M. *Picart*, measuring an arc of one degree, and 22′ 55″ of the meridian between *Amiens* and *Malvoisine*, found an arc of one degree to be 57060 *Paris* toises. M. *Cassini*, the father, measured the distance upon the meridian from the town of *Collioure* in *Roussillon* to the Observatory of *Paris;* and his son added the distance from the Observatory to the Citadel of *Dunkirk*. The whole distance was 486156½ toises and the difference of the latitudes of *Collioure* and *Dunkirk* was 8 degrees, and 31′

$11\frac{5}{8}''$. Hence an arc of one degree appears to be 57061 *Paris* toises. And from these measures we conclude that the circumference of the earth is 123249600, and its semi-diameter 19615800 *Paris* feet, upon the supposition that the earth is of a spherical figure.

In the latitude of *Paris* a heavy body falling in a second of time describes 15 *Paris* feet, 1 inch, $1\frac{7}{9}$ line, as above, that is, 2173 lines $\frac{7}{9}$. The weight of the body is diminished by the weight of the ambient air. Let us suppose the weight lost thereby to be $\frac{1}{11000}$ part of the whole weight; then that heavy body falling *in vacuo* will describe a height of 2174 lines in one second of time.

A body in every sidereal day of 23^h. 56′ 4″ uniformly revolving in a circle at the distance of 19615800 feet from the centre, in one second of time describes an arc of 1433,46 feet; the versed sine of which is 0,05236561 feet, or 7,54064 lines. And therefore the force with which bodies descend in the latitude of *Paris* is to the centrifugal force of bodies in the equator arising from the diurnal motion of the earth as 2174 to 7,54064.

The centrifugal force of bodies in the equator is to the centrifugal force with which bodies recede directly from the earth in the latitude of *Paris* 48° 50′ 10″ in the duplicate proportion of the radius to the cosine of the latitude, that is, as 7,54064 to 3,267. Add this force to the force with which bodies descend by their weight in the latitude of *Paris*, and a body, in the latitude of *Paris*, falling by its whole undiminished force of gravity, in the time of one second, will describe 2177,267 lines, or 15 *Paris* feet, 1 inch, and 5,267 lines. And the total force of gravity in that latitude will be to the centrifugal force of bodies in the equator of the earth as 2177,267 to 7,54064, or as 289 to 1.

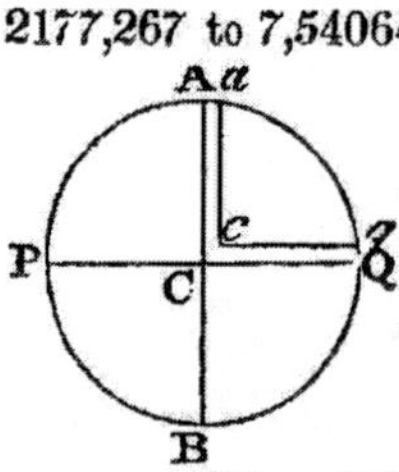

Wherefore if APBQ represent the figure of the earth, now no longer spherical, but generated by the rotation of an ellipsis about its lesser axis PQ; and ACQ*qca* a canal full of water, reaching from the pole Q*q* to the centre C*c*, and thence rising to the equator A*a*; the weight of the water in the leg of the canal AC*ca* will be to the weight of water in the other leg QC*cq* as 289 to 288, because the centrifugal force arising from the circular motion sustains and takes off one of the 289 parts of the weight (in the one leg), and the weight of 288 in the other sustains the rest. But by computation (from Cor. 2, Prop. XCI, Book I) I find, that, if the matter of the earth was all uniform, and without any motion, and its axis PQ were to the diameter AB as 100 to 101, the force of gravity in the place Q towards the earth would be to the force of gravity in the same place Q towards a sphere described about the centre C with the radius PC, or QC, as 126 to 125. And, by the same argument, the force of gravity in the place A towards the spheroid generated by the rotation of the ellipsis APBQ about the axis AB is to the force of gravity in the same place A, towards the sphere described about the centre C with the radius AC, as 125 to 126. But the force of gravity in the place A towards the earth is a mean proportional betwixt the forces of gravity towards the spheroid and this sphere; because the sphere, by having its diameter PQ diminished in the proportion of 101 to 100, is transformed into the figure of the earth; and this figure, by having a third diameter perpendicular to the two diameters AB and PQ diminished in the same proportion, is converted into the said spheroid; and the force of gravity in A, in either case, is diminished nearly in the same proportion. Therefore the force of gravity in A towards the sphere described about the centre C with the radius AC, is to the force of gravity in A towards the earth as 126 to $125\frac{1}{2}$. And the force of gravity in the place Q towards the sphere described about the centre C with the radius QC, is to the force of gravity in the place A towards the sphere described about the centre C, with the radius AC, in the proportion of the diameters (by Prop. LXXII, Book I), that is, as 100 to 101. If, therefore, we compound those three proportions 126 to 125, 126 to $125\frac{1}{2}$, and 100 to 101, into one, the force of gravity in the place Q towards the earth will be to the force of gravity in the place A towards the earth as $126 \times 126 \times 100$ to $125 \times 125\frac{1}{2} \times 101$; or as 501 to 500.

Now since (by Cor. 3, Prop. XCI, Book I) the force of gravity in either leg of the canal AC*ca*, or QC*cq*, is as the distance of the places from the centre of the earth, if those legs are conceived to be divided by transverse, parallel, and equidistant surfaces, into parts proportional to the wholes, the weights of any number of parts in the one leg AC*ca* will be to the weights of the same number of parts in the other leg as their magnitudes and the accelerative forces of their gravity conjunctly, that is, as 101 to 100, and 500 to 501, or as 505 to 501. And therefore if the centrifugal force of every part in the leg AC*ca*, arising from the diurnal motion, was to the weight of the same part as 4 to 505, so that from the weight of every part, conceived to be divided into 505 parts, the centrifugal force might take off four of those parts, the weights would remain equal in each leg, and therefore the fluid would rest in an equilibrium. But the centrifugal force of every part is to the weight of the same part as 1 to 289; that is, the centrifugal force, which should be $\frac{4}{505}$ parts of the weight, is only $\frac{1}{289}$ part thereof. And, therefore, I say, by the rule of proportion, that if the centrifugal force $\frac{4}{505}$ make the height of the water in the leg AC*ca* to exceed the height of the water in the leg QC*cq* by one $\frac{1}{100}$ part of its whole height, the centrifugal force $\frac{1}{289}$ will make the excess of the height in the leg AC*ca* only $\frac{1}{289}$ part of the height of the water in the other leg QC*cq*; and therefore the diameter of the earth at the equator, is to its diameter from pole to pole as 230 to 229. And since the mean semi-diameter of the earth, according to *Picart's* mensuration, is 19615800 *Paris* feet, or 3923,16 miles (reckoning 5000 feet to a mile), the earth will be higher at the equator than at the poles by 85472 feet, or $17\frac{1}{10}$ miles. And its height at the equator will be about 19658600 feet, and at the poles 19573000 feet.

If, the density and periodic time of the diurnal revolution remaining the same, the planet was greater or less than the earth, the proportion of the centrifugal force to that of gravity, and therefore also of the diameter betwixt the poles to the diameter at the equator, would likewise remain the same. But if the diurnal motion was accelerated or retarded in any proportion, the centrifugal force would be augmented or diminished nearly in the same duplicate proportion; and therefore the difference of the diameters will be increased or diminished in the same duplicate ratio very nearly. And if the density of the planet was augmented or diminished in any proportion, the force of gravity tending towards it would also be augmented or diminished in the same proportion: and the difference of the diameters contrariwise would be diminished in proportion as the force of gravity is augmented, and augmented in proportion as the force of gravity is diminished. Wherefore, since the earth, in respect of the fixed stars, revolves in $23^h.\ 56'$, but Jupiter in $9^h.\ 56'$, and the squares of their periodic times are as 29 to 5, and their densities as 400 to $94\frac{1}{2}$, the difference of the diameters of Jupiter will be to its lesser diameter as $\frac{29}{5} \times \frac{400}{94\frac{1}{2}} \times \frac{1}{229}$ to 1, or as 1 to $9\frac{1}{3}$, nearly. Therefore the diameter of Jupiter from east to west is to its diameter from pole to pole nearly as $10\frac{1}{3}$ to $9\frac{1}{3}$. Therefore since its greatest diameter is $37''$, its lesser diameter lying between the poles will be $33''\ 25'''$. Add thereto about $3''$ for the irregular refraction of light, and the apparent diameters of this planet will become $40''$ and $36''\ 25'''$; which are to each other as $11\frac{1}{6}$ to $10\frac{1}{6}$, very nearly. These things are so upon the supposition that the body of Jupiter is uniformly dense. But now if its body be denser towards the plane of the equator than towards the poles, its diameters may be to each other as 12 to 11, or 13 to 12, or perhaps as 14 to 13.

And *Cassini* observed in the year 1691, that the diameter of Jupiter reaching from east to west is greater by about a fifteenth part than the other diameter. Mr. *Pound* with his 123 feet telescope, and an excellent micrometer, measured the diameters of Jupiter in the year 1719, and found them as follow.

The Times.			Greatest diam.	Lesser diam.	The diam. to each other.		
	Day.	Hours.	Parts.	Parts.			
January	28	6	13,40	12,28	As 12	to	11
March	6	7	13,12	12,20	$13\frac{3}{4}$	to	$12\frac{3}{4}$
March	9	7	13,12	12,08	$12\frac{2}{3}$	to	$11\frac{2}{3}$
April	9	9	12,32	11,48	$14\frac{1}{2}$	to	$13\frac{1}{2}$

So that the theory agrees with the phænomena; for the planets are more heated by the sun's rays towards their equators, and therefore are a little more condensed by that heat than towards their poles.

Moreover, that there is a diminution of gravity occasioned by the diurnal rotation of the earth, and therefore the earth rises higher there than it does at the poles (supposing that its matter is uniformly dense), will appear by the experiments of pendulums related under the following Proposition.

PROPOSITION XX. PROBLEM IV.

To find and compare together the weights of bodies in the different regions of our earth.

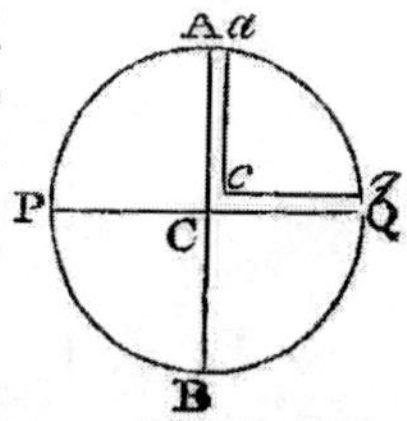

Because the weights of the unequal legs of the canal of water ACQ*qca* are equal; and the weights of the parts proportional to the whole legs, and alike situated in them, are one to another as the weights of the wholes, and therefore equal betwixt themselves; the weights of equal parts, and alike situated in the legs, will be reciprocally as the legs, that is, reciprocally as 230 to 229. And the case is the same in all homogeneous equal bodies alike situated in the legs of the canal. Their weights are reciprocally as the legs, that is, reciprocally as the distances of the bodies from the centre of the earth. Therefore if the bodies are situated in the uppermost parts of the canals, or on the surface of the earth, their weights will be one to another reciprocally as their distances from the centre. And, by the same argument, the weights in all other places round the whole surface of the earth are reciprocally as the distances of the places from the centre; and, therefore, in the hypothesis of the earth's being a spheroid are given in proportion.

Whence arises this Theorem, that the increase of weight in passing from the equator to the poles is nearly as the versed sine of double the latitude; or, which comes to the same thing, as the square of the right sine of the latitude; and the arcs of the degrees of latitude in the meridian increase nearly in the same proportion. And, therefore, since the latitude of *Paris* is 48° 50′, that of places under the equator 00° 00′, and that of places under the poles 90°; and the versed sines of double those arcs are 11334,00000 and 20000, the radius being 10000; and the force of gravity at the pole is to the force of gravity at the equator as 230 to 229; and the excess of the force of gravity at the pole to the force of gravity at the equator as 1 to 229; the excess of the force of gravity in the latitude of *Paris* will be to the force of gravity at the equator as $1 \times \frac{11334}{20000}$ to 229, or as 5667 to 2290000. And therefore the whole forces of gravity in those places will be one to the other as 2295667 to 2290000. Wherefore since the lengths of pendulums vibrating in equal times are as the forces of gravity, and in the latitude of *Paris*, the length of a pendulum vibrating seconds is 3 *Paris* feet, and $8\frac{1}{3}$ lines, or rather because of the weight of the air, $8\frac{5}{9}$ lines, the length of a pendulum vibrating in the same time under the equator will be shorter by 1,087 lines. And by a like calculus the following table is made.

Latitude of the place.	Length of the pendulum		Measure of one degree in the meridian.
Deg.	Feet.	Lines.	Toises.
0	3 .	7,468	56637
5	3 .	7,482	56642
10	3 .	7,526	56659
15	3 .	7,596	56687
20	3 .	7,692	56724
25	3 .	7,812	56769
30	3 .	7,948	56823
35	3 .	8,099	56882
40	3 .	8,261	56945
1	3 .	8,294	56958
2	3 .	8,327	56971
3	3 .	8,361	56984
4	3 .	8,394	56997
45	3 .	8,428	57010
6	3 .	8,461	57022
7	3 .	8,494	57035
8	3 .	8,528	57048
9	3 .	8,561	57061
50	3 .	8,594	57074
55	3 .	8,756	57137
60	3 .	8,907	57196
65	3 .	9,044	57250
70	3 .	9,162	57295
75	3 .	9,258	57332
80	3 .	9,329	57360
85	3 .	9,372	57377
90	3 .	9,387	57382

By this table, therefore, it appears that the inequality of degrees is so small, that the figure of the earth, in geographical matters, may be considered as spherical; especially if the earth be a little denser towards the plane of the equator than towards the poles.

Now several astronomers, sent into remote countries to make astronomical observations, have found that pendulum clocks do accordingly move slower near the equator than in our climates. And, first of all, in the year 1672, M. *Richer* took notice of it in the island of *Cayenne;* for when, in the month of *August*, he was observing the transits of the fixed stars over the meridian, he found his clock to go slower than it ought in respect of the mean motion of the sun at the rate of 2′ 28″ a day. Therefore, fitting up a simple pendulum to vibrate in seconds, which were measured by an excellent clock, he observed the length of that simple pendulum; and this he did over and over every week for ten months together. And upon his return to *France*, comparing the length of that pendulum with the length of the pendulum at *Paris* (which was 3 *Paris* feet and $8\frac{3}{5}$ lines), he found it shorter by $1\frac{1}{4}$ line.

Afterwards, our friend Dr. *Halley*, about the year 1677, arriving at the island of St. *Helena*, found his pendulum clock to go slower there than at *London*, without marking the difference. But he shortened the rod of his clock by more than the $\frac{1}{8}$ of an inch, or $1\frac{1}{2}$ line; and to effect this, because the length of the screw at the lower end of the rod was not sufficient, he interposed a wooden ring betwixt the nut and the ball.

Then, in the year 1682, M. *Varin* and M. *des Hayes* found the length of a simple pendulum vibrating in seconds at the Royal Observatory of *Paris* to be 3 feet and $8\frac{5}{9}$ lines. And by the same method in the island of *Goree*, they found the length of an isochronal pendulum to be 3 feet and $6\frac{5}{9}$ lines, differing from the former by two lines. And in the same year, going to the islands of *Guadaloupe* and *Martinico*, they found that the length of an isochronal pendulum in those islands was 3 feet and $6\frac{1}{2}$ lines.

After this, M. *Couplet*, the son, in the month of *July* 1697, at the Royal Observatory of *Paris*, so fitted his pendulum clock to the mean motion of the sun, that for a considerable time together the clock agreed with the motion of the sun. In *November* following, upon his arrival at *Lisbon*, he found his clock to go slower than before at the rate of 2′ 13″ in 24 hours. And next *March* coming to *Paraiba*, he found his clock to go slower than at *Paris*, and at the rate 4′ 12″ in 24 hours; and he affirms, that the pendulum vibrating in seconds was shorter at *Lisbon* by $2\frac{1}{2}$ lines, and at *Paraiba* by $3\frac{2}{3}$ lines, than at *Paris*. He had done better to have reckoned those differences $1\frac{1}{3}$ and $2\frac{5}{9}$; for these differences correspond to the differences of the times 2′ 13″ and 4′ 12″. But this gentleman's observations are so gross, that we cannot confide in them.

In the following years, 1699, and 1700, M. *des Hayes*, making another voyage to *America*, determined that in the island of *Cayenne* and *Granada* the length of the pendulum vibrating in seconds was a small matter less than 3 feet and $6\frac{1}{2}$ lines; that in the island of St. *Christophers* it was 3 feet and $6\frac{3}{4}$ lines; and in the island of St. *Domingo* 3 feet and 7 lines.

And in the year 1704, P. *Feuillé*, at *Puerto Bello* in *America*, found that the length of the pendulum vibrating in seconds was 3 *Paris feet*, and only $5\frac{7}{12}$ lines, that is, almost 3 lines shorter than at *Paris*; but the observation was faulty. For afterward, going to the island of *Martinico*, he found the length of the isochronal pendulum there 3 *Paris* feet and $5\frac{10}{12}$ lines.

Now the latitude of *Paraiba* is 6° 38′ south; that of *Puerto Bello* 9° 33′ north; and the latitudes of the islands *Cayenne, Goree, Gaudaloupe, Martinico, Granada*, St. *Christophers*, and St. *Domingo*, are respectively 4° 55′, 14° 40″, 15° 00′, 14° 44′, 12° 06′, 17° 19′, and 19° 48′, north. And the excesses of the length of the pendulum at *Paris* above the lengths of the isochronal pendulums observed in those latitudes are a little greater than by the table of the lengths of the pendulum before computed. And therefore the earth is a little higher under the equator than by the preceding calculus, and a little denser at the centre than in mines near the surface, unless, perhaps, the heats of the torrid zone have a little extended the length of the pendulums.

For M. *Picart* has observed, that a rod of iron, which in frosty weather in the winter season was one foot long, when heated by fire, was lengthened into one foot and $\frac{1}{4}$ line. Afterward M. *de la Hire* found that a rod of iron, which in the like winter season was 6 feet long, when exposed to the heat of the summer sun, was extended into 6 feet and $\frac{2}{3}$ line. In the former case the heat was greater than in the latter; but in the latter it was greater than the heat of the external parts of a human body; for metals exposed to the summer sun acquire a very considerable degree of heat. But the rod of a pendulum clock is never exposed to the heat of the summer sun, nor ever acquires a heat equal to that of the external parts of a human body; and, therefore, though the 3 feet rod of a pendulum clock will indeed be a little longer in the summer than in the winter season, yet the difference will scarcely amount to $\frac{1}{4}$ line. Therefore the total difference of the lengths of isochronal pendulums in different climates cannot be ascribed to the difference of heat; nor indeed to the mistakes of the *French* astronomers. For although there is not a perfect agreement betwixt their observations, yet the errors are so small that they may be neglected; and in this they all agree, that isochronal pendulums are shorter under the equator than at the Royal Observatory of *Paris*, by a difference not less than $1\frac{1}{4}$ line, nor greater than $2\frac{2}{3}$ lines. By the observations of M. *Richer*, in the island of *Cayenne*, the difference was $1\frac{1}{4}$ line. That difference being corrected by those of M. *des Hayes*, becomes $1\frac{1}{2}$ line or $1\frac{3}{4}$ line. By the less accurate observations of others, the same was made about two lines. And this disagreement might arise partly from the errors of the observations, partly from the dissimilitude of the internal parts of the earth, and the height of mountains; partly from the different heats of the air.

I take an iron rod of 3 feet long to be shorter by a sixth part of one line in winter time with us here in *England* than in the summer. Because of the great heats under the equator, subduct this quantity from the difference of one line and a quarter observed by M. *Richer*, and there will remain one line $\frac{1}{12}$, which agrees very well with $1\frac{87}{1000}$ line collected, by the theory a little before. M. *Richer* repeated his observations, made in the island of *Cayenne*, every week for ten months together, and compared the lengths of the pendulum which he had there noted in the iron rods with the lengths thereof which he observed in *France*. This diligence and care seems to have been wanting to the other observers. If this gentleman's observations are to be depended on, the earth is higher under the equator than at the poles, and that by an excess of about 17 miles; as appeared above by the theory.

PROPOSITION XXI. THEOREM XVII.

That the equinoctial points go backward, and that the axis of the earth, by a nutation in every annual revolution, twice vibrates towards the ecliptic, and as often returns to its former position.

The proposition appears from Cor. 20, Prop. LXVI, Book I; but that motion of nutation must be very small, and, indeed, scarcely perceptible.

PROPOSITION XXII. THEOREM XVIII.

That all the motions of the moon, and all the inequalities of those motions, follow from the principles which we have laid down.

That the greater planets, while they are carried about the sun, may in the mean time carry other lesser planets, revolving about them; and that those lesser planets must move in ellipses which have their foci in the centres of the greater, appears from Prop. LXV, Book I. But then their motions will be several ways disturbed by the action of the sun, and they will suffer such inequalities as are observed in our moon. Thus our moon (by Cor. 2, 3, 4, and 5, Prop. LXVI, Book I) moves faster, and, by a radius drawn to the earth, describes an area greater for the time, and has its orbit less curved, and therefore approaches nearer to the earth in the syzygies than in the quadratures, excepting in so far as these effects are hindered by the motion of eccentricity; for (by Cor. 9, Prop. LXVI, Book I) the eccentricity is greatest when the apogeon of the moon is in the syzygies, and least when the same is in the quadratures; and upon this account the perigeon moon is swifter, and nearer to us, but the apogeon moon slower, and farther from us, in the syzygies than in the quadratures. Moreover, the apogee goes forward, and the nodes backward; and this is done not with a regular but an unequal motion. For (by Cor. 7 and 8, Prop. LXVI, Book I) the apogee goes more swiftly forward in its syzygies, more slowly backward in its quadratures; and, by the excess of its progress above its regress, advances yearly *in consequentia.* But, contrariwise, the nodes (by Cor. 11, Prop. LXVI, Book I) are quiescent in their syzygies, and go fastest back in their quadratures. Farther, the greatest latitude of the moon (by Cor. 10, Prop. LXVI, Book I) is greater in the quadratures of the moon than in its syzygies. And (by Cor. 6, Prop. LXVI, Book I) the mean motion of the moon is slower in the perihelion of the earth than in its aphelion. And these are the principal inequalities (of the moon) taken notice of by astronomers.

But there are yet other inequalities not observed by former astronomers, by which the motions of the moon are so disturbed, that to this day we have not been able to bring them under any certain rule. For the velocities or horary motions of the apogee and nodes of the moon, and their equations, as well as the difference betwixt the greatest eccentricity in the syzygies, and the least eccentricity in the quadratures, and that inequality which we call the variation, are (by Cor. 14, Prop. LXVI, Book I) in the course of the year augmented and diminished in the triplicate proportion of the sun's apparent diameter. And besides (by Cor. 1 and 2, Lem. 10, and Cor. 16, Prop. LXVI, Book I) the variation is augmented and diminished nearly in the duplicate proportion of the time between the quadratures. But in astronomical calculations, this inequality is commonly thrown into and confounded with the equation of the moon's centre.

PROPOSITION XXIII. PROBLEM V.

To derive the unequal motions of the satellites of Jupiter and Saturn from the motions of our moon.

From the motions of our moon we deduce the corresponding motions of the moons or satellites of Jupiter in this manner, by Cor. 16, Prop. LXVI, Book I. The mean motion of the nodes of the outmost satellite of Jupiter is to the mean motion of the nodes of our moon in a proportion compounded of the duplicate proportion of the periodic times of the earth about the sun to the periodic times of Jupiter about the sun, and the simple proportion of the periodic time of the satellite about Jupiter to the periodic time of our moon about the earth; and, therefore, those nodes, in the space of a hundred years, are carried 8° 24′ backward, or *in antecedentia*. The mean motions of the nodes of the inner satellites are to the mean motion of the nodes of the outmost as their periodic times to the periodic time of the former, by the same Corollary, and are thence given. And the motion of the apsis of every satellite *in consequentia* is to the motion of its nodes *in antecedentia* as the motion of the apogee of our moon to the motion of its nodes (by the same Corollary), and is thence given. But the motions of the apsides thus found must be diminished in the proportion of 5 to 9, or of about 1 to 2, on account of a cause which I cannot here descend to explain. The greatest equations of the nodes, and of the apsis of every satellite, are to the greatest equations of the nodes, and apogee of our moon respectively, as the motions of the nodes and apsides of the satellites, in the time of one revolution of the former equations, to the motions of the nodes and apogee of our moon, in the time of one revolution of the latter equations. The variation of a satellite seen from Jupiter is to the variation of our moon in the same proportion as the whole motions of their nodes respectively during the times in which the satellite and our moon (after parting from) are revolved (again) to the sun, by the same Corollary; and therefore in the outmost satellite the variation does not exceed 5″ 12‴.

PROPOSITION XXIV. THEOREM XIX.

That the flux and reflux of the sea arise from the actions of the sun and moon.

By Cor. 19 and 20, Prop. LXVI, Book I, it appears that the waters of the sea ought twice to rise and twice to fall every day, as well lunar as solar; and that the greatest height of the waters in the open and deep seas ought to follow the appulse of the luminaries to the meridian of the place by a less interval than 6 hours; as happens in all that eastern tract of the *Atlantic* and *Æthiopic* seas between *France* and the *Cape of Good Hope;* and on the coasts of *Chili* and *Peru* in the *South Sea;* in all which shores the flood falls out about the second, third, or fourth hour, unless where the motion propagated from the deep ocean is by the shallowness of the channels, through which it passes to some particular places, retarded to the fifth, sixth, or seventh hour, and even later. The hours I reckon from the appulse of each luminary to the meridian of the place, as well under as above the horizon; and by the hours of the lunar day I understand the 24th parts of that time which the moon, by its apparent diurnal motion, employs to come about again to the meridian of the place which it left the day before. The force of the sun or moon in raising the sea is greatest in the appulse of the luminary to the meridian of the place; but the force impressed upon the sea at that time continues a little while after the impression, and is afterwards increased by a new though less force still acting upon it. This makes the sea rise higher and higher, till this new force becoming too weak to raise it any more, the sea rises to its greatest height. And this will come to pass, perhaps, in one or two hours, but more frequently near the shores in about three hours, or even more, where the sea is shallow.

The two luminaries excite two motions, which will not appear distinctly, but between them will arise one mixed motion compounded out of both. In the conjunction or opposition of the luminaries their forces will be conjoined, and bring on the greatest flood and ebb. In the quadratures the sun will raise the waters which the moon depresses, and depress the waters which the moon raises, and from the difference of their forces the smallest of all tides will follow. And because (as experience tells us) the force of the moon is greater than that of the sun, the greatest height of the waters will happen about the third lunar hour. Out of the syzygies and quadratures, the greatest tide, which by the single force of the moon ought to fall out at the third lunar hour, and by the single force of the sun at the third solar hour, by the compounded forces of both must fall out in an intermediate time that aproaches nearer to the third hour of the moon than to that of the sun. And, therefore, while the moon is passing from the syzygies to the quadratures, during which time the 3d hour of the sun precedes the 3d hour of the moon, the greatest height of the waters will also precede the 3d hour of the moon, and that, by the greatest interval, a little after the octants of the moon; and, by like intervals, the greatest tide will follow the 3d lunar hour, while the moon is passing from the quadratures to the syzygies. Thus it happens in the open sea; for in the mouths of rivers the greater tides come later to their height.

But the effects of the luminaries depend upon their distances from the earth; for when they are less distant, their effects are greater, and when more distant, their effects are less, and that in the triplicate proportion of their apparent diameter. Therefore it is that the sun, in the winter time, being then in its perigee, has a greater effect, and makes the tides in the syzygies something greater, and those in the quadratures something less than in the summer season; and every month the moon, while in the perigee, raises greater tides than at the distance of 15 days before or after, when it is in its apogee. Whence it comes to pass that two highest tides do not follow one the other in two immediately succeeding syzygies.

The effect of either luminary doth likewise depend upon its declination or distance from the equator; for if the luminary was placed at the pole, it would constantly attract all the parts of the waters without any intension or remission of its action, and could cause no reciprocation of motion. And, therefore, as the luminaries decline from the equator towards either pole, they will, by degrees, lose their force, and on this account will excite lesser tides in the solstitial than in the equinoctial syzygies. But in the solstitial quadratures they will raise greater tides than in the quadratures about the equinoxes; because the force of the moon, then situated in the equator, most exceeds the force of the sun. Therefore the greatest tides fall out in those syzygies, and the least in those quadratures, which happen about the time of both equinoxes: and the greatest tide in the syzygies is always succeeded by the least tide in the quadratures, as we find by experience. But, because the sun is less distant from the earth in winter than in summer, it comes to pass that the greatest and least tides more frequently appear before than after the vernal equinox, and more frequently after than before the autumnal.

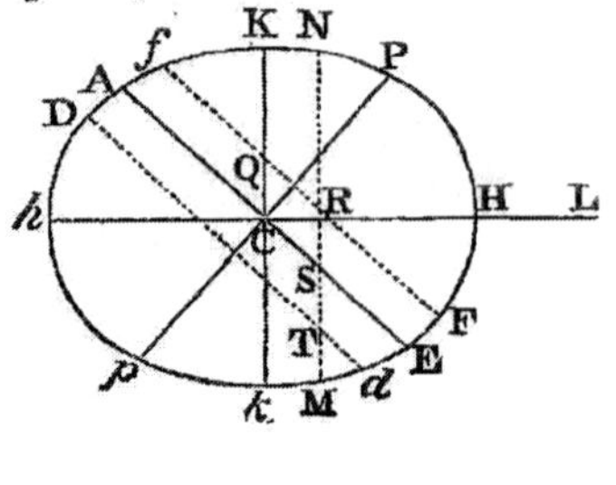

Moreover, the effects of the luminaries depend upon the latitudes of places. Let $ApEP$ represent the earth covered with deep waters; C its centre; P, p its poles; AE the equator; F any place without the equator; Ff the parallel of the place; Dd the correspondent parallel on the other side of the equator; L the place of the moon three hours before; H the place of the earth directly under it; *h* the opposite place; K, *k* the places at 90 degrees distance; CH, C*h*, the greatest heights of the sea from the centre of the earth; and CK, C*k*, its least heights: and if with the axes H*h*, K*k*, an ellipsis is described, and by the revolution of that ellipsis about its longer axis H*h* a spheroid HPK*hpk* is formed, this spheroid will nearly represent the figure of the sea; and CF, C*f*, CD, C*d*, will represent the heights of the sea in the places F*f*, D*d*. But farther; in the said revolution of the ellipsis any point N describes the circle NM cutting the parallels F*f*, D*d*, in any places RT, and the equator AE in S; CN will represent the height of the sea in all those places R, S, T, situated in this circle. Wherefore, in the diurnal revolution of any place F, the greatest flood will be in F, at the third hour after the appulse of the moon to the meridian above the horizon; and afterwards the greatest ebb in Q, at the third hour after the setting of the moon; and then the greatest flood in *f*, at the third hour after the appulse of the moon to the meridian under the horizon; and, lastly, the greatest ebb in Q, at the third hour after the rising of the moon; and the latter flood in *f* will be less than the preceding flood in F. For the whole sea is divided into two hemispherical floods, one in the hemisphere KH*k* on the north side, the other in the opposite hemisphere K*hk*, which we may therefore call the northern and the southern floods. These floods, being always opposite the one to the other, come by turns to the meridians of all places, after an interval of 12 lunar hours. And seeing the northern countries partake more of the northern flood, and the southern countries more of the southern flood, thence arise tides, alternately greater and less in all places without the equator, in which the luminaries rise and set. But the greatest tide will happen when the moon declines towards the vertex of the place, about the third hour after the appulse of the moon to the meridian above the horizon; and when the moon changes its declination *to the other side of the equator*, that which was the greater tide will be changed into a lesser. And the greatest difference of the floods will fall out about the times of the solstices; especially if the ascending node of the moon is about the first of Aries. So it is found by experience that the morning tides in winter exceed those of the evening, and the evening tides in summer exceed those of the morning; at *Plymouth* by the height of one foot, but at *Bristol* by the height of 15 inches, according to the observations of *Colepress* and *Sturmy*.

But the motions which we have been describing suffer some alteration from that force of reciprocation, which the waters, being once moved, retain a little while *by their vis insita*. Whence it comes to pass that the tides may continue for some time, though the actions of the luminaries should cease. This power of retaining the impressed motion lessens the difference of the alternate tides, and makes those tides which immediately succeed after the syzygies greater, and those which follow next after the quadratures less. And hence it is that the alternate tides at *Plymouth* and *Bristol* do not differ much more one from the other than by the height of a foot or 15 inches, and that the greatest tides of all at those ports are not the first but the third after the syzygies. And, besides, all the motions are retarded in their passage through shallow channels, so that the greatest tides of all, in some straits and mouths of rivers, are the fourth or even the fifth after the syzygies.

Farther, it may happen that the tide may be propagated from the ocean through different channels towards the same port, and may pass quicker through some channels than through others; in which case the same tide, divided into two or more succeeding one another, may compound new motions of different kinds. Let us suppose two equal tides flowing towards the same port from different places, the one preceding the other by 6 hours; and suppose the first tide to happen at the third hour of the appulse of the moon to the meridian of the port. If the moon at the time of the appulse to the meridian was in the equator, every 6 hours alternately there would arise equal floods, which, meeting with as many equal ebbs, would so balance one the other, that for that day, the water would stagnate and remain quiet. If the moon then declined from the equator, the tides in the ocean would be alternately greater and less, as was said; and from thence two greater and two lesser tides would be alternately propagated towards that port. But the two greater floods would make the greatest height of the waters to fall out in the middle time betwixt both; and the greater and lesser floods would make the waters to rise to a mean height in the middle time between them, and in the middle time between the two lesser floods the waters would rise to their least height. Thus in the space of 24 hours the waters would come, not twice, as commonly, but once only to their greatest, and once only to their least height; and their greatest height, if the moon declined towards the elevated pole, would happen at the 6th or 30th hour after the appulse of the moon to the meridian; and when the moon changed its declination, this flood would be changed into an ebb. An example of all which Dr. *Halley* has given us, from the observations of seamen in the port of *Batsham*, in the kingdom of *Tunquin*, in the latitude of 20° 50′ north. In that port, on the day which follows after the passage of the moon over the equator, the waters stagnate: when the moon declines to the north, they begin to flow and ebb, not twice, as in other ports, but once only every day; and the flood happens at the setting, and the greatest ebb at the rising of the moon. This tide increases with the declination of the moon till the 7th or 8th day; then for the 7 or 8 days following it decreases at the same rate as it had increased before, and ceases when the moon changes its declination, crossing over the equator to the south. After which the flood is immediately changed into an ebb; and thenceforth the ebb happens at the setting and the flood at the rising of the moon; till the moon, again passing the equator, changes its declination. There are two inlets to this port and the neighboring channels, one from the seas of *China*, between the continent and the island of *Leuconia;* the other from the *Indian* sea, between the continent and the island of *Borneo.* But whether there be really two tides propagated through the said channels, one from the *Indian* sea in the space of 12 hours, and one from the sea of *China* in the space of 6 hours, which therefore happening at the 3d and 9th lunar hours, by being compounded together, produce those motions; or whether there be any other circumstances in the state of those seas, I leave to be determined by observations on the neighbouring shores.

Thus I have explained the causes of the motions of the moon and of the sea. Now it is fit to subjoin something concerning the quantity of those motions.

PROPOSITION XXV. PROBLEM VI.

To find the forces with which the sun disturbs the motions of the moon.

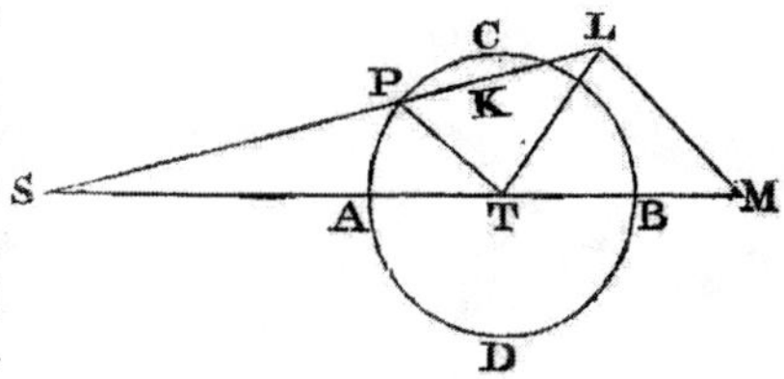

Let S represent the sun, T the earth, P the moon, CADB the moon's orbit. In SP take SK equal to ST; and let SL be to SK in the duplicate proportion of SK to SP: draw LM parallel to PT; and if ST or SK is supposed to represent the accelerated force of gravity of the earth towards the sun, SL will represent the accelerative force of gravity of the moon towards the sun. But that force is compounded of the parts SM and LM, of which the force LM, and that part of SM which is represented by TM, disturb the motion of the moon, as we have shewn in Prop. LXVI, Book I, and its Corollaries. Forasmuch as the earth and moon are revolved about their common centre of gravity, the motion of the earth about that centre will be also disturbed by the like forces; but we may consider the sums both of the forces and of the motions as in the moon, and represent the sum of the forces by the lines TM and ML, which are analogous to them both. The force ML (in its mean quantity) is to the centripetal force by which the moon may be retained in its orbit revolving about the earth at rest, at the distance PT, in the duplicate proportion of the periodic time of the moon about the earth to the periodic time of the earth about the sun (by Cor. 17, Prop. LXVI, Book I); that is, in the duplicate proportion of $27^d. 7^h. 43'$ to $365^d. 6^h. 9'$; or as 1000 to 178725; or as 1 to $178\frac{29}{40}$. But in the

4th Prop. of this Book we found, that, if both earth and moon were revolved about their common centre of gravity, the mean distance of the one from the other would be nearly $60\frac{1}{2}$ mean semi-diameters of the earth; and the force by which the moon may be kept revolving in its orbit about the earth in rest at the distance PT of $60\frac{1}{2}$ semi-diameters of the earth, is to the force by which it may be revolved in the same time, at the distance of 60 semi-diameters, as $60\frac{1}{2}$ to 60: and this force is to the force of gravity with us very nearly as 1 to 60×60. Therefore the mean force ML is to the force of gravity on the surface of our earth as $1 \times 60\frac{1}{2}$ to $60 \times 60 \times 60 \times 178\frac{29}{40}$, or as 1 to 638092,6; whence by the proportion of the lines TM, ML, the force TM is also given; and these are the forces with which the sun disturbs the motions of the moon. Q.E.I.

PROPOSITION XXVI. PROBLEM VII.

To find the horary increment of the area which the moon, by a radius drawn to the earth, describes in a circular orbit.

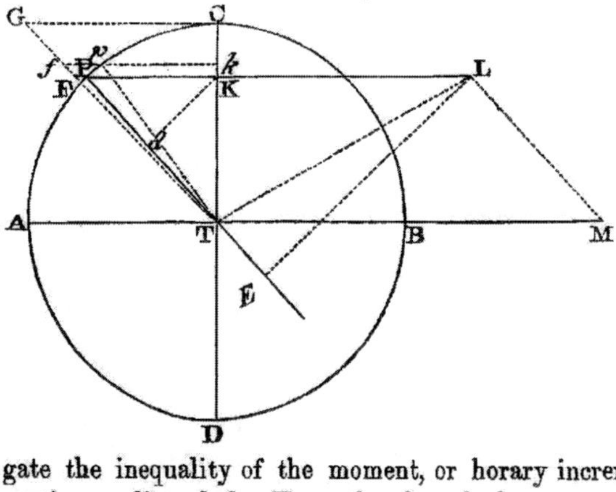

We have above shown that the area which the moon describes by a radius drawn to the earth is proportional to the time of description, excepting in so far as the moon's motion is disturbed by the action of the sun; and here we propose to investigate the inequality of the moment, or horary increment *of that area or motion so disturbed.* To render the calculus more easy, we shall suppose the orbit of the moon to be circular, and neglect all inequalities but that only which is now under consideration; and, because of the immense distance of the sun, we shall farther suppose that the lines SP and ST are parallel. By this means, the force LM will be always reduced to its mean quantity TP, as well as the force TM to its mean quantity 3PK. These forces (by Cor. 2 of the Laws of Motion) compose the force TL; and this force, by letting fall the perpendicular LE upon the radius TP, is resolved into the forces TE, EL; of which the force TE, acting constantly in the direction of the radius TP, neither accelerates nor retards the description of the area TPC made by that radius TP; but EL, *acting on the radius* TP in a perpendicular direction, accelerates or retards *the description of the area* in proportion as it accelerates or retards the moon.

That acceleration of the moon, in its passage from the quadrature C to the conjunction A, is in every moment of time as the *generating* accelerative force EL, that is, as $\frac{3PK \times TK}{TP}$. Let the time be represented by the mean motion of the moon, or (which comes to the same thing) by the angle CTP, or even by the arc CP. At right angles upon CT erect CG equal to CT; and, supposing the quadrantal arc AC to be divided into an infinite number of equal parts P*p*, &c., these *parts* may represent the like *infinite* number of the equal parts of time. Let fall *pk* perpendicular on CT, and draw TG meeting with KP, *kp* produced in F and *f*; then will FK be equal to TK, and K*k* be to PK as P*p* to T*p*, that is, in a given proportion; and therefore FK × K*k*, or the area FK*kf*, will be as $\frac{3PK \times TK}{TP}$, that is, as EL; and compounding, the whole area GCKF will be as the sum of all the forces EL impressed upon the moon in the whole time CP; and therefore also as the velocity generated by that sum, that is, as the acceleration of the description of the area CTP, or as the increment of the moment *thereof*. The force by which the moon may in its periodic time CADB of $27^d. 7^h. 43'$ be retained revolving about the earth in rest at the distance TP, would cause a body falling in the time CT to describe the length $\frac{1}{2}$CT, and at the same time to acquire a velocity equal to that with which the moon is moved in its orbit. This appears from Cor. 9, Prop. IV., Book I. But since K*d*, drawn perpendicular on TP, is *but* a third part of EL, and *equal* to the half of TP, or ML, in the octants, the force EL in the octants, where it is greatest, will exceed the force ML in the proportion of 3 to 2; and therefore will be to that force by which the moon in its periodic time may be retained revolving about the earth at rest as 100 to $\frac{2}{3} \times 17872\frac{1}{2}$, or 11915; and in the time CT will generate a velocity equal to $\frac{100}{11915}$ parts of the velocity of the moon; but in the time CPA will generate a greater velocity in the proportion of CA to CT or TP. Let the greatest force EL in the octants be represented by the area FK × K*k*, or by the rectangle $\frac{1}{2}$TP × P*p*, which is equal thereto; and the velocity which that greatest force can generate in any time CP will be to the velocity which any other lesser force EL can generate in the same time as the rectangle $\frac{1}{2}$TP × CP to the area KCGF; but the velocities generated in the whole time CPA will be one to the other as the rectangle $\frac{1}{2}$TP × CA to the triangle TCG, or as the quadrantal arc CA to the radius TP; and therefore the latter velocity generated in the whole time will be $\frac{100}{11915}$ parts of the velocity of the moon. To this velocity of the moon, which is proportional to the mean moment of the area (supposing this mean moment to be represented by the number 11915), we add and subtract the half of the other velocity; the sum 11915 + 50, or 11965, will represent the greatest moment of the area in the syzygy A; and the difference 11915 — 50, or 11865, the least moment thereof in the quadratures. Therefore the areas which in equal times are described in the syzygies and quadratures are one to the other as 11965 to 11865. And if to the least moment 11865 we add a moment which shall be to 100, the difference of the two former moments, as the trapezium FKCG to the triangle TCG, or, which comes to the same thing, as the square of the sine PK to the square of the radius TP (that is, as Pd to TP), the sum will represent the moment of the area when the moon is in any intermediate place P.

But these things take place only in the hypothesis that the sun and the earth are at rest, and that the synodical revolution of the moon is finished in $27^{d}.\ 7^{h}.\ 43'$. But since the moon's synodical period is really $29^{d}.\ 12^{h}.\ 44'$, the increments of the moments must be enlarged in the same proportion as the time is, that is, in the proportion of 1080853 to 1000000. Upon which account, the whole increment, which was $\frac{100}{11915}$ parts of the mean moment, will now become $\frac{100}{11023}$ parts thereof; and therefore the moment of the area in the quadrature of the moon will be to the moment thereof in the syzygy as 11023 — 50 to 11023 + 50; or as 10973 to 11073; and to the moment thereof, when the moon is in any intermediate place P, as 10973 to 10973 + Pd; that is, supposing TP = 100.

The area, therefore, which the moon, by a radius drawn to the earth, describes in the several little equal parts of time, is nearly as the sum of the number 219,46, and the versed sine of the double distance of the moon from the nearest quadrature, considered in a circle which hath unity for its radius. Thus it is when the variation in the octants is in its mean quantity. But if the variation there is greater or less, that versed sine must be augmented or diminished in the same proportion.

PROPOSITION XXVII. PROBLEM VIII.

From the horary motion of the moon to find its distance from the earth.

The area which the moon, by a radius drawn to the earth, describes in every moment of time, is as the horary motion of the moon and the square of the distance of the moon from the earth conjunctly. And therefore the distance of the moon from the earth is in a proportion compounded of the subduplicate proportion of the area directly, and the subduplicate proportion of the horary motion inversely. Q.E.I.

Cor. 1. Hence the apparent diameter of the moon is given; for it is reciprocally as the distance of the moon from the earth. Let astronomers try how accurately this rule agrees with the phænomena.

Cor. 2. Hence also the orbit of the moon may be more exactly defined from the phænomena than hitherto could be done.

PROPOSITION XXVIII. PROBLEM IX.

To find the diameters of the orbit, in which, without eccentricity, the moon would move.

The curvature of the orbit which a body describes, if attracted in lines perpendicular to the orbit, is as the force of attraction directly, and the square of the velocity inversely. I estimate the curvatures of lines compared one with another according to the evanescent proportion of the sines or tangents of their angles of contact to equal radii, supposing those radii to be infinitely diminished. But the attraction of the moon towards the earth in the syzygies is the excess of its gravity towards the earth above the force of the sun 2PK (see Fig. Prop. XXV), by which force the accelerative gravity of the moon towards the sun exceeds the accelerative gravity of the earth towards the sun, or is exceeded by it. But in the quadratures that attraction is the sum of the gravity of the moon towards the earth, and the sun's force KT, by which the moon is attracted towards the earth. And these attractions, putting N for $\frac{AT + CT}{2}$, are nearly as $\frac{178725}{AT^2} - \frac{2000}{CT \times N}$ and $\frac{178725}{CT^2} + \frac{1000}{AT \times N}$, or as $178725N \times CT^2 - 2000AT^2 \times CT$, and $178725N \times AT^2 + 1000CT^2 \times AT$. For if the accelerative gravity of the moon towards the earth be represented by the number 178725, the mean force ML, which in the quadratures is PT or TK, and draws the moon towards the earth, will be 1000, and the mean force TM in the syzygies will be 3000; from which, if we subtract the mean force ML, there will remain 2000, the force by which the moon in the syzygies is drawn from the earth, and which we above called 2PK. But the velocity of the moon in the syzygies A and B is to its velocity in the quadratures C and D as CT to AT, and the moment of the area, which the moon by a radius drawn to the earth describes in the syzygies, to the moment of that area *described* in the quadratures conjunctly; that is, as 11073CT to 10973AT. Take this ratio twice inversely, and the former ratio once directly, and the curvature of the orb of the moon in the syzygies will be to the curvature thereof in the quadratures as $120406729 \times 178725AT^2 \times CT^2 \times N - 120406729 \times 2000AT^4 \times CT$ to $122611329 \times 178725AT^2 \times CT^2 \times N + 122611329 \times 1000CT^4 \times AT$, that is, as $2151969AT \times CT \times N - 24081AT^3$ to $2191371AT \times CT \times N + 12261CT^3$.

Because the figure of the moon's orbit is unknown, let us, in its stead, assume the ellipsis DBCA, in the centre of which we suppose the earth to be situated, and the greater axis DC to lie between the quadratures as the lesser AB between the syzygies. But since the plane of this ellipsis is revolved about the earth by an angular motion, and the orbit, whose curvature we now examine, should be described in a plane void of such motion,

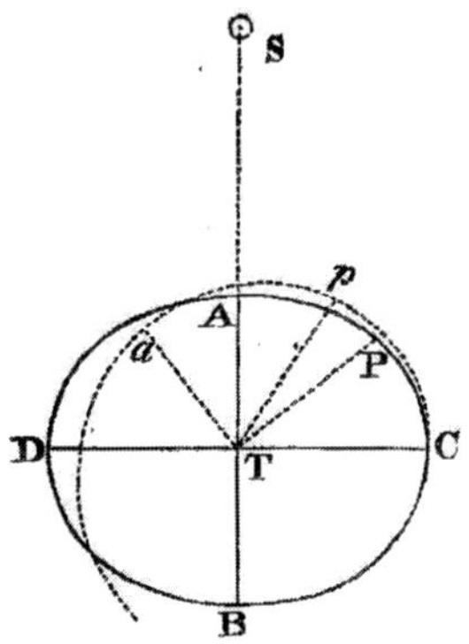

we are to consider the figure which the moon, while it is revolved in that ellipsis, describes in this plane, that is to say, the figure C*pa*, the several points *p* of which are found by assuming any point P in the ellipsis, which may represent the place of the moon, and drawing T*p* equal to TP in such manner that the angle PT*p* may be equal to the apparent motion of the sun from the time of the last quadrature in C; or (which comes to the same thing) that the angle CT*p* may be to the angle CTP as the time of the synodic revolution of the moon to the time of the periodic revolution thereof, or as $29^d. 12^h. 44'$ to $27^d. 7^h. 43'$. If, therefore, in this proportion we take the angle CT*a* to the right angle CTA, and make T*a* of equal length with TA, we shall have *a* the lower and C the upper apsis of this orbit C*pa*. But, by computation, I find that the difference betwixt the curvature of this orbit C*pa* at the vertex *a*, and the curvature of a circle described about the centre T with the interval TA, is to the difference between the curvature of the ellipsis at the vertex A, and the curvature of the same circle, in the duplicate proportion of the angle CTP to the angle CT*p*; and that the curvature of the ellipsis in A is to the curvature of that circle in the duplicate proportion of TA to TC; and the curvature of that circle to the curvature of a circle described about the centre T with the interval TC as TC to TA; but that the curvature of this *last arch* is to the curvature of the ellipsis in C in the duplicate proportion of TA to TC; and that the difference betwixt the curvature of the ellipsis in the vertex C, and the curvature of this last circle, is to the difference betwixt the curvature of the figure C*pa*, at the vertex C, and the curvature of this same *last* circle, in the duplicate proportion of the angle CT*p* to the angle CTP; all which proportions are easily drawn from the sines of the angles of contact, and of the differences of those angles. But, by comparing those proportions together, we find the curvature of the figure C*pa* at *a* to be to its curvature at C as $AT^3 - \frac{16824}{100000}CT^2 AT$ to $CT^3 + \frac{16824}{100000}AT^2 \times CT$; where the number $\frac{16824}{100000}$ represents the difference of the squares of the angles CTP and CT*p*, applied to the square of the lesser angle CTP; or (which is all one) the difference of the squares of the times $27^d. 7^h. 43'$, and $29^d. 12^h. 44'$, applied to the square of the time $27^d. 7^h. 43'$.

Since, therefore, *a* represents the syzygy of the moon, and C its quadrature, the proportion now found must be the same with that proportion of the curvature of the moon's orb in the syzygies to the curvature thereof in the quadratures, which we found above. Therefore, in order to find the proportion of CT to AT, let us multiply the extremes and the means, and the terms which come out, applied to AT $\times$ CT, become $2062{,}79CT^4 - 2151969N \times CT^3 + 368676N \times AT \times CT^2 + 36342\,AT^2 \times CT^2 - 362047N \times AT^2 \times CT + 2191371N \times AT^3 + 4051{,}4AT^4 = 0$. Now if for the half sum N of the terms AT and CT we put 1, and x for their half difference, then CT will be $= 1 + x$, and AT $= 1 - x$. And substituting those values in the equation, after resolving thereof, we shall find $x = 0{,}00719$; and from thence the semi-diameter CT $= 1{,}00719$, and the semi-diameter AT $= 0{,}99281$, which numbers are nearly as $70\frac{1}{24}$, and $69\frac{1}{24}$. Therefore the moon's distance from the earth in the syzygies is to its distance in the quadratures (setting aside the consideration of eccentricity) as $69\frac{1}{24}$ to $70\frac{1}{24}$; or, in round numbers, as 69 to 70.

PROPOSITION XXIX. PROBLEM X.

To find the variation of the moon.

This inequality is owing partly to the elliptic figure of the moon's orbit, partly to the inequality of the moments of the area which the moon by a radius drawn to the earth describes. If the moon P revolved in the ellipsis DBCA about the earth quiescent in the centre of the ellipsis, and by the radius TP, drawn to the earth, described the area CTP, proportional to the time *of description;* and the greatest semi-diameter CT of the ellipsis was to the least TA as 70 to 69; the tangent of the angle CTP would be to the tangent of the angle of the mean motion, computed from the quadrature C, as the semi-diameter TA of the ellipsis to its semi-diameter TC, or as 69 to 70. But the description of the area CTP, as the moon advances from the quadrature to the syzygy, ought to be in such manner accelerated, that the moment of the area in the moon's syzygy may be to the moment thereof in its quadrature as 11073 to 10973; and that the excess of the moment in any intermediate place P above the moment in the quadrature may be as the square of the sine of the angle CTP; which we may effect with accuracy enough, if we diminish the tangent of the angle CTP in the subduplicate proportion of the number 10973 to the number 11073, that is, in proportion of the number 68,6877 to the number 69. Upon which account the tangent of the angle CTP will now be to the tangent of the mean motion as 68,6877 to 70; and the angle CTP in the octants, where the mean motion is 45°, will be found 44° 27′ 28″, which subtracted from 45°, the angle of the mean motion, leaves the greatest variation 32′ 32″. Thus it would be, if the moon, in passing from the quadrature to the syzygy, described an angle CTA of 90 degrees only. But because of the motion of the earth, by which the sun is apparently transferred *in consequentia*, the moon, before it overtakes the sun, describes an angle CT*a*, greater than a right angle, in the proportion of the time of the synodic revolution of the moon to the time of its periodic revolution, that is, in the proportion of $29^d. 12^h. 44'$ to $27^d. 7^h. 43'$. Whence it comes to pass that all the angles about the centre T are dilated in the same proportion; and the greatest variation, which otherwise would be *but* 32′ 32″, now augmented in the said proportion, becomes 35′ 10″.

And this is its magnitude in the mean distance of the sun from the earth, neglecting the differences which may arise from the curvature of the *orbis magnus*, and the stronger action of the sun upon the moon when horned and new, than when gibbous and full. In other distances of the sun from the earth, the greatest variation is in a proportion compounded of the duplicate proportion of the time of the synodic revolution of the moon (the time of the year being given) directly, and the triplicate proportion of the distance of the sun from the earth inversely. And, therefore, in the apogee of the sun, the greatest variation is 33′ 14″, and in its perigee 37′ 11″, if the eccentricity of the sun is to the transverse semi-diameter of the *orbis magnus* as $16\frac{15}{16}$ to 1000.

Hitherto we have investigated the variation in an orb not eccentric, in which, to wit, the moon in its octants is always in its mean distance from the earth. If the moon, on account of its eccentricity, is more or less removed from the earth than if placed in this orb, the variation may be something greater, or something less, than according to this rule. But I leave the excess or defect to the determination of astronomers from the phænomena.

PROPOSITION XXX. PROBLEM XI.

To find the horary motion of the nodes of the moon in a circular orbit.

Let S represent the sun, T the earth, P the moon, NP*n* the orbit of the moon, N*pn* the orthographic projection of the orbit upon the plane of the ecliptic; N, *n* the nodes, *n*TN*m* the line of the nodes produced indefi-

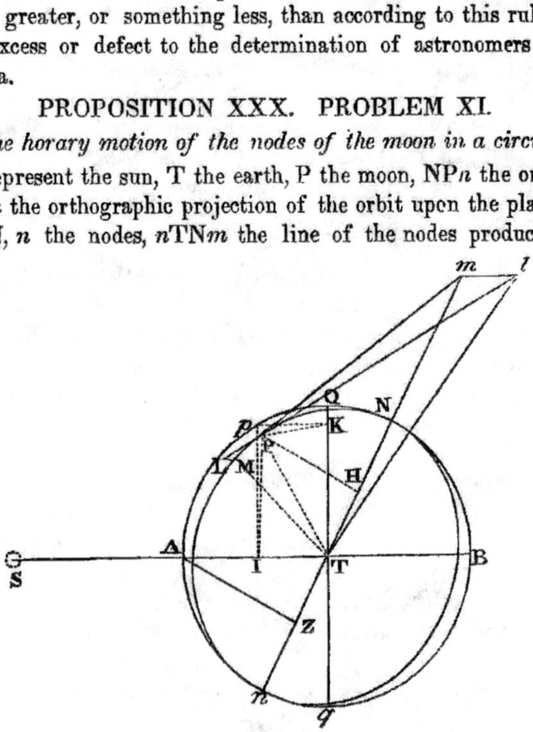

nitely; PI, PK perpendiculars upon the lines ST, Qq; Pp a perpendicular upon the plane of the ecliptic; A, B the moon's syzygies in the plane of the ecliptic; AZ a perpendicular let fall upon Nn, the line of the nodes; Q, q the quadratures of the moon in the plane of the ecliptic, and pK a perpendicular on the line Qq lying between the quadratures. The force of the sun to disturb the motion of the moon (by Prop. XXV) is twofold, one proportional to the line LM, the other to the line MT, in the scheme of that Proposition; and the moon by the former force is drawn towards the earth, by the latter towards the sun, in a direction parallel to the right line ST joining the earth and the sun. The former force LM acts in the direction of the plane of the moon's orbit, and therefore makes no change upon the situation thereof, and is upon that account to be neglected; the latter force MT, by which the plane of the moon's orbit is disturbed, is the same with the force 3PK or 3IT. And this force (by Prop. XXV) is to the force by which the moon may, in its periodic time, be uniformly revolved in a circle about the earth at rest, as 3IT to the radius of the circle multiplied by the number 178,725, or as IT to the radius thereof multiplied by 59,575. But in this calculus, and all that follows, I consider all the lines drawn from the moon to the sun as parallel to the line which joins the earth and the sun; because what inclination there is almost as much diminishes all effects in some cases as it augments them in others; and we are now inquiring after the mean motions of the nodes, neglecting such niceties as are of no moment, and would only serve to render the calculus more perplexed.

Now suppose PM to represent an arc which the moon describes in the least moment of time, and ML a little line, the half of which the moon, by the impulse of the said force 3IT, would describe in the same time; and joining PL, MP, let them be produced to m and l, where they cut the plane of the ecliptic, and upon Tm let fall the perpendicular PH. Now, since the right line ML is parallel to the plane of the ecliptic, and therefore can never meet with the right line ml which lies in that plane, and yet both those right lines lie in one common plane LMPml, they will be parallel, and upon that account the triangles LMP, lmP will be similar. And seeing MPm lies in the plane of the orbit, in which the moon did move while in the place P, the point m will fall upon the line Nn, which passes through the nodes N, n, of that orbit. And because the force by which the half of the little line LM is generated, if the whole had been together, and at once impressed in the point P, would have generated that whole line, and caused the moon to move in the arc whose chord is LP; that is to say, would have transferred the moon from the plane MPmT into the plane LPlT; therefore the angular motion of the nodes generated by that force will be equal to the angle mTl. But ml is to mP as ML to MP; and since MP, because of the time given, is also given, ml will be as the rectangle ML $\times$ mP, that is, as the rectangle IT $\times$ mP. And if Tml is a right angle, the angle mTl will be as $\frac{ml}{\mathrm{T}m}$, and therefore as $\frac{\mathrm{IT} \times \mathrm{P}m}{\mathrm{T}m}$, that is (because T$m$ and mP, TP and PH are proportional), as $\frac{\mathrm{IT} \times \mathrm{PH}}{\mathrm{TP}}$; and, therefore, because TP is given, as IT $\times$ PH. But if the angle Tml or STN is oblique, the angle mTl will be yet less, in proportion of the sine of the angle STN to the radius, or AZ to AT. And therefore the velocity of the nodes is as IT $\times$ PH $\times$ AZ, or as the solid content of the sines of the three angles TPI, PTN, and STN.

If these are right angles, as happens when the nodes are in the quadratures, and the moon in the syzygy, the little line ml will be removed to an infinite distance, and the angle mTl will become equal to the angle mPl. But in this case the angle mPl is to the angle PTM, which the moon in the same time by its apparent motion describes about the earth, as 1 to 59,575. For the angle mPl is equal to the angle LPM, that is, to the angle of the moon's deflexion from a rectilinear path; which angle, if the gravity of the moon should have then ceased, the said force of the sun 3IT would by itself have generated in that given time; and the angle PTM is equal to the angle of the moon's deflexion from a rectilinear path; which angle, if the force of the sun 3IT should have then ceased, the force alone by which the moon is retained in its orbit would have generated in the same time. And these forces (as we have above shewn) are the one to the other as 1 to 59,575. Since, therefore, the mean horary motion of the moon (in respect of the fixed stars) is $32' \; 56'' \; 27''' \; 12\frac{1}{2}^{\text{iv}}$. the horary motion of the node in this case will be $33'' \; 10''' \; 33^{\text{iv}}. \; 12^{\text{v}}$. But in other cases the horary motion will be to $33'' \; 10''' \; 33^{\text{iv}}. \; 12^{\text{v}}$. as the solid content of the sines of the three angles TPI, PTN, and STN (or of the distances of the moon from the quadrature, of the moon from the node, and of the node from the sun) to the cube of the radius. And as often as the sine of any angle is changed from positive to negative, and from negative to positive, so often must the regressive be changed into a progressive, and the progressive into a regressive motion. Whence it comes to pass that the nodes are progressive as often as the moon happens to be placed between either quadrature, and the node nearest to that quadrature. In other cases they are regressive, and by the excess of the regress above the progress, they are monthly transferred *in antecedentia.*

Cor. 1. Hence if from P and M, the extreme points of a least arc PM, on the line Qq joining the quadratures we let fall the perpendiculars PK, Mk, and produce the same till they cut the line of the nodes Nn in D and d, the horary motion of the nodes will be as the area MPDd, and the square of the line AZ conjunctly. For let PK, PH, and AZ, be the three said sines, viz., PK the sine of the distance of the moon from the quadra-

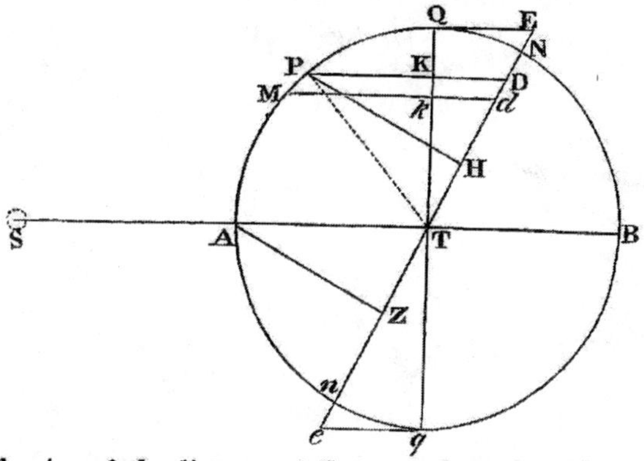

ture, PH the sine of the distance of the moon from the node, and AZ the sine of the distance of the node from the sun; and the velocity of the node will be as the solid content of PK × PH × AZ. But PT is to PK as PM to K*k*; and, therefore, because PT and PM are given, K*k* will be as PK. Likewise AT is to PD as AZ to PH, and therefore PH is as the rectangle PD × AZ; and, by compounding those proportions, PK × PH is as the solid content K*k* × PD × AZ, and PK × PH × AZ as K*k* × PD × AZ^2; that is, as the area PD*d*M and AZ^2 conjunctly. Q.E.D.

Cor. 2. In any given position of the nodes their mean horary motion is half their horary motion in the moon's syzygies; and therefore is to 16″ 35‴ 16^{iv}. 36^{v}. as the square of the sine of the distance of the nodes from the syzygies to the square of the radius, or as AZ^2 to AT^2. For if the moon, by an uniform motion, describes the semi-circle QA*q*, the sum of all the areas PD*d*M, during the time of the moon's passage from Q to M, will make up the area QM*d*E, terminating at the tangent QE of the circle; and by the time that the moon has arrived at the point *n*, that sum will make up the whole area EQA*n* described by the line PD: but when the moon proceeds from *n* to *q*, the line PD will fall without the circle, and describe the area *nqe*, terminating at the tangent *qe* of the circle, which area, because the nodes were before regressive, but are now progressive, must be subducted from the former area, and, being itself equal to the area QEN, will leave the semi-circle NQA*n*. While, therefore, the moon describes a semi-circle, the sum of all the areas PD*d*M will be the area of that semi-circle; and while the moon describes a complete circle, the sum of those areas will be the area of the whole circle. But the area PD*d*M, when the moon is in the syzygies, is the rectangle of the arc PM into the radius PT; and the sum of all the areas, *every one* equal to this area, in the time that the moon describes a complete circle, is the rectangle of the whole circumference into the radius of the circle; and this rectangle, being double the area of the circle, will be double the quantity of the former sum.

If, therefore, the nodes went on with that velocity uniformly continued which they acquire in the moon's syzygies, they would describe a space double of that which they describe in fact; and, therefore, the mean motion, by which, if uniformly continued, they would describe the same space with that which they do in fact describe by an unequal motion, is *but* one-half of that motion which they are possessed of in the moon's syzygies. Wherefore since their greatest horary motion, if the nodes are in the quadratures, is 33″ 10‴ 33^{iv}. 12^{v}. their mean horary motion in this case will be 16″ 35‴ 16^{iv}. 36^{v}. And seeing the horary motion of the nodes is every where as AZ^2 and the area PD*d*M conjunctly, and, therefore, in the moon's syzygies, the horary motion of the nodes is as AZ^2 and the area PD*d*M conjunctly, that is (because the area PD*d*M described in the syzygies is given), as AZ^2, therefore the mean motion also will be as AZ^2; and, therefore, when the nodes are without the quadratures, this motion will be to 16″ 35‴ 16^{iv}. 36^{v}. as AZ^2 to AT^2. Q.E.D.

PROPOSITION XXXI. PROBLEM XII.

To find the horary motion of the nodes of the moon in an elliptic orbit.

Let Q*pmaq* represent an ellipsis described with the greater axis Q*q*, and the lesser axis *ab*; QA*q*B a circle circumscribed; T the earth in the common centre of both; S the sun; *p* the moon moving in this ellipsis; and

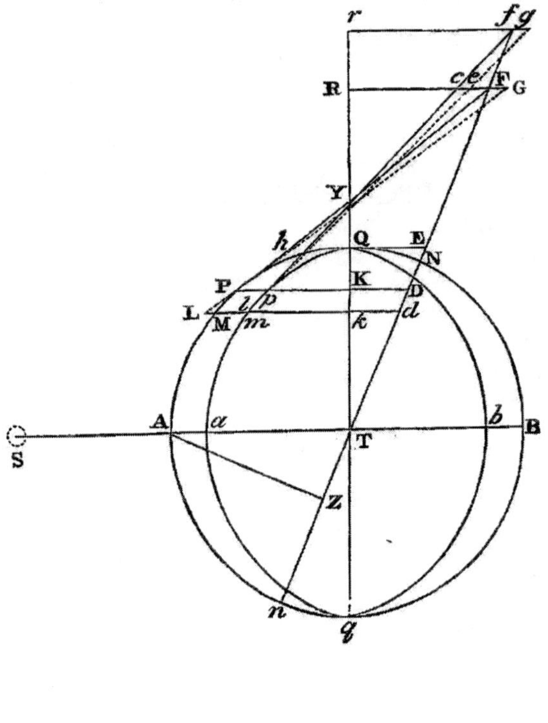

cY to fY × cp, or as fp to fY, and cY to cp; that is, if ph parallel to TN meet FP in h, as Fh to FY and FY to FP; that is, as Fh to FP or Dp to DP, and therefore as the area Dpmd to the area DPMd. And, therefore, seeing (by Corol. 1, Prop. XXX) the latter area and AZ^2 conjunctly are proportional to the horary motion of the nodes in the circle, the former area and AZ^2 conjunctly will be proportional to the horary motion of the nodes in the ellipsis. Q.E.D.

Cor. Since, therefore, in any given position of the nodes, the sum of all the areas pDdm, in the time while the moon is carried from the quadrature to any place m, is the area mpQEd terminated at the tangent of the ellipsis QE; and the sum of all those areas, in one entire revolution, is the area of the whole ellipsis; the mean motion of the nodes in the ellipsis will be to the mean motion of the nodes in the circle as the ellipsis to the circle; that is, as Ta to TA, or 69 to 70. And, therefore, since (by Corol 2, Prop. XXX) the mean horary motion of the nodes in the circle is to $16''\ 35'''\ 16^{iv}.\ 36^{v}.$ as AZ^2 to AT^2, if we take the angle $16''\ 21'''\ 3^{iv}.\ 30^{v}.$ to the angle $16''\ 35'''\ 16^{iv}.\ 36^{v}.$ as 69 to 70, the mean horary motion of the nodes in the ellipsis will be to $16''\ 21'''\ 3^{iv}.\ 30^{v}.$ as AZ^2 to AT^2; that is, as the square of the sine of the distance of the node from the sun to the square of the radius.

But the moon, by a radius drawn to the earth, describes the area in the syzygies with a greater velocity than it does that in the quadratures, and upon that account the time is contracted in the syzygies, and prolonged in the quadratures; and together with the time the motion of the nodes is likewise augmented or diminished. But the moment of the area in the quadrature of the moon was to the moment thereof in the syzygies as 10973 to 11073; and therefore the mean moment in the octants is to the excess in the syzygies, and to the defect in the quadratures, as 11023, the half sum of those numbers, to their half difference 50. Wherefore since the time of the moon in the several little equal parts of its orbit is reciprocally as its velocity, the mean time in the octants will be to the excess of the time in the quadratures, and to the defect *of the time* in the syzygies arising from this cause, nearly as 11023 to 50. But, reckoning from the quadratures to the syzygies, I find that the excess of the moments of the area, in the several places above the least moment in the quadratures, is nearly as the square of the sine of the moon's distance from the quadratures; and therefore the difference betwixt the moment in any place, and the mean moment in the octants, is as the difference betwixt the square of the sine of the moon's distance from the quadratures, and the square of the sine of 45 degrees, or half the square of the radius; and the increment of the time in the several places between the octants and quadratures, and the decrement thereof between the octants and syzygies, is in the same proportion. But the motion of the nodes, while the moon describes the several little equal parts of its orbit, is accelerated or retarded

cY to fY × cp, or as fp to fY, and cY to cp; that is, if ph parallel to TN meet FP in h, as Fh to FY and FY to FP; that is, as Fh to FP or Dp to DP, and therefore as the area Dpmd to the area DPMd. And, therefore, seeing (by Corol. 1, Prop. XXX) the latter area and AZ^2 conjunctly are proportional to the horary motion of the nodes in the circle, the former area and AZ^2 conjunctly will be proportional to the horary motion of the nodes in the ellipsis. Q.E.D.

COR. Since, therefore, in any given position of the nodes, the sum of all the areas pDdm, in the time while the moon is carried from the quadrature to any place m, is the area mpQEd terminated at the tangent of the ellipsis QE; and the sum of all those areas, in one entire revolution, is the area of the whole ellipsis; the mean motion of the nodes in the ellipsis will be to the mean motion of the nodes in the circle as the ellipsis to the circle; that is, as Ta to TA, or 69 to 70. And, therefore, since (by Corol 2, Prop. XXX) the mean horary motion of the nodes in the circle is to 16″ 35‴ 16^{iv}. 36^{v}. as AZ^2 to AT^2, if we take the angle 16″ 21‴ 3^{iv}. 30^{v}. to the angle 16″ 35‴ 16^{iv}. 36^{v}. as 69 to 70, the mean horary motion of the nodes in the ellipsis will be to 16″ 21‴ 3^{iv}. 30^{v}. as AZ^2 to AT^2; that is, as the square of the sine of the distance of the node from the sun to the square of the radius.

But the moon, by a radius drawn to the earth, describes the area in the syzygies with a greater velocity than it does that in the quadratures, and upon that account the time is contracted in the syzygies, and prolonged in the quadratures; and together with the time the motion of the nodes is likewise augmented or diminished. But the moment of the area in the quadrature of the moon was to the moment thereof in the syzygies as 10973 to 11073; and therefore the mean moment in the octants is to the excess in the syzygies, and to the defect in the quadratures, as 11023, the half sum of those numbers, to their half difference 50. Wherefore since the time of the moon in the several little equal parts of its orbit is reciprocally as its velocity, the mean time in the octants will be to the excess of the time in the quadratures, and to the defect *of the time* in the syzygies arising from this cause, nearly as 11023 to 50. But, reckoning from the quadratures to the syzygies, I find that the excess of the moments of the area, in the several places above the least moment in the quadratures, is nearly as the square of the sine of the moon's distance from the quadratures; and therefore the difference betwixt the moment in any place, and the mean moment in the octants, is as the difference betwixt the square of the sine of the moon's distance from the quadratures, and the square of the sine of 45 degrees, or half the square of the radius; and the increment of the time in the several places between the octants and quadratures, and the decrement thereof between the octants and syzygies, is in the same proportion. But the motion of the nodes, while the moon describes the several little equal parts of its orbit, is accelerated or retarded in the duplicate proportion of the time; for that motion, while the moon describes PM, is (*cæteris paribus*) as ML, and ML is in the duplicate proportion of the time. Wherefore the motion of the nodes in the syzygies, in the time while the moon describes given little parts of its orbit, is diminished in the duplicate proportion of the number 11073 to the number 11023; and the decrement is to the remaining motion as 100 to 10973; but to the whole motion as 100 to 11073 nearly. But the decrement in the places between the octants and syzygies, and the increment in the places between the octants and quadratures, is to this decrement nearly as the whole motion in these places to the whole motion in the syzygies, and the difference betwixt the square of the sine of the moon's distance from the quadrature, and the half square of the radius, to the half square of the radius conjunctly. Wherefore, if the nodes are in the quadratures, and we take two places, one on one side, one on the other, equally distant from the octant and other two distant by the same interval, one from the syzygy, the other from the quadrature, and from the decrements of the motions in the two places between the syzygy and octant we subtract the increments of the motions in the two other places between the octant and the quadrature, the remaining decrement will be equal to the decrement in the syzygy, as will easily appear by computation; and therefore the mean decrement, which ought to be subducted from the mean motion of the nodes, is the fourth part of the decrement in the syzygy. The whole horary motion of the nodes in the syzygies (when the moon by a radius drawn to the earth was supposed to describe an area proportional to the time) was $32'' \, 42''' \, 7^{iv}$. And we have shewn that the decrement of the motion of the nodes, in the time while the moon, now moving with greater velocity, describes the same space, was to this motion as 100 to 11073; and therefore this decrement is $17''' \, 43^{iv}. \, 11^{v}$. The fourth part of which $4''' \, 25^{iv}. \, 48^{v}$. subtracted from the mean horary motion above found, $16'' \, 21''' \, 3^{iv}. \, 30^{v}$. leaves $16'' \, 16''' \, 37^{iv}. \, 42^{v}$. their correct mean horary motion.

If the nodes are without the quadratures, and two places are considered, one on one side, one on the other, equally distant from the syzygies, the sum of the motions of the nodes, when the moon is in those places, will be to the sum of their motions, when the moon is in the same places and the nodes in the quadratures, as AZ^2 to AT^2. And the decrements of the motions arising from the causes but now explained will be mutually as the motions themselves, and therefore the remaining motions will be mutually betwixt themselves as AZ^2 to AT^2; and the mean motions will be as the remaining motions. And, therefore, in any given position of the nodes, their correct mean horary motion is to $16'' \, 16''' \, 37^{iv}. \, 42^{v}$. as AZ^2 to AT^2; that is, as the square of the sine of the distance of the nodes from the syzygies to the square of the radius.

PROPOSITION XXXII. PROBLEM XIII.

To find the mean motion of the nodes of the moon.

The yearly mean motion is the sum of all the mean horary motions throughout the course of the year. Suppose that the node is in N, and that, after every hour is elapsed, it is drawn back again to its former place; so that, notwithstanding its proper motion, it may constantly remain in the same situation with respect to the fixed stars; while in the mean time the sun S, by the motion of the earth, is seen to leave the node, and to proceed till it completes its apparent annual course by an uniform motion. Let A*a* represent a given least arc, which the right line TS always drawn to the sun, by its intersection with the circle NA*n*, describes in the least given moment of time; and the mean horary motion (from what we have above shewn) will be as AZ^2, that is (because AZ and ZY are proportional), as the rectangle of AZ into ZY, that is, as the area AZY*a*; and the sum of all the mean horary motions from the beginning will be as the sum of all the areas *a*YZA, that is, as the area NAZ. But the greatest AZY*a* is equal to the rectangle of the arc A*a* into the radius of the circle; and therefore the sum of all these rectangles in the whole circle will be to the like sum of all the greatest rectangles as the area of the whole circle to the rectangle of the whole circumference into the radius, that is, as 1 to 2. But the horary motion corresponding to that greatest rectangle was 16″ 16‴ 37^{iv}. 42^{v}. and this motion in the complete course of the sidereal year, 365^{d}. 6^{h}. 9′, amounts to 39° 38′ 7″ 50‴, and therefore the half thereof, 19° 49′ 3″ 55‴, is the mean motion of the nodes corresponding to the whole circle. And the motion of the nodes, in the time while the sun is carried from N to A, is to 19° 49′ 3″ 55‴ as the area NAZ to the whole circle.

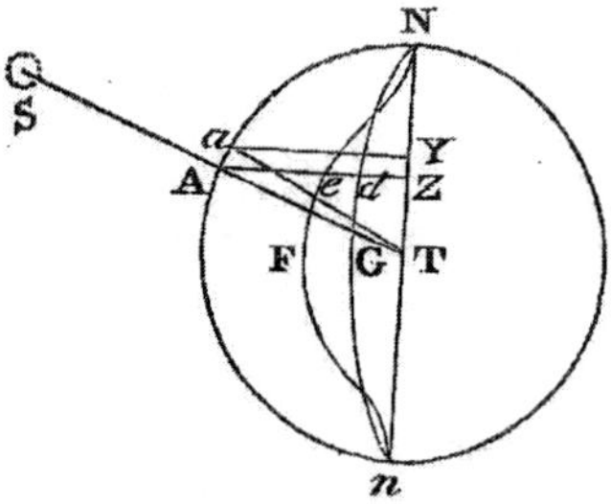

Thus it would be if the node was after every hour drawn back again to its former place, that so, after a complete revolution, the sun at the year's end would be found again in the same node which it had left when the year begun. But, because of the motion of the node in the mean time, the sun must needs meet the node sooner; and now it remains that we compute the abbreviation of the time. Since, then, the sun, in the course of the year, travels 360 degrees, and the node in the same time by its greatest motion would be carried 39° 38′ 7″ 50‴, or 39,6355 degrees; and the mean motion of the node in any place N is to its mean motion in its quadratures as AZ^2 to AT^2; the motion of the sun will be to the motion of the node in N as $360AT^2$ to $39,6355AZ^2$; that is, as $9,0827646AT^2$ to AZ^2. Wherefore if we suppose the circumference N*An* of the whole circle to be divided into little equal parts, such as A*a*, the time in which the sun would describe the little arc A*a*, if the circle was quiescent, will be to the time of which it would describe the same arc, supposing the circle together with the nodes to be revolved about the centre T, reciprocally as $9,0827646AT^2$ to $9,0827646AT^2 + AZ^2$; for the time is reciprocally as the velocity with which the little arc is described, and this velocity is the sum of the velocities of both sun and node. If, therefore, the sector NTA represent the time in which the sun by itself, without the motion of the node, would describe the arc NA, and the indefinitely small part AT*a* of the sector represent the little moment of the time in which it would describe the least arc A*a*; and (letting fall *a*Y perpendicular upon N*n*) if in AZ we take *d*Z of such length that the rectangle of *d*Z into ZY may be to the least part AT*a* of the sector as AZ^2 to $9,0827646AT^2 + AZ^2$, that is to say, that *d*Z may be to $\frac{1}{2}AZ$ as AT^2 to $9,0827646AT^2 + AZ^2$; the rectangle of *d*Z into ZY will represent the decrement of the time arising from the motion of the node, while the arc A*a* is described; and if the curve N*d*G*n* is the locus where the point *d* is always found, the curvilinear area N*d*Z will be as the whole decrement *of time* while the whole arc NA is described; and, therefore, the excess of the sector NAT above the area N*d*Z will be as the whole time. But because the motion of the node in a less time is less in proportion of the time, the area A*a*YZ must also be diminished in the same proportion; which may be done by taking in AZ the line *e*Z of such length, that it may be to the length of AZ as AZ^2 to $9,0827646AT^2 + AZ^2$; for so the rectangle of *e*Z into ZY will be to the area AZY*a* as the decrement of the time in which the arc A*a* is described to the whole time in which it would have been described, if the node had been quiescent; and, therefore, that rectangle will be as the decrement of the motion of the node. And if the curve N*e*F*n* is the locus of the point *e*, the whole area N*e*Z, which is the sum of all the decrements *of that motion*, will be as the whole decrement *thereof* during the time in which the arc AN is described; and the remaining area NA*e* will be as the remaining motion, which is the true motion of the node, during the time in which the whole arc NA is described by the joint motions of both sun and node. Now the area of the semi-circle is to the area of the figure N*e*F*n* found by the method of infinite series nearly as 793 to 60. But the motion corresponding *or proportional* to the whole circle was 19° 49′ 3″ 55‴; and therefore the motion corresponding to double the figure N*e*F*n* is 1° 29′ 58″ 2‴, which taken from the former motion leaves 18° 19′ 5″ 53‴, the whole motion of the node with respect to the fixed stars in the interval between two of its conjunctions with the sun; and this motion subducted from the annual motion of the sun 360°, leaves 341° 40′ 54″ 7‴, the motion of the sun in the interval between the same conjunctions. But as this motion is to the annual motion 360°, so is the motion of the node but just now found 18° 19′ 5″ 53‴ to its annual motion, which will therefore be 19° 18′ 1″ 23‴; and this is the mean motion of the nodes in the sidereal year. By astronomical tables, it is 19° 21′ 21″ 50‴. The difference is less than $\frac{1}{300}$ part of the whole motion, and seems to arise from the eccentricity of the moon's orbit, and its inclination to the plane of the ecliptic. By the eccentricity of this orbit the motion of the nodes is too much accelerated; and, on the other hand, by the inclination of the orbit, the motion of the nodes is something retarded, and reduced to its just velocity.

PROPOSITION XXXIII. PROBLEM XIV.

To find the true motion of the nodes of the moon.

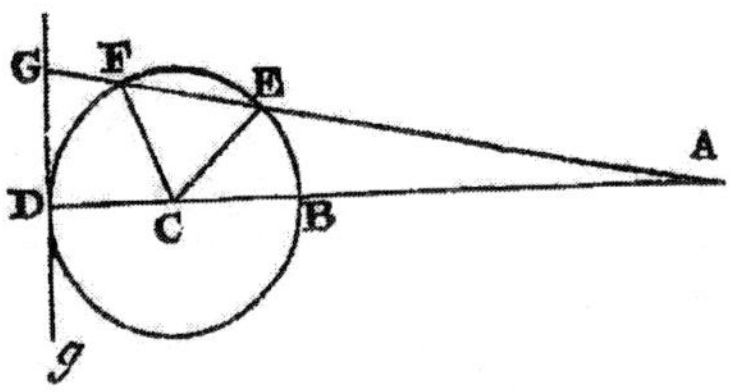

In the time which is as the area NTA—N*d*Z (in the preceding Fig.) that motion is as the area NA*e*, and is thence given; but because the calculus is too difficult, it will be better to use the following construction of the Problem. About the centre C, with any interval CD, describe the circle BEFD; produce DC to A so as AB may be to AC as the mean motion to half the mean true motion when the nodes are in their quadratures (that is, as 19° 18′ 1″ 23‴ to 19° 49′ 3″ 55‴; and therefore BC to AC as the difference of those motions 0° 31′ 2″ 32‴ to the latter motion 19° 49′ 3″ 55‴, that is, as 1 to $38\frac{3}{10}$). Then through the point D draw the indefinite line G*g*, touching the circle in D; and if we take the angle BCE, or BCF, equal to the double distance of the sun from the place of the node, as found by the mean motion, and drawing AE or AF cutting the perpendicular DG in G, we take another angle which shall be to the whole motion of the node in the interval between its syzygies (that is, to 9° 11′ 3″) as the tangent DG to the whole circumference of the circle BED, and add this *last* angle (for which the angle DAG may be used) to the mean motion of the nodes, while they are passing from the quadratures to the syzygies, and subtract it from their mean motion while they are passing from the syzygies to the quadratures, we shall have their true motion; for the true motion so found will nearly agree with the true motion which comes out from assuming the times as the area NTA—N*d*Z, and the motion of the node as the area NA*e*; as whoever will please to examine and make the computations will find: and this is the semi-menstrual equation of the motion of the nodes. But there is also a menstrual equation, but which is by no means necessary for finding of the moon's latitude; for since the variation of the inclination of the moon's orbit to the plane of the ecliptic is liable to a twofold inequality, the one semi-menstrual, the other menstrual, the menstrual inequality *of this variation*, and the menstrual equation of the nodes, so moderate and correct each other, that in computing the latitude of the moon both may be neglected.

Cor. From this and the preceding Prop. it appears that the nodes are quiescent in their syzygies, but regressive in their quadratures, by an hourly motion of 16″ 19‴ 26^{iv}.; and that the equation of the motion of the nodes in the octants is 1° 30′; all which exactly agree with the phænomena of the heavens.

SCHOLIUM.

Mr. *Machin*, Astron., Prof. Gresh., and Dr. *Henry Pemberton*, separately found out the motion of the nodes by a different method. Mention has been made of this method in another place. Their several papers, both of which I have seen, contained two Propositions, and exactly agreed with each other in both of them. Mr. *Machin's* paper coming first to my hands, I shall here insert it.

OF THE MOTION OF THE MOON'S NODES.

"PROPOSITION I.

" *The mean motion of the sun from the node is defined by a geometric mean proportional between the mean motion of the sun and that mean motion with which the sun recedes with the greatest swiftness from the node in the quadratures.*

"Let T be the earth's place, N*n* the line of the moon's nodes at any given time, KTM a perpendicular thereto, TA a right line revolving about the centre with the same angular velocity with which the sun and the node recede from one another, in such sort that the angle between the quiescent right line N*n* and the revolving line TA may be always equal to the distance of the places of the sun and node. Now if any right line TK be divided into parts TS and SK, and those parts be taken as the mean horary motion of the sun to the mean horary motion of the node in the quadratures, and there be taken the right line TH, a mean proportional between the part TS and the whole TK, this right line will be proportional to the sun's mean motion from the node.

"For let there be described the circle NK*n*M from the centre T and with the radius TK, and about the same centre, with the semi-axis TH

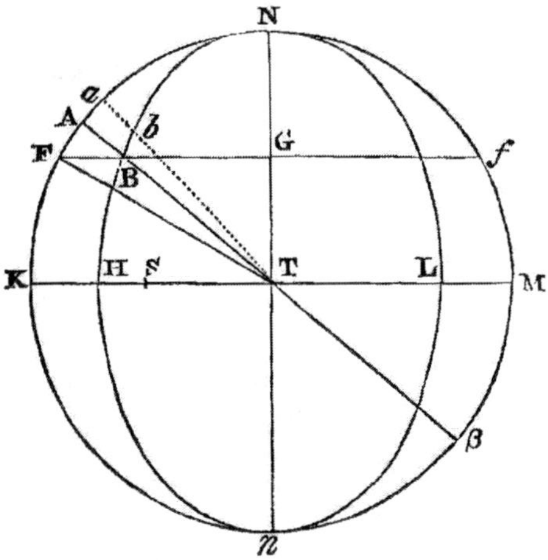

and TN, let there be described an ellipsis NHnL; and in the time in which the sun recedes from the node through the arc Na, if there be drawn, the right line Tba, the area of the sector NTa will be the exponent of the sum of the motions of the sun and node in the same time. Let, therefore, the extremely small arc aA be that which the right line Tba, revolving according to the aforesaid law, will uniformly describe in a given particle of time, and the extremely small sector TAa will be as the sum of the velocities with which the sun and node are carried two different ways in that time. Now the sun's velocity is almost uniform, its inequality being so small as scarcely to produce the least inequality in the mean motion of the nodes. The other part of this sum, namely, the mean quantity of the velocity of the node, is increased in the recess from the syzygies in a duplicate ratio of the sine of its distance from the sun (by Cor. Prop. XXXI, of this Book), and, being greatest in its quadratures with the sun in K, is in the same ratio to the sun's velocity as SK to TS, that is, as (the difference of the squares of TK and TH, or) the rectangle KHM to TH^2. But the ellipsis NBH divides the sector ATa, the exponent of the sum of these two velocities, into two parts ABba and BTb, proportional to the velocities. For produce BT to the circle in β, and from the point B let fall upon the greater axis the perpendicular BG, which being produced both ways may meet the circle in the points F and f; and because the space ABba is to the sector TBb as the rectangle ABβ to BT^2 (that rectangle being equal to the difference of the squares of TA and TB, because the right line Aβ is equally cut in T, and unequally in B), therefore when the space ABba is the greatest of all in K, this ratio will be the same as the ratio of the rectangle KHM to HT^2. But the greatest mean velocity of the node was shewn above to be in that very ratio to the velocity of the sun; and therefore in the quadratures the sector ATa is divided into parts proportional to the velocities. And because the rectangle KHM is to HT^2 as FBf to BG^2, and the rectangle ABβ is equal to the rectangle FBf, therefore the little area ABba, where it is greatest, is to the remaining sector TBb as the rectangle ABβ to BG^2. But the ratio of these little areas always was as the rectangle ABβ to BT^2; and therefore the little area ABba in the place A is less than its correspondent little area in the quadratures in the duplicate ratio of BG to BT, that is, in the duplicate ratio of the sine of the sun's distance from the node. And therefore the sum of all the little areas ABba, to wit, the space ABN, will be as the motion of the node in the time in which the sun hath been going over the arc NA since he left the node; and the remaining space, namely, the elliptic sector NTB, will be as the sun's mean motion in the same time. And because the mean annual motion of the node is that motion which it performs in the time that the sun completes one period of its course, the mean motion of the node from the sun will be to the mean motion of the sun itself as the area of the circle to the area of the ellipsis; that is, as the right line TK to the right line TH, which is a mean proportional between TK and TS; or, which comes to the same as the mean proportional TH to the right line TS.

"PROPOSITION II.

" *The mean motion of the moon's nodes being given, to find their true motion.*

" Let the angle A be the distance of the sun from the mean place of the node, or the sun's mean motion from the node. Then if we take the angle B, whose tangent is to the tangent of the angle A as TH to TK, that is,

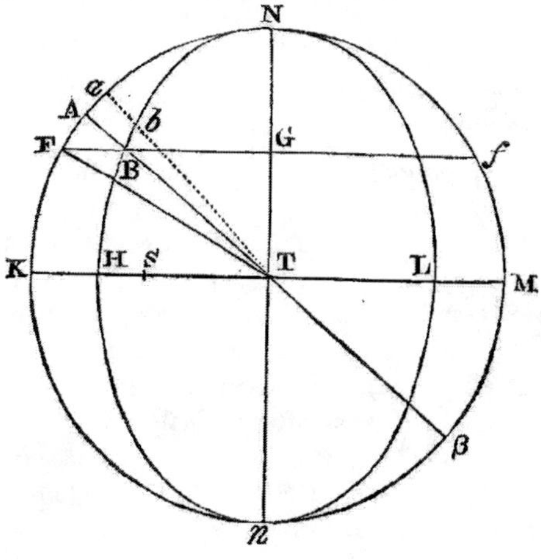

in the sub-duplicate ratio of the mean horary motion of the sun to the mean horary motion of the sun from the node, when the node is in the quadrature, that angle B will be the distance of the sun from the node's true place. For join FT, and, by the demonstration of the last Proposition, the angle FTN will be the distance of the sun from the mean place of the node, and the angle ATN the distance from the true place, and the tangents of these angles are between themselves as TK to TH.

"Cor. Hence the angle FTA is the equation of the moon's nodes; and the sine of this angle, where it is greatest in the octants, is to the radius as KH to TK + TH. But the sine of this equation in any other place A is to the greatest sine as the sine of the sums of the angles FTN + ATN to the radius; that is, nearly as the sine of double the distance of the sun from the mean place of the node (namely, 2FTN) to the radius.

"SCHOLIUM.

"If the mean horary motion of the nodes in the quadratures be 16″ 16‴ 37^{iv}. 42^{v}. that is, in a whole sidereal year, 39° 38′ 7″ 50‴, TH will be to TK in the subduplicate ratio of the number 9,0827646 to the number 10,0827646, that is, as 18,6524761 to 19,6524761. And, therefore, TH is to HK as 18,6524761 to 1; that is, as the motion of the sun in a sidereal year to the mean motion of the node 19° 18′ 1″ 23⅔‴.

"But if the mean motion of the moon's nodes in 20 Julian years is 386° 50′ 15″, as is collected from the observations made use of in the theory of the moon, the mean motion of the nodes in one sidereal year will be 19° 20′ 31″ 58‴. and TH will be to HK as 360° to 19° 20′ 31′ 58‴; that is, as 18,61214 to 1: and from hence the mean horary motion of the nodes in the quadratures will come out 16″ 18‴ 48^{iv}. And the greatest equation of the nodes in the octants will be 1° 29′ 57″."

PROPOSITION XXXIV. PROBLEM XV.

To find the horary variation of the inclination of the moon's orbit to the plane of the ecliptic.

Let A and *a* represent the syzygies; Q and *q* the quadratures; N and *n* the nodes; P the place of the moon in its orbit; *p* the orthographic projection of that place upon the plane of the ecliptic; and *m*T*l* the momentaneous motion of the nodes as above. If upon T*m* we let fall the perpendicular PG, and joining *p*G we produce it till it meet T*l* in *g*, and join also P*g*, the angle PG*p* will be the inclination of the moon's orbit to the plane of the ecliptic when the moon is in P; and the angle P*gp* will be the inclination of the same after a small moment of time is elapsed; and therefore the angle GP*g* will be the momentaneous variation of the inclination. But this angle GP*g* is to the angle GT*g* as TG to PG and P*p* to PG conjunctly. And, therefore, if for the moment of time we as-

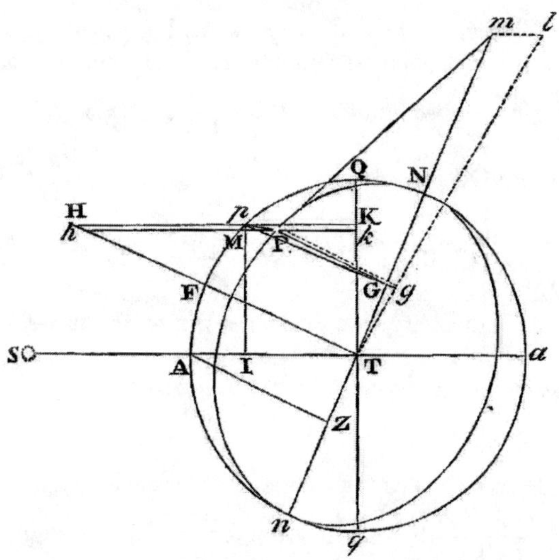

sume an hour, since the angle GTg (by Prop. XXX) is to the angle 33″ 10‴ 33^{iv}. as IT × PG × AZ to AT³, the angle GPg (or the horary variation of the inclination) will be to the angle 33″ 10‴ 33^{iv}. as IT × AZ × TG × $\frac{Pp}{PG}$ to AT³. Q.E.I.

And thus it would be if the moon was uniformly revolved in a circular orbit. But if the orbit is elliptical, the mean motion of the nodes will be diminished in proportion of the lesser axis to the greater, as we have shewn above; and the variation of the inclination will be also diminished in the same proportion.

Cor. 1. Upon Nn erect the perpendicular TF, and let pM be the horary motion of the moon in the plane of the ecliptic; upon QT let fall the perpendiculars pK, Mk, and produce them till they meet TF in H and h; then IT will be to AT as Kk to Mp; and TG to Hp as TZ to AT; and, therefore, IT × TG will be equal to $\frac{Kk \times Hp \times TZ}{Mp}$, that is, equal to the area HpMh multiplied into the ratio $\frac{TZ}{Mp}$: and therefore the horary variation of the inclination will be to 33″ 10‴ 33^{iv}. as the area HpMh multiplied into AZ × $\frac{TZ}{Mp}$ × $\frac{Pp}{PG}$ to AT³.

Cor. 2. And, therefore, if the earth and nodes were after every hour drawn back from their new and instantly restored to their old places, so as their situation might continue given for a whole periodic month together, the whole variation of the inclination during that month would be to 33″

$10''' \; 33^{iv}$. as the aggregate of all the areas HpMh, generated in the time of one revolution of the point p (with due regard in summing to their proper signs + —), multiplied into $AZ \times TZ \times \frac{Pp}{PG}$ to $Mp \times AT^3$; that is, as the whole circle QAqa multiplied into $AZ \times TZ \times \frac{Pp}{PG}$ to $Mp \times AT^3$, that is, as the circumference QAqa multiplied into $AZ \times TZ \times \frac{Pp}{PG}$ to $2Mp \times AT^2$.

Cor. 3. And, therefore, in a given position of the nodes, the mean horary variation, from which, if uniformly continued through the whole month, that menstrual variation might be generated, is to $33'' \; 10''' \; 33^{iv}$. as $AZ \times TZ \times \frac{Pp}{PG}$ to $2AT^2$, or as $Pp \times \frac{AZ \times TZ}{\frac{1}{2}AT}$ to $PG \times 4AT$; that is (because Pp is to PG as the sine of the aforesaid inclination to the radius, and $\frac{AZ \times TZ}{\frac{1}{2}AT}$ to 4AT as the sine of double the angle ATn to four times the radius), as the sine of the same inclination multiplied into the sine of double the distance of the nodes from the sun to four times the square of the radius.

Cor. 4. Seeing the horary variation of the inclination, when the nodes are in the quadratures, is (by this Prop.) to the angle $33'' \; 10''' \; 33^{iv}$. as $IT \times AZ \times TG \times \frac{Pp}{PG}$ to AT^3, that is, as $\frac{IT \times TG}{\frac{1}{2}AT} \times \frac{Pp}{PG}$ to 2AT, that is, as the sine of double the distance of the moon from the quadratures multiplied into $\frac{Pp}{PG}$ to twice the radius, the sum of all the horary variations during the time that the moon, in this situation of the nodes, passes from the quadrature to the syzygy (that is, in the space of $177\frac{1}{6}$ hours) will be to the sum of as many angles $33'' \; 10''' \; 33^{iv}$. or 5878'', as the sum of all the sines of double the distance of the moon from the quadratures multiplied into $\frac{Pp}{PG}$ to the sum of as many diameters; that is, as the diameter multiplied into $\frac{Pp}{PG}$ to the circumference; that is, if the inclination be 5° 1', as $7 \times \frac{874}{10000}$ to 22, or as 278 to 10000. And, therefore, the whole variation, composed out of the sum of all the horary variations in the aforesaid time, is 163'', or 2' 43''.

PROPOSITION XXXV. PROBLEM XVI.

To a given time to find the inclination of the moon's orbit to the plane of the ecliptic.

Let AD be the sine of the greatest inclination, and AB the sine of the least. Bisect BD in C; and round the centre C, with the interval BC, describe the circle BGD. In AC take CE in the same proportion to EB

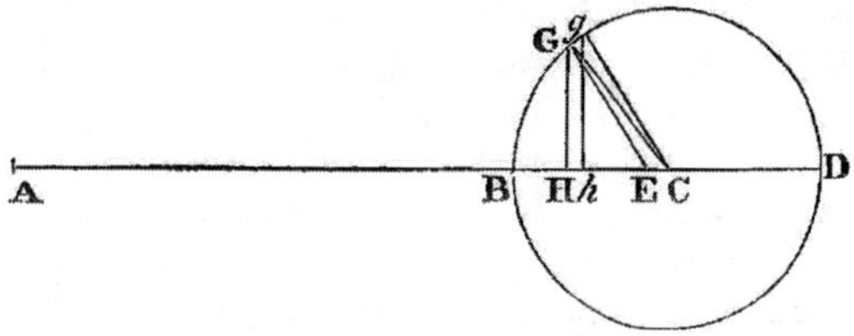

as EB to twice BA. And if to the time given we set off the angle AEG equal to double the distance of the nodes from the quadratures, and upon AD let fall the perpendicular GH, AH will be the sine of the inclination required.

For GE^2 is equal to $GH^2 + HE^2 = BHD + HE^2 = HBD + HE^2 - BH^2 = HBD + BE^2 - 2BH \times BE = BE^2 + 2EC \times BH = 2EC \times AB + 2EC \times BH = 2EC \times AH$; wherefore since 2EC is given, GE^2 will be as AH. Now let AE*g* represent double the distance of the nodes from the quadratures, in a given moment of time after, and the arc G*g*, on account of the given angle GE*g*, will be as the distance GE. But H*h* is to G*g* as GH to GC, and, therefore, H*h* is as the rectangle GH × G*g*, or GH × GE, that is, as $\frac{GH}{GE} \times GE^2$, or $\frac{GH}{GE} \times AH$; that is, as AH and the sine of the angle AEG conjunctly. If, therefore, in any one case, AH be the sine of inclination, it will increase by the same increments as the sine of inclination doth, by Cor. 3 of the preceding Prop. and therefore will always continue equal to that sine. But when the point G falls upon either point B or D, AH is equal to this sine, and therefore remains always equal thereto. Q.E.D.

In this demonstration I have supposed that the angle BEG, representing double the distance of the nodes from the quadratures, increaseth uniformly; for I cannot descend to every minute circumstance of inequality. Now suppose that BEG is a right angle, and that G*g* is in this case the horary increment of double the distance of the nodes from the sun; then, by Cor. 3 of the last Prop. the horary variation of the inclination in the same case will be to $33'' \ 10''' \ 33^{iv}$. as the rectangle of AH, the sine of the inclination, into the sine of the right angle BEG, double the distance of the nodes from the sun, to four times the square of the radius; that is, as AH, the sine of the mean inclination, to four times the radius; that is, seeing the mean inclination is about 5° $8\frac{1}{2}$, as its sine 896 to 40000, the quadruple of the radius, or as 224 to 10000. But the whole variation corresponding to BD, the difference of the sines, is to this horary variation as the diameter BD to the arc G*g*, that is, conjunctly as the diameter BD to the semi-circumference BGD, and as the time of $2079\frac{7}{10}$ hours, in which the node proceeds from the quadratures to the syzygies, to one hour, that is, as 7 to 11, and $2079\frac{7}{10}$ to 1. Wherefore, compounding all these proportions, we shall have the whole variation BD to 33″ 10‴ 33^{iv}. as 224 × 7 × $2079\frac{7}{10}$ to 110000, that is, as 29645 to 1000; and from thence that variation BD will come out 16′ $23\frac{1}{2}$″.

And this is the greatest variation of the inclination, abstracting from the situation of the moon in its orbit; for if the nodes are in the syzygies, the inclination suffers no change from the various positions of the moon. But if the nodes are in the quadratures, the inclination is less when the moon is in the syzygies than when it is in the quadratures by a difference of 2′ 43″, as we shewed in Cor. 4 of the preceding Prop.; and the whole mean variation BD, diminished by 1′ $21\frac{1}{2}$″, the half of this excess, becomes 15′ 2″, when the moon is in the quadratures; and increased by the same, becomes 17′ 45″ when the moon is in the syzygies. If, therefore, the moon be in the syzygies, the whole variation in the passage of the nodes from the quadratures to the syzygies will be 17′ 45″; and, therefore, if the inclination be 5° 17′ 20″, when the nodes are in the syzygies, it will be 4° 59′ 35″ when the nodes are in the quadratures and the moon in the syzygies. The truth of all which is confirmed by observations.

Now if the inclination of the orbit should be required when the moon is in the syzygies, and the nodes any where between them and the quadratures, let AB be to AD as the sine of 4° 59′ 35″ to the sine of 5° 17′ 20″, and take the angle AEG equal to double the distance of the nodes from the quadratures; and AH will be the sine of the inclination desired. To this inclination of the orbit the inclination of the same is equal, when the moon is 90° distant from the nodes. In other situations of the moon, this menstrual inequality, to which the variation of the inclination is obnoxious in the calculus of the moon's latitude, is balanced, and in a manner took off, by the menstrual inequality of the motion of the nodes (as we said before), and therefore may be neglected in the computation of the said latitude.

SCHOLIUM.

By these computations of the lunar motions I was willing to shew that by the theory of gravity the motions of the moon could be calculated from their physical causes. By the same theory I moreover found that the annual equation of the mean motion of the moon arises from the various dilatation which the orbit of the moon suffers from the action of the sun, according to Cor. 6, Prop. LXVI, Book 1. The force of this action is greater in the perigeon sun, and dilates the moon's orbit; in the apogeon sun it is less, and permits the orbit to be again contracted. The moon moves slower in the dilated and faster in the contracted orbit; and the annual equation, by which this inequality is regulated, vanishes in the apogee and perigee of the sun. In the mean distance of the sun from the earth it arises to about 11′ 50″; in other distances of the sun it is proportional to the equation of the sun's centre, and is added to the mean motion of the moon, while the earth is passing from its aphelion to its perihelion, and subducted while the earth is in the opposite semi-circle. Taking for the radius of the *orbis magnus* 1000, and $16\frac{7}{8}$ for the earth's eccentricity, this equation, when of the greatest magnitude, by the theory of gravity comes out 11′ 49″. But the eccentricity of the earth seems to be something greater, and with the eccentricity this equation will be augmented in the same proportion. Suppose the eccentricity $16\frac{11}{12}$, and the greatest equation will be 11′ 51″.

Farther; I found that the apogee and nodes of the moon move faster in the perihelion of the earth, where the force of the sun's action is greater, than in the aphelion thereof, and that in the reciprocal triplicate proportion of the earth's distance from the sun; and hence arise annual equations of those motions proportional to the equation of the sun's centre. Now the motion of the sun is in the reciprocal duplicate proportion of the earth's distance from the sun; and the greatest equation of the centre which this inequality generates is 1° 56′ 20″, corresponding to the abovementioned eccentricity of the sun, $16\frac{11}{12}$. But if the motion of the sun had been in the reciprocal triplicate proportion of the distance, this inequality would have generated the greatest equation 2° 54′ 30″; and therefore the greatest equations which the inequalities of the motions of the moon's apogee and nodes do generate are to 2° 54′ 30″ as the mean diurnal motion of the moon's apogee and the mean diurnal motion of its nodes are to the mean diurnal motion of the sun. Whence the greatest equation of the mean motion of the apogee comes out 19′ 43″, and the greatest equation of the mean motion of the nodes 9′ 24″. The former equation is added, and the latter subducted, while the earth is passing from its perihelion to its aphelion, and contrariwise when the earth is in the opposite semi-circle.

By the theory of gravity I likewise found that the action of the sun upon the moon is something greater when the transverse diameter of the moon's orbit passeth through the sun than when the same is perpendicular upon the line which joins the earth and the sun; and therefore the moon's orbit is something larger in the former than in the latter case. And hence arises another equation of the moon's mean motion, depending upon the situation of the moon's apogee in respect of the sun, which is in its greatest quantity when the moon's apogee is in the octants of the sun, and vanishes when the apogee arrives at the quadratures or syzygies; and it is added to the mean motion while the moon's apogee is passing from the quadrature of the sun to the syzygy, and subducted while the apogee is passing from the syzygy to the quadrature. This equation, which I shall call the semi-annual, when greatest in the octants of the apogee, arises to about 3′ 45″, so far as I could collect from the phænomena: and this is its quantity in the mean distance of the sun from the earth. But it is increased and diminished in the reciprocal triplicate proportion of the sun's distance, and therefore is nearly 3′ 34″ when that distance is greatest, and 3′ 56″ when least. But when the moon's apogee is without the octants, it becomes less, and is to its greatest quantity as the sine of double the distance of the moon's apogee from the nearest syzygy or quadrature to the radius.

By the same theory of gravity, the action of the sun upon the moon is something greater when the line of the moon's nodes passes through the sun than when it is at right angles with the line which joins the sun and the earth; and hence arises another equation of the moon's mean motion, which I shall call the second semi-annual; and this is greatest when the nodes are in the octants of the sun, and vanishes when they are in the syzygies or quadratures; and in other positions of the nodes is proportional to the sine of double the distance of either node from the nearest syzygy or quadrature. And it is added to the mean motion of the moon, if the sun is *in antecedentia*, to the node which is nearest to him, and subducted if *in consequentia;* and in the octants, where it is of the greatest magnitude, it arises to 47″ in the mean distance of the sun from the earth, as I find from the theory of gravity. In other distances of the sun, this equation, greatest in the octants of the nodes, is reciprocally as the cube of the sun's distance from the earth; and therefore in the sun's perigee it comes to about 49″, and in its apogee to about 45″.

By the same theory of gravity, the moon's apogee goes forward at the greatest rate when it is either in conjunction with or in opposition to the sun, but in its quadratures with the sun it goes backward; and the eccentricity comes, in the former case, to its greatest quantity; in the latter to its least, by Cor. 7, 8, and 9, Prop. LXVI, Book I. And those inequalities, by the Corollaries we have named, are very great, and generate the principal which I call the semi-annual equation of the apogee; and this semi-annual equation in its greatest quantity comes to about 12° 18′, as nearly as I could collect from the phænomena. Our countryman, *Horrox*, was the first who advanced the theory of the moon's moving in an ellipsis about the earth placed in its lower focus. Dr. *Halley* improved the notion, by putting the centre of the ellipsis in an epicycle whose centre is uniformly revolved about the earth; and from the motion in this epicycle the mentioned inequalities in the progress and regress of the apogee, and in the quantity of eccentricity, do arise. Suppose the mean distance of the moon from the earth to be divided into 100000 parts, and let T represent the earth, and TC the moon's mean eccentricity of 5505 such parts. Produce TC to B, so as CB may be the sine of the greatest semi-annual equation 12° 18′ to the radius TC; and the circle BDA described about the centre C, with the interval CB, will be the epicycle spoken of, in which the centre of the moon's orbit is placed, and revolved according to the order of the letters BDA. Set off the angle BCD equal to twice the annual argument, or twice the distance of the sun's true place from the place of the moon's apogee once equated, and CTD will be the semi-annual equation of the moon's apogee, and TD the eccentricity of its orbit, tending to the place of the apogee now twice equated. But, having the moon's mean motion, the place of its apogee, and its eccentricity, as well as the longer axis of its orbit 200000, from these *data* the true place of the moon in its orbit, together with its distance from the earth, may be determined by the methods commonly known.

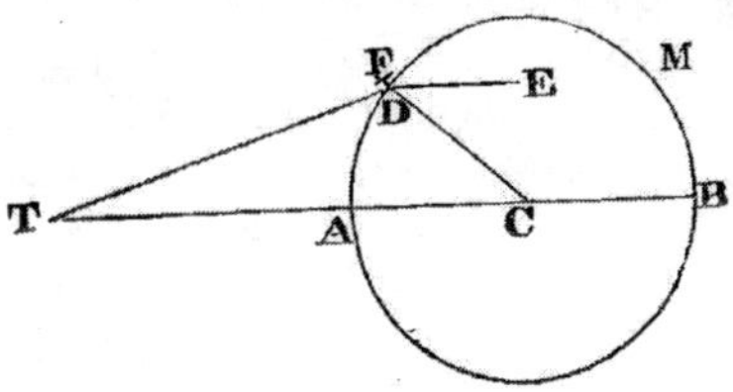

In the perihelion of the earth, where the force of the sun is greatest, the centre of the moon's orbit moves faster about the centre C than in the aphelion, and that in the reciprocal triplicate proportion of the sun's distance from the earth. But, because the equation of the sun's centre is included in the annual argument, the centre of the moon's orbit moves faster in its epicycle BDA, in the reciprocal duplicate proportion of the sun's distance from the earth. Therefore, that it may move yet faster in the reciprocal simple proportion of the distance, suppose that from D, the centre of the orbit, a right line DE is drawn, tending towards the moon's apogee once equated, that is, parallel to TC; and set off the angle EDF equal to the excess of the aforesaid annual argument above the distance of the moon's apogee from the sun's perigee *in consequentia;* or, which comes to the same thing, take the angle CDF equal to the complement of the sun's true anomaly to 360°; and let DF be to DC as twice the eccentricity of the *orbis magnus* to the sun's mean distance from the earth, and the sun's mean diurnal motion from the moon's apogee to the sun's mean diurnal motion from its own apogee conjunctly, that is, as $33\frac{7}{8}$ to 1000, and 52′ 27″ 16‴ to 59′ 8″ 10‴ conjunctly, or as 3 to 100; and imagine the centre of the moon's orbit placed in the point F to be revolved in an epicycle whose centre is D, and radius DF, while the point D moves in the circumference of the circle DABD: for by this means the centre of the moon's orbit comes to describe a certain curve line about the centre C, with a velocity which will be almost reciprocally as the cube of the sun's distance from the earth, as it ought to be.

The calculus of this motion is difficult, but may be rendered more easy by the following approximation. Assuming, as above, the moon's mean distance from the earth of 100000 parts, and the eccentricity TC of 5505 such parts, the line CB or CD will be found $1172\frac{3}{4}$, and DF $35\frac{1}{5}$ of those parts; and this line DF at the distance TC subtends the angle at the earth, which the removal of the centre of the orbit from the place D to the place F generates in the motion of this centre; and double this line DF in a parallel position, at the distance of the upper focus of the moon's orbit from the earth, subtends at the earth the same angle as DF did before, which that removal generates in the motion of this upper focus; but at the distance of the moon from the earth this double line 2DF at the upper focus, in a parallel position to the first line DF, subtends an angle at the moon, which the said removal generates in the motion of the moon, which angle may be therefore called the second equation of the moon's centre; and this equation, in the mean distance of the moon from the earth, is nearly as the sine of the angle which that line DF contains with the line drawn from the point F to the moon, and when in its greatest quantity amounts to 2′ 25″. But the angle which the line DF contains with the line drawn from the point F to the moon is found either by subtracting the angle EDF from the mean anomaly of the moon, or by adding the distance of the moon from the sun to the distance of the moon's apogee from the apogee of the sun; and as the radius to the sine of the angle thus found, so is 2′ 25″ to the second equation of the centre: to be added, if the forementioned sum be less than a semi-circle; to be subducted, if greater. And from the moon's place in its orbit thus corrected, its longitude may be found in the syzygies of the luminaries.

The atmosphere of the earth to the height of 35 or 40 miles refracts the sun's light. This refraction scatters and spreads the light over the earth's shadow; and the dissipated light near the limits of the shadow dilates the shadow. Upon which account, to the diameter of the shadow, as it comes out by the parallax, I add 1 or $1\frac{1}{3}$ minute in lunar eclipses.

But the theory of the moon ought to be examined and proved from the phænomena, first in the syzygies, then in the quadratures, and last of all in the octants; and whoever pleases to undertake the work will find it not amiss to assume the following mean motions of the sun and moon at the Royal Observatory of *Greenwich*, to the last day of *December* at noon, *anno* 1700, O.S. viz. The mean motion of the sun ♑ 20° 43′ 40″, and of its apogee ♋ 7° 44′ 30″; the mean motion of the moon ♒ 15° 21′ 00″; of its apogee, ♓ 8° 20′ 00″; and of its ascending node ♌ 27° 24′ 20″; and the difference of meridians betwixt the Observatory at *Greenwich* and the Royal Observatory at *Paris*, $0^h. 9' 20''$: but the mean motion of the moon and of its apogee are not yet obtained with sufficient accuracy.

PROPOSITION XXXVI. PROBLEM XVII.

To find the force of the sun to move the sea.

The sun's force ML or PT to disturb the motions of the moon, was (by Prop. XXV.) in the moon's quadratures, to the force of gravity with us, as 1 to 638092,6; and the force TM — LM or 2PK in the moon's syzygies is double that quantity. But, descending to the surface of the earth, these forces are diminished in proportion of the distances from the centre of the earth, that is, in the proportion of $60\frac{1}{2}$ to 1; and therefore the former force on the earth's surface is to the force of gravity as 1 to 38604600; and by this force the sea is depressed in such places as are 90 degrees distant from the sun. But by the other force, which is twice as great, the sea is raised not only in the places directly under the sun, but in those also which are directly opposed to it; and the sum of these forces is to the force of gravity as 1 to 12868200. And because the same force excites the same motion, whether it depresses the waters in those places which are 90 degrees distant from the sun, or raises them in the places which are directly under and directly opposed to the sun, the aforesaid sum will be the total force of the sun to disturb the sea, and will have the same effect as if the whole was employed in raising the sea in the places directly under and directly opposed to the sun, and did not act at all in the places which are 90 degrees removed from the sun.

And this is the force of the sun to disturb the sea in any given place, where the sun is at the same time both vertical, and in its mean distance from the earth. In other positions of the sun, its force to raise the sea is as the versed sine of double its altitude above the horizon of the place directly, and the cube of the distance from the earth reciprocally.

Cor. Since the centrifugal force of the parts of the earth, arising from the earth's diurnal motion, which is to the force of gravity as 1 to 289, raises the waters under the equator to a height exceeding that under the poles by 85472 *Paris* feet, as above, in Prop. XIX., the force of the sun, which we have now shewed to be to the force of gravity as 1 to 12868200, and therefore is to that centrifugal force as 289 to 12868200, or as 1 to 44527, will be able to raise the waters in the places directly under and directly opposed to the sun to a height exceeding that in the places which are 90 degrees removed from the sun only by one *Paris* foot and $113\frac{1}{30}$ inches; for this measure is to the measure of 85472 feet as 1 to 44527.

PROPOSITION XXXVII. PROBLEM XVIII.

To find the force of the moon to move the sea.

The force of the moon to move the sea is to be deduced from its proportion to the force of the sun, and this proportion is to be collected from the proportion of the motions of the sea, which are the effects of those forces. Before the mouth of the river *Avon*, three miles below *Bristol*, the height of the ascent of the water in the vernal and autumnal syzygies of the luminaries (by the observations of *Samuel Sturmy*) amounts to about 45 feet, but in the quadratures to 25 only. The former of those heights arises from the sum of the aforesaid forces, the latter from their difference. If, therefore, S and L are supposed to represent respectively the forces of the sun and moon while they are in the equator, as well as in their mean distances from the earth, we shall have L + S to L — S as 45 to 25, or as 9 to 5.

At *Plymouth* (by the observations of *Samuel Colepress*) the tide in its mean height rises to about 16 feet, and in the spring and autumn the height thereof in the syzygies may exceed that in the quadratures by more than 7 or 8 feet. Suppose the greatest difference of those heights to be 9 feet, and L + S will be to L — S as $20\frac{1}{2}$ to $11\frac{1}{2}$, or as 41 to 23; a proportion that agrees well enough with the former. But because of the great tide at *Bristol*, we are rather to depend upon the observations of *Sturmy;* and, therefore, till we procure something that is more certain, we shall use the proportion of 9 to 5.

But because of the reciprocal motions of the waters, the greatest tides do not happen at the times of the syzygies of the luminaries, but, as we have said before, are the third in order after the syzygies; or (reckoning from the syzygies) follow next after the third appulse of the moon to the meridian of the place after the syzygies; or rather (as *Sturmy* observes) are the third after the day of the new or full moon, or rather nearly after the twelfth hour from the new or full moon, and therefore fall nearly upon the forty-third hour after the new or full of the moon. But in this port they fall out about the seventh hour after the appulse of the moon to the meridian of the place; and therefore follow next after the appulse of the moon to the meridian, when the moon is distant from the sun, or from opposition with the sun by about 18 or 19 degrees *in consequentia.* So the summer and winter seasons come not to their height in the solstices themselves, but when the sun is advanced beyond the solstices by about a tenth part of its whole course, that is, by about 36 or 37 degrees. In like manner, the greatest tide is raised after the appulse of the moon to the meridian of the place, when the moon has passed by the sun, *or the opposition thereof,* by about a tenth part of the whole motion from *one greatest* tide to *the next following greatest* tide. Suppose that distance about $18\frac{1}{2}$ degrees; and the sun's force in this distance of the moon from the syzygies and quadratures will be of less moment to augment and diminish that part of the motion of the sea which proceeds from the motion of the moon than in the syzygies and quadratures themselves in the proportion of the radius to the co-sine of double this distance, or of an angle of 37 degrees; that is, in proportion of 10000000 to 7986355; and, therefore, in the preceding analogy, in place of S we must put 0,7986355S.

But farther; the force of the moon in the quadratures must be diminished, on account of its declination from the equator; for the moon in those quadratures, or rather in $18\frac{1}{2}$ degrees past the quadratures, declines from the equator by about 23° 13′; and the force of either luminary to move the sea is diminished as it declines from the equator nearly in the duplicate proportion of the co-sine of the declination; and therefore the force of the moon in those quadratures is only 0,8570327L; whence we have L+0,7986355S to 0,8570327L — 0,7986355S as 9 to 5.

Farther yet; the diameters of the orbit in which the moon should move, setting aside the consideration of eccentricity, are one to the other as 69 to 70; and therefore the moon's distance from the earth in the syzygies is to its distance in the quadratures, *cæteris paribus*, as 69 to 70; and its distances, when $18\frac{1}{2}$ degrees advanced beyond the syzygies, where the greatest tide was excited, and when $18\frac{1}{2}$ degrees passed by the quadratures, where the least tide was produced, are to its mean distance as 69,098747 and 69,897345 to $69\frac{1}{2}$. But the force of the moon to move the sea is in the reciprocal triplicate proportion of its distance; and therefore its forces, in the greatest and least of those distances, are to its force in its mean distance as 0,9830427 and 1,017522 to 1. From whence we have 1,017522L × 0,7986355S to 0,9830427 × 0,8570327L — 0,7986355S as 9 to 5; and S to L as 1 to 4,4815. Wherefore since the force of the sun is to the force of gravity as 1 to 12868200, the moon's force will be to the force of gravity as 1 to 2871400.

COR. 1. Since the waters excited by the sun's force rise to the height of a foot and $11\frac{1}{30}$ inches, the moon's force will raise the same to the height of 8 feet and $7\frac{5}{22}$ inches; and the joint forces of both will raise the same to the height of $10\frac{1}{2}$ feet; and when the moon is in its perigee to the height of $12\frac{1}{2}$ feet, and more, especially when the wind sets the same way as the tide. And a force of that quantity is abundantly sufficient to excite all the motions of the sea, and agrees well with the proportion of those motions; for in such seas as lie free and open from east to west, as in the *Pacific* sea, and in those tracts of the *Atlantic* and *Ethiopic* seas which lie without the tropics, the waters commonly rise to 6, 9, 12, or 15 feet; but in the *Pacific* sea, which is of a greater depth, as well as of a larger extent, the tides are said to be greater than in the *Atlantic* and *Ethiopic* seas; for to have a full tide raised, an extent of sea from east to west is required of no less than 90 degrees. In the *Ethiopic* sea, the waters rise to a less height within the tropics than in the temperate zones, because of the narrowness of the sea between *Africa* and the southern parts of *America*. In the middle of the open sea the waters cannot rise without falling together, and at the same time, upon both the eastern and western shores, when, notwithstanding, in our narrow seas, they ought to fall on those shores by alternate turns; upon which account there is commonly but a small flood and ebb in such islands as lie far distant from the continent. On the contrary, in some ports, where to fill and empty the bays alternately the waters are with great violence forced in and out through shallow channels, the flood and ebb must be greater than ordinary; as at *Plymouth* and *Chepstow Bridge* in *England*, at the mountains of St. *Michael*, and the town of *Avranches*, in *Normandy*, and at *Cambaia* and *Pegu* in the *East Indies*. In these places the sea is hurried in and out with such violence, as sometimes to lay the shores under water, sometimes to leave them dry for many miles. Nor is this force of the influx and efflux to be broke till it has raised and depressed the waters to 30, 40, or 50 feet and above. And a like account is to be given of long and shallow channels or straits, such as the *Magellanic* straits, and those channels which environ *England*. The tide in such ports and straits, by the violence of the influx and efflux, is augmented above measure. But on such shores as lie towards the deep and open sea with a steep descent, where the waters may freely rise and fall without that precipitation of influx and efflux, the proportion of the tides agrees with the forces of the sun and moon.

Cor. 2. Since the moon's force to move the sea is to the force of gravity as 1 to 2871400, it is evident that this force is far less than to appear sensibly in statical or hydrostatical experiments, or even in those of pendulums. It is in the tides only that this force shews itself by any sensible effect.

Cor. 3. Because the force of the moon to move the sea is to the like force of the sun as 4,4815 to 1, and those forces (by Cor. 14, Prop. LXVI, Book 1) are as the densities of the bodies of the sun and moon and the cubes of their apparent diameters conjunctly, the density of the moon will be to the density of the sun as 4,4815 to 1 directly, and the cube of the moon's diameter to the cube of the sun's diameter inversely; that is (seeing the mean apparent diameters of the moon and sun are $31' \, 16\frac{1}{2}''$, and $32' \, 12''$), as 4891 to 1000. But the density of the sun was to the density of the earth as 1000 to 4000; and therefore the density of the moon is to the density of the earth as 4891 to 4000, or as 11 to 9. Therefore the body of the moon is more dense and more earthly than the earth itself.

Cor. 4. And since the true diameter of the moon (from the observations of astronomers) is to the true diameter of the earth as 100 to 365, the mass of matter in the moon will be to the mass of matter in the earth as 1 to 39,788.

Cor. 5. And the accelerative gravity on the surface of the moon will be about three times less than the accelerative gravity on the surface of the earth.

COR. 6. And the distance of the moon's centre from the centre of the earth will be to the distance of the moon's centre from the common centre of gravity of the earth and moon as 40,788 to 39,788.

COR. 7. And the mean distance of the centre of the moon from the centre of the earth will be (in the moon's octants) nearly $60\frac{2}{5}$ of the greatest semi-diameters of the earth; for the greatest semi-diameter of the earth was 19658600 *Paris* feet, and the mean distance of the centres of the earth and moon, consisting of $60\frac{2}{5}$ such semi-diameters, is equal to 1187379440 feet. And this distance (by the preceding Cor.) is to the distance of the moon's centre from the common centre of gravity of the earth and moon as 40,788 to 39,788; which latter distance, therefore, is 1158268534 feet. And since the moon, in respect of the fixed stars, performs its revolution in $27^d. 7^h. 43\frac{4}{9}'$, the versed sine of that angle which the moon in a minute of time describes is 12752341 to the radius 1000,000000,000000; and as the radius is to this versed sine, so are 1158268534 feet to 14,7706353 feet. The moon, therefore, falling towards the earth by that force which retains it in its orbit, would in one minute of time describe 14,7706353 feet; and if we augment this force in the proportion of $178\frac{29}{40}$ to $177\frac{29}{40}$, we shall have the total force of gravity at the orbit of the moon, by Cor. Prop. III; and the moon falling by this force, in one minute of time would describe 14,8538067 feet. And at the 60th part of the distance of the moon from the earth's centre, that is, at the distance of 197896573 feet from the centre of the earth, a body falling by its weight, would, in one second of time, likewise describe 14,8538067 feet. And, therefore, at the distance of 19615800, which compose one mean semi-diameter of the earth, a heavy body would describe in falling 15,11175, or 15 feet, 1 inch, and $4\frac{1}{11}$ lines, in the same time. This will be the descent of bodies in the latitude of 45 degrees. And by the foregoing table, to be found under Prop. XX, the descent in the latitude of *Paris* will be a little greater by an excess of about $\frac{2}{3}$ parts of a line. Therefore, by this computation, heavy bodies in the latitude of *Paris* falling *in vacuo* will describe 15 *Paris* feet, 1 inch, $4\frac{25}{33}$ lines, very nearly, in one second of time. And if the gravity be diminished by taking away a quantity equal to the centrifugal force arising in that latitude from the earth's diurnal motion, heavy bodies falling there will describe in one second of time 15 feet, 1 inch, and $1\frac{1}{2}$ line. And with this velocity heavy bodies do really fall in the latitude of *Paris*, as we have shown above in Prop. IV and XIX.

COR. 8. The mean distance of the centres of the earth and moon in the syzygies of the moon is equal to 60 of the greatest semi-diameters of the earth, subducting only about one 30th part of a semi-diameter; and in the moon's quadratures the mean distance of the same centres is $60\frac{5}{6}$ such semi-diameters of the earth; for these two distances are to the mean distance of the moon in the octants as 69 and 70 to $69\frac{1}{2}$, by Prop. XXVIII.

Cor. 9. The mean distance of the centres of the earth and moon in the syzygies of the moon is 60 mean semi-diameters of the earth, and a 10th part of one semi-diameter; and in the moon's quadratures the mean distance of the same centres is 61 mean semi-diameters of the earth, subducting one 30th part of one semi-diameter.

Cor. 10. In the moon's syzygies its mean horizontal parallax in the latitudes of 0, 30, 38, 45, 52, 60, 90 degrees is 57′ 20″, 57′ 16″, 57′ 14″, 57′ 12″, 57′ 10″, 57′ 8″, 57′ 4″, respectively.

In these computations I do not consider the magnetic attraction of the earth, whose quantity is very small and unknown: if this quantity should ever be found out, and the measures of degrees upon the meridian, the lengths of isochronous pendulums in different parallels, the laws of the motions of the sea, and the moon's parallax, with the apparent diameters of the sun and moon, should be more exactly determined from phænomena: we should then be enabled to bring this calculation to a greater accuracy.

PROPOSITION XXXVIII. PROBLEM XIX.

To find the figure of the moon's body.

If the moon's body were fluid like our sea, the force of the earth to raise that fluid in the nearest and remotest parts would be to the force of the moon by which our sea is raised in the places under and opposite to the moon as the accelerative gravity of the moon towards the earth to the accelerative gravity of the earth towards the moon, and the diameter of the moon to the diameter of the earth conjunctly; that is, as 39,788 to 1, and 100 to 365 conjunctly, or as 1081 to 100. Wherefore, since our sea, by the force of the moon, is raised to $8\frac{3}{5}$ feet, the lunar fluid would be raised by the force of the earth to 93 feet; and upon this account the figure of the moon would be a spheroid, whose greatest diameter produced would pass through the centre of the earth, and exceed the diameters perpendicular thereto by 186 feet. Such a figure, therefore, the moon affects, and must have put on from the beginning. Q.E.I.

Cor. Hence it is that the same face of the moon always respects the earth; nor can the body of the moon possibly rest in any other position, but would return always by a libratory motion to this situation; but those librations, however, must be exceedingly slow, because of the weakness of the forces which excite them; so that the face of the moon, which should be always obverted to the earth, may, for the reason assigned in Prop. XVII. be turned towards the other focus of the moon's orbit, without being immediately drawn back, and converted again towards the earth.

LEMMA I.

If APEp *represent the earth uniformly dense, marked with the centre* C, *the poles* P, p, *and the equator* AE; *and if about the centre* C, *with the radius* CP, *we suppose the sphere* Pape *to be described, and* QR *to denote the plane on which a right line, drawn from the centre of the sun to the centre of the earth, insists at right angles; and further suppose that the several particles of the whole exterior earth* PapAPepE, *without the height of the said sphere, endeavour to recede towards this side and that side from the plane* QR, *every particle by a force proportional to its distance from that plane; I say, in the first place, that the whole force and efficacy of all the particles that are situate in* AE, *the circle of the equator, and disposed uniformly without the globe, encompassing the same after the manner of a ring, to wheel the earth about its centre, is to the whole force and efficacy of as many particles in that point* A *of the equator which is at the greatest distance from the plane* QR, *to wheel the earth about its centre with a like circular motion, as* 1 *to* 2. *And that circular motion will be performed about an axis lying in the common section of the equator and the plane* QR.

For let there be described from the centre K, with the diameter IL, the semi-circle INL. Suppose the semi-circumference INL to be divided into innumerable equal parts, and from the several parts N to the diameter

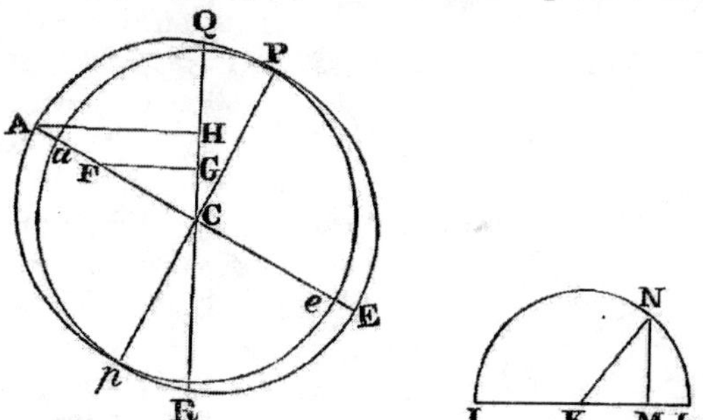

IL let fall the sines NM. Then the sums of the squares of all the sines NM will be equal to the sums of the squares of the sines KM, and both sums together will be equal to the sums of the squares of as many semi-diameters KN; and therefore the sum of the squares of all the sines NM will be but half so great as the sum of the squares of as many semi-diameters KN.

Suppose now the circumference of the circle AE to be divided into the like number of little equal parts, and from every such part F a perpendicular FG to be let fall upon the plane QR, as well as the perpendicular AH from the point A. Then the force by which the particle F recedes from the plane QR will (by supposition) be as that perpendicular FG; and this force multiplied by the distance CG will represent the power of the particle F to turn the earth round its centre. And, therefore, the power of a particle in the place F will be to the power of a particle in the place A as $FG \times GC$ to $AH \times HC$; that is, as FC^2 to AC^2: and therefore the whole power of all the particles F, in their proper places F, will be to the power of the like number of particles in the place A as the sum of all the FC^2 to the sum of all the AC^2, that is (by what we have demonstrated before), as 1 to 2. Q.E.D.

And because the action of those particles is exerted in the direction of lines perpendicularly receding from the plane QR, and that equally from each side of this plane, they will wheel about the circumference of the circle of the equator, together with the adherent body of the earth, round an axis which lies as well in the plane QR as in that of the equator.

LEMMA II.

The same things still supposed, I say, in the second place, that the total force or power of all the particles situated every where about the sphere to turn the earth about the said axis is to the whole force of the like number of particles, uniformly disposed round the whole circumference of the equator AE *in the fashion of a ring, to turn the whole earth about with the like circular motion, as* 2 *to* 5.

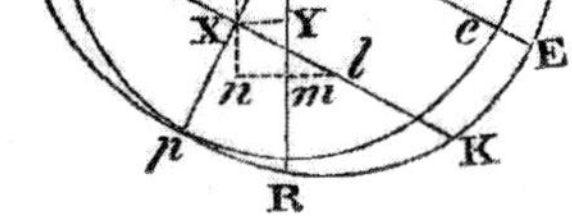

For let IK be any lesser circle parallel to the equator AE, and let L*l* be any two equal particles in this circle, situated without the sphere P*ape*; and if upon the plane QR, which is at right angles with a radius drawn to the sun, we let fall the perpendiculars LM, *lm*, the total forces by which these particles recede from the plane QR will be proportional to the perpendiculars LM, *lm*. Let the right line L*l* be drawn parallel to the plane P*ape*, and bisect the same in X; and through the point X draw N*n* parallel to the plane QR, and meeting the perpendiculars LM, *lm*, in N and *n*; and upon the plane QR let fall the perpendicular XY. And the contrary forces of the particles L and *l* to wheel about the earth contrariwise are as $LM \times MC$, and $lm \times mC$; that is, as $LN \times MC + NM \times MC$, and $ln \times mC - nm \times mC$; or $LN \times MC + NM \times MC$, and $LN \times mC - NM \times mC$, and $LN \times Mm - NM \times \overline{MC + mC}$, the difference of the two, is the force of both taken together to turn the earth round. The affirmative part of this difference $LN \times Mm$, or $2LN \times NX$, is to $2AH \times HC$, the force of two particles of the same size situated in A, as LX^2 to AC^2; and the negative part NM

$\times \overline{MC + mC}$, or $2XY \times CY$, is to $2AH \times HC$, the force of the same two particles situated in A, as CX^2 to AC^2. And therefore the difference of the parts, that is, the force of the two particles L and *l*, taken together, to wheel the earth about, is to the force of two particles, equal to the former and situated in the place A, to turn in like manner the earth round, as $LX^2 - CX^2$ to AC^2. But if the circumference IK of the circle IK is supposed to be divided into an infinite number of little equal parts L, all the LX^2 will be to the like number of IX^2 as 1 to 2 (by Lem. 1); and to the same number of AC^2 as IX^2 to $2AC^2$; and the same number of CX^2 to as many AC^2 as $2CX^2$ to $2AC^2$. Wherefore the united forces of all the particles in the circumference of the circle IK are to the joint forces of as many particles in the place A as $IX^2 - 2CX^2$ to $2AC^2$; and therefore (by Lem. 1) to the united forces of as many particles in the circumference of the circle AE as $IX^2 - 2CX^2$ to AC^2.

Now if P*p*, the diameter of the sphere, is conceived to be divided into an infinite number of equal parts, upon which a like number of circles IK are supposed to insist, the matter in the circumference of every circle IK will be as IX^2; and therefore the force of that matter to turn the earth about will be as IX^2 into $IX^2 - 2CX^2$; and the force of the same matter, if it was situated in the circumference of the circle AE, would be as IX^2 into AC^2. And therefore the force of all the particles of the whole matter situated without the sphere in the circumferences of all the circles is to the force of the like number of particles situated in the circumference of the greatest circle AE as all the IX^2 into $IX^2 - 2CX^2$ to as many IX^2 into AC^2; that is, as all the $AC^2 - CX^2$ into $AC^2 - 3CX^2$ to as many $AC^2 - CX^2$ into AC^2; that is, as all the $AC^4 - 4AC^2 \times CX^2 + 3CX^4$ to as many $AC^4 - AC^2 \times CX^2$; that is, as the whole fluent quantity, whose fluxion is $AC^4 - 4AC^2 \times CX^2 + 3CX^4$, to the whole fluent quantity, whose fluxion is $AC^4 - AC^2 \times CX^2$; and, therefore, by the method of fluxions, as $AC^4 \times CX - \frac{4}{3}AC^2 \times CX^3 + \frac{3}{5}CX^5$ to $AC^4 \times CX - \frac{1}{3}AC^2 \times CX^3$; that is, if for CX we write the whole C*p*, or AC, as $\frac{4}{15}AC^5$ to $\frac{2}{3}AC^5$; that is, as 2 to 5. Q.E.D.

LEMMA III.

The same things still supposed, I say, in the third place, that the motion of the whole earth about the axis above-named arising from the motions of all the particles, will be to the motion of the aforesaid ring about the same axis in a proportion compounded of the proportion of the matter in the earth to the matter in the ring; and the proportion of three squares of the quadrantal arc of any circle to two squares of its diameter, that is, in the proportion of the matter to the matter, and of the number 925275 *to the number* 1000000.

For the motion of a cylinder revolved about its quiescent axis is to the motion of the inscribed sphere revolved together with it as any four equal squares to three circles inscribed in three of those squares; and the motion of this cylinder is to the motion of an exceedingly thin ring surrounding both sphere and cylinder in their common contact as double the matter in the cylinder to triple the matter in the ring; and this motion of the ring, uniformly continued about the axis of the cylinder, is to the uniform motion of the same about its own diameter performed in the same periodic time as the circumference of a circle to double its diameter.

HYPOTHESIS II.

If the other parts of the earth were taken away, and the remaining ring was carried alone about the sun in the orbit of the earth by the annual motion, while by the diurnal motion it was in the mean time revolved about its own axis inclined to the plane of the ecliptic by an angle of $23\frac{1}{2}$ degrees, the motion of the equinoctial points would be the same, whether the ring were fluid, or whether it consisted of a hard and rigid matter.

PROPOSITION XXXIX. PROBLEM XX.

To find the precession of the equinoxes.

The middle horary motion of the moon's nodes in a circular orbit, when the nodes are in the quadratures, was $16'' \, 35''' \, 16^{iv}. \, 36^{v}.$; the half of which, $8'' \, 17''' \, 38^{iv}. \, 18^{v}.$ (for the reasons above explained) is the mean horary motion of the nodes in such an orbit, which motion in a whole sidereal year becomes $20° \, 11' \, 46''$. Because, therefore, the nodes of the moon in such an orbit would be yearly transferred $20° \, 11' \, 46''$ *in antecedentia;* and, if there were more moons, the motion of the nodes of every one (by Cor. 16, Prop. LXVI, Book 1) would be as its periodic time; if upon the surface of the earth a moon was revolved in the time of a sidereal day, the annual motion of the nodes of this moon would be to $20° \, 11' \, 46''$ as $23^h. \, 56'$, the sidereal day, to $27^d. \, 7^h. \, 43'$, the periodic time of our moon, that is, as 1436 to 39343. And the same thing would happen to the nodes of a ring of moons encompassing the earth, whether these moons did not mutually touch each the other, or whether they were molten, and formed into a continued ring, or whether that ring should become rigid and inflexible.

Let us, then, suppose that this ring is in quantity of matter equal to the whole exterior earth P*ap*AP*ep*E, which lies without the sphere P*ape* (see fig. Lem. II); and because this sphere is to that exterior earth as aC^2 to $AC^2 - aC^2$, that is (seeing PC or aC the least semi-diameter of the earth is to AC the greatest semi-diameter of the same as 229 to 230), as 52441 to 459; if this ring encompassed the earth round the equator, and both together were revolved about the diameter of the ring, the motion of the ring (by Lem. III) would be to the motion of the inner sphere as 459 to 52441 and 1000000 to 925275 conjunctly, that is, as 4590 to 485223; and therefore the motion of the ring would be to the sum of the motions of both ring and sphere as 4590 to 489813. Wherefore if the ring adheres to the sphere, and communicates its motion to the sphere, by which its nodes or equinoctial points recede, the motion remaining in the ring will be to its former motion as 4590 to 489813; upon which account the motion of the equinoctial points will be diminished in the same proportion. Wherefore the annual motion of the equinoctial points of the body, composed of both ring and sphere, will be to the motion 20° 11′ 46″ as 1436 to 39343 and 4590 to 489813 conjunctly, that is, as 100 to 292369. But the forces by which the nodes of a number of moons (as we explained above), and therefore by which the equinoctial points of the ring recede (that is, the forces 3IT, in fig. Prop. XXX), are in the several particles as the distances of those particles from the plane QR; and by these forces the particles recede from that plane: and therefore (by Lem. II) if the matter of the ring was spread all over the surface of the sphere, after the fashion of the figure *PapAPepE*, in order to make up that exterior part of the earth, the total force or power of all the particles to wheel about the earth round any diameter of the equator, and therefore to move the equinoctial points, would become less than before in the proportion of 2 to 5. Wherefore the annual regress of the equinoxes now would be to 20° 11′ 46″ as 10 to 73092; that is, would be 9″ 56‴ 50^{iv}.

But because the plane of the equator is inclined to that of the ecliptic, this motion is to be diminished in the proportion of the sine 91706 (which is the co-sine of 23½ deg.) to the radius 100000; and the remaining motion will now be 9″ 7‴ 20^{iv}. which is the annual precession of the equinoxes arising from the force of the sun.

But the force of the moon to move the sea was to the force of the sun nearly as 4,4815 to 1; and the force of the moon to move the equinoxes is to that of the sun in the same proportion. Whence the annual precession of the equinoxes proceeding from the force of the moon comes out 40″ 52‴ 52^{iv}. and the total annual precession arising from the united forces of both will be 50″ 00‴ 12^{iv}. the quantity of which motion agrees with the phænomena; for the precession of the equinoxes, by astronomical observations, is about 50″ yearly.

If the height of the earth at the equator exceeds its height at the poles by more than 17⅙ miles, the matter thereof will be more rare near the surface than at the centre; and the precession of the equinoxes will be augmented by the excess of height, and diminished by the greater rarity.

And now we have described the system of the sun, the earth, moon, and planets, it remains that we add something about the comets.

LEMMA IV.

That the comets are higher than the moon, and in the regions of the planets.

As the comets were placed by astronomers above the moon, because they were found to have no diurnal parallax, so their annual parallax is a convincing proof of their descending into the regions of the planets; for all the comets which move in a direct course according to the order of the signs, about the end of their appearance become more than ordinarily slow or retrograde, if the earth is between them and the sun; and more than ordinarily swift, if the earth is approaching to a heliocentric opposition with them; whereas, on the other hand, those which move against the order of the signs, towards the end of their appearance appear swifter than they ought to be, if the earth is between them and the sun; and slower, and perhaps retrograde, if the earth is in the other side of its orbit. And these appearances proceed chiefly from the diverse situations which the earth acquires in the course of its motion, after the same manner as it happens to the planets, which appear sometimes retrograde, sometimes more slowly, and sometimes more swiftly, progressive, according as the motion of the earth falls in with that of the planet, or is directed the contrary way. If the earth move the same way with the comet, but, by an angular motion about the sun, so much swifter that right lines drawn from the earth to the comet converge towards the parts beyond the comet, the comet seen from the earth, because of its slower motion, will appear retrograde; and even if the earth is slower than the comet, the motion of the earth being subducted, the motion of the comet will at least appear retarded; but if the earth tends the contrary way to that of the comet, the motion of the comet will from thence appear accelerated; and from this apparent acceleration, or retardation, or regressive motion, the distance of the comet may be inferred in this manner. Let ♈QA, ♈QB, ♈QC, be three observed longitudes of the comet about the time of its first appearing, and ♈QF its last observed longitude before its disappearing. Draw the right line ABC, whose parts AB, BC, intercepted between the right lines QA and QB, QB and QC, may be one to the other as the two times between the three first observations. Produce AC to G, so as AG may be to AB as the time between the first and last observation to the time between the first and second; and join QG. Now if the comet did move uniformly in a right line, and the earth either stood still, or was likewise carried forwards in a right line by an uniform motion, the angle ♈QG would be the

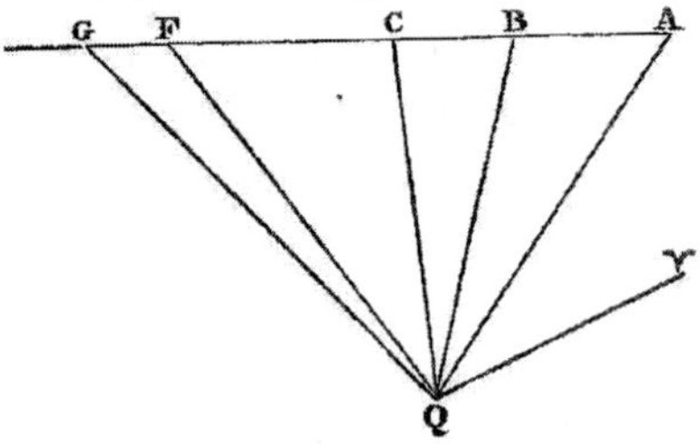

longitude of the comet at the time of the last observation. The angle, therefore, FQG, which is the difference of the longitude, proceeds from the inequality of the motions of the comet and the earth; and this angle, if the earth and comet move contrary ways, is added to the angle ♈QG, and accelerates the apparent motion of the comet; but if the comet move the same way with the earth, it is subtracted, and either retards the motion of the comet, or perhaps renders it retrograde, as we have but now explained. This angle, therefore, proceeding chiefly from the motion of the earth, is justly to be esteemed the parallax of the comet; neglecting, to wit, some little increment or decrement that may arise from the unequal motion of the comet in its orbit; and from this parallax we thus deduce the distance of the comet. Let S represent the sun, *ac*T the *orbis magnus*, *a* the earth's place in the first observation, *c* the place of the earth in the third observation, T the place of the earth in the last observation, and T♈ a right line drawn to the beginning of Aries. Set off the angle ♈TV equal to the angle ♈QF, that is, equal to the longitude of the comet at the time when the earth is in T; join *ac*, and produce it to *g*, so as *ag* may be to *ac* as AG to AC; and *g* will be the place at which the earth would have arrived in the time of the last observation, if it had continued to move uniformly in the right line *ac*. Wherefore, if we draw *g*♈ parallel to T♈, and make the angle ♈*g*V equal to the angle ♈QG, this angle ♈*g*V will be equal to the longitude of the comet seen from the place *g*, and the angle TV*g* will be the parallax which arises from the earth's being transferred from the place *g* into the place T; and therefore V will be the place of the comet in the plane of the ecliptic. And this place V is commonly lower than the orb of Jupiter.

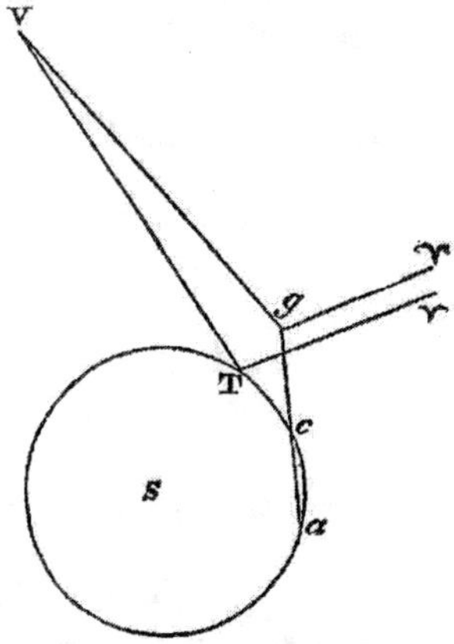

The same thing may be deduced from the incurvation of the way of the comets; for these bodies move almost in great circles, while their velocity is great; but about the end of their course, when that part of their apparent motion which arises from the parallax bears a greater proportion to their whole apparent motion, they commonly deviate from those circles, and when the earth goes to one side, they deviate to the other; and this deflexion, because of its corresponding with the motion of the earth, must arise chiefly from the parallax; and the quantity thereof is so considerable, as, by my computation, to place the disappearing comets a good deal lower than Jupiter. Whence it follows that when they approach nearer to us in their perigees and perihelions they often descend below the orbs of Mars and the inferior planets.

The near approach of the comets is farther confirmed from the light of their heads; for the light of a celestial body, illuminated by the sun, and receding to remote parts, is diminished in the quadruplicate proportion of the distance; to wit, in one duplicate proportion, on account of the increase of the distance from the sun, and in another duplicate proportion, on account of the decrease of the apparent diameter. Wherefore if both the quantity of light and the apparent diameter of a comet are given, its distance will be also given, by taking the distance of the comet to the distance of a planet in the direct proportion of their diameters and the reciprocal subduplicate proportion of their lights. Thus, in the comet of the year 1682, Mr. *Flamsted* observed with a telescope of 16 feet, and measured with a micrometer, the least diameter of its head, 2′ 00; but the nucleus or star in the middle of the head scarcely amounted to the tenth part of this measure; and therefore its diameter was only 11″ or 12″; but in the light and splendor of its head it surpassed that of the comet in the year 1680, and might be compared with the stars of the first or second magnitude. Let us suppose that Saturn with its ring was about four times more lucid; and because the light of the ring was almost equal to the light of the globe within, and the apparent diameter of the globe is about 21″, and therefore the united light of both globe and ring would be equal to the light of a globe whose diameter is 30″, it follows that the distance of the comet was to the distance of Saturn as 1 to $\sqrt{4}$ inversely, and 12″ to 30 directly; that is, as 24 to 30, or 4 to 5. Again; the comet in the month of *April* 1665, as *Hevelius* informs us, excelled almost all the fixed stars in splendor, and even Saturn itself, as being of a much more vivid colour; for this comet was more lucid than that other which had appeared about the end of the preceding year, and had been compared to the stars of the first magnitude. The diameter of its head was about 6′; but the nucleus, compared with the planets by means of a telescope, was plainly less than Jupiter; and sometimes judged less, sometimes judged equal, to the globe of Saturn within the ring. Since, then, the diameters of the heads of the comets seldom exceed 8′ or 12′, and the diameter of the nucleus or central star is but about a tenth or perhaps fifteenth part of the diameter of the head, it appears that these stars are generally of about the same apparent magnitude with the planets. But in regard that their light may be often compared with the light of Saturn, yea, and sometimes exceeds it, it is evident that all comets in their perihelions must either be placed below or not far above Saturn; and they are much mistaken who remove them almost as far as the fixed stars; for if it was so, the comets could receive no more light from our sun than our planets do from the fixed stars.

So far we have gone, without considering the obscuration which comets suffer from that plenty of thick smoke which encompasseth their heads, and through which the heads always shew dull, as through a cloud; for by how much the more a body is obscured by this smoke, by so much the more near it must be allowed to come to the sun, that it may vie with the planets in the quantity of light which it reflects. Whence it is probable that the comets descend far below the orb of Saturn, as we proved before from their parallax. But, above all, the thing is evinced from their tails, which must be owing either to the sun's light reflected by a smoke arising from them, and dispersing itself through the æther, or to the light of their own heads. In the former case, we must shorten the distance of the comets, lest we be obliged to allow that the smoke arising from their heads is propagated through such a vast extent of space, and with such a velocity and expansion as will seem altogether incredible; in the latter case, the whole light of both head and tail is to be ascribed to the central nucleus. But, then, if we suppose all this light to be united and condensed within the disk of the nucleus, certainly the nucleus will by far exceed Jupiter itself in splendor, especially when it emits a very large and lucid tail. If, therefore, under a less apparent diameter, it reflects more light, it must be much more illuminated by the sun, and therefore much nearer to it; and the same argument will bring down the heads of comets sometimes within the orb of Venus, viz., when, being hid under the sun's rays, they emit such huge and splendid tails, like beams of fire, as sometimes they do; for if all that light was supposed to be gathered together into one star, it would sometimes exceed not one Venus only, but a great many such united into one.

Lastly; the same thing is inferred from the light of the heads, which increases in the recess of the comets from the earth towards the sun, and decreases in their return from the sun towards the earth; for so the comet of the year 1665 (by the observations of *Hevelius*), from the time that it was first seen, was always losing of its apparent motion, and therefore had already passed its perigee; but yet the splendor of its head was daily increasing, till, being hid under the sun's rays, the comet ceased to appear. The comet of the year 1683 (by the observations of the same *Hevelius*), about the end of *July*, when it first appeared, moved at a very slow rate, advancing only about 40 or 45 minutes in its orb in a day's time; but from that time its diurnal motion was continually upon the increase, till *September* 4, when it arose to about 5 degrees; and therefore, in all this interval of time, the comet was approaching to the earth. Which is likewise proved from the diameter of its head, measured with a micrometer; for, *August* 6, *Hevelius* found it only 6′ 05″, including the *coma*, which, *September* 2, he observed to be 9′ 07″, and therefore its head appeared far less about the beginning than towards the end of the motion; though about the beginning, because nearer to the sun, it appeared far more lucid than towards the end, as the same *Hevelius* declares. Wherefore in all this interval of time, on account of its recess from the sun, it decreased in splendor, notwithstanding its access towards the earth. The comet of the year 1618, about the middle of *December*, and that of the year 1680, about the end of the same month, did both move with their greatest velocity, and were therefore then in their perigees; but the greatest splendor of their heads was seen two weeks before, when they had just got clear of the sun's rays; and the greatest splendor of their tails a little more early, when yet nearer to the sun. The head of the former comet (according to the observations of *Cysatus*), *December* 1, appeared greater than the stars of the first magnitude; and, *December* 16 (then in the perigee), it was but little diminished in magnitude, but in the splendor and brightness of its light a great deal. *January* 7, *Kepler*, being uncertain about the head, left off observing. *December* 12, the head of the latter comet was seen and observed by Mr. *Flamsted*, when but 9 degrees distant from the sun; which is scarcely to be done in a star of the third magnitude. *December* 15 and 17, it appeared as a star of the third magnitude, its lustre being diminished by the brightness of the clouds near the setting sun. *December* 26, when it moved with the greatest velocity, being almost in its perigee, it was less than the mouth of *Pegasus*, a star of the third magnitude. *January* 3, it appeared as a star of the fourth. *January* 9, as one of the fifth. *January* 13, it was hid by the splendor of the moon, then in her increase. *January* 25, it was scarcely equal to the stars of the seventh magnitude. If we compare equal intervals of time on one side and on the other from the perigee, we shall find that the head of the comet, which at both intervals of time was far, but yet equally, removed from the earth, and should have therefore shone with equal splendor, appeared brightest on the side of the perigee towards the sun, and disappeared on the other. Therefore, from the great difference of light in the one situation and in the other, we conclude the great vicinity of the sun and comet in the former; for the light of comets uses to be regular, and to appear greatest when the heads move fastest, and are therefore in their perigees; excepting in so far as it is increased by their nearness to the sun.

Cor. 1. Therefore the comets shine by the sun's light, which they reflect.

Cor. 2. From what has been said, we may likewise understand why comets are so frequently seen in that hemisphere in which the sun is, and so seldom in the other. If they were visible in the regions far above Saturn, they would appear more frequently in the parts opposite to the sun; for such as were in those parts would be nearer to the earth, whereas the presence of the sun must obscure and hide those that appear in the hemisphere in which he is. Yet, looking over the history of comets, I find that four or five times more have been seen in the hemisphere towards the sun than in the opposite hemisphere; besides, without doubt, not a few, which have been hid by the light of the sun: for comets descending into our parts neither emit tails, nor are so well illuminated by the sun, as to discover themselves to our naked eyes, until they are come nearer to us than Jupiter. But the far greater part of that spherical space, which is described about the sun with so small an interval, lies on that side of the earth which regards the sun; and the comets in that greater part are commonly more strongly illuminated, as being for the most part nearer to the sun.

Cor. 3. Hence also it is evident that the celestial spaces are void of resistance; for though the comets are carried in oblique paths, and sometimes contrary to the course of the planets, yet they move every way with the greatest freedom, and preserve their motions for an exceeding long time, even where contrary to the course of the planets. I am out in my judgment if they are not a sort of planets revolving in orbits returning into themselves with a perpetual motion; for, as to what some writers contend, that they are no other than meteors, led into this opinion by the perpetual changes that happen to their heads, it seems to have no foundation; for the heads of comets are encompassed with huge atmospheres, and the lowermost parts of these atmospheres must be the densest; and therefore it is in the clouds only, not in the bodies of the comets themselves, that these changes are seen. Thus the earth, if it was viewed from the planets, would, without all doubt, shine by the light of its clouds, and the solid body would scarcely appear through the surrounding clouds. Thus also the belts of Jupiter are formed in the clouds of that planet, for they change their position one to another, and the solid body of Jupiter is hardly to be seen through them; and much more must the bodies of comets be hid under their atmospheres, which are both deeper and thicker.

PROPOSITION XL. THEOREM XX.

That the comets move in some of the conic sections, having their foci in the centre of the sun; and by radii drawn to the sun describe areas proportional to the times.

This proposition appears from Cor. 1, Prop. XIII, Book 1, compared with Prop. VIII, XII, and XIII, Book III.

Cor. 1. Hence if comets are revolved in orbits returning into themselves, those orbits will be ellipses; and their periodic times be to the periodic times of the planets in the sesquiplicate proportion of their principal axes. And therefore the comets, which for the most part of their course are higher than the planets, and upon that account describe orbits with greater axes, will require a longer time to finish their revolutions. Thus if the axis of a comet's orbit was four times greater than the axis of the orbit of Saturn, the time of the revolution of the comet would be to the time of the revolution of Saturn, that is, to 30 years, as $4\sqrt{4}$ (or 8) to 1, and would therefore be 240 years.

Cor. 2. But their orbits will be so near to parabolas, that parabolas may be used for them without sensible error.

Cor. 3. And, therefore, by Cor. 7, Prop. XVI, Book 1, the velocity of every comet will always be to the velocity of any planet, supposed to be revolved at the same distance in a circle about the sun, nearly in the subduplicate proportion of double the distance of the planet from the centre of the sun to the distance of the comet from the sun's centre, very nearly. Let us suppose the radius of the *orbis magnus*, or the greatest semidiameter of the ellipsis which the earth describes, to consist of 100000000 parts; and then the earth by its mean diurnal motion will describe 1720212 of those parts, and $71675\frac{1}{2}$ by its horary motion. And therefore the comet, at the same mean distance of the earth from the sun, with a velocity which is to the velocity of the earth as $\surd 2$ to 1, would by its diurnal motion describe 2432747 parts, and $101364\frac{1}{2}$ parts by its horary motion. But at greater or less distances both the diurnal and horary motion will be to this diurnal and horary motion in the reciprocal subduplicate proportion of the distances, and is therefore given.

Cor. 4. Wherefore if the *latus rectum* of the parabola is quadruple of the radius of the *orbis magnus*, and the square of that radius is supposed to consist of 100000000 parts, the area which the comet will daily describe by a radius drawn to the sun will be $1216373\frac{1}{2}$ parts, and the horary area will be $50682\frac{1}{4}$ parts. But, if the *latus rectum* is greater or less in any proportion, the diurnal and horary area will be less or greater in the subduplicate of the same proportion reciprocally.

LEMMA V.

To find a curve line of the parabolic kind which shall pass through any given number of points.

Let those points be A, B, C, D, E, F, &c., and from the same to any right line HN, given in position, let fall as many perpendiculars AH, BI, CK, DL, EM, FN, &c.

$b \quad 2b \quad 3b \quad 4b \quad 5b$

$c \quad 2c \quad 3c \quad 4c$

$d \quad 2d \quad 3d$

$e \quad 2e$

f

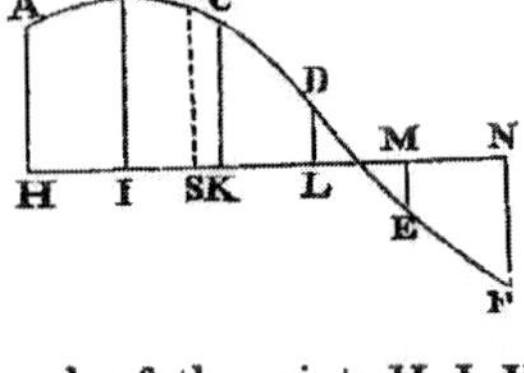

Case 1. If HI, IK, KL, &c., the intervals of the points H, I, K, L, M, N, &c., are equal, take b, $2b$, $3b$, $4b$, $5b$, &c., the first differences of the perpendiculars AH, BI, CK, &c.; their second differences c, $2c$, $3c$, $4c$, &c.; their third, d, $2d$, $3d$, &c., that is to say, so as AH — BI may be $= b$, BI

— CK = 2*b*, CK — DL = 3*b*, DL + EM = 4*b*, — EM + FN = 5*b*, &c.; then *b* — 2*b* = *c*, &c., and so on to the last difference, which is here *f*. Then, erecting any perpendicular RS, which may be considered as an ordinate of the curve required, in order to find the length of this ordinate, suppose the intervals HI, IK, KL, LM, &c., to be units, and let AH = *a*, — HS = *p*, $\frac{1}{2}p$ into — IS = *q*, $\frac{1}{3}q$ into + SK = *r*, $\frac{1}{4}r$ into + SL = *s*, $\frac{1}{5}s$ into + SM = *t*; proceeding, to wit, to ME, the last perpendicular but one, and prefixing negative signs before the terms HS, IS, &c., which lie from S towards A; and affirmative signs before the terms SK, SL, &c., which lie on the other side of the point S; and, observing well the signs, RS will be = *a* + *bp* + *cq* + *dr* + *es* + *ft*, + &c.

CASE 2. But if HI, IK, &c., the intervals of the points H, I, K, L, &c., are unequal, take *b*, 2*b*, 3*b*, 4*b*, 5*b*, &c., the first differences of the perpendiculars AH, BI, CK, &c., divided by the intervals between those perpendiculars; *c*, 2*c*, 3*c*, 4*c*, &c., their second differences, divided by the intervals between every two; *d*, 2*d*, 3*d*, &c., their third differences, divided by the intervals between every three; *e*, 2*e*, &c., their fourth differences, divided by the intervals between every four; and so forth; that is, in such manner, that *b* may be $= \frac{AH - BI}{HI}$, $2b = \frac{BI - CK}{IK}$, $3b = \frac{CK - DL}{KL}$, &c., then $c = \frac{b - 2b}{HK}$, $2c = \frac{2b - 3b}{IL}$, $3c = \frac{3b - 4b}{KM}$, &c., then $d = \frac{c - 2c}{HL}$, $2d = \frac{2c - 3c}{IM}$, &c. And those differences being found, let AH be = *a*, — HS = *p*, *p* into — IS = *q*, *q* into + SK = *r*, *r* into + SL = *s*, *s* into + SM = *t*; proceeding, to wit, to ME, the last perpendicular but one; and the ordinate RS will be = *a* + *bp* + *cq* + *dr* + *es* + *ft*, + &c.

COR. Hence the areas of all curves may be nearly found; for if some number of points of the curve to be squared are found, and a parabola be supposed to be drawn through those points, the area of this parabola will be nearly the same with the area of the curvilinear figure proposed to be squared: but the parabola can be always squared geometrically by methods vulgarly known.

LEMMA VI.

Certain observed places of a comet being given, to find the place of the same to any intermediate given time.

Let HI, IK, KL, LM (in the preceding Fig.), represent the times between the observations; HA, IB, KC, LD, ME, five observed longitudes of the comet; and HS the given time between the first observation and the longitude required. Then if a regular curve ABCDE is supposed to be drawn through the points A, B, C, D, E, and the ordinate RS is found out by the preceding lemma, RS will be the longitude required.

After the same method, from five observed latitudes, we may find the latitude to a given time.

If the differences of the observed longitudes are small, suppose of 4 or 5 degrees, three or four observations will be sufficient to find a new longitude and latitude; but if the differences are greater, as of 10 or 20 degrees, five observations ought to be used.

LEMMA VII.

Through a given point P *to draw a right line* BC, *whose parts* PB, PC, *cut off by two right lines* AB, AC, *given in position, may be one to the other in a given proportion.*

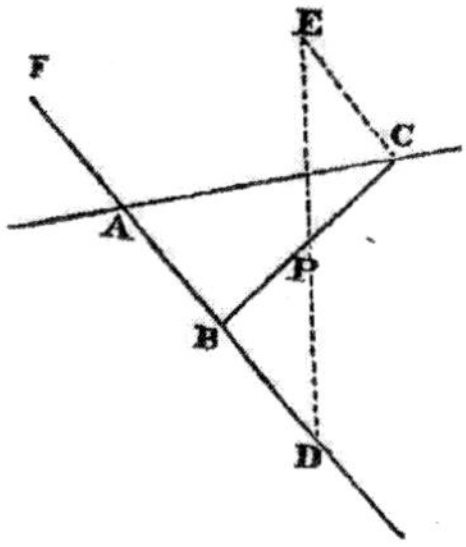

From the given point P suppose any right line PD to be drawn to either of the right lines given, as AB; and produce the same towards AC, the other given right line, as far as E, so as PE may be to PD in the given proportion. Let EC be parallel to AD. Draw CPB, and PC will be to PB as PE to PD. Q.E.F.

LEMMA VIII.

Let ABC *be a parabola, having its focus in* S. *By the chord* AC *bisected in* I *cut off the segment* ABCI, *whose diameter is* Iμ *and vertex* μ. *In* Iμ *produced take* μO *equal to one half of* Iμ. *Join* OS, *and produce it to* ξ, *so as* Sξ *may be equal to* 2SO. *Now, supposing a comet to revolve in the arc* CBA, *draw* ξB, *cutting* AC *in* E; *I say, the point* E *will cut off from the chord* AC *the segment* AE, *nearly proportional to the time.*

For if we join EO, cutting the parabolic arc ABC in Y, and draw μX touching the same arc in the vertex μ, and meeting EO in X, the curvilinear area AEXμA will be to the curvilinear area ACYμA as AE to AC; and, therefore, since the triangle ASE is to the triangle ASC in the same proportion, the whole area ASEXμA will be to the whole area ASCYμA as

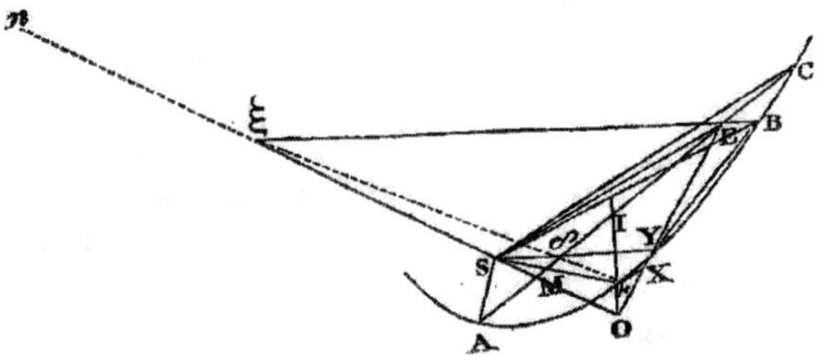

AE to AC. But, because ξO is to SO as 3 to 1, and EO to XO in the same proportion, SX will be parallel to EB; and, therefore, joining BX, the triangle SEB will be equal to the triangle XEB. Wherefore if to the area ASEXμA we add the triangle EXB, and from the sum subduct the triangle SEB, there will remain the area ASBXμA, equal to the area ASEXμA, and therefore in proportion to the area ASCYμA as AE to AC. But the area ASBYμA is nearly equal to the area ASBXμA; and this area ASBYμA is to the area ASCYμA as the time of description of the arc AB to the time of description of the whole arc AC; and, therefore, AE is to AC nearly in the proportion of the times. Q.E.D.

COR. When the point B falls upon the vertex μ of the parabola, AE is to AC accurately in the proportion of the times.

SCHOLIUM.

If we join $\mu\xi$ cutting AC in δ, and in it take ξn in proportion to μB as 27MI to 16Mμ, and draw Bn, this Bn will cut the chord AC, in the proportion of the times, more accurately than before; but the point n is to be taken beyond or on this side the point ξ, according as the point B is more or less distant from the principal vertex of the parabola than the point μ.

LEMMA IX.

The right lines Iμ *and* μM, *and the length* $\frac{AI^2}{4S\mu}$, *are equal among themselves.*

For 4Sμ is the latus rectum of the parabola belonging to the vertex μ.

LEMMA X.

Produce Sμ *to* N *and* P, *so as* μN *may be one third of* μI, *and* SP *may be to* SN *as* SN *to* Sμ; *and in the time that a comet would describe the arc* AμC, *if it was supposed to move always forwards with the velocity which it hath in a height equal to* SP, *it would describe a length equal to the chord* AC.

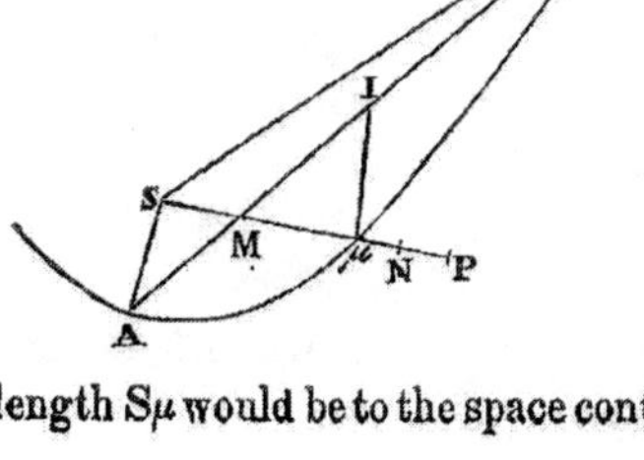

For if the comet with the velocity which it hath in μ was in the said time supposed to move uniformly forward in the right line which touches the parabola in μ, the area which it would describe by a radius drawn to the point S would be equal to the parabolic area ASCμA; and therefore the space contained under the length described in the tangent and the length Sμ would be to the space contained under the lengths AC and SM as the area ASCμA to the triangle ASC, that is, as SN to SM. Wherefore AC is to the length described in the tangent as Sμ to SN. But since the velocity of the comet in the height SP (by Cor. 6, Prop. XVI., Book I) is to the velocity of the same in the height Sμ in the reciprocal subduplicate proportion of SP to Sμ, that is, in the proportion of Sμ to SN, the length described with this velocity will be to the length in the same time described in the tangent as Sμ to SN. Wherefore since AC, and the length described with this new velocity, are in the same proportion to the length described in the tangent, they must be equal betwixt themselves. Q.E.D.

Cor. Therefore a comet, with that velocity which it hath in the height Sμ + $\frac{2}{3}$Iμ, would in the same time describe the chord AC nearly.

LEMMA XI.

If a comet void of all motion was let fall from the height SN, *or* Sμ + $\frac{1}{3}$Iμ, *towards the sun, and was still impelled to the sun by the same force uniformly continued by which it was impelled at first, the same, in one half of that time in which it might describe the arc* AC *in its own orbit, would in descending describe a space equal to the length* Iμ.

For in the same time that the comet would require to describe the parabolic arc AC, it would (by the last Lemma), with that velocity which it hath in the height SP, describe the chord AC; and, therefore (by Cor. 7, Prop. XVI, Book 1), if it was in the same time supposed to revolve by the force of its own gravity in a circle whose semi-diameter was SP, it would describe an arc of that circle, the length of which would be to the chord of the parabolic arc AC in the subduplicate proportion of 1 to 2. Wherefore if with that weight, which in the height SP it hath towards the sun, it should fall from that height towards the sun, it would (by Cor. 9, Prop. XVI, Book 1) in half the said time describe a space equal to the square of half the said chord applied to quadruple the height SP, that is, it would describe the space $\frac{AI^2}{4SP}$. But since the weight of the comet towards the sun in the height SN is to the weight of the same towards the sun in the height SP as SP to Sμ, the comet, by the weight which it hath in the height SN, in falling from that height towards the sun, would in the same time describe the space $\frac{AI^2}{4S\mu}$; that is, a space equal to the length Iμ or μM. Q.E.D.

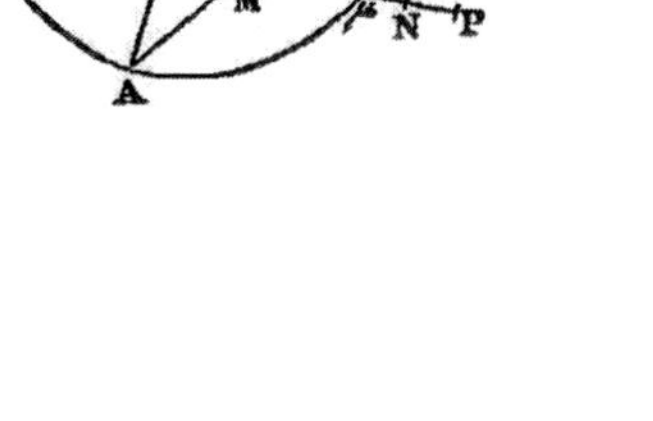

PROPOSITION XLI. PROBLEM XXI.

From three observations given to determine the orbit of a comet moving in a parabola.

This being a Problem of very great difficulty, I tried many methods of resolving it; and several of those Problems, the composition whereof I have given in the first Book, tended to this purpose. But afterwards I contrived the following solution, which is something more simple.

Select three observations distant one from another by intervals of time nearly equal; but let that interval of time in which the comet moves more slowly be somewhat greater than the other; so, to wit, that the difference of the times may be to the sum of the times as the sum of the

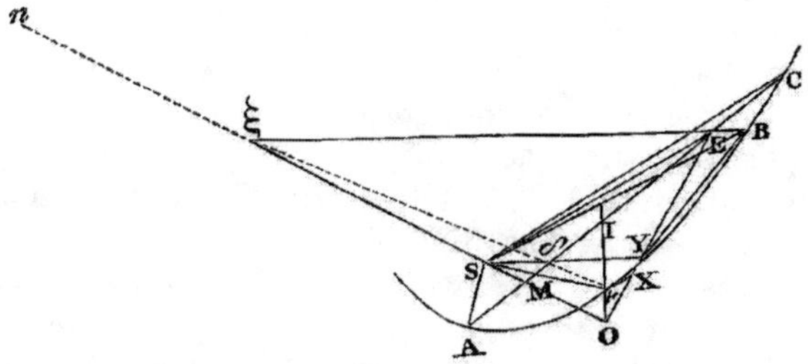

times to about 600 days; or that the point E may fall upon M nearly, and may err therefrom rather towards I than towards A. If such direct observations are not at hand, a new place of the comet must be found, by Lem. VI.

Let S represent the sun; T, t, τ, three places of the earth in the *orbis magnus;* TA, tB, τC, three observed longitudes of the comet; V the time between the first observation and the second; W the time between the second and the third; X the length which in the whole time V + W

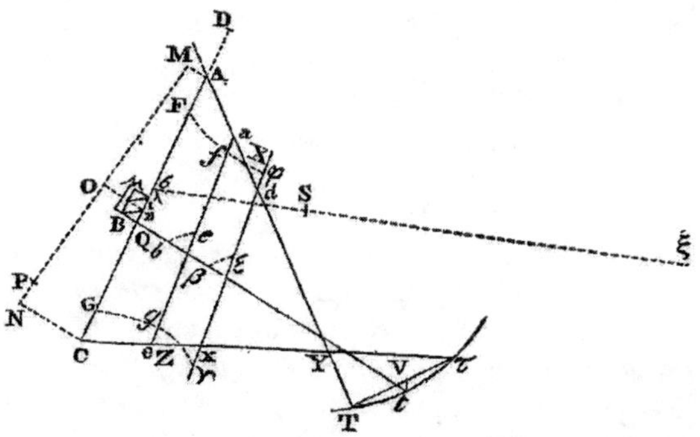

the comet might describe with that velocity which it hath in the mean distance of the earth from the sun, which length is to be found by Cor. 3,

Prop. XL, Book III; and tV a perpendicular upon the chord $T\tau$. In the mean observed longitude tB take at pleasure the point B, for the place of the comet in the plane of the ecliptic; and from thence, towards the sun S, draw the line BE, which may be to the perpendicular tV as the content under SB and St^2 to the cube of the hypothenuse of the right angled triangle, whose sides are SB, and the tangent of the latitude of the comet in the second observation to the radius tB. And through the point E (by Lemma VII) draw the right line AEC, whose parts AE and EC, terminating in the right lines TA and τC; may be one to the other as the times V and W: then A and C will be nearly the places of the comet in the plane of the ecliptic in the first and third observations, if B was its place rightly assumed in the second.

Upon AC, bisected in I, erect the perpendicular Ii. Through B draw the obscure line Bi parallel to AC. Join the obscure line Si, cutting AC in λ, and complete the parallelogram $iI\lambda\mu$. Take $I\sigma$ equal to $3I\lambda$; and through the sun S draw the obscure line $\sigma\xi$ equal to $3S\sigma + 3i\lambda$. Then, cancelling the letters A, E, C, I, from the point B towards the point ξ, draw the new obscure line BE, which may be to the former BE in the duplicate proportion of the distance BS to the quantity $S\mu + \frac{1}{3}i\lambda$. And through the point E draw again the right line AEC by the same rule as before; that is, so as its parts AE and EC may be one to the other as the times V and W between the observations. Thus A and C will be the places of the comet more accurately.

Upon AC, bisected in I, erect the perpendiculars AM, CN, IO, of which AM and CN may be the tangents of the latitudes in the first and third observations, to the radii TA and τC. Join MN, cutting IO in O. Draw the rectangular parallelogram $iI\lambda\mu$, as before. In IA produced take ID equal to $S\mu + \frac{2}{3}i\lambda$. Then in MN, towards N, take MP, which may be to the above found length X in the subduplicate proportion of the mean distance of the earth from the sun (or of the semi-diameter of the *orbis magnus*) to the distance OD. If the point P fall upon the point N; A, B, and C, will be three places of the comet, through which its orbit is to be described in the plane of the ecliptic. But if the point P falls not upon the point N, in the right line AC take CG equal to NP, so as the points G and P may lie on the same side of the line NC.

By the same method as the points E, A, C, G, were found from the assumed point B, from other points b and β assumed at pleasure, find out the new points e, a, c, g; and $\varepsilon, \alpha, \kappa, \gamma$. Then through G, g, and γ, draw the circumference of a circle $Gg\gamma$, cutting the right line τC in Z: and Z will be one place of the comet in the plane of the ecliptic. And in AC, ac, $\alpha\kappa$, taking AF, af, $\alpha\phi$, equal respectively to CG, cg, $\kappa\gamma$; through the points F, f, and ϕ, draw the circumference of a circle $Ff\phi$, cutting the right line AT in X; and the point X will be another place of the comet in the plane of the ecliptic. And at the points X and Z, erecting the tangents of the latitudes of the comet to the radii TX and τZ, two places of the comet in its own orbit will be determined. Lastly, if (by Prop. XIX., Book 1) to the focus S a parabola is described passing through those two places, this parabola will be the orbit of the comet. Q.E.I.

The demonstration of this construction follows from the preceding Lemmas, because the right line AC is cut in E in the proportion of the times, by Lem. VII., as it ought to be, by Lem. VIII.; and BE, by Lem. XI., is a portion of the right line BS or Bξ in the plane of the ecliptic, intercepted between the arc ABC and the chord AEC; and MP (by Cor. Lem. X.) is the length of the chord of that arc, which the comet should describe in its proper orbit between the first and third observation, and therefore is equal to MN, providing B is a true place of the comet in the plane of the ecliptic.

But it will be convenient to assume the points B, b, β, not at random, but nearly true. If the angle AQt, at which the projection of the orbit in the plane of the ecliptic cuts the right line tB, is rudely known, at that angle with Bt draw the obscure line AC, which may be to $\frac{4}{3}$Tτ in the subduplicate proportion of SQ to St; and, drawing the right line SEB so as its part EB may be equal to the length Vt, the point B will be determined, which we are to use for the first time. Then, cancelling the right line AC, and drawing anew AC according to the preceding construction, and, moreover, finding the length MP, in tB take the point b, by this rule, that, if TA and τC intersect each other in Y, the distance Yb may be to the distance YB in a proportion compounded of the proportion of MP to MN, and the subduplicate proportion of SB to Sb. And by the same method you may find the third point β, if you please to repeat the operation the third time; but if this method is followed, two operations generally will be sufficient; for if the distance Bb happens to be very small, after the points F, f, and G, g, are found, draw the right lines Ff and Gg, and they will cut TA and τC in the points required, X and Z.

EXAMPLE.

Let the comet of the year 1680 be proposed. The following table shews the motion thereof, as observed by *Flamsted*, and calculated afterwards by him from his observations, and corrected by Dr. *Halley* from the same observations.

	Time.		Sun's	Comet's	
	Appar.	True.	Longitude.	Longitude.	Lat. N.
	h. ″	h. ′ ″	° ′ ″	° ′ ″	° ′ ″
1680, *Dec.* 12	4.46	4.46. 0	♑ 1.51.23	♑ 6.32.30	8.28. 0
21	6.32½	6.36.59	11.06.44	♒ 5.08.12	21.42.13
24	6.12	6.17.52	14.09.26	18.49.23	25.23. 5
26	5.14	5 20.44	16.09.22	28.24.13	27.00.52
29	7.55	8.03.02	19.19.43	♓ 13.10.41	28.09.58
30	8.02	8.10.26	20.21.09	17.38.20	28.11.53
1681, *Jan.* 5	5.51	6.01.38	26.22.18	♈ 8.48.53	26.15. 7
9	6.49	7.00.53	♒ 0.29.02	18.44.04	24.11.56
10	5.54	6.06.10	1.27.43	20.40.50	23.43.52
13	6.56	7.08.55	4.33.20	25.59.48	22.17.28
25	7.44	7.58.42	16.45.36	♉ 9.35. 0	17.56.30
30	8.07	8.21.53	21.49.58	13.19.51	16.42.18
Feb. 2	6.20	6.34.51	24.46.59	15.13.53	16.04. 1
5	6.50	7.04.41	27.49.51	16.59.06	15.27. 3

To these you may add some observations of mine.

	Ap.	Comet's	
	Time	Longitude.	Lat. N.
	h. ′	° ′ ″	° ′ ″
1681, *Feb.* 25	8.30	♉ 26.18.35	12.46.46
27	8.15	27.04.30	12.36.12
Mar. 1	11. 0	27.52.42	12.23.40
2	8. 0	28.12.48	12.19.38
5	11.30	29.18. 0	12.03.16
7	9.30	♊ 0. 4. 0	11.57. 0
9	8.30	0.43. 4	11.45.52

These observations were made by a telescope of 7 feet, with a micrometer and threads placed in the focus of the telescope; by which instruments we determined the positions both of the fixed stars among themselves, and of the comet in respect of the fixed stars. Let A represent the star of the fourth magnitude in the left heel of *Perseus* (*Bayer's o*), B the following star of the third magnitude in the left foot (*Bayer's* ζ), C a star of the sixth magnitude (*Bayer's n*) in the heel of the same foot, and D, E, F, G, H, I, K, L, M, N, O, Z, α, β, γ, δ, other smaller stars in the same foot; and let p, P, Q, R, S, T, V, X, represent the places of the comet in the observations above set down; and, reckoning the distance AB of $80\frac{7}{12}$ parts, AC was $52\frac{1}{4}$ of those parts; BC, $58\frac{5}{6}$; AD, $57\frac{5}{12}$; BD, $82\frac{6}{11}$; CD, $23\frac{2}{3}$; AE, $29\frac{4}{7}$; CE, $57\frac{1}{2}$; DE, $49\frac{11}{12}$; AI, $27\frac{7}{12}$; BI, $52\frac{1}{6}$; CI, $36\frac{7}{12}$; DI, $53\frac{5}{11}$; AK, $38\frac{2}{3}$; BK, 43; CK, $31\frac{5}{9}$; FK, 29; FB, 23; FC, $36\frac{1}{4}$; AH, $18\frac{6}{7}$; DH, $50\frac{7}{8}$; BN, $46\frac{5}{12}$; CN, $31\frac{1}{3}$; BL, $45\frac{5}{12}$; NL, $31\frac{5}{7}$. HO was to HI as 7 to 6, and, produced, did pass between the stars D and E, so as the distance of the star D from this right line was $\frac{1}{6}$CD. LM was to LN as 2 to 9, and, produced, did pass through the star H. Thus were the positions of the fixed stars determined in respect of one another.

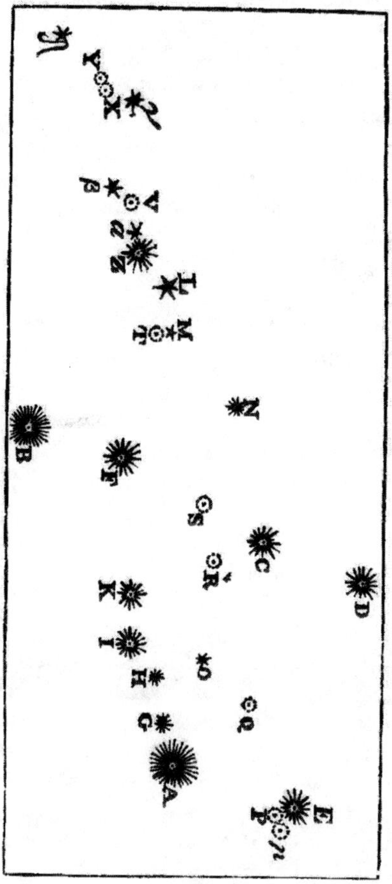

Mr. *Pound* has since observed a second time the positions of these fixed stars amongst themselves, and collected their longitudes and latitudes according to the following table.

The fixed stars.	Their Longitudes	Latitude North.	The fixed stars.	Their Longitudes	Latitude North.
	° ′ ″	° ′ ″		° ′ ″	° ′ ″
A	♉ 26.41.50	12. 8.36	L	♉ 29.33.34	12. 7.48
B	28.40.23	11.17.54	M	29.18.54	12. 7.20
C	27.58.30	12.40.25	N	28.48.29	12.31. 9
E	26.27.17	12.52. 7	Z	29.44.48	11.57.13
F	28.28.37	11.52.22	α	29.52. 3	11.55.48
G	26.56. 8	12. 4.58	β	♊ 0. 8.23	11.48.56
H	27.11.45	12. 2. 1	γ	0.40.10	11.55.18
I	27.25. 2	11.53.11	δ	1. 3.20	11.30.42
K	27.42. 7	11.53.26			

The positions of the comet to these fixed stars were observed to be as follow:

Friday, *February* 25, O.S. at $8\frac{1}{2}^h$. P. M. the distance of the comet in p from the star E was less than $\frac{3}{13}$AE, and greater than $\frac{1}{5}$AE, and therefore nearly equal to $\frac{3}{14}$AE; and the angle ApE was a little obtuse, but almost right. For from A, letting fall a perpendicular on pE, the distance of the comet from that perpendicular was $\frac{1}{5}p$E.

The same night, at $9\frac{1}{2}^h$., the distance of the comet in P from the star E was greater than $\frac{1}{4\frac{1}{2}}$ AE, and less than $\frac{1}{5\frac{1}{4}}$ AE, and therefore nearly equal to $\frac{1}{4\frac{7}{8}}$ of AE, or $\frac{8}{39}$AE. But the distance of the comet from the perpendicular let fall from the star A upon the right line PE was $\frac{4}{5}$PE.

Sunday, *February* 27, $8\frac{1}{4}^h$. P. M. the distance of the comet in Q from the star O was equal to the distance of the stars O and H; and the right line QO produced passed between the stars K and B. I could not, by reason of intervening clouds, determine the position of the star to greater accuracy.

Tuesday, *March* 1, 11^h. P. M. the comet in R lay exactly in a line between the stars K and C, so as the part CR of the right line CRK was a little greater than $\frac{1}{3}$CK, and a little less than $\frac{1}{3}$CK + $\frac{1}{8}$CR, and therefore = $\frac{1}{3}$CK + $\frac{1}{15}$ CR, or $\frac{16}{45}$CK.

Wednesday, *March* 2, 8^h. P. M. the distance of the comet in S from the star C was nearly $\frac{4}{9}$FC; the distance of the star F from the right line CS produced was $\frac{1}{24}$FC; and the distance of the star B from the same right line was five times greater than the distance of the star F; and the right line NS produced passed between the stars H and I five or six times nearer to the star H than to the star I.

Saturday, *March* 5, $11\frac{1}{2}^h$. P. M. when the comet was in T, the right line MT was equal to $\frac{1}{2}$ML, and the right line LT produced passed between B and F four or five times nearer to F than to B, cutting off from BF a fifth or sixth part thereof towards F: and MT produced passed on the outside of the space BF towards the star B four times nearer to the star B than to the star F. M was a very small star, scarcely to be seen by the telescope; but the star L was greater, and of about the eighth magnitude.

Monday, *March* 7, $9\frac{1}{2}^h$. P. M. the comet being in V, the right line Vα produced did pass between B and F, cutting off, from BF towards F, $\frac{1}{16}$ of BF, and was to the right line Vβ as 5 to 4. And the distance of the comet from the right line $\alpha\beta$ was $\frac{1}{2}$Vβ.

Wednesday, *March* 9, $8\frac{1}{2}^h$. P. M. the comet being in X, the right line γX was equal to $\frac{1}{4}\gamma\delta$; and the perpendicular let fall from the star δ upon the right γX was $\frac{2}{5}$ of $\gamma\delta$.

The same night, at 12^h. the comet being in Y, the right line γY was equal to $\frac{1}{3}$ of $\gamma\delta$, or a little less, as perhaps $\frac{5}{16}$ of $\gamma\delta$; and a perpendicular let fall from the star δ on the right line γY was equal to about $\frac{1}{6}$ or $\frac{1}{7}$ $\gamma\delta$. But the comet being then extremely near the horizon, was scarcely discernible, and therefore its place could not be determined with that certainty as in the foregoing observations.

From these observations, by constructions of figures and calculations, I deduced the longitudes and latitudes of the comet; and Mr. *Pound*, by correcting the places of the fixed stars, hath determined more correctly the places of the comet, which correct places are set down above. Though my micrometer was none of the best, yet the errors in longitude and latitude (as derived from my observations) scarcely exceed one minute. The comet (according to my observations), about the end of its motion, began to decline sensibly towards the north, from the parallel which it described about the end of *February*.

Now, in order to determine the orbit of the comet out of the observations above described, I selected those three which *Flamsted* made, *Dec.* 21, *Jan.* 5, and *Jan.* 25; from which I found S*t* of 9842,1 parts, and V*t* of 455, such as the semi-diameter of the *orbis magnus* contains 10000. Then for the first observation, assuming *t*B of 5657 of those parts, I found SB 9747, BE for the first time 412, Sμ 9503, *i*λ 413, BE for the second time 421, OD 10186, X 8528,4, PM 8450, MN 8475, NP 25; from whence, by the second operation, I collected the distance *tb* 5640; and by this operation I at last deduced the distances TX 4775 and τZ 11322. From which, limiting the orbit, I found its descending node in ♋, and ascending node in ♑ 1° 53′; the inclination of its plane to the plane of the ecliptic 61° 20$\frac{1}{3}$′; the vertex thereof (or the perihelion of the comet) distant from the node 8° 38′, and in ♐ 27° 43′, with latitude 7° 34′ south; its *latus rectum* 236,8; and the diurnal area described by a radius drawn to the sun 93585, supposing the square of the semi-diameter of the *orbis magnus* 100000000; that the comet in this orbit moved directly according to the order of the signs, and on *Dec.* 8^d. 00^h. 04′ P. M. was in the vertex or perihelion of its orbit. All which I determined by scale and compass, and the chords of angles, taken from the table of natural sines, in a pretty large figure, in which, to wit, the radius of the *orbis magnus* (consisting of 10000 parts) was equal to 16$\frac{1}{3}$ inches of an *English* foot.

Lastly, in order to discover whether the comet did truly move in the orbit so determined, I investigated its places in this orbit partly by arithmetical operations, and partly by scale and compass, to the times of some of the observations, as may be seen in the following table:—

The Comet's	Dist. from sun.	Longitude computed.	Latitud. computed.	Longitude observed.	Latitude observed	Dif Lo.	Dif. Lat.
Dec. 12	2792	♑ 6°.32′	8°.18$\frac{1}{2}$	♑ 6°.31$\frac{1}{3}$	8°.26	+1	− 7$\frac{1}{2}$
29	8403	♓ 13 .13$\frac{2}{3}$	28. 00	♓ 13 .11$\frac{3}{4}$	28 .10$\frac{1}{12}$	+2	−10$\frac{1}{12}$
Feb. 5	16669	♉ 17 .00	15. 29$\frac{2}{3}$	♉ 16 .59$\frac{7}{8}$	15 .27$\frac{2}{5}$	+0	+ 2$\frac{1}{4}$
Mar. 5	21737	29 .19$\frac{3}{4}$	12. 4	29 .20$\frac{6}{7}$	12 . 3$\frac{1}{2}$	−1	+ $\frac{1}{2}$

But afterwards Dr. *Halley* did determine the orbit to a greater accuracy by an arithmetical calculus than could be done by linear descriptions; and, retaining the place of the nodes in ♋ and ♑ 1° 53′, and the inclination of the plane of the orbit to the ecliptic 61° 20⅓′, as well as the time of the comet's being in perihelio, *Dec.* 8^{d}. 00^{h}. 04′, he found the distance of the perihelion from the ascending node measured in the comet's orbit 9° 20′, and the *latus rectum* of the parabola 2430 parts, supposing the mean distance of the sun from the earth to be 100000 parts; and from these *data*, by an accurate arithmetical calculus, he computed the places of the comet to the times of the observations as follows:—

True time.		The Comet's Dist from the sun.	Longitude computed.	Latitude computed.	Errors in Long.	Errors in Lat.
	d. h. ′ ″		° ′ ″	° ′ ″	′ ″	′ ″
Dec.	12. 4.46.	28028	♑ 6.29.25	8.26. 0 bor.	−3. 5	−2. 0
	21. 6.37.	61076	♒ 5. 6.30	21.43.20	−1.42	+ 1. 7
	24. 6.18.	70008	18.48.20	25.22.40	−1. 3	−0.25
	26. 5.20.	75576	28.22.45	27. 1.36	−1.28	+ 0.44
	29. 8. 3.	84021	♓ 13.12.40	28.10.10	+ 1.59	+ 0.12
	30. 8.10.	86661	17.40. 5	28.11.20	+ 1.45	−0.33
Jan.	5. 6. 1.$\frac{1}{2}$	101440	♈ 8.49.49	26.15.15	+ 0.56	+ 0. 8
	9. 7. 0.	110959	18.44.36	24.12.54	+ 0.32	+ 0.58
	10. 6. 6.	113162	20.41. 0	23.44.10	+ 0.10	+ 0.18
	13. 7. 9.	120000	26. 0.21	22.17.30	+ 0.33	+ 0. 2
	25. 7.59.	145370	♉ 9.33.40	17.57.55	−1.20	+ 1.25
	30. 8.22.	155303	13.17.41	16.42. 7	−2.10	−0.11
Feb.	2. 6.35.	160951	15.11.11	16. 4.15	−2.42	+ 0.14
	5. 7. 4.$\frac{1}{2}$	166686	16.58.55	15.29.13	−0.41	+ 2. 0
	25. 8.41.	202570	26.15.46	12.48. 0	−2.49	+ 1.10
Mar.	5.11.39.	216205	29.18.35	12. 5.40	+ 0.35	+ 2.14

This comet also appeared in the *November* before, and at *Coburg*, in *Saxony*, was observed by Mr. *Gottfried Kirch*, on the 4th of that month, on the 6th and 11th O. S.; from its positions to the nearest fixed stars observed with sufficient accuracy, sometimes with a two feet, and sometimes with a ten feet telescope; from the difference of longitudes of *Coburg* and *London*, 11°; and from the places of the fixed stars observed by Mr. *Pound*, Dr. *Halley* has determined the places of the comet as follows:—

Nov. 3, 17^h. 2′, apparent time at *London*, the comet was in ♌ 29 deg. 51′, with 1 deg. 17′ 45″ latitude north.

November 5, 15^h. 58′ the comet was in ♍ 3° 23′, with 1° 6′ north lat.

November 10, 16^h. 31′, the comet was equally distant from two stars in ♌, which are σ and τ in *Bayer;* but it had not quite touched the right line that joins them, but was very little distant from it. In *Flamsted's* catalogue this star σ was then in ♍ 14° 15′, with 1 deg. 41′ lat. north nearly, and τ in ♍ 17° 3½′ with 0 deg. 34′ lat. south; and the middle point between those stars was ♍ 15° 39¼′, with 0° 33½′ lat. north. Let the distance of the comet from that right line be about 10′ or 12′; and the difference of the longitude of the comet and that middle point will be 7′; and the difference of the latitude nearly 7½′; and thence it follows that the comet was in ♍ 15° 32′, with about 26′ lat. north.

The first observation from the position of the comet with respect to certain small fixed stars had all the exactness that could be desired; the second also was accurate enough. In the third observation, which was the least accurate, there might be an error of 6 or 7 minutes, but hardly greater. The longitude of the comet, as found in the first and most accurate observation, being computed in the aforesaid parabolic orbit, comes out ♌ 29° 30′ 22″, its latitude north 1° 25′ 7″, and its distance from the sun 115546.

Moreover, Dr. *Halley*, observing that a remarkable comet had appeared four times at equal intervals of 575 years (that is, in the month of *September* after *Julius Cæsar* was killed; *An. Chr.* 531, in the consulate of *Lampadius* and *Orestes; An. Chr.* 1106, in the month of *February;* and at the end of the year 1680; and that with a long and remarkable tail, except when it was seen after *Cæsar's* death, at which time, by reason of the inconvenient situation of the earth, the tail was not so conspicuous), set himself to find out an elliptic orbit whose greater axis should be 1382957 parts, the mean distance of the earth from the sun containing 10000 such; in which orbit a comet might revolve in 575 years; and, placing the ascending node in ♋ 2° 2′, the inclination of the plane of the orbit to the plane of the ecliptic in an angle of 61° 6′ 48″, the perihelion of the comet in this plane in ♐ 22° 44′ 25″, the equal time of the perihelion *December* 7^d. 23^h. 9′, the distance of the perihelion from the ascending node in the plane of the ecliptic 9° 17′ 35″, and its conjugate axis 18481,2, he computed the motions of the comet in this elliptic orbit. The places of the comet, as deduced from the observations, and as arising from computation made in this orbit, may be seen in the following table.

True time.	Longitude observed.	Latit'de North obs.	Longitude comp.	Latitude computed.	Errors in Long.	Errors in Lat.
d. h. ′	° ′ ″	° ′ ″	° ′ ″	° ′ ″	′ ″	′ ″
Nov. 3.16.47	♌ 29.51. 0	1.17.45	♌ 29.51.22	1.17.32 N	+ 0.22	— 0.13
5.15.37	♍ 3.23. 0	1. 6. 0	♍ 3.24.32	1. 6. 9	+ 1.32	+ 0. 9
10.16.18	15.32. 0	0.27. 0	15.33. 2	0.25. 7	+ 1. 2	— 1.53
16.17.00			♎ 8.16.45	0.53. 7 S		
18.21.34			18.52.15	1.26.54		
20.17. 0			28.10.36	1.53.35		
23.17. 5			♏ 13.22.42	2.29. 0		
Dec. 12. 4.46	♑ 6.32.30	8.28. 0	♑ 6.31.20	8.29. 6 N	— 1.10	+ 1. 6
21. 6.37	♒ 5. 8.12	21.42.13	♒ 5. 6.14	21.44.42	— 1.58	+ 2.29
24. 6.18	18.49.23	25.23. 5	18.47.30	25.23.35	— 1.53	+ 0.30
26. 5.21	28.24.13	27. 0.52	28.21.42	27. 2. 1	— 2.31	+ 1. 9
29. 8. 3	♓ 13.10.41	28. 9.58	♓ 13.11.14	28.10.38	+ 0.33	+ 0.40
30. 8.10	17.38. 0	28.11.53	17.38.27	28.11.37	+ 0. 7	— 0.16
Jan. 5. 6. 1½	♈ 8.48.53	26.15. 7	♈ 8.48.51	26.14.57	— 0. 2	— 0.10
9. 7. 1	18.44. 4	24.11.56	18.43.51	24.12.17	— 0.13	+ 0.21
10. 6. 6	20.40.50	23.43.32	20.40.23	23.43.25	— 0.27	— 0. 7
13. 7. 9	25.59.48	22.17.28	26. 0. 8	22.16.32	+ 0.20	— 0.56
25. 7.59	♉ 9.35. 0	17.56.30	♉ 9.34.11	17.56. 6	— 0.49	— 0.24
30. 8.22	13.19.51	16.42.18	13.18.28	16.40. 5	— 1.23	— 2.13
Feb. 2. 6.35	15.13.53	16. 4. 1	15.11.59	16. 2.17	— 1.54	— 1.54
5. 7. 4½	16.59. 6	15.27. 3	16.59.17	15.27. 0	+ 0.11	— 0. 3
25. 8.41	26.18.35	12.46.46	26.16.59	12.45.22	— 1.36	— 1.24
Mar. 1.11.10	27.52.42	12.23.40	27.51.47	12.22.28	— 0.55	— 1.12
5.11.39	29.18. 0	12. 3.16	29.20.11	12. 2.50	+ 2.11	— 0.26
9. 8.38	♊ 0.43. 4	11.45.52	♊ 0.42.43	11.45.35	— 0.21	— 0.17

The observations of this comet from the beginning to the end agree as perfectly with the motion of the comet in the orbit just now described as the motions of the planets do with the theories from whence they are calculated; and by this agreement plainly evince that it was one and the same comet that appeared all that time, and also that the orbit of that comet is here rightly defined.

In the foregoing table we have omitted the observations of *Nov.* 16, 18, 20, and 23, as not sufficiently accurate, for at those times several persons had observed the comet. *Nov.* 17, O. S. *Ponthæus* and his companions, at 6ʰ. in the morning at *Rome* (that is, 5ʰ. 10′ at *London*), by threads directed to the fixed stars, observed the comet in ♎ 8° 30′, with latitude 0° 40′ south. Their observations may be seen in a treatise which *Ponthæus* published concerning this comet. *Cellius*, who was present, and communicated his observations in a letter to *Cassini*, saw the comet at the same hour in ♎ 8° 30′, with latitude 0° 30′ south. It was likewise seen by *Galletius* at the same hour at *Avignon* (that is, at 5ʰ. 42′ morning at *London*) in ♎ 8° without latitude. But by the theory the comet was at that time in ♎ 8° 16′ 45″, and its latitude was 0° 53′ 7″ south.

Nov. 18, at 6ʰ. 30′ in the morning at *Rome* (that is, at 5ʰ. 40′ at *London*), *Ponthæus* observed the comet in ♎ 13° 30′, with latitude 1° 20′

south; and *Cellius* in ♎ 13° 30′, with latitude 1° 00′ south. But at 5^{h}. 30′ in the morning at *Avignon*, *Galletius* saw it in ♎ 13° 00′, with latitude 1° 00′ south. In the University of *La Fleche*, in *France*, at 5^{h}. in the morning (that is, at 5^{h}. 9′ at *London*), it was seen by *P. Ango*, in the middle between two small stars, one of which is the middle of the three which lie in a right line in the southern hand of Virgo, *Bayers* ψ; and the other is the outmost of the wing, *Bayer's* θ. Whence the comet was then in ♎ 12° 46′ with latitude 50′ south. And I was informed by Dr. *Halley*, that on the same day at *Boston* in *New England*, in the latitude of 42½ deg. at 5^{h}. in the morning (that is, at 9^{h}. 44′ in the morning at *London*), the comet was seen near ♎ 14°, with latitude 1° 30′ south.

Nov. 19, at $4\frac{1}{2}^{h}$. at *Cambridge*, the comet (by the observation of a young man) was distant from *Spica* ♍ about 2° towards the north west. Now the spike was at that time in ♎ 19° 23′ 47″, with latitude 2° 1′ 59″ south. The same day, at 5^{h}. in the morning, at *Boston* in *New England*, the comet was distant from *Spica* ♍ 1°, with the difference of 40′ in latitude. The same day, in the island of *Jamaica*, it was about 1° distant from *Spica* ♍. The same day, Mr. *Arthur Storer*, at the river *Patuxent*, near *Hunting Creek*, in *Maryland*, in the confines of *Virginia*, in lat. $38\frac{1}{2}°$, at 5 in the morning (that is, at 10^{h}. at *London*), saw the comet above *Spica* ♍, and very nearly joined with it, the distance between them being about ¾ of one deg. And from these observations compared, I conclude, that at 9^{h}. 44′ at *London* the comet was in ♎ 18° 50′, with about 1° 25′ latitude south. Now by the theory the comet was at that time in ♎ 18° 52′ 15″, with 1° 26′ 54″ lat. south.

Nov. 20, *Montenari*, professor of astronomy at *Padua*, at 6^{h}. in the morning at *Venice* (that is, 5^{h}. 10′ at *London*), saw the comet in ♎ 23°, with latitude 1° 30′ south. The same day, at *Boston*, it was distant from *Spica* ♍ by about 4° of longitude east, and therefore was in ♎ 23° 24′ nearly.

Nov. 21, *Ponthœus* and his companions, at $7\frac{1}{4}^{h}$. in the morning, observed the comet in ♎ 27° 50′, with latitude 1° 16′ south; *Cellius*, in ♎ 28°; *P. Ango* at 5^{h}. in the morning, in ♎ 27° 45′; *Montenari* in ♎ 27° 51′. The same day, in the island of *Jamaica*, it was seen near the beginning of ♏, and of about the same latitude with *Spica* ♍, that is, 2° 2′. The same day, at 5^{h}. morning, at *Ballasore*, in the *East Indies* (that is, at 11^{h}. 20′ of the night preceding at *London*), the distance of the comet from *Spica* ♍ was taken 7° 35′ to the east. It was in a right line between the spike and the balance, and therefore was then in ♎ 26° 58′, with about 1° 11′ lat. south; and after 5^{h}. 40′ (that is, at 5^{h}. morning at *London*), it was in ♎ 28° 12′. with 1° 16′ lat. south. Now by the theory the comet was then in ♎ 28° 10′ 36″, with 1° 53′ 35″ lat. south.

Nov. 22, the comet was seen by *Montenari* in ♏ 2° 33′; but at *Boston* in *New England*, it was found in about ♏ 3°, and with almost the same latitude as before, that is, 1° 30′. The same day, at 5^h. morning at *Ballasore*, the comet was observed in ♏ 1° 50′; and therefore at 5^h. morning at *London*, the comet was in ♏ 3° 5′ nearly. The same day, at 6½^h. in the morning at *London*, Dr. *Hook* observed it in about ♏ 3° 30′, and that in the right line which passeth through *Spica* ♍ and *Cor Leonis;* not, indeed, exactly, but deviating a little from that line towards the north. *Montenari* likewise observed, that this day, and some days after, a right line drawn from the comet through *Spica* passed by the south side of *Cor Leonis* at a very small distance therefrom. The right line through *Cor Leonis* and *Spica* ♍ did cut the ecliptic in ♍ 3° 46′ at an angle of 2° 51′; and if the comet had been in this line and in ♏ 3°, its latitude would have been 2° 26′; but since *Hook* and *Montenari* agree that the comet was at some small distance from this line towards the north, its latitude must have been something less. On the 20th, by the observation of *Montenari*, its latitude was almost the same with that of *Spica* ♍, that is, about 1° 30′. But by the agreement of *Hook, Montenari,* and *Ango,* the latitude was continually increasing, and therefore must now, on the 22d, be sensibly greater than 1° 30′; and, taking a mean between the extreme limits but now stated, 2° 26′ and 1° 30′, the latitude will be about 1° 58′. *Hook* and *Montenari* agree that the tail of the comet was directed towards *Spica* ♍, declining a little from that star towards the south according to *Hook*, but towards the north according to *Montenari;* and, therefore, that declination was scarcely sensible; and the tail, lying nearly parallel to the equator, deviated a little from the opposition of the sun towards the north.

Nov. 23, O. S. at 5^h. morning, at *Nuremberg* (that is, at 4½^h. at *London*), Mr. *Zimmerman* saw the comet in ♏ 8° 8′, with 2° 31′ south lat. its place being collected by taking its distances from fixed stars.

Nov. 24, before sun-rising, the comet was seen by *Montenari* in ♏ 12° 52′ on the north side of the right line through *Cor Leonis* and *Spica* ♍, and therefore its latitude was something less than 2° 38′; and since the latitude, as we said, by the concurring observations of *Montenari, Ango,* and *Hook*, was continually increasing, therefore, it was now, on the 24th, something greater than 1° 58′; and, taking the mean quantity, may be reckoned 2° 18′, without any considerable error. *Ponthæus* and *Galletius* will have it that the latitude was now decreasing; and *Cellius*, and the observer in *New England*, that it continued the same, viz., of about 1°, or 1½°. The observations of *Ponthæus* and *Cellius* are more rude, especially those which were made by taking the azimuths and altitudes; as are also the observations of *Galletius.* Those are better which were made by taking the position of the comet to the fixed stars by *Montenari, Hook, Ango,* and the observer in *New England*, and sometimes by

Ponthæus and *Cellius*. The same day, at 5^h. morning, at *Ballasore*, the comet was observed in ♏ 11° 45′; and, therefore, at 5^h. morning at *London*, was in ♏ 13° nearly. And, by the theory, the comet was at that time in ♏ 13° 22′ 42″.

Nov. 25, before sunrise, *Montenari* observed the comet in ♏ 17¾ nearly; and *Cellius* observed at the same time that the comet was in a right line between the bright star in the right thigh of Virgo and the southern scale of Libra; and this right line cuts the comet's way in ♏ 18° 36′. And, by the theory, the comet was in ♏ 18⅓° nearly.

From all this it is plain that these observations agree with the theory, so far as they agree with one another; and by this agreement it is made clear that it was one and the same comet that appeared all the time from *Nov.* 4 to *Mar.* 9. The path of this comet did twice cut the plane of the ecliptic, and therefore was not a right line. It did cut the ecliptic not in opposite parts of the heavens, but in the end of Virgo and beginning of Capricorn, including an arc of about 98°; and therefore the way of the comet did very much deviate from the path of a great circle; for in the month of *Nov.* it declined at least 3° from the ecliptic towards the south; and in the month of *Dec.* following it declined 29° from the ecliptic towards the north; the two parts of the orbit in which the comet descended towards the sun, and ascended again from the sun, declining one from the other by an apparent angle of above 30°, as observed by *Montenari*. This comet travelled over 9 signs, to wit, from the last deg. of ♌ to the beginning of ♊, beside the sign of ♌, through which it passed before it began to be seen; and there is no other theory by which a comet can go over so great a part of the heavens with a regular motion. The motion of this comet was very unequable; for about the 20th of *Nov.* it described about 5° a day. Then its motion being retarded between *Nov.* 26 and *Dec.* 12, to wit, in the space of 15½ days, it described only 40°. But the motion thereof being afterwards accelerated, it described near 5° a day, till its motion began to be again retarded. And the theory which justly corresponds with a motion so unequable, and through so great a part of the heavens, which observes the same laws with the theory of the planets, and which accurately agrees with accurate astronomical observations, cannot be otherwise than true.

And, thinking it would not be improper, I have given a true representation of the orbit which this comet described, and of the tail which it emitted in several places, in the annexed figure; protracted in the plane of the trajectory. In this scheme ABC represents the trajectory of the comet, D the sun DE the axis of the trajectory, DF the line of the nodes, GH the intersection of the sphere of the *orbis magnus* with the plane of the trajectory, I the place of the comet *Nov.* 4, *Ann.* 1680; K the place of the same *Nov.* 11; L the place of the same *Nov.* 19; M its place *Dec.* 12; N

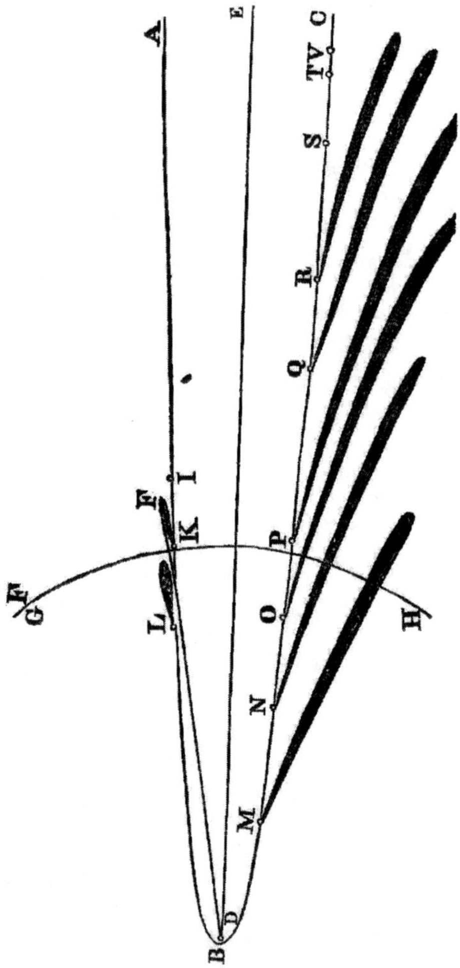

its place *Dec.* 21; O its place *Dec.* 29; P its place *Jan.* 5 following; Q its place *Jan.* 25; R its place *Feb.* 5; S its place *Feb.* 25; T its place *March* 5; and V its place *March* 9. In determining the length of the tail, I made the following observations.

Nov. 4 and 6, the tail did not appear; *Nov.* 11, the tail just begun to shew itself, but did not appear above ½ deg. long through a 10 feet telescope; *Nov.* 17, the tail was seen by *Ponthæus* more than 15° long; *Nov.* 18, in *New-England*, the tail appeared 30° long, and directly opposite to the sun, extending itself to the planet Mars, which was then in ♍, 9° 54′; *Nov.* 19, in *Maryland*, the tail was found 15° or 20° long; *Dec.* 10 (by the observation of Mr. *Flamsted*), the tail passed through the middle of the distance intercepted between the tail of the Serpent of *Ophiuchus* and the star δ in the south wing of *Aquila*, and did terminate near the stars A, ω, b, in *Bayer's* tables. Therefore the end of the tail was in ♑ $19\frac{1}{2}°$, with latitude about $34\frac{1}{4}°$ north; *Dec.* 11, it ascended to the head of *Sagitta* (*Bayer's* α, β), terminating in ♑ 26° 43′, with latitude 38° 34′ north; *Dec.* 12, it passed through the middle of *Sagitta*, nor did it reach much farther; terminating in ♒ 4°, with latitude $42\frac{1}{2}°$ north nearly. But these things are to be understood of the length of the brighter part of the tail; for with a more faint light, observed, too, perhaps, in a serener sky, at *Rome*, *Dec.* 12, 5^h. 40′, by the observation of *Ponthæus*, the tail arose to 10° above the rump of the Swan, and the side thereof towards the west and towards the north was 45′ distant from this star. But about that time the tail was 3° broad towards the upper end; and therefore the middle thereof was 2° 15′ distant from that star towards the south, and the upper end was ♓ in 22°, with latitude 61° north; and thence the tail was about 70° long; *Dec.* 21, it extended almost to *Cassiopeia's* chair, equally distant from β and from *Schedir*, so as its distance from either of the two was equal to the distance of the one from the other, and therefore did terminate in ♈ 24°, with latitude $47\frac{1}{2}°$; *Dec.* 29, it reached to a contact with *Scheat* on its left, and exactly filled up the space between the two stars in the northern foot of *Andromeda*, being 54° in length; and therefore terminated in ♉ 19°, with 35° of latitude; *Jan.* 5, it touched the star π in the breast of *Andromeda* on its right side, and the star μ of the girdle on its left; and, according to our observations, was 40° long; but it was curved, and the convex side thereof lay to the south; and near the head of the comet it made an angle of 4° with the circle which passed through the sun and the comet's head; but towards the other end it was inclined to that circle in an angle of about 10° or 11°; and the chord of the tail contained with that circle an angle of 8°. *Jan.* 13, the tail terminated between *Alamech* and *Algol*, with a light that was sensible enough; but with a faint light it ended over against the star κ in *Perseus's* side. The distance of the end of the tail from the circle passing through the sun and the comet was 3° 50′; and the inclination of the chord of the tail to that circle was $8\frac{1}{2}°$. *Jan.* 25 and 26, it shone with a faint light to the length of 6° or 7°; and for a night or two after, when there was a very clear sky, it extended to the length of 12°, or something more, with a light that was very faint and very hardly to be seen; but the axis thereof was exactly directed to the bright star in the eastern shoulder of *Auriga*, and therefore deviated from the opposition of the sun towards the north by an angle of 10°. Lastly, *Feb.* 10, with a telescope I observed the tail 2° long; for that fainter light which I spoke of did not appear through the glasses. But *Ponthæus* writes, that, on *Feb.* 7, he saw the tail 12° long. *Feb.* 25, the comet was without a tail, and so continued till it disappeared.

Now if one reflects upon the orbit described, and duly considers the other appearances of this comet, he will be easily satisfied that the bodies of comets are solid, compact, fixed, and durable, like the bodies of the planets; for if they were nothing else but the vapours or exhalations of the earth, of the sun, and other planets, this comet, in its passage by the neighbourhood of the sun, would have been immediately dissipated; for the heat of the sun is as the density of its rays, that is, reciprocally as the square of the distance of the places from the sun. Therefore, since on *Dec.* 8, when the comet was in its perihelion, the distance thereof from the centre of the sun was to the distance of the earth from the same as about 6 to 1000, the sun's heat on the comet was at that time to the heat of the summer-sun with us as 1000000 to 36, or as 28000 to 1. But the heat of boiling water is about 3 times greater than the heat which dry earth acquires from the summer-sun, as I have tried; and the heat of red-hot iron (if my conjecture is right) is about three or four times greater than the heat of boiling water. And therefore the heat which dry earth on the comet, while in its perihelion, might have conceived from the rays of the sun, was about 2000 times greater than the heat of red-hot iron. But by so fierce a heat, vapours and exhalations, and every volatile matter, must have been immediately consumed and dissipated.

This comet, therefore, must have conceived an immense heat from the sun, and retained that heat for an exceeding long time; for a globe of iron of an inch in diameter, exposed red-hot to the open air, will scarcely lose all its heat in an hour's time; but a greater globe would retain its heat longer in the proportion of its diameter, because the surface (in proportion to which it is cooled by the contact of the ambient air) is in that proportion less in respect of the quantity of the included hot matter; and therefore a globe of red hot iron equal to our earth, that is, about 40000000 feet in diameter, would scarcely cool in an equal number of days, or in above 50000 years. But I suspect that the duration of heat may, on account of some latent causes, increase in a yet less proportion than that of the diameter; and I should be glad that the true proportion was investigated by experiments.

It is farther to be observed, that the comet in the month of *December*, just after it had been heated by the sun, did emit a much longer tail, and much more splendid, than in the month of *November* before, when it had not yet arrived at its perihelion; and, universally, the greatest and most fulgent tails always arise from comets immediately after their passing by the neighbourhood of the sun. Therefore the heat received by the comet conduces to the greatness of the tail: from whence, I think, I may infer, that the tail is nothing else but a very fine vapour, which the head or nucleus of the comet emits by its heat.

But we have had three several opinions about the tails of comets; for some will have it that they are nothing else but the beams of the sun's light transmitted through the comets' heads, which they suppose to be transparent; others, that they proceed from the refraction which light suffers in passing from the comet's head to the earth; and, lastly, others, that they are a sort of clouds or vapour constantly rising from the comets' heads, and tending towards the parts opposite to the sun. The first is the opinion of such as are yet unacquainted with optics; for the beams of the sun are seen in a darkened room only in consequence of the light that is reflected from them by the little particles of dust and smoke which are always flying about in the air; and, for that reason, in air impregnated with thick smoke, those beams appear with great brightness, and move the sense vigorously; in a yet finer air they appear more faint, and are less easily discerned; but in the heavens, where there is no matter to reflect the light, they can never be seen at all. Light is not seen as it is in the beam, but as it is thence reflected to our eyes; for vision can be no otherwise produced than by rays falling upon the eyes; and, therefore, there must be some reflecting matter in those parts where the tails of the comets are seen: for otherwise, since all the celestial spaces are equally illuminated by the sun's light, no part of the heavens could appear with more splendor than another. The second opinion is liable to many difficulties. The tails of comets are never seen variegated with those colours which commonly are inseparable from refraction; and the distinct transmission of the light of the fixed stars and planets to us is a demonstration that the æther or celestial medium is not endowed with any refractive power: for as to what is alleged, that the fixed stars have been sometimes seen by the *Egyptians* environed with a *Coma* or *Capitlitium*, because that has but rarely happened, it is rather to be ascribed to a casual refraction of clouds; and so the radiation and scintillation of the fixed stars to the refractions both of the eyes and air; for upon laying a telescope to the eye, those radiations and scintillations immediately disappear. By the tremulous agitation of the air and ascending vapours, it happens that the rays of light are alternately turned aside from the narrow space of the pupil of the eye; but no such thing can have place in the much wider aperture of the object-glass of a telescope; and hence it is that a scintillation is occasioned in the former case, which ceases in the latter; and this cessation in the latter case is a demonstration of the regular transmission of light through the heavens, without any sensible refraction. But, to obviate an objection that may be made from the appearing of no tail in such comets as shine but with a faint light, as if the secondary rays were then too weak to affect the eyes, and for that reason it is that the tails of the fixed stars do not appear, we are to consider, that by the means of telescopes the light of the fixed stars may be augmented above an hundred fold, and yet no tails are seen; that the light of the planets is yet more copious without any tail; but that comets are seen sometimes with huge tails, when the light of their heads is but faint and dull. For so it happened in the comet of the year 1680, when in the month of *December* it was scarcely equal in light to the stars of the second magnitude, and yet emitted a notable tail, extending to the length of 40°, 50°, 60°, or 70°, and upwards; and afterwards, on the 27th and 28th of *January*, when the head appeared but as a star of the 7th magnitude, yet the tail (as we said above), with a light that was sensible enough, though faint, was stretched out to 6 or 7 degrees in length, and with a languishing light that was more difficultly seen, even to 12°, and upwards. But on the 9th and 10th of *February*, when to the naked eye the head appeared no more, through a telescope I viewed the tail of 2° in length. But farther; if the tail was owing to the refraction of the celestial matter, and did deviate from the opposition of the sun, according to the figure of the heavens, that deviation in the same places of the heavens should be always directed towards the same parts. But the comet of the year 1680, *December* 28^d. $8\frac{1}{2}^h$. P. M. at *London*, was seen in ♓ 8° 41′, with latitude north 28° 6′; while the sun was in ♑ 18° 26′. And the comet of the year 1577, *December* 29^d. was in ♓ 8° 41′, with latitude north 28° 40′, and the sun, as before, in about ♑ 18° 26′. In both cases the situation of the earth was the same, and the comet appeared in the same place of the heavens; yet in the former case the tail of the comet (as well by my observations as by the observations of others) deviated from the opposition of the sun towards the north by an angle of $4\frac{1}{2}$ degrees; whereas in the latter there was (according to the observations of *Tycho*) a deviation of 21 degrees towards the south. The refraction, therefore, of the heavens being thus disproved, it remains that the *phænomena* of the tails of comets must be derived from some reflecting matter.

And that the tails of comets do arise from their heads, and tend towards the parts opposite to the sun, is farther confirmed from the laws which the tails observe. As that, lying in the planes of the comets' orbits which pass through the sun, they constantly deviate from the opposition of the sun towards the parts which the comets' heads in their progress along these orbits have left. That to a spectator, placed in those planes, they appear in the parts directly opposite to the sun; but, as the spectator recedes from those planes, their deviation begins to appear, and daily becomes greater. That the deviation, *cæteris paribus*, appears less when the tail is more oblique to the orbit of the comet, as well as when the head of the comet approaches nearer to the sun, especially if the angle of deviation is estimated near the head of the comet. That the tails which have no deviation appear straight, but the tails which deviate are likewise bended into a certain curvature. That this curvature is greater when the deviation is greater; and is more sensible when the tail, *cæteris paribus*, is longer; for in the shorter tails the curvature is hardly to be perceived. That the angle of deviation is less near the comet's head, but greater towards the other end of the tail; and that because the convex side of the tail regards the parts from which the deviation is made, and which lie in a right line drawn out infinitely from the sun through the comet's head. And that the tails that are long and broad, and shine with a stronger light, appear more resplendent and more exactly defined on the convex than on the concave side. Upon which accounts it is plain that the *phænomena* of the tails of comets depend upon the motions of their heads, and by no means upon the places of the heavens in which their heads are seen; and that, therefore, the tails of comets do not proceed from the refraction of the heavens, but from their own heads, which furnish the matter that forms the tail. For, as in our air, the smoke of a heated body ascends either perpendicularly if the body is at rest, or obliquely if the body is moved obliquely, so in the heavens, where all bodies gravitate towards the sun, smoke and vapour must (as we have already said) ascend from the sun, and either rise perpendicularly if the smoking body is at rest, or obliquely if the body, in all the progress of its motion, is always leaving those places from which the upper or higher parts of the vapour had risen before; and that obliquity will be least where the vapour ascends with most velocity, to wit, near the smoking body, when that is near the sun. But, because the obliquity varies, the column of vapour will be incurvated; and because the vapour in the preceding sides is something more recent, that is, has ascended something more late from the body, it will therefore be something more dense on that side, and must on that account reflect more light, as well as be better defined. I add nothing concerning the sudden uncertain agitation of the tails of comets, and their irregular figures, which authors sometimes describe, because they may arise from the mutations of our air, and the motions of our clouds, in part obscuring those tails; or, perhaps, from parts of the *Via Lactea*, which might have been confounded with and mistaken for parts of the tails of the comets as they passed by.

But that the atmospheres of comets may furnish a supply of vapour great enough to fill so immense spaces, we may easily understand from the rarity of our own air; for the air near the surface of our earth possesses a space 850 times greater than water of the same weight; and therefore a cylinder of air 850 feet high is of equal weight with a cylinder of water of the same breadth, and but one foot high. But a cylinder of air reaching to the top of the atmosphere is of equal weight with a cylinder of water about 33 feet high: and, therefore, if from the whole cylinder of air the lower part of 850 feet high is taken away, the remaining upper part will be of equal weight with a cylinder of water 32 feet high: and from thence (and by the hypothesis, confirmed by many experiments, that the compression of air is as the weight of the incumbent atmosphere, and that the force of gravity is reciprocally as the square of the distance from the centre of the earth) raising a calculus, by Cor. Prop. XXII, Book II, I found, that, at the height of one semi-diameter of the earth, reckoned from the earth's surface, the air is more rare than with us in a far greater proportion than of the whole space within the orb of Saturn to a spherical space of one inch in diameter; and therefore if a sphere of our air of but one inch in thickness was equally rarefied with the air at the height of one semi-diameter of the earth from the earth's surface, it would fill all the regions of the planets to the orb of Saturn, and far beyond it. Wherefore since the air at greater distances is immensely rarefied, and the *coma* or atmosphere of comets is ordinarily about ten times higher, reckoning from their centres, than the surface of the nucleus, and the tails rise yet higher, they must therefore be exceedingly rare; and though, on account of the much thicker atmospheres of comets, and the great gravitation of their bodies towards the sun, as well as of the particles of their air and vapours mutually one towards another, it may happen that the air in the celestial spaces and in the tails of comets is not so vastly rarefied, yet from this computation it is plain that a very small quantity of air and vapour is abundantly sufficient to produce all the appearances of the tails of comets; for that they are, indeed, of a very notable rarity appears from the shining of the stars through them. The atmosphere of the earth, illuminated by the sun's light, though but of a few miles in thickness, quite obscures and extinguishes the light not only of all the stars, but even of the moon itself; whereas the smallest stars are seen to shine through the immense thickness of the tails of comets, likewise illuminated by the sun, without the least diminution of their splendor. Nor is the brightness of the tails of most comets ordinarily greater than that of our air, an inch or two in thickness, reflecting in a darkened room the light of the sun-beams let in by a hole of the window-shutter.

And we may pretty nearly determine the time spent during the ascent of the vapour from the comet's head to the extremity of the tail, by drawing a right line from the extremity of the tail to the sun, and marking the place where that right line intersects the comet's orbit; for the vapour that is now in the extremity of the tail, if it has ascended in a right line from the sun, must have begun to rise from the head at the time when the head was in the point of intersection. It is true, the vapour does not rise in a right line from the sun, but, retaining the motion which it had from the comet before its ascent, and compounding that motion with its motion of ascent, arises obliquely; and, therefore, the solution of the Problem will be more exact, if we draw the line which intersects the orbit parallel to the length of the tail; or rather (because of the curvilinear motion of the comet) diverging a little from the line or length of the tail. And by means of this principle I found that the vapour which, *January* 25, was in the extremity of the tail, had begun to rise from the head before *December* 11, and therefore had spent in its whole ascent 45 days; but that the whole tail which appeared on *December* 10 had finished its ascent in the space of the two days then elapsed from the time of the comet's being in its perihelion. The vapour, therefore, about the beginning and in the neighbourhood of the sun rose with the greatest velocity, and afterwards continued to ascend with a motion constantly retarded by its own gravity; and the higher it ascended, the more it added to the length of the tail; and while the tail continued to be seen, it was made up of almost all that vapour which had risen since the time of the comet's being in its perihelion; nor did that part of the vapour which had risen first, and which formed the extremity of the tail, cease to appear, till its too great distance, as well from the sun, from which it received its light, as from our eyes, rendered it invisible. Whence also it is that the tails of other comets which are short do not rise from their heads with a swift and continued motion, and soon after disappear, but are permanent and lasting columns of vapours and exhalations, which, ascending from the heads with a slow motion of many days, and partaking of the motion of the heads which they had from the beginning, continue to go along together with them through the heavens. From whence again we have another argument proving the celestial spaces to be free, and without resistance, since in them not only the solid bodies of the planets and comets, but also the extremely rare vapours of comets' tails, maintain their rapid motions with great freedom, and for an exceeding long time.

Kepler ascribes the ascent of the tails of the comets to the atmospheres of their heads; and their direction towards the parts opposite to the sun to the action of the rays of light carrying along with them the matter of the comets' tails; and without any great incongruity we may suppose, that, in so free spaces, so fine a matter as that of the æther may yield to the action of the rays of the sun's light, though those rays are not able sensibly to move the gross substances in our parts, which are clogged with so palpable a resistance. Another author thinks that there may be a sort of particles of matter endowed with a principle of levity, as well as others are with a power of gravity; that the matter of the tails of comets may be of the former sort, and that its ascent from the sun may be owing to its levity; but, considering that the gravity of terrestrial bodies is as the matter of the bodies, and therefore can be neither more nor less in the same quantity of matter, I am inclined to believe that this ascent may rather proceed from the rarefaction of the matter of the comets' tails. The ascent of smoke in a chimney is owing to the impulse of the air with which it is entangled. The air rarefied by heat ascends, because its specific gravity is diminished, and in its ascent carries along with it the smoke with which it is engaged; and why may not the tail of a comet rise from the sun after the same manner? For the sun's rays do not act upon the mediums which they pervade otherwise than by reflection and refraction; and those reflecting particles heated by this action, heat the matter of the æther which is involved with them. That matter is rarefied by the heat which it acquires, and because, by this rarefaction, the specific gravity with which it tended towards the sun before is diminished, it will ascend therefrom, and carry along with it the reflecting particles of which the tail of the comet is composed. But the ascent of the vapours is further promoted by their circumgyration about the sun, in consequence whereof they endeavour to recede from the sun, while the sun's atmosphere and the other matter of the heavens are either altogether quiescent, or are only moved with a slower circumgyration derived from the rotation of the sun. And these are the causes of the ascent of the tails of the comets in the neighbourhood of the sun, where their orbits are bent into a greater curvature, and the comets themselves are plunged into the denser and therefore heavier parts of the sun's atmosphere: upon which account they do then emit tails of an huge length; for the tails which then arise, retaining their own proper motion, and in the mean time gravitating towards the sun, must be revolved in ellipses about the sun in like manner as the heads are, and by that motion must always accompany the heads, and freely adhere to them. For the gravitation of the vapours towards the sun can no more force the tails to abandon the heads, and descend to the sun, than the gravitation of the heads can oblige them to fall from the tails. They must by their common gravity either fall together towards the sun, or be retarded together in their common ascent therefrom; and, therefore (whether from the causes already described, or from any others), the tails and heads of comets may easily acquire and freely retain any position one to the other, without disturbance or impediment from that common gravitation.

The tails, therefore, that rise in the perihelion positions of the comets will go along with their heads into far remote parts, and together with the heads will either return again from thence to us, after a long course of years, or rather will be there rarefied, and by degrees quite vanish away; for afterwards, in the descent of the heads towards the sun, new short tails will be emitted from the heads with a slow motion; and those tails by degrees will be augmented immensely, especially in such comets as in their perihelion distances descend as low as the sun's atmosphere; for all vapour in those free spaces is in a perpetual state of rarefaction and dilatation; and from hence it is that the tails of all comets are broader at their upper extremity than near their heads. And it is not unlikely but that the vapour, thus perpetually rarefied and dilated, may be at last dissipated and scattered through the whole heavens, and by little and little be attracted towards the planets by its gravity, and mixed with their atmosphere; for as the seas are absolutely necessary to the constitution of our earth, that from them, the sun, by its heat, may exhale a sufficient quantity of vapours, which, being gathered together into clouds, may drop down in rain, for watering of the earth, and for the production and nourishment of vegetables; or, being condensed with cold on the tops of mountains (as some philosophers with reason judge), may run down in springs and rivers; so for the conservation of the seas, and fluids of the planets, comets seem to be required, that, from their exhalations and vapours condensed, the wastes of the planetary fluids spent upon vegetation and putrefaction, and converted into dry earth, may be continually supplied and made up; for all vegetables entirely derive their growths from fluids, and afterwards, in great measure, are turned into dry earth by putrefaction; and a sort of slime is always found to settle at the bottom of putrefied fluids; and hence it is that the bulk of the solid earth is continually increased; and the fluids, if they are not supplied from without, must be in a continual decrease, and quite fail at last. I suspect, moreover, that it is chiefly from the comets that spirit comes, which is indeed the smallest but the most subtle and useful part of our air, and so much required to sustain the life of all things with us.

The atmospheres of comets, in their descent towards the sun, by running out into the tails, are spent and diminished, and become narrower, at least on that side which regards the sun; and in receding from the sun, when they less run out into the tails, they are again enlarged, if *Hevelius* has justly marked their appearances. But they are seen least of all just after they have been most heated by the sun, and on that account then emit the longest and most resplendent tails; and, perhaps, at the same time, the nuclei are environed with a denser and blacker smoke in the lowermost parts of their atmosphere; for smoke that is raised by a great and intense heat is commonly the denser and blacker. Thus the head of that comet which we have been describing, at equal distances both from the sun and from the earth, appeared darker after it had passed by its perihelion than it did before; for in the month of *December* it was commonly compared with the stars of the third magnitude, but in *November* with those of the first or second; and such as saw both appearances have described the first as of another and greater comet than the second. For, *November* 19, this comet appeared to a young man at *Cambridge*, though with a pale and dull light, yet equal to *Spica Virginis;* and at that time it shone with greater brightness than it did afterwards. And *Montenari*, *November* 20, st. vet. observed it larger than the stars of the first magnitude, its tail being then 2 degrees long. And Mr. *Storer* (by letters which have come into my hands) writes, that in the month of *December*, when the tail appeared of the greatest bulk and splendor, the head was but small, and far less than that which was seen in the month of *November* before sun-rising; and, conjecturing at the cause of the appearance, he judged it to proceed from there being a greater quantity of matter in the head at first, which was afterwards gradually spent.

And, which farther makes for the same purpose, I find, that the heads of other comets, which did put forth tails of the greatest bulk and splendor, have appeared but obscure and small. For in *Brazil, March* 5, 1668, 7[h]. P. M., *St. N. P. Valentinus Estancius* saw a comet near the horizon, and towards the south west, with a head so small as scarcely to be discerned, but with a tail above measure splendid, so that the reflection thereof from the sea was easily seen by those who stood upon the shore; and it looked like a fiery beam extended 23° in length from the west to south, almost parallel to the horizon. But this excessive splendor continued only three days, decreasing apace afterwards; and while the splendor was decreasing, the bulk of the tail increased: whence in *Portugal* it is said to have taken up one quarter of the heavens, that is, 45 degrees, extending from west to east with a very notable splendor, though the whole tail was not seen in those parts, because the head was always hid under the horizon: and from the increase of the bulk and decrease of the splendor of the tail, it appears that the head was then in its recess from the sun, and had been very near to it in its perihelion, as the comet of 1680 was. And we read, in the *Saxon* Chronicle, of a like comet appearing in the year 1106, *the star whereof was small and obscure* (as that of 1680), *but the splendour of its tail was very bright, and like a huge fiery beam stretched out in a direction between the east and north*, as *Hevelius* has it also from *Simeon*, the monk of *Durham.* This comet appeared in the beginning of *February*, about the evening, and towards the south west part of heaven; from whence, and from the position of the tail, we infer that the head was near the sun. *Matthew Paris* says, *It was distant from the sun by about a cubit, from three of the clock* (rather six) *till nine, putting forth a long tail.* Such also was that most resplendent comet described by *Aristotle*, lib. 1, *Meteor.* 6. *The head whereof could not be seen, because it had set before the sun, or at least was hid under the sun's rays; but next day it was seen as well as might be; for, having left the sun but a very little way, it set immediately after it. And the scattered light of the head, obscured by the too great splendour* (of the tail) *did not yet appear. But afterwards* (as *Aristotle says*) *when the splendour* (of the tail) *was now diminished* (the head of), *the comet recovered its native brightness; and the splendour* (of its tail) *reached now to a third part of the heavens* (that is, to 60°). *This appearance was in the winter season* (an. 4, Olymp. 101), *and, rising to* Orion's *girdle, it there vanished away.* It is true that the comet of 1618, which came out directly from under the sun's rays with a very large tail, seemed to equal, if not to exceed, the stars of the first magnitude; but, then, abundance of other comets have appeared yet greater than this, that put forth shorter tails; some of which are said to have appeared as big as Jupiter, others as big as Venus, or even as the moon.

We have said, that comets are a sort of planets revolved in very eccentric orbits about the sun; and as, in the planets which are without tails, those are commonly less which are revolved in lesser orbits, and nearer to the sun, so in comets it is probable that those which in their perihelion approach nearer to the sun are generally of less magnitude, that they may not agitate the sun too much by their attractions. But as to the transverse diameters of their orbits, and the periodic times of their revolutions, I leave them to be determined by comparing comets together which after long intervals of time return again in the same orbit. In the mean time, the following Proposition may give some light in that inquiry.

PROPOSITION XLII. PROBLEM XXII.

To correct a comet's trajectory found as above.

OPERATION 1. Assume that position of the plane of the trajectory which was determined according to the preceding proposition; and select three places of the comet, deduced from very accurate observations, and at great distances one from the other. Then suppose A to represent the time between the first observation and the second, and B the time between the second and the third; but it will be convenient that in one of those times the comet be in its perigeon, or at least not far from it. From those apparent places find, by trigonometric operations, the three true places of the comet in that assumed plane of the trajectory; then through the places found, and about the centre of the sun as the focus, describe a conic section by arithmetical operations, according to Prop. XXI., Book 1. Let the areas of this figure which are terminated by radii drawn from the sun to the places found be D and E; to wit, D the area between the first observation and the second, and E the area between the second and third; and let T represent the whole time in which the whole area D + E should be described with the velocity of the comet found by Prop. XVI., Book 1.

OPER. 2. Retaining the inclination of the plane of the trajectory to the plane of the ecliptic, let the longitude of the nodes of the plane of the trajectory be increased by the addition of 20 or 30 minutes, which call P. Then from the aforesaid three observed places of the comet let the three true places be found (as before) in this new plane; as also the orbit passing through those places, and the two areas of the same described between the two observations, which call d and e; and let t be the whole time in which the whole area $d + e$ should be described.

OPER. 3. Retaining the longitude of the nodes in the first operation, let the inclination of the plane of the trajectory to the plane of the ecliptic be increased by adding thereto 20′ or 30′, which call Q. Then from the aforesaid three observed apparent places of the comet let the three true places be found in this new plane, as well as the orbit passing through them, and the two areas of the same described between the observation, which call δ and ε; and let τ be the whole time in which the whole area $\delta + \varepsilon$ should be described.

Then taking C to 1 as A to B; and G to 1 as D to E; and g to 1 as d to e; and γ to 1 as δ to ε; let S be the true time between the first observation and the third; and, observing well the signs + and —, let such numbers m and n be found out as will make $2G - 2C, = mG - mg + nG - n\gamma$; and $2T - 2S = mT - mt + nT - n\tau$. And if, in the first operation, I represents the inclination of the plane of the trajectory to the plane of the ecliptic, and K the longitude of either node, then $I + nQ$ will be the true inclination of the plane of the trajectory to the plane of the ecliptic, and $K + mP$ the true longitude of the node. And, lastly, if in the first, second, and third operations, the quantities R, r, and ρ, represent the parameters of the trajectory, and the quantities $\frac{1}{L}, \frac{1}{l}, \frac{1}{\lambda}$, the transverse diameters of the same, then $R + mr - mR + n\rho - nR$ will be the true parameter, and $\frac{1}{L + ml - mL + n\lambda - nL}$ will be the true transverse diameter of the trajectory which the comet describes; and from the transverse diameter given the periodic time of the comet is also given. Q.E.I. But the periodic times of the revolutions of comets, and the transverse diameters of their orbits, cannot be accurately enough determined but by comparing comets together which appear at different times. If, after equal intervals of time, several comets are found to have described the same orbit, we may thence conclude that they are all but one and the same comet revolved in the same orbit; and then from the times of their revolutions the transverse diameters of their orbits will be given, and from those diameters the elliptic orbits themselves will be determined.

To this purpose the trajectories of many comets ought to be computed, supposing those trajectories to be parabolic; for such trajectories will always nearly agree with the *phænomena*, as appears not only from the parabolic trajectory of the comet of the year 1680, which I compared above with the observations, but likewise from that of the notable comet which appeared in the year 1664 and 1665, and was observed by *Hevelius*, who, from his own observations, calculated the longitudes and latitudes thereof, though with little accuracy. But from the same observations Dr. *Halley* did again compute its places; and from those new places determined its trajectory, finding its ascending node in ♊ 21° 13′ 55″; the inclination of the orbit to the plane of the ecliptic 21° 18′ 40″; the distance of its perihelion from the node, estimated in the comet's orbit, 49° 27′ 30″, its perihelion in ♌ 8° 40′ 30″, with heliocentric latitude south

16° 01′ 45″; the comet to have been in its perihelion *November* 24^{d}. 11^{h}. 52′ P.M. equal time at *London*, or 13^{h}. 8′ at *Dantzick*, O. S.; and that the *latus rectum* of the parabola was 410286 such parts as the sun's mean distance from the earth is supposed to contain 100000. And how nearly the places of the comet computed in this orbit agree with the observations, will appear from the annexed table, calculated by Dr. *Halley*.

Appar. Time at Dantzick.	The observed Distances of the Comet from		The observed Places.	The Places computed in the Orb.
December d. h. ′		° ′ ″	° ′ ″	° ′ ″
3.18.29½	The Lion's heart	46.24.20	Long. ♎ 7.01.00	♎ 7. 1.29
	The Virgin's spike	22.52.10	Lat. S. 21.39. 0	21.38.50
4.18. 1½	The Lion's heart	46. 2.45	Long. ♎ 6.15. 0	♎ 6.16. 5
	The Virgin's spike	23.52.40	Lat. S. 22.24. 0	22.24. 0
7.17.48	The Lion's heart	44.48. 0	Long. ♎ 3. 6. 0	♎ 3. 7.33
	The Virgin's spike	27.56.40	Lat. S. 25.22. 0	25.21.40
17.14.43	The Lion's heart	53.15.15	Long. ♌ 2.56. 0	♌ 2.56. 0
	Orion's right shoulder	45.43.30	Lat. S. 49.25. 0	49.25. 0
19. 9.25	Procyon	35.13.50	Long. ♊ 28.40.30	♊ 28.43. 0
	Bright star of Whale's jaw	52.56. 0	Lat. S. 45.48. 0	45.46. 0
20. 9.53½	Procyon	40.49. 0	Long. ♊ 13.03. 0	♊ 13. 5. 0
	Bright star of Whale's jaw	40.04. 0	Lat. S. 39.54. 0	39.53. 0
21. 9. 9½	Orion's right shoulder	26.21.25	Long. ♊ 2.16. 0	♊ 2.18.30
	Bright star of Whale's jaw	29.28. 0	Lat. S. 33.41. 0	33.39.40
22. 9. 0	Orion's right shoulder	29.47. 0	Long. ♉ 24.24. 0	♉ 24.27. 0
	Bright star of Whale's jaw	20.29.30	Lat. S. 27.45. 0	27.46. 0
26. 7.58	The bright star of Aries	23.20. 0	Long. ♉ 9. 0. 0	♉ 9. 2.28
	Aldebaran	26.44. 0	Lat. S. 12.36. 0	12.34.13
27. 6.45	The bright star of Aries	20.45. 0	Long. ♉ 7. 5.40	♉ 7. 8.45
	Aldebaran	28.10. 0	Lat. S. 10.23. 0	10.23.13
28. 7.39	The bright star of Aries	18.29. 0	Long. ♉ 5.24.45	♉ 5.27.52
	Palilicium	29.37. 0	Lat. S. 8.22.50	8.23.37
31. 6.45	Andromeda's girdle	30.48.10	Long. ♉ 2. 7.40	♉ 2. 8.20
	Palilicium	32.53.30	Lat. S. 4.13. 0	4.16.25
Jan. 1665 7. 7.37½	Andromeda's girdle	25.11. 0	Long. ♈ 28.24.47	♈ 28.24. 0
	Palilicium	37.12.25	Lat. N. 0.54. 0	0.53. 0
13. 7. 0	Andromeda's head	28. 7.10	Long. ♈ 27. 6.54	♈ 27. 6.39
	Palilicium	38.55.20	Lat. N. 3. 6.50	3. 7.40
24. 7.29	Andromeda's girdle	20.32.15	Long. ♈ 26.29.15	♈ 26.28.50
	Palilicium	40. 5. 0	Lat. N. 5.25.50	5.26. 0
Feb. 7. 8.37			Long. ♈ 27. 4.46	♈ 27.24.55
			Lat. N. 7. 3.29	7. 3.15
22. 8.46			Long. ♈ 28.29.46	♈ 28.29.58
			Lat. N. 8.12.36	8.10.25
March. 1. 8.16			Long. ♈ 29.18 15	♈ 29.18.20
			Lat. N. 8.36.26	8.36.12
7. 8.37			Long. ♉ 0. 2.48	♉ 0. 2.42
			Lat. N. 8.56.30	8.56.56

In *February*, the beginning of the year 1665, the first star of Aries, which I shall hereafter call γ, was in ♈ 28° 30′ 15″, with 7° 8′ 58″ north lat.; the second star of Aries was in ♈ 29° 17′ 18″, with 8° 28′ 16″ north lat.; and another star of the seventh magnitude, which I call A, was in ♈ 28° 24′ 45″, with 8° 28′ 33″ north lat. The comet *Feb.* 7ᵈ. 7ʰ. 30′ at *Paris* (that is, *Feb.* 7ᵈ. 8ʰ. 37′ at *Dantzick*) O. S. made a triangle with those stars γ and A, which was right-angled in γ; and the distance of the comet from the star γ was equal to the distance of the stars γ and A, that is, 1° 19′ 46″ of a great circle; and therefore in the parallel of the latitude of the star γ it was 1° 20′ 26″. Therefore if from the longitude of the star γ there be subducted the longitude 1° 20′ 26″, there will remain the longitude of the comet ♈ 27° 9′ 49″. M. *Auzout*, from this observation of his, placed the comet in ♈ 27° 0′, nearly; and, by the scheme in which Dr. *Hooke* delineated its motion, it was then in ♈ 26° 59′ 24″. I place it in ♈ 27° 4′ 46″, taking the middle between the two extremes.

From the same observations, M. *Auzout* made the latitude of the comet at that time 7° and 4′ or 5′ to the north; but he had done better to have made it 7° 3′ 29″, the difference of the latitudes of the comet and the star γ being equal to the difference of the longitude of the stars γ and A.

February 22ᵈ. 7ʰ. 30′ at *London*, that is, *February* 22ᵈ. 8ʰ. 46′ at *Dantzick*, the distance of the comet from the star A, according to Dr. *Hooke's* observation, as was delineated by himself in a scheme, and also by the observations of M. *Auzout*, delineated in like manner by M. *Petit*, was a fifth part of the distance between the star A and the first star of Aries, or 15′ 57″; and the distance of the comet from a right line joining the star A and the first of Aries was a fourth part of the same fifth part, that is, 4′; and therefore the comet was in ♈ 28° 29′ 46″, with 8° 12′ 36″ north lat.

March 1, 7ʰ. 0′ at *London*, that is, *March* 1, 8ʰ. 16′ at *Dantzick*, the comet was observed near the second star in Aries, the distance between them being to the distance between the first and second stars in Aries, that is, to 1° 33′, as 4 to 45 according to Dr. *Hooke*, or as 2 to 23 according to M. *Gottignies*. And, therefore, the distance of the comet from the second star in Aries was 8′ 16″ according to Dr. *Hooke*, or 8′ 5″ according to M. *Gottignies*; or, taking a mean between both, 8′ 10″. But, according to M. *Gottignies*, the comet had gone beyond the second star of Aries about a fourth or a fifth part of the space that it commonly went over in a day, to wit, about 1′ 35″ (in which he agrees very well with M. *Auzout*); or, according to Dr. *Hooke*, not quite so much, as perhaps only 1′. Wherefore if to the longitude of the first star in Aries we add 1′, and 8′ 10″ to its latitude, we shall have the longitude of the comet ♈ 29° 18′, with 8° 36′ 26″ north lat.

March 7, 7ʰ. 30′ at *Paris* (that is, *March* 7, 8ʰ. 37′ at *Dantzick*), from the observations of M. *Auzout*, the distance of the comet from the second star in Aries was equal to the distance of that star from the star

A, that is, 52,′29″; and the difference of the longitude of the comet and the second star in Aries was 45′ or 46′, or, taking a mean quantity, 45′ 30″; and therefore the comet was in ♉ 0° 2′ 48″. From the scheme of the observations of M. *Auzout*, constructed by M. *Petit*, *Hevelius* collected the latitude of the comet 8° 54′. But the engraver did not rightly trace the curvature of the comet's way towards the end of the motion; and *Hevelius*, in the scheme of M. *Auzout's* observations which he constructed himself, corrected this irregular curvature, and so made the latitude of the comet 8° 55′ 30″. And, by farther correcting this irregularity, the latitude may become 8° 56′, or 8° 57′.

This comet was also seen *March* 9, and at that time its place must have been in ♉ 0° 18′, with 9° 3½′ north lat. nearly.

This comet appeared three months together, in which space of time it travelled over almost six signs, and in one of the days thereof described almost 20 deg. Its course did very much deviate from a great circle, bending towards the north, and its motion towards the end from retrograde became direct; and, notwithstanding its course was so uncommon, yet by the table it appears that the theory, from beginning to end, agrees with the observations no less accurately than the theories of the planets usually do with the observations of them; but we are to subduct about 2′ when the comet was swiftest, which we may effect by taking off 12″ from the angle between the ascending node and the perihelion, or by making that angle 49° 27′ 18″. The annual parallax of both these comets (this and the preceding) was very conspicuous, and by its quantity demonstrates the annual motion of the earth in the *orbis magnus*.

This theory is likewise confirmed by the motion of that comet, which in the year 1683 appeared retrograde, in an orbit whose plane contained almost a right angle with the plane of the ecliptic, and whose ascending node (by the computation of Dr. *Halley*) was in ♍ 23° 23′; the inclination of its orbit to the ecliptic 83° 11′; its perihelion in ♊ 25° 29′ 30″; its perihelion distance from the sun 56020 of such parts as the radius of the *orbis magnus* contains 100000; and the time of its perihelion *July* 2^d. 3^h. 50′. And the places thereof, computed by Dr. *Halley* in this orbit, are compared with the places of the same observed by Mr. *Flamsted*, in the following table:—

1683 Eq. time.	Sun's place	Comet's Long. com.	Lat.Nor. comput.	Comet's Long. obs'd	Lat.Nor. observ'd	Diff. Long.	Diff. Lat.
d. h. ′	° ′ ″	° ′ ″	° ′ ″	° ′ ″	° ′ ″	′ ″	′ ″
July 13.12.55	♌ 1.02.30	♋ 13.05.42	29.28.13	♋ 13. 6.42	29.28.20	+ 1.00	+ 0.07
15.11.15	2.53.12	11.37.48	29.34. 0	11.39.43	29.34.50	+ 1.55	+ 0.50
17.10.20	4.45.45	10. 7. 6	29.33.30	10. 8.40	29.34. 0	+ 1.34	+ 0.30
23.13.40	10.38.21	5.10.27	28.51.42	5.11.30	28.50.28	+ 1.03	— 1.14
25.14. 5	12.35.28	3.27.53	24.24.47	3.27. 0	28.23.40	— 0.53	— 1. 7
31. 9.42	18.09.22	♊ 27.55. 3	26.22.52	♊ 27.54.24	26.22.25	— 0.39	— 0.27
31.14.55	18.21.53	27.41. 7	26.16.57	27.41. 8	26.14.50	+ 0. 1	— 2. 7
Aug. 2.14.56	20.17.16	25.29.32	25.16.19	25.28.46	25.17.28	— 0.46	+ 1. 9
4.10.49	22.02.50	23.18.20	24.10.49	23.16.55	24.12.19	— 1.25	+ 1.30
6.10. 9	23.56.45	20.42.23	22.47. 5	20.40.32	22.49. 5	— 1.51	+ 2. 0
9.10.26	26.50.52	16. 7.57	20. 6.37	16. 5.55	20. 6.10	— 2. 2	— 0.27
15.14. 1	♍ 2.47.13	3.30.48	11.37.33	3.26.18	11.32. 1	— 4.30	— 5.32
16.15.10	3.48. 2	0.43. 7	9.34.16	0.41.55	9.34.13	— 1.12	— 0. 3
18.15.44	5.45.33	♉ 24.52.53	5.11.15	♉ 24.49. 5	5. 9.11	— 3.48	— 2. 4
			South.		South		
22.14.44	9.35.49	11. 7.14	5.16.58	11.07.12	5.16.58	— 0. 2	— 0. 3
23.15.52	10.36.48	7. 2.18	8.17. 9	7. 1.17	8.16.41	— 1. 1	— 0.28
26.16. 2	13.31.10	♈ 24.45.31	16.38. 0	♈ 24.44.00	16.38.20	— 1.31	+ 0.20

This theory is yet farther confirmed by the motion of that retrograde comet which appeared in the year 1682. The ascending node of this (by Dr. *Halley's* computation) was in ♉ 21° 16′ 30″; the inclination of its orbit to the plane of the ecliptic 17° 56′ 00″; its perihelion in ♒ 2° 52′ 50″; its perihelion distance from the sun 58328 parts, of which the radius of the *orbis magnus* contains 100000; the equal time of the comet's being in its perihelion *Sept.* 4^d. 7^h. 39′. And its places, collected from Mr. *Flamsted's* observations, are compared with its places computed from our theory in the following table:—

1682 App. Time.	Sun's place	Comet's Lon. comp.	Lat.Nor comp.	Com. Long. observed.	Lat.Nor observ.	Diff. Long.	Diff. Lat.
d. h. ′	° ′ ″	° ′ ″	° ′ ″	° ′ ″	° ′ ″	′ ″	′ ″
Aug 19.16.38	♍ 7. 0. 7	♌ 18.14.28	25.50. 7	♌ 18.14.40	25.49.55	— 0.12	+ 0.12
20.15.38	7.55.52	24.46.23	26.14.42	24.46.22	26.12.52	+ 0. 1	+ 1.50
21. 8.21	8.36.14	29.37.15	26.20. 3	29.38.02	26.17.37	— 0.47	+ 2.26
22. 8. 8	9.33.55	♍ 6.29.53	26. 8.42	♍ 6.30. 3	26. 7.12	— 0.10	+ 1.30
29.08.20	16.22.40	♎ 12.37.54	18.37.47	♎ 12.37.49	18.34. 5	+ 0. 5	+ 3.42
30. 7.45	17.19.41	15.36. 1	17.26.43	15.35.18	17.27.17	+ 0.43	— 0.34
Sept. 1. 7.33	19.16. 9	20.30.53	15.13. 0	20.27. 4	15. 9.49	+ 3.49	+ 3.11
4. 7.22	22.11.28	25.42. 0	12.23.48	25.40.58	12.22. 0	+ 1. 2	+ 1.48
5. 7.32	23.10.29	27. 0.46	11.33.08	26.59.24	11.33.51	+ 1.22	— 0.43
8. 7.16	26. 5.58	29.58.44	9.26.46	29.58.45	9.26.43	— 0. 1	+ 0. 3
9. 7.26	27. 5. 9	♏ 0.44.10	8.49.10	♏ 0.44. 4	8.48.25	+ 0. 6	+ 0.45

This theory is also confirmed by the retrograde motion of the comet that appeared in the year 1723. The ascending node of this comet (according to the computation of Mr. *Bradley*, Savilian Professor of Astronomy at *Oxford*) was in ♈ 14° 16′. The inclination of the orbit to the plane of the ecliptic 49° 59′. Its perihelion was in ♉ 12° 15′ 20″. Its perihelion distance from the sun 998651 parts, of which the radius of the *orbis magnus* contains 1000000, and the equal time of its perihelion *September* 16^d.

$16^h. 10'$. The places of this comet computed in this orbit by Mr. *Bradley*, and compared with the places observed by himself, his uncle Mr. *Pound*, and Dr. *Halley*, may be seen in the following table.

1723 Eq. time.	Comet's Long. obs.	Lat.Nor obs.	Comet's Lon. com.	Lat.Nor comp.	Diff. Lon.	Diff. Lat.
d. h. ′	° ′ ″	° ′ ″	° ′ ″	° ′ ″	″	″
Oct. 9.8. 5	♒ 7.22.15	5. 2. 0	♒ 7.21.26	5. 2.47	+ 49	— 47
10.6.21	6.41.12	7.44.13	6.41.42	7.43.18	— 50	+ 55
12.7.22	5 39.58	11.55. 0	5.40.19	11.54.55	— 21	+ 5
14.8.57	4.59.49	14.43.50	5. 0.37	14.44. 1	— 48	— 11
15.6.35	4.47.41	15.40.51	4.47.45	15.40.55	— 4	— 4
21.6.22	4. 2.32	19.41.49	4. 2.21	19.42. 3	+ 11	— 14
22.6.24	3.59. 2	20. 8.12	3.59.10	20. 8.17	— 8	— 5
24.8. 2	3.55.29	20.55.18	3.55.11	20.55. 9	+ 18	+ 9
29.8.56	3.56.17	22.20.27	3.56.42	22.20.10	— 25	+ 17
30.6.20	3.58. 9	22.32.28	3.58.17	22.32.12	— 8	+ 16
Nov. 5.5.53	4.16.30	23.38.33	4.16.23	23.38. 7	+ 7	+ 26
8.7. 6	4:29.36	24. 4.30	4.29.54	24. 4.40	— 18	— 10
14.6.20	5. 2.16	24.48.46	5. 2.51	24.48.16	— 35	+ 30
20.7.45	5.42.20	25.24.45	5.43.13	25.25.17	— 53	— 32
Dec. 7.6.45	8. 4.13	26.54.18	8. 3.55	26.53.42	+ 18	+ 36

From these examples it is abundantly evident that the motions of comets are no less accurately represented by our theory than the motions of the planets commonly are by the theories of them; and, therefore, by means of this theory, we may enumerate the orbits of comets, and so discover the periodic time of a comet's revolution in any orbit; whence, at last, we shall have the transverse diameters of their elliptic orbits and their aphelion distances.

That retrograde comet which appeared in the year 1607 described an orbit whose ascending node (according to Dr. *Halley's* computation) was in ♉ 20° 21′; and the inclination of the plane of the orbit to the plane of the ecliptic 17° 2′; whose perihelion was in ♒ 2° 16′; and its perihelion distance from the sun 58680 of such parts as the radius of the *orbis magnus* contains 100000; and the comet was in its perihelion *October* $16^d. 3^h. 50'$; which orbit agrees very nearly with the orbit of the comet which was seen in 1682. If these were not two different comets, but one and the same, that comet will finish one revolution in the space of 75 years; and the greater axis of its orbit will be to the greater axis of the *orbis magnus* as $\sqrt[3]{:75 \times 75}$ to 1, or as 1778 to 100, nearly. And the aphelion distance of this comet from the sun will be to the mean distance of the earth from the sun as about 35 to 1; from which data it will be no hard matter to determine the elliptic orbit of this comet. But these things are to be supposed on condition, that, after the space of 75 years, the same comet shall return again in the same orbit. The other comets seem to ascend to greater heights, and to require a longer time to perform their revolutions.

But, because of the great number of comets, of the great distance of their aphelions from the sun, and of the slowness of their motions in the aphelions, they will, by their mutual gravitations, disturb each other; so that their eccentricities and the times of their revolutions will be sometimes a little increased, and sometimes diminished. Therefore we are not to expect that the same comet will return exactly in the same orbit, and in the same periodic times: it will be sufficient if we find the changes no greater than may arise from the causes just spoken of.

And hence a reason may be assigned why comets are not comprehended within the limits of a zodiac, as the planets are; but, being confined to no bounds, are with various motions dispersed all over the heavens; namely, to this purpose, that in their aphelions, where their motions are exceedingly slow, receding to greater distances one from another, they may suffer less disturbance from their mutual gravitations: and hence it is that the comets which descend the lowest, and therefore move the slowest in their aphelions, ought also to ascend the highest.

The comet which appeared in the year 1680 was in its perihelion less distant from the sun than by a sixth part of the sun's diameter; and because of its extreme velocity in that proximity to the sun, and some density of the sun's atmosphere, it must have suffered some resistance and retardation; and therefore, being attracted something nearer to the sun in every revolution, will at last fall down upon the body of the sun. Nay, in its aphelion, where it moves the slowest, it may sometimes happen to be yet farther retarded by the attractions of other comets, and in consequence of this retardation descend to the sun. So fixed stars, that have been gradually wasted by the light and vapours emitted from them for a long time, may be recruited by comets that fall upon them; and from this fresh supply of new fuel those old stars, acquiring new splendor, may pass for new stars. Of this kind are such fixed stars as appear on a sudden, and shine with a wonderful brightness at first, and afterwards vanish by little and little. Such was that star which appeared in *Cassiopeia's* chair; which *Cornelius Gemma* did not see upon the 8th of *November*, 1572, though he was observing that part of the heavens upon that very night, and the sky was perfectly serene; but the next night (*November* 9) he saw it shining much brighter than any of the fixed stars, and scarcely inferior to *Venus* in splendor. *Tycho Brahe* saw it upon the 11th of the same month, when it shone with the greatest lustre; and from that time he observed it to decay by little and little; and in 16 months' time it entirely disappeared. In the month of *November*, when it first appeared, its light was equal to that of *Venus*. In the month of *December* its light was a little diminished, and was now become equal to that of *Jupiter*. In *January* 1573 it was less than *Jupiter*, and greater than *Sirius*; and about the end of *February* and the beginning of *March* became equal to that star. In the months of *April* and *May* it was equal to a star of the second magnitude; in *June*, *July*, and *August*, to a star of the third magnitude; in *September*, *October*, and *November*, to those of the fourth magnitude; in *December* and *January* 1574 to those of the fifth; in *February* to those of the sixth magnitude; and in *March* it entirely vanished. Its colour at the beginning was clear, bright, and inclining to white; afterwards it turned a little yellow; and in *March* 1573 it became ruddy, like *Mars* or *Aldebaran:* in *May* it turned to a kind of dusky whiteness, like that we observe in *Saturn;* and that colour it retained ever after, but growing always more and more obscure. Such also was the star in the right foot of *Serpentarius*, which *Kepler's* scholars first observed *September* 30, O.S. 1604, with a light exceeding that of *Jupiter*, though the night before it was not to be seen; and from that time it decreased by little and little, and in 15 or 16 months entirely disappeared. Such a new star appearing with an unusual splendor is said to have moved *Hipparchus* to observe, and make a catalogue of, the fixed stars. As to those fixed stars that appear and disappear by turns, and increase slowly and by degrees, and scarcely ever exceed the stars of the third magnitude, they seem to be of another kind, which revolve about their axes, and, having a light and a dark side, shew those two different sides by turns. The vapours which arise from the sun, the fixed stars, and the tails of the comets, may meet at last with, and fall into, the atmospheres of the planets by their gravity, and there be condensed and turned into water and humid spirits; and from thence, by a slow heat, pass gradually into the form of salts, and sulphurs, and tinctures, and mud, and clay, and sand, and stones, and coral, and other terrestrial substances.

GENERAL SCHOLIUM.

The hypothesis of vortices is pressed with many difficulties. That every planet by a radius drawn to the sun may describe areas proportional to the times of description, the periodic times of the several parts of the vortices should observe the duplicate proportion of their distances from the sun; but that the periodic times of the planets may obtain the sesquiplicate proportion of their distances from the sun, the periodic times of the parts of the vortex ought to be in the sesquiplicate proportion of their distances. That the smaller vortices may maintain their lesser revolutions about *Saturn*, *Jupiter*, and other planets, and swim quietly and undisturbed in the greater vortex of the sun, the periodic times of the parts of the sun's vortex should be equal; but the rotation of the sun and planets about their axes, which ought to correspond with the motions of their vortices, recede far from all these proportions. The motions of the comets are exceedingly regular, are governed by the same laws with the motions of the planets, and can by no means be accounted for by the hypothesis of vortices; for comets are carried with very eccentric motions through all parts of the heavens indifferently, with a freedom that is incompatible with the notion of a vortex.

Bodies projected in our air suffer no resistance but from the air. Withdraw the air, as is done in Mr. *Boyle's* vacuum, and the resistance ceases; for in this void a bit of fine down and a piece of solid gold descend with equal velocity. And the parity of reason must take place in the celestial spaces above the earth's atmosphere; in which spaces, where there is no air to resist their motions, all bodies will move with the greatest freedom; and the planets and comets will constantly pursue their revolutions in orbits given in kind and position, according to the laws above explained; but though these bodies may, indeed, persevere in their orbits by the mere laws of gravity, yet they could by no means have at first derived the regular position of the orbits themselves from those laws.

The six primary planets are revolved about the sun in circles concentric with the sun, and with motions directed towards the same parts, and almost in the same plane. Ten moons are revolved about the earth, Jupiter and Saturn, in circles concentric with them, with the same direction of motion, and nearly in the planes of the orbits of those planets; but it is not to be conceived that mere mechanical causes could give birth to so many regular motions, since the comets range over all parts of the heavens in very eccentric orbits; for by that kind of motion they pass easily through the orbs of the planets, and with great rapidity; and in their aphelions, where they move the slowest, and are detained the longest, they recede to the greatest distances from each other, and thence suffer the least disturbance from their mutual attractions. This most beautiful system of the sun, planets, and comets, could only proceed from the counsel and dominion of an intelligent and powerful Being. And if the fixed stars are the centres of other like systems, these, being formed by the like wise counsel, must be all subject to the dominion of One; especially since the light of the fixed stars is of the same nature with the light of the sun, and from every system light passes into all the other systems: and lest the systems of the fixed stars should, by their gravity, fall on each other mutually, he hath placed those systems at immense distances one from another.

This Being governs all things, not as the soul of the world, but as Lord over all; and on account of his dominion he is wont to be called *Lord God* παντοκράτωρ, or *Universal Ruler;* for *God* is a relative word, and has a respect to servants; and *Deity* is the dominion of God not over his own body, as those imagine who fancy God to be the soul of the world, but over servants. The Supreme God is a Being eternal, infinite, absolutely perfect; but a being, however perfect, without dominion, cannot be said to be Lord God; for we say, my God, your God, the God of *Israel*, the God of Gods, and Lord of Lords; but we do not say, my Eternal, your Eternal, the Eternal of *Israel*, the Eternal of Gods; we do not say, my Infinite, or my Perfect: these are titles which have no respect to servants. The word *God** usually signifies *Lord;* but every lord is not a God. It is the dominion of a spiritual being which constitutes a God: a true, supreme, or imaginary dominion makes a true, supreme, or imaginary God. And from his true dominion it follows that the true God is a living, intelligent, and powerful Being; and, from his other perfections, that he is supreme, or most perfect. He is eternal and infinite, omnipotent and omniscient; that is, his duration reaches from eternity to eternity; his presence from infinity to infinity; he governs all things, and knows all things that are or can be done. He is not eternity or infinity, but eternal and infinite; he is not duration or space, but he endures and is present. He endures for ever, and is every where present; and by existing always and every where, he constitutes duration and space. Since every particle of space is *always*, and every indivisible moment of duration is *every where*, certainly the Maker and Lord of all things cannot be *never* and *no where.* Every soul that has perception is, though in different times and in different organs of sense and motion, still the same indivisible person. There are given successive parts in duration, co-existent parts in space, but neither the one nor the other in the person of a man, or his thinking principle; and much less can they be found in the thinking substance of God. Every man, so far as he is a thing that has perception, is one and the same man during his whole life, in all and each of his organs of sense. God is the same God, always and every where. He is omnipresent not *virtually* only, but also *substantially;* for virtue cannot subsist without substance. In him† are all things contained and moved; yet neither affects the other: God suffers nothing from the motion of bodies; bodies find no resistance from the omnipresence of God. It is allowed by all that the Supreme God exists necessarily; and by the same necessity he exists *always* and *every where.* Whence also he is all similar, all eye, all ear, all brain, all arm, all power to perceive, to understand, and to act; but in a manner not at all human, in a manner not at all corporeal, in a manner utterly unknown to us. As a blind man has no idea of colours, so have we no idea of the manner by

* Dr. *Pocock* derives the Latin word *Deus* from the *Arabic du* (in the oblique case *di),* which signifies *Lord.* And in this sense princes are called *gods, Psal.* lxxxii. ver. 6; and *John* x. ver. 35. And *Moses* is called a *god* to his brother *Aaron,* and a *god* to *Pharaoh (Exod.* iv. ver. 16; and vii. ver. 1). And in the same sense the souls of dead princes were formerly, by the Heathens, called *gods,* but falsely, because of their want of dominion.

† This was the opinion of the Ancients. So *Pythagoras,* in *Cicer. de Nat. Deor.* lib. i. *Thales, Anaxagoras, Virgil,* Georg. lib. iv. ver. 220; and Æneid, lib. vi. ver. 721. *Philo Allegor,* at the beginning of lib. i. *Aratus,* in his Phænom. at the beginning. So also the sacred writers; as St. *Paul, Acts,* xvii. ver. 27, 28. St. *John's* Gosp. chap. xiv. ver. 2. *Moses,* in *Deut.* iv. ver. 39; and x. ver. 14. *David, Psal.* cxxxix. ver. 7, 8, 9. *Solomon,* 1 *Kings,* viii. ver. 27. *Job,* xxii. ver. 12, 13, 14. *Jeremiah,* xxiii. ver. 23, 24. The Idolaters supposed the sun, moon, and stars, the souls of men, and other parts of the world, to be parts of the Supreme God, and therefore to be worshipped; but erroneously.

which the all-wise God perceives and understands all things. He is utterly void of all body and bodily figure, and can therefore neither be seen, nor heard, nor touched; nor ought he to be worshipped under the representation of any corporeal thing. We have ideas of his attributes, but what the real substance of any thing is we know not. In bodies, we see only their figures and colours, we hear only the sounds, we touch only their outward surfaces, we smell only the smells, and taste the savours; but their inward substances are not to be known either by our senses, or by any reflex act of our minds: much less, then, have we any idea of the substance of God. We know him only by his most wise and excellent contrivances of things, and final causes; we admire him for his perfections; but we reverence and adore him on account of his dominion: for we adore him as his servants; and a god without dominion, providence, and final causes, is nothing else but Fate and Nature. Blind metaphysical necessity, which is certainly the same always and every where, could produce no variety of things. All that diversity of natural things which we find suited to different times and places could arise from nothing but the ideas and will of a Being necessarily existing. But, by way of allegory, God is said to see, to speak, to laugh, to love, to hate, to desire, to give, to receive, to rejoice, to be angry, to fight, to frame, to work, to build; for all our notions of God are taken from the ways of mankind by a certain similitude, which, though not perfect, has some likeness, however. And thus much concerning God; to discourse of whom from the appearances of things, does certainly belong to Natural Philosophy.

Hitherto we have explained the phænomena of the heavens and of our sea by the power of gravity, but have not yet assigned the cause of this power. This is certain, that it must proceed from a cause that penetrates to the very centres of the sun and planets, without suffering the least diminution of its force; that operates not according to the quantity of the surfaces of the particles upon which it acts (as mechanical causes use to do), but according to the quantity of the solid matter which they contain, and propagates its virtue on all sides to immense distances, decreasing always in the duplicate proportion of the distances. Gravitation towards the sun is made up out of the gravitations towards the several particles of which the body of the sun is composed; and in receding from the sun decreases accurately in the duplicate proportion of the distances as far as the orb of Saturn, as evidently appears from the quiescence of the aphelions of the planets; nay, and even to the remotest aphelions of the comets, if those aphelions are also quiescent. But hitherto I have not been able to discover the cause of those properties of gravity from phænomena, and I frame no hypotheses; for whatever is not deduced from the phænomena is to be called an hypothesis; and hypotheses, whether metaphysical or physical, whether of occult qualities or mechanical, have no place in experimental philosophy. In this philosophy particular propositions are inferred from the phænomena, and afterwards rendered general by induction. Thus it was that the impenetrability, the mobility, and the impulsive force of bodies, and the laws of motion and of gravitation, were discovered. And to us it is enough that gravity does really exist, and act according to the laws which we have explained, and abundantly serves to account for all the motions of the celestial bodies, and of our sea.

And now we might add something concerning a certain most subtle Spirit which pervades and lies hid in all gross bodies; by the force and action of which Spirit the particles of bodies mutually attract one another at near distances, and cohere, if contiguous; and electric bodies operate to greater distances, as well repelling as attracting the neighbouring corpuscles; and light is emitted, reflected, refracted, inflected, and heats bodies; and all sensation is excited, and the members of animal bodies move at the command of the will, namely, by the vibrations of this Spirit, mutually propagated along the solid filaments of the nerves, from the outward organs of sense to the brain, and from the brain into the muscles. But these are things that cannot be explained in few words, nor are we furnished with that sufficiency of experiments which is required to an accurate determination and demonstration of the laws by which this electric and elastic Spirit operates.

END OF THE MATHEMATICAL PRINCIPLES.

THE SYSTEM OF THE WORLD.

THE

SYSTEM OF THE WORLD.

It was the ancient opinion of not a few, in the earliest ages of philosophy, that the fixed stars stood immoveable in the highest parts of the world; that under the fixed stars the planets were carried about the sun; that the earth, as one of the planets, described an annual course about the sun, while by a diurnal motion it was in the mean time revolved about its own axis; and that the sun, as the common fire which served to warm the whole, was fixed in the centre of the universe.

This was the philosophy taught of old by *Philolaus, Aristarchus* of *Samos, Plato* in his riper years, and the whole sect of the *Pythagoreans;* and this was the judgment of *Anaximander*, more ancient than any of them; and of that wise king of the *Romans, Numa Pompilius,* who, as a symbol of the figure of the world with the sun in the centre, erected a temple in honour of *Vesta,* of a round form, and ordained perpetual fire to be kept in the middle of it.

The *Egyptians* were early observers of the heavens; and from them, probably, this philosophy was spread abroad among other nations; for from them it was, and the nations about them, that the *Greeks,* a people of themselves more addicted to the study of philology than of nature, derived their first, as well as soundest, notions of philosophy; and in the vestal ceremonies we may yet trace the ancient spirit of the *Egyptians;* for it was their way to deliver their mysteries, that is, their philosophy of things above the vulgar way of thinking, under the veil of religious rites and hieroglyphic symbols.

It is not to be denied but that *Anaxagoras, Democritus,* and others, did now and then start up, who would have it that the earth possessed the centre of the world, and that the stars of all sorts were revolved towards the west about the earth quiescent in the centre, some at a swifter, others at a slower rate.

However, it was agreed on both sides that the motions of the celestial bodies were performed in spaces altogether free and void of resistance. The whim of solid orbs was of a later date, introduced by *Eudoxus, Calippus,* and *Aristotle;* when the ancient philosophy began to decline, and to give place to the new prevailing fictions of the Greeks.

But, above all things, the phænomena of comets can by no means consist with the notion of solid orbs. The *Chaldeans*, the most learned astronomers of their time, looked upon the comets (which of ancient times before had been numbered among the celestial bodies) as a particular sort of planets, which, describing very eccentric orbits, presented themselves to our view only by turns, viz., once in a revolution, when they descended into the lower parts of their orbits.

And as it was the unavoidable consequence of the hypothesis of solid orbs, while it prevailed, that the comets should be thrust down below the moon, so no sooner had the late observations of astronomers restored the comets to their ancient places in the higher heavens, but these celestial spaces were at once cleared of the incumbrance of solid orbs, which by these observations were broke into pieces, and discarded for ever.

Whence it was that the planets came to be retained within any certain bounds in these free spaces, and to be drawn off from the rectilinear courses, which, left to themselves, they should have pursued, into regular revolutions in curvilinear orbits, are questions which we do not know how the ancients explained; and probably it was to give some sort of satisfaction to this difficulty that solid orbs were introduced.

The later philosophers pretend to account for it either by the action of certain vortices, as *Kepler* and *Des Cartes;* or by some other principle of impulse or attraction, as *Borelli*, *Hooke*, and others of our nation; for, from the laws of motion, it is most certain that these effects must proceed from the action of some force or other.

But our purpose is only to trace out the quantity and properties of this force from the phænomena (p. 218), and to apply what we discover in some simple cases as principles, by which, in a mathematical way, we may estimate the effects thereof in more involved cases; for it would be endless and impossible to bring every particular to direct and immediate observation.

We said, *in a mathematical way*, to avoid all questions about the nature or quality of this force, which we would not be understood to determine by any hypothesis; and therefore call it by the general name of a centripetal force, as it is a force which is directed towards some centre; and as it regards more particularly a body in that centre, we call it circum-solar, circum-terrestrial, circum-jovial; and in like manner in respect of other central bodies.

That by means of centripetal forces the planets may be retained in certain orbits, we may easily understand, if we consider the motions of projectiles (p. 75, 76, 77); for a stone projected is by the pressure of its own weight forced out of the rectilinear path, which by the projection alone it should have pursued, and made to describe a curve line in the air; and through that crooked way is at last brought down to the ground; and the greater the velocity is with which it is projected, the farther it goes before it falls to the earth. We may therefore suppose the velocity to be so increased, that it would describe an arc of 1, 2, 5, 10, 100, 1000 miles before it arrived at the earth, till at last, exceeding the limits of the earth, it should pass quite by without touching it.

Let AFB represent the surface of the earth, C its centre, VD, VE, VF, the curve lines which a body would describe, if projected in an horizontal direction from the top of an high mountain successively with more and

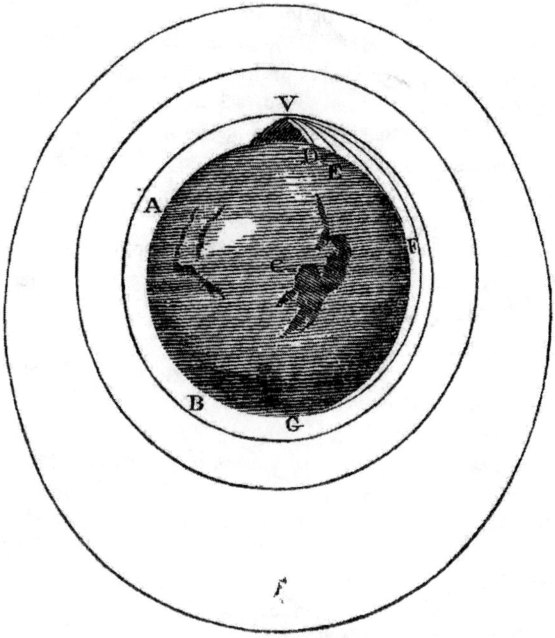

more velocity (p. 400); and, because the celestial motions are scarcely retarded by the little or no resistance of the spaces in which they are performed, to keep up the parity of cases, let us suppose either that there is no air about the earth, or at least that it is endowed with little or no power of resisting; and for the same reason that the body projected with a less velocity describes the lesser arc VD, and with a greater velocity the greater arc VE, and, augmenting the velocity, it goes farther and farther to F and G, if the velocity was still more and more augmented, it would reach at last quite beyond the circumference of the earth, and return to the mountain from which it was projected.

And since the areas which by this motion it describes by a radius drawn to the centre of the earth are (by Prop. 1, Book 1, *Princip. Math.*) proportional to the times in which they are described, its velocity, when it returns to the mountain, will be no less than it was at first; and, retaining the same velocity, it will describe the same curve over and over, by the same law.

But if we now imagine bodies to be projected in the directions of lines parallel to the horizon from greater heights, as of 5, 10, 100, 1000, or more miles, or rather as many semi-diameters of the earth, those bodies, according to their different velocity, and the different force of gravity in different heights, will describe arcs either concentric with the earth, or variously eccentric, and go on revolving through the heavens in those trajectories, just as the planets do in their orbs.

As when a stone is projected obliquely, that is, any way but in the perpendicular direction, the perpetual deflection thereof towards the earth from the right line in which it was projected is a proof of its gravitation to the earth, no less certain than its direct descent when only suffered to fall freely from rest; so the deviation of bodies moving in free spaces from rectilinear paths, and perpetual deflection therefrom towards any place, is a sure indication of the existence of some force which from all quarters impels those bodies towards that place.

And as, from the supposed existence of gravity, it necessarily follows that all bodies about the earth must press downwards, and therefore must either descend directly to the earth, if they are let fall from rest, or at least perpetually deviate from right lines towards the earth, if they are projected obliquely; so from the supposed existence of a force directed to any centre, it will follow, by the like necessity, that all bodies upon which this force acts must either descend directly to that centre, or at least deviate perpetually towards it from right lines, if otherwise they should have moved obliquely in these right lines.

And how from the motions given we may infer the forces, or from the forces given we may determine the motions, is shewn in the two first Books of our *Principles of Philosophy*.

If the earth is supposed to stand still, and the fixed stars to be revolved in free spaces in the space of 24 hours, it is certain the forces by which the fixed stars are retained in their orbs are not directed to the earth, but to the centres of the several orbs, that is, of the several parallel circles, which the fixed stars, declining to one side and the other from the equator, describe daily; also that by radii drawn to the centres of those orbs the fixed stars describe areas exactly proportional to the times of description. Then, because the periodic times are equal (by Cor. III, Prop. IV, Book 1), it follows that the centripetal forces are as the radii of the several orbs, and that they will perpetually revolve in the same orbs. And the like consequences may be drawn from the supposed diurnal motion of the planets.

That forces should be directed to no body on which they physically depend, but to innumerable imaginary points in the axis of the earth, is an hypothesis too incongruous. It is more incongruous still that those forces should increase exactly in proportion of the distances from this axis; for this is an indication of an increase to immensity, or rather to infinity; whereas the forces of natural things commonly decrease in receding from the fountain from which they flow. But, what is yet more absurd, neither are the areas described by the same star proportional to the times, nor are its revolutions performed in the same orb; for as the star recedes from the neighbouring pole, both areas and orb increase; and from the increase of the area it is demonstrated that the forces are not directed to the axis of the earth. And this difficulty (Cor. 1, Prop. II) arises from the twofold motion that is observed in the fixed stars, one diurnal round the axis of the earth, the other exceedingly slow round the axis of the ecliptic. And the explication thereof requires a composition of forces so perplexed and so variable, that it is hardly to be reconciled with any physical theory.

That there are centripetal forces actually directed to the bodies of the sun, of the earth, and other planets, I thus infer.

The moon revolves about our earth, and by radii drawn to its centre (p. 390) describes areas nearly proportional to the times in which they are described, as is evident from its velocity compared with its apparent diameter; for its motion is slower when its diameter is less (and therefore its distance greater), and its motion is swifter when its diameter is greater.

The revolutions of the satellites of Jupiter about that planet are more regular (p. 386); for they describe circles concentric with Jupiter by equable motions, as exactly as our senses can distinguish.

And so the satellites of Saturn are revolved about this planet with motions nearly (p. 387) circular and equable, scarcely disturbed by any eccentricity hitherto observed.

That Venus and Mercury are revolved about the sun, is demonstrable from their moon-like appearances (p. 388): when they shine with a full face, they are in those parts of their orbs which in respect of the earth lie beyond the sun; when they appear half full, they are in those parts which lie over against the sun; when horned, in those parts which lie between the earth and the sun; and sometimes they pass over the sun's disk, when directly interposed between the earth and the sun.

And Venus, with a motion almost uniform, describes an orb nearly circular and concentric with the sun.

But Mercury, with a more eccentric motion, makes remarkable approaches to the sun, and goes off again by turns; but it is always swifter as it is near to the sun, and therefore by a radius drawn to the sun still describes areas proportional to the times.

Lastly, that the earth describes about the sun, or the sun about the earth, by a radius from the one to the other, areas exactly proportional to the times, is demonstrable from the apparent diameter of the sun compared with its apparent motion.

These are astronomical experiments; from which it follows, by Prop. I,

II, III, in the first Book of our *Principles*, and their Corollaries (p. 212, 213, 214), that there are centripetal forces actually directed (either accurately or without considerable error) to the centres of the earth, of Jupiter, of Saturn, and of the sun. In Mercury, Venus, Mars, and the lesser planets, where experiments are wanting, the arguments from analogy must be allowed in their place.

That those forces (p. 212, 213, 214) decrease in the duplicate proportion of the distances from the centre of every planet, appears by Cor. VI, Prop. IV, Book 1; for the periodic times of the satellites of Jupiter are one to another (p. 386, 387) in the sesquiplicate proportion of their distances from the centre of this planet.

This proportion has been long ago observed in those satellites; and Mr. *Flamsted*, who had often measured their distances from Jupiter by the micrometer, and by the eclipses of the satellites, wrote to me, that it holds to all the accuracy that possibly can be discerned by our senses. And he sent me the dimensions of their orbits taken by the micrometer, and reduced to the mean distance of Jupiter from the earth, or from the sun, together with the times of their revolutions, as follows:—

The greatest elongation of the satelites from the centre of Jupiter as seen from the sun.			The periodic times of their revolutions.			
	′ ″	″	d	h	′	″
1st	1 48	or 108	1	18	28	36
2d	3 01	or 181	3	13	17	54
3d	4 46	or 286	7	03	59	36
4th	8 13½	or 493½	16	18	5	13

Whence the sesquiplicate proportion may be easily seen. For example; the $16^d. 18^h. 05' 13''$ is to the time $1^d. 18^h. 28' 36''$ as $493\frac{1}{2}'' \times \sqrt{493\frac{1}{2}''}$ to $108'' \times \sqrt{108''}$, neglecting those small fractions which, in observing, cannot be certainly determined.

Before the invention of the micrometer, the same distances were determined in semi-diameters of Jupiter thus:—

Distance of the	1st	2d	3d	4th
By Galileo, . . .	6	10	16	28
" Simon Marius .	6	10	16	26
" Cassini . . .	5	8	13	23
" Borelli, more exactly . . .	5⅔	8⅔	14	24⅔

After the invention of the micrometer:—

By Townley . . .	5,51	8,78	13,47	24,72
" Flamsted . . .	5,31	8,85	13,98	24,23
More accurately by the eclipses . .	5,578	8,876	14,159	24,903

And the periodic times of those satellities, by the observations of Mr. *Flamsted*, are $1^d. 18^h. 28' 36''$ | $3^d. 13^h. 17' 54''$ | $7^d. 3^h. 59' 36''$ | $16^d. 18^h. 5' 13''$, as above.

And the distances thence computed are 5,578 | 8,878 | 14,168 | 24,968, accurately agreeing with the distances by observation.

Cassini assures us (p. 388, 389) that the same proportion is observed in the circum-saturnal planets. But a longer course of observations is required before we can have a certain and accurate theory of those planets.

In the circum-solar planets, Mercury and Venus, the same proportion holds with great accuracy, according to the dimensions of their orbs, as determined by the observations of the best astronomers.

That Mars is revolved about the sun is demonstrated from the phases which it shews, and the proportion of its apparent diameters (p. 388, 389, and 390); for from its appearing full near conjunction with the sun, and gibbous in its quadratures, it is certain that it surrounds the sun.

And since its diameter appears about five times greater when in opposition to the sun than when in conjunction therewith, and its distance from the earth is reciprocally as its apparent diameter, that distance will be about five times less when in opposition to than when in conjunction with the sun; but in both cases its distance from the sun will be nearly about the same with the distance which is inferred from its gibbous appearance in the quadratures. And as it encompasses the sun at almost equal distances, but in respect of the earth is very unequally distant, so by radii drawn to the sun it describes areas nearly uniform; but by radii drawn to the earth, it is sometimes swift, sometimes stationary, and sometimes retrograde.

That Jupiter, in a higher orb than Mars, is likewise revolved about the sun, with a motion nearly equable, as well in distance as in the areas described, I infer thus.

Mr. *Flamsted* assured me, by letters, that all the eclipses of the innermost satellite which hitherto have been well observed do agree with his theory so nearly, as never to differ therefrom by two minutes of time; that in the outmost the error is little greater; in the outmost but one, scarcely three times greater; that in the innermost but one the difference is indeed much greater, yet so as to agree as nearly with his computations as the moon does with the common tables; and that he computes those eclipses only from the mean motions corrected by the equation of light discovered and introduced by Mr. *Romer*. Supposing, then, that the theory differs by a less error than that of $2'$ from the motion of the outmost satellite as hitherto described, and taking as the periodic time $16^d. 18^h. 5' 13''$ to $2'$ in time, so is the whole circle or $360°$ to the arc $1' 48''$, the error of Mr. *Flamsted's* computation, reduced to the satellite's orbit, will be less than $1' 48''$; that is, the longitude of the satellite, as seen from the centre of Jupiter, will be determined with a less error than $1' 48''$. But when the satellite is in the middle of the shadow, that longitude is the same with the heliocentric longitude of Jupiter; and, therefore, the hypothesis which Mr. *Flamsted* follows, viz., the *Copernican*, as improved by *Kepler*, and (as to the motion of Jupiter) lately corrected by himself, rightly represents that longitude within a less error than 1′ 48″; but by this longitude, together with the geocentric longitude, which is always easily found, the distance of Jupiter from the sun is determined; which must, therefore, be the very same with that which the hypothesis exhibits. For that greatest error of 1′ 48″ that can happen in the heliocentric longitude is almost insensible, and quite to be neglected, and perhaps may arise from some yet undiscovered eccentricity of the satellite; but since both longitude and distance are rightly determined, it follows of necessity that Jupiter, by radii drawn to the sun, describes areas so conditioned as the hypothesis requires, that is, proportional to the times.

And the same thing may be concluded of Saturn from his satellite, by the observations of Mr. *Huygens* and Dr. *Halley;* though a longer series of observations is yet wanting to confirm the thing, and to bring it under a sufficiently exact computation.

For if Jupiter was viewed from the sun, it would never appear retrograde nor stationary, as it is seen sometimes from the earth, but always to go forward with a motion nearly uniform (p. 389). And from the very great inequality of its apparent geocentric motion, we infer (by Prop. III Cor. IV) that the force by which Jupiter is turned out of a rectilinear course, and made to revolve in an orb, is not directed to the centre of the earth. And the same argument holds good in Mars and in Saturn. Another centre of these forces is therefore to be looked for (by Prop. II and III, and the Corollaries of the latter), about which the areas described by radii intervening may be equable; and that this is the sun, we have proved already in Mars and Saturn nearly, but accurately enough in Jupiter. It may be alledged that the sun and planets are impelled by some other force equally and in the direction of parallel lines; but by such a force (by Cor. VI of the Laws of Motion) no change would happen in the situation of the planets one to another, nor any sensible effect follow: but our business is with the causes of sensible effects. Let us, therefore, neglect every such force as imaginary and precarious, and of no use in the phænomena of the heavens; and the whole remaining force by which Jupiter is impelled will be directed (by Prop. III, Cor. I) to the centre of the sun.

The distances of the planets from the sun come out the same, whether, with *Tycho*, we place the earth in the centre of the system, or the sun with *Copernicus:* and we have already proved that these distances are true in Jupiter.

Kepler and *Bullialdus* have, with great care (p. 388), determined the distances of the planets from the sun; and hence it is that their tables agree best with the heavens. And in all the planets, in Jupiter and Mars, in Saturn and the earth, as well as in Venus and Mercury, the cubes of their distances are as the squares of their periodic times; and therefore (by Cor. VI, Prop. IV) the centripetal circum-solar force throughout all the planetary regions decreases in the duplicate proportion of the distances from the sun. In examining this proportion, we are to use the mean distances, or the transverse semi-axes of the orbits (by Prop. XV), and to neglect those little fractions, which, in defining the orbits, may have arisen from the insensible errors of observation, or may be ascribed to other causes which we shall afterwards explain. And thus we shall always find the said proportion to hold exactly; for the distances of Saturn, Jupiter, Mars, the Earth, Venus, and Mercury, from the sun, drawn from the observations of astronomers, are, according to the computation of *Kepler*, as the numbers 951000, 519650, 152350, 100000, 72400, 38806; by the computation of *Bullialdus*, as the numbers 954198, 522520, 152350, 100000, 72398, 38585; and from the periodic times they come out 953806, 520116, 152399, 100000, 72333, 38710. Their distances, according to *Kepler* and *Bullialdus*, scarcely differ by any sensible quantity, and where they differ most the distances drawn from the periodic times, fall in between them.

That the circum-terrestrial force likewise decreases in the duplicate proportion of the distances, I infer thus.

The mean distance of the moon from the centre of the earth, is, in semi-diameters of the earth, according to *Ptolemy*, *Kepler* in his *Ephemerides*, *Bullialdus*, *Hevelius*, and *Ricciolus*, 59; according to *Flamsted*, $59\frac{1}{3}$; according to *Tycho*, $56\frac{1}{2}$; to *Vendelin*, 60; to *Copernicus*, $60\frac{1}{3}$; to *Kircher*, $62\frac{1}{2}$ (p. 391, 392, 393).

But *Tycho*, and all that follow his tables of refraction, making the refractions of the sun and moon (altogether against the nature of light) to exceed those of the fixed stars, and that by about four or five minutes in the horizon, did thereby augment the horizontal parallax of the moon by about the like number of minutes; that is, by about the 12th or 15th part of the whole parallax. Correct this error, and the distance will become 60 or 61 semi-diameters of the earth, nearly agreeing with what others have determined.

Let us, then, assume the mean distance of the moon 60 semi-diameters of the earth, and its periodic time in respect of the fixed stars $27^d. 7^h. 43'$, as astronomers have determined it. And (by Cor. VI, Prop. IV) a body revolved in our air, near the surface of the earth supposed at rest, by means of a centripetal force which should be to the same force at the distance of the moon in the reciprocal duplicate proportion of the distances from the centre of the earth, that is, as 3600 to 1, would (secluding the resistance of the air) complete a revolution in $1^h. 24' 27''$.

Suppose the circumference of the earth to be 123249600 *Paris* feet, as has been determined by the late mensuration of the *French* (vide p. 406); then the same body, deprived of its circular motion, and falling by the impulse of the same centripetal force as before, would, in one second of time, describe $15\frac{1}{12}$ *Paris* feet.

This we infer by a calculus formed upon Prop. XXXVI, and it agrees with what we observe in all bodies about the earth. For by the experiments of pendulums, and a computation raised thereon, Mr. *Huygens* has demonstrated that bodies falling by all that centripetal force with which (of whatever nature it is) they are impelled near the surface of the earth, do, in one second of time, describe $15\frac{1}{12}$ *Paris* feet.

But if the earth is supposed to move, the earth and moon together (by Cor. IV of the Laws of Motion, and Prop. LVII) will be revolved about their common centre of gravity. And the moon (by Prop. LX) will in the same periodic time, $27^d. 7^h. 43'$, with the same circum-terrestrial force diminished in the duplicate proportion of the distance, describe an orbit whose semi-diameter is to the semi-diameter of the former orbit, that is, to 60 semi-diameters of the earth, as the sum of both the bodies of the earth and moon to the first of two mean proportionals between this sum and the body of the earth; that is, if we suppose the moon (on account of its mean apparent diameter $31\frac{1}{2}'$) to be about $\frac{1}{42}$ of the earth, as 43 to $\sqrt[3]{\overline{42 + 43}|^2}$, or as about 128 to 127. And therefore the semi-diameter of the orbit, that is, the distance between the centres of the moon and earth, will in this case be $60\frac{1}{2}$ semi-diameters of the earth, almost the same with that assigned by *Copernicus*, which the *Tychonic* observations by no means disprove; and, therefore, the duplicate proportion of the decrement of the force holds good in this distance. I have neglected the increment of the orbit which arises from the action of the sun as inconsiderable; but if that is subducted, the true distance will remain about $60\frac{4}{9}$ semi-diameters of the earth.

But farther (p. 390); this proportion of the decrement of the forces is confirmed from the eccentricity of the planets, and the very slow motion of their apses; for (by the Corollaries of Prop. XLV) in no other proportion could the circum-solar planets once in every revolution descend to their least and once ascend to their greatest distance from the sun, and the places of those distances remain immoveable. A small error from the duplicate proportion would produce a motion of the apses considerable in every revolution, but in many enormous.

But now, after innumerable revolutions, hardly any such motion has been perceived in the orbs of the circum-solar planets. Some astronomers affirm that there is no such motion; others reckon it no greater than what may easily arise from the causes hereafter to be assigned, and is of no moment in the present question.

We may even neglect the motion of the moon's apsis (p. 390, 391), which is far greater than in the circum-solar planets, amounting in every revolution to three degrees; and from this motion it is demonstrable that the circum-terrestrial force decreases in no less than the duplicate, but far less than the triplicate proportion of the distance; for if the duplicate proportion was gradually changed into the triplicate, the motion of the apsis would thereby increase to infinity; and, therefore, by a very small mutation, would exceed the motion of the moon's apsis. This slow motion arises from the action of the circum-solar force, as we shall afterwards explain. But, secluding this cause, the apsis or apogeon of the moon will be fixed, and the duplicate proportion of the decrease of the circum-terrestrial force in different distances from the earth will accurately take place.

Now that this proportion has been established, we may compare the forces of the several planets among themselves (p. 391).

In the mean distance of Jupiter from the earth, the greatest elongation of the outmost satellite from Jupiter's centre (by the observations of Mr. *Flamsted*) is 8′ 13″; and therefore the distance of the satellite from the centre of Jupiter is to the mean distance of Jupiter from the centre of the sun as 124 to 52012, but to the mean distance of Venus from the centre of the sun as 124 to 7234; and their periodic times are $16\frac{3}{4}^{d}$. and $224\frac{2}{3}^{d}$.; and from hence (according to Cor. II, Prop. IV), dividing the distances by the squares of the times, we infer that the force by which the satellite is impelled towards Jupiter is to the force by which Venus is impelled towards the sun as 442 to 143; and if we diminish the force by which the satellite is impelled in the duplicate proportion of the distance 124 to 7234, we shall have the circum-jovial force in the distance of Venus from the sun to the circum-solar force by which Venus is impelled as $\frac{13}{100}$ to 143, or as 1 to 1100; wherefore at equal distances the circum-solar force is 1100 times greater than the circum-jovial.

And, by the like computation, from the periodic time of the satellite of Saturn 15^{d}. 22^{h}. and its greatest elongation from Saturn, while that planet is in its mean distance from us, 3′ 20″, it follows that the distance of this satellite from Saturn's centre is to the distance of Venus from the sun as $92\frac{2}{5}$ to 7234; and from thence that the absolute circum-solar force is 2360 times greater than the absolute circum-saturnal.

From the regularity of the heliocentric and irregularity of the geocentric motions of Venus, of Jupiter, and the other planets, it is evident (by Cor. IV, Prop. III) that the circum-terrestrial force, compared with the circum-solar, is very small.

Ricciolus and *Vendelin* have severally tried to determine the sun's parallax from the moon's dichotomies observed by the telescope, and they agree that it does not exceed half a minute.

Kepler, from *Tycho's* observations and his own, found the parallax of

Mars insensible, even in opposition to the sun, when that parallax is something greater than the sun's.

Flamsted attempted the same parallax with the micrometer in the perigeon position of Mars, but never found it above 25″; and thence concluded the sun's parallax at most 10″.

Whence it follows that the distance of the moon from the earth bears no greater proportion to the distance of the earth from the sun than 29 to 10000; nor to the distance of Venus from the sun than 29 to 7233.

From which distances, together with the periodic times, by the method above explained, it is easy to infer that the absolute circum-solar force is greater than the absolute circum-terrestrial force at least 229400 times.

And though we were only certain, from the observations of *Ricciolus* and *Vendelin*, that the sun's parallax was less than half a minute, yet from this it will follow that the absolute circum-solar force exceeds the absolute circum-terrestrial force 8500 times.

By the like computations I happened to discover an analogy, that is observed between the forces and the bodies of the planets; but, before I explain this analogy, the apparent diameters of the planets in their mean distances from the earth must be first determined.

Mr. *Flamsted* (p. 387), by the micrometer, measured the diameter of Jupiter 40″ or 41″; the diameter of Saturn's ring 50″; and the diameter of the sun about 32′ 13″ (p. 387).

But the diameter of Saturn is to the diameter of the ring, according to Mr. *Huygens* and Dr. *Halley*, as 4 to 9; according to *Galletius*, as 4 to 10; and according to *Hooke* (by a telescope of 60 feet), as 5 to 12. And from the mean proportion, 5 to 12, the diameter of Saturn's body is inferred about 21″.

Such as we have said are the apparent magnitudes; but, because of the unequal refrangibility of light, all lucid points are dilated by the telescope, and in the focus of the object-glass possess a circular space whose breadth is about the 50th part of the aperture of the glass.

It is true, that towards the circumference the light is so rare as hardly to move the sense; but towards the middle, where it is of greater density, and is sensible enough, it makes a small lucid circle, whose breadth varies according to the splendor of the lucid point, but is generally about the 3d, or 4th, or 5th part of the breadth of the whole.

Let ABD represent the circle of the whole light; PQ the small circle of the denser and clearer light; C the centre of both; CA, CB, semi-diameters of the greater circle containing a right angle at C; ACBE the square comprehended under these semi-diameters; AB the diagonal of that square; EGH an hyperbola with the centre C and asymptotes CA, CB; PG a perpendicular erected from any point P of the line BC, and meeting the hyperbola in G, and the right lines AB, AE, in K and F: and the density of the light in any place P, will, by my computation, be as the line FG, and therefore at the centre infinite, but near the circumference very small. And the whole light within the small circle PQ is to the whole without as the area of the quadrilateral figure CAKP to the trian-

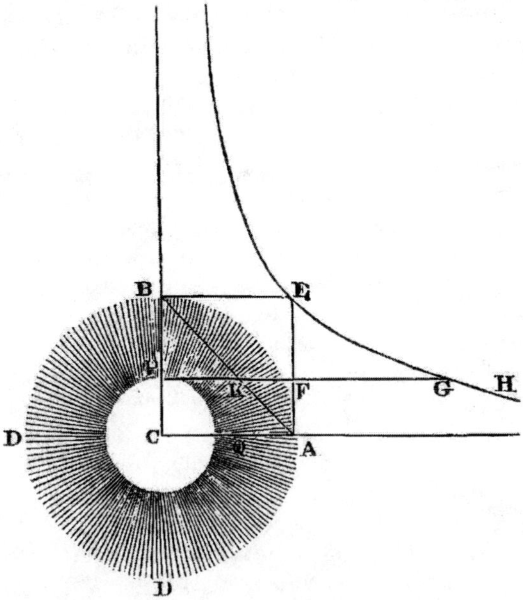

gle PKB. And we are to understand the small circle PQ to be there terminated, where FG, the density of the light, begins to be less than what is required to move the sense.

Hence it was, that, at the distance of 191382 feet, a fire of 3 feet in diameter, through a telescope of 3 feet, appeared to Mr. *Picart* of 8″ in breadth, when it should have appeared only of 3″ 14‴; and hence it is that the brighter fixed stars appear through the telescope as of 5″ or 6″ in diameter, and that with a good full light; but with a fainter light they appear to run out to a greater breadth. Hence, likewise, it was that *Hevelius*, by diminishing the aperture of the telescope, did cut off a great part of the light towards the circumference, and brought the disk of the star to be more distinctly defined, which, though hereby diminished, did yet appear as of 5″ or 6″ in diameter. But Mr. *Huygens*, only by clouding the eye-glass with a little smoke, did so effectually extinguish this scattered light, that the fixed stars appeared as mere points, void of all sensible breadth. Hence also it was that Mr. *Huygens*, from the breadth of bodies interposed to intercept the whole light of the planets, reckoned their diameters greater than others have measured them by the micrometer; for the scattered light, which could not be seen before for the stronger light of the planet, when the planet is hid, appears every way farther spread. Lastly, from hence it is that the planets appear so small in the disk of the sun, being lessened by the dilated light. For to *Hevelius*, *Galletius*, and Dr. *Halley*, Mercury did not seem to exceed 12″ or 15″; and Venus appeared to Mr. *Crabtrie* only 1′ 3″; to *Horrox* but 1′ 12″; though by the mensurations of *Hevelius* and *Hugenius* without the sun's disk, it ought to have been seen at least 1′ 24″. Thus the apparent diameter of the moon, which in 1684, a few days both before and after the sun's eclipse, was measured at the observatory of *Paris* 31′ 30″, in the eclipse itself did not seem to exceed 30′ or 30′ 05″; and therefore the diameters of the planets are to be diminished when without the sun, and to be augmented when within it, by some seconds. But the errors seem to be less than usual in the mensurations that are made by the micrometer. So from the diameter of the shadow, determined by the eclipses of the satellites, Mr. *Flamsted* found that the semi-diameter of Jupiter was to the greatest elongation of the outmost satellite as 1 to 24,903. Wherefore since that elongation is 8′ 13″, the diameter of Jupiter will be 39½″; and, rejecting the scattered light, the diameter found by the micrometer 40″ or 41″ will be reduced to 39½″; and the diameter of Saturn 21″ is to be diminished by the like correction, and to be reckoned 20″, or something less. But (if I am not mistaken) the diameter of the sun, because of its stronger light, is to be diminished something more, and to be reckoned about 32′, or 32′ 6″.

That bodies so different in magnitude should come so near to an analogy with their forces, is not without some mystery (p. 400).

It may be that the remoter planets, for want of heat, have not those metallic substances and ponderous minerals with which our earth abounds; and that the bodies of Venus and Mercury, as they are more exposed to the sun's heat, are also harder baked, and more compact.

For, from the experiment of the burning-glass, we see that the heat increases with the density of light; and this density increases in the reciprocal duplicate proportion of the distance from the sun; from whence the sun's heat in Mercury is proved to be sevenfold its heat in our summer seasons. But with this heat our water boils; and those heavy fluids, quicksilver and the spirit of vitriol, gently evaporate, as I have tried by the thermometer; and therefore there can be no fluids in Mercury but what are heavy, and able to bear a great heat, and from which substances of great density may be nourished.

And why not, if God has placed different bodies at different distances from the sun, so as the denser bodies always possess the nearer places, and each body enjoys a degree of heat suitable to its condition, and proper for its nourishment? From this consideration it will best appear that the weights of all the planets are one to another as their forces.

But I should be glad the diameters of the planets were more accurately measured; and that may be done, if a lamp, set at a great distance, is made to shine through a circular hole, and both the hole and the light of the lamp are so diminished that the spectrum may appear through the telescope just like the planet, and may be defined by the same measure: then the diameter of the hole will be to its distance from the objective glass as the true diameter of the planet to its distance from us. The light of the lamp may be diminished by the interposition either of pieces of cloth, or of smoked glass.

Of kin to the analogy we have been describing, there is another observed between the forces and the bodies attracted (p. 395, 396, 397). Since the action of the centripetal force upon the planets decreases in the duplicate proportion of the distance, and the periodic time increases in the sesquiplicate thereof, it is evident that the actions of the centripetal force, and therefore the periodic times, would be equal in equal planets at equal distances from the sun; and in equal distances of unequal planets the total actions of the centripetal force would be as the bodies of the planets; for if the actions were not proportional to the bodies to be moved, they could not equally retract these bodies from the tangents of their orbs in equal times: nor could the motions of the satellites of Jupiter be so regular, if it was not that the circum-solar force was equally exerted upon Jupiter and all its satellites in proportion of their several weights. And the same thing is to be said of Saturn in respect of its satellites, and of our earth in respect of the moon, as appears from Cor. II and III, Prop. LXV. And, therefore, at equal distances, the actions of the centripetal force are equal upon all the planets in proportion of their bodies, or of the quantities of matter in their several bodies; and for the same reason must be the same upon all the particles of the same size of which the planet is composed; for if the action was greater upon some sort of particles than upon others than in proportion to their quantity of matter, it would be also greater or less upon the whole planets not in proportion to the quantity only, but likewise of the sort of the matter more copiously found in one and more sparingly in another.

In such bodies as are found on our earth of very different sorts, I examined this analogy with great accuracy (p. 343, 344).

If the action of the circum-terrestrial force is proportional to the bodies to be moved, it will (by the Second Law of Motion) move them with equal velocity in equal times, and will make all bodies let fall to descend through equal spaces in equal times, and all bodies hung by equal threads to vibrate in equal times. If the action of the force was greater, the times would be less; if that was less, these would be greater.

But it has been long ago observed by others, that (allowance being made for the small resistance of the air) all bodies descend through equal spaces in equal times; and, by the help of pendulums, that equality of times may be distinguished to great exactness.

I tried the thing in gold, silver, lead, glass, sand, common salt, wood, water, and wheat. I provided two equal wooden boxes. I filled the one with wood, and suspended an equal weight of gold (as exactly as I could) in the centre of oscillation of the other. The boxes, hung by equal threads of 11 feet, made a couple of pendulums perfectly equal in weight and figure, and equally exposed to the resistance of the air: and, placing the one by the other, I observed them to play together forwards and backwards for a long while, with equal vibrations. And therefore (by Cor. I and VI, Prop. XXIV, Book II) the quantity of matter in the gold was to the quantity of matter in the wood as the action of the motive force upon all the gold to the action of the same upon all the wood; that is, as the weight of the one to the weight of the other.

And by these experiments, in bodies of the same weight, could have discovered a difference of matter less than the thousandth part of the whole.

Since the action of the centripetal force upon the bodies attracted is, at equal distances, proportional to the quantities of matter in those bodies, reason requires that it should be also proportional to the quantity of matter in the body attracting.

For all action is mutual, and (p. 83, 93, by the Third Law of Motion) makes the bodies mutually to approach one to the other, and therefore must be the same in both bodies. It is true that we may consider one body as attracting, another as attracted; but this distinction is more mathematical than natural. The attraction is really common of either to other, and therefore of the same kind in both.

And hence it is that the attractive force is found in both. The sun attracts Jupiter and the other planets; Jupiter attracts its satellites; and, for the same reason, the satellites act as well one upon another as upon Jupiter, and all the planets mutually one upon another.

And though the mutual actions of two planets may be distinguished and considered as two, by which each attracts the other, yet, as those actions are intermediate, they do not make two but one operation between two terms. Two bodies may be mutually attracted each to the other by the contraction of a cord interposed. There is a double cause of action, to wit, the disposition of both bodies, as well as a double action in so far as the action is considered as upon two bodies; but as betwixt two bodies it is but one single one. It is not one action by which the sun attracts Jupiter, and another by which Jupiter attracts the sun; but it is one action by which the sun and Jupiter mutually endeavour to approach each the other. By the action with which the sun attracts Jupiter, Jupiter and the sun endeavours to come nearer together (by the Third Law of Motion); and by the action with which Jupiter attracts the sun, likewise Jupiter and the sun endeavor to come nearer together. But the sun is not attracted towards Jupiter by a twofold action, nor Jupiter by a twofold action towards the sun; but it is one single intermediate action, by which both approach nearer together.

Thus iron draws the load-stone (p. 93), as well as the load-stone draws the iron; for all iron in the neighbourhood of the load-stone draws other iron. But the action betwixt the load-stone and iron is single, and is considered as single by the philosophers. The action of iron upon the load-stone, is, indeed, the action of the load-stone betwixt itself and the iron, by which both endeavour to come nearer together: and so it manifestly appears; for if you remove the load-stone, the whole force of the iron almost ceases.

In this sense it is that we are to conceive one single action to be exerted betwixt two planets, arising from the conspiring natures of both; and this action standing in the same relation to both, if it is proportional to the quantity of matter in the one, it will be also proportional to the quantity of matter in the other.

Perhaps it may be objected, that, according to this philosophy (p. 398), all bodies should mutually attract one another, contrary to the evidence of experiments in terrestrial bodies; but I answer, that the experiments in terrestrial bodies come to no account; for the attraction of homogeneous spheres near their surfaces are (by Prop. LXXII) as their diameters. Whence a sphere of one foot in diameter, and of a like nature to the earth, would attract a small body placed near its surface with a force 20000000 times less than the earth would do if placed near its surface; but so small a force could produce no sensible effect. If two such spheres were distant but by $\frac{1}{4}$ of an inch, they would not, even in spaces void of

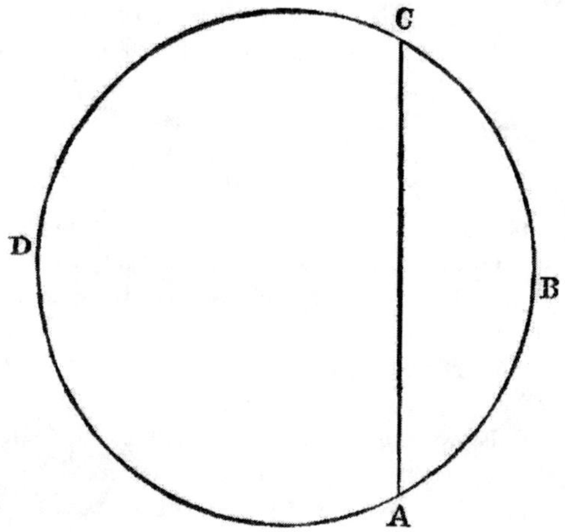

resistance, come together by the force of their mutual attraction in less than a month's time; and less spheres will come together at a rate yet slower, viz., in the proportion of their diameters. Nay, whole mountains will not be sufficient to produce any sensible effect. A mountain of an hemispherical figure, three miles high, and six broad, will not, by its attraction, draw the pendulum two minutes out of the true perpendicular; and it is only in the great bodies of the planets that these forces are to be perceived, unless we may reason about smaller bodies in manner following.

Let ABCD (p. 93) represent the globe of the earth cut by any plane AC into two parts ACB, and ACD. The part ACB bearing upon the part ACD presses it with its whole weight; nor can the part ACD sustain this pressure and continue unmoved, if it is not opposed by an equal contrary pressure. And therefore the parts equally press each other by their weights, that is, equally attract each other, according to the third Law of Motion; and, if separated and let go, would fall towards each other with velocities reciprocally as the bodies. All which we may try and see in the load-stone, whose attracted part does not propel the part attracting, but is only stopped and sustained thereby.

Suppose now that ACB represents some small body on the earth's surface; then, because the mutual attractions of this particle, and of the remaining part ACD of the earth towards each other, are equal, but the attraction of the particle towards the earth (or its weight) is as the matter of the particle (as we have proved by the experiment of the pendulums), the attraction of the earth towards the particle will likewise be as the matter of the particle; and therefore the attractive forces of all terrestrial bodies will be as their several quantities of matter.

The forces (p. 396), which are as the matter in terrestrial bodies of all forms, and therefore are not mutable with the forms, must be found in all sorts of bodies whatsoever, celestial as well as terrestrial, and be in all proportional to their quantities of matter, because among all there is no difference of substance, but of modes and forms only. But in the celestial bodies the same thing is likewise proved thus. We have shewn that the action of the circum-solar force upon all the planets (reduced to equal distances) is as the matter of the planets; that the action of the circum-jovial force upon the satellites of Jupiter observes the same law; and the same thing is to be said of the attraction of all the planets towards every planet: but thence it follows (by Prop. LXIX) that their attractive forces are as their several quantities of matter.

As the parts of the earth mutually attract one another, so do those of all the planets. If Jupiter and its satellites were brought together, and formed into one globe, without doubt they would continue mutually to attract one another as before. And, on the other hand, if the body of Jupiter was broke into more globes, to be sure, these would no less attract one another than they do the satellites now. From these attractions it is that the bodies of the earth and all the planets effect a spherical figure, and their parts cohere, and are not dispersed through the æther. But we have before proved that these forces arise from the universal nature of matter (p. 398), and that, therefore, the force of any whole globe is made up of the several forces of all its parts. And from thence it follows (by Cor. III, Prop. LXXIV) that the force of every particle decreases in the duplicate proportion of the distance from that particle; and (by Prop. LXXIII and LXXV) that the force of an entire globe, reckoning from the surface outwards, decreases in the duplicate, but, reckoning inwards, in the simple proportion of the distances from the centres, if the matter of the globe be uniform. And though the matter of the globe, reckoning from the centre towards the surface, is not uniform (p. 398, 399), yet the decrease in the duplicate proportion of the distance outwards would (by Prop. LXXVI) take place, provided that difformity is similar in places round about at equal distances from the centre. And two such globes will (by the same Proposition) attract one the other with a force decreasing in the duplicate proportion of the distance between their centres.

Wherefore the absolute force of every globe is as the quantity of matter which the globe contains; but the motive force by which every globe is attracted towards another, and which, in terrestrial bodies, we commonly call their weight, is as the content under the quantities of matter in both globes applied to the square of the distance between their centres (by Cor. IV, Prop. LXXVI), to which force the quantity of motion, by which each globe in a given time will be carried towards the other, is proportional. And the accelerative force, by which every globe according to its quantity of matter is attracted towards another, is as the quantity of matter in that other globe applied to the square of the distance between the centres of the two (by Cor. II, Prop. LXXVI); to which force, the velocity by which the attracted globe will, in a given time, be carried towards the other is proportional. And from these principles well understood, it will be now easy to determine the motions of the celestial bodies among themselves.

From comparing the forces of the planets one with another, we have above seen that the circum-solar does more than a thousand times exceed all the rest; but by the action of a force so great it is unavoidable but that all bodies within, nay, and far beyond, the bounds of the planetary system must descend directly to the sun, unless by other motions they are impelled towards other parts: nor is our earth to be excluded from the number of such bodies; for certainly the moon is a body of the same nature with the planets, and subject to the same attractions with the other planets, seeing it is by the circum-terrestrial force that it is retained in its orbit. But that the earth and moon are equally attracted towards the sun, we have above proved; we have likewise before proved that all bodies are subject to the said common laws of attraction. Nay, supposing any of those bodies to be deprived of its circular motion about the sun, by having its distance from the sun, we may find (by Prop. XXXVI) in what space of time it would in its descent arrive at the sun; to wit, in half that periodic time in which the body might be revolved at one half of its former distance; or in a space of time that is to the periodic time of the planet as 1 to $4\sqrt{2}$; as that Venus in its descent would arrive at the sun in the space of 40 days, Jupiter in the space of two years and one month, and the earth and moon together in the space of 66 days and 19 hours. But, since no such thing happens, it must needs be, that those bodies are moved towards other parts (p. 75), nor is every motion sufficient for this purpose. To hinder such a descent, a due proportion of velocity is required. And hence depends the force of the argument drawn from the retardation of the motions of the planets. Unless the circum-solar force decreased in the duplicate ratio of their increasing slowness, the excess thereof would force those bodies to descend to the sun; for instance, if the motion (*cæteris paribus*) was retarded by one half, the planet would be retained in its orb by one fourth of the former circum-solar force, and by the excess of the other three fourths would descend to the sun. And therefore the planets (Saturn, Jupiter, Mars, Venus, and Mercury) are not really retarded in their perigees, nor become really stationary, or regressive with slow motions. All these are but apparent, and the absolute motions, by which the planets continue to revolve in their orbits, are always direct, and nearly equable. But that such motions are performed about the sun, we have already proved; and therefore the sun, as the centre of the absolute motions, is quiescent. For we can by no means allow quiescence to the earth, lest the planets in their perigees should indeed be truly retarded, and become truly stationary and regressive, and so for want of motion should descend to the sun. But farther; since the planets (Venus, Mars, Jupiter, and the rest) by radii drawn to the sun describe regular orbits, and areas (as we have shewn) nearly and to sense proportional to the times, it follows (by Prop. III, and Cor. III, Prop. LXV) that the sun is moved with no notable force, unless perhaps with such as all the planets are equally moved with, according to their several quantities of matter, in parallel lines, and so the whole system is transferred in right lines. Reject that translation of the whole system, and the sun will be almost quiescent in the centre thereof. If the sun was revolved about the earth, and carried the other planets round about itself, the earth ought to attract the sun with a great force, but the circum-solar planets with no force producing any sensible effect, which is contrary to Cor. III, Prop. LXV. Add to this, that if hitherto the earth, because of the gravitation of its parts, has been placed by most authors in the lowermost region of the universe; now, for better reason, the sun possessed of a centripetal force exceeding our terrestrial gravitation a thousand times and more, ought to be depressed into the lowermost place, and to be held for the centre of the system. And thus the true disposition of the whole system will be more fully and more exactly understood.

Because the fixed stars are quiescent one in respect of another (p. 401, 402), we may consider the sun, earth, and planets, as one system of bodies carried hither and thither by various motions among themselves; and the common centre of gravity of all (by Cor. IV of the Laws of Motion) will either be quiescent, or move uniformly forward in a right line: in which case the whole system will likewise move uniformly forward in right lines. But this is an hypothesis hardly to be admitted; and, therefore, setting it aside, that common centre will be quiescent: and from it the sun is never far removed. The common centre of gravity of the sun and Jupiter falls on the surface of the sun; and though all the planets were placed towards the same parts from the sun with Jupiter the common centre of the sun and all of them would scarcely recede twice as far from the sun's centre; and, therefore, though the sun, according to the various situation of the planets, is variously agitated, and always wandering to and fro with a slow motion of libration, yet it never recedes one entire diameter of its own body from the quiescent centre of the whole system. But from the weights of the sun and planets above determined, and the situation of all among themselves, their common centre of gravity may be found; and, this being given, the sun's place to any supposed time may be obtained.

About the sun thus librated the other planets are revolved in elliptic orbits (p. 403), and, by radii drawn to the sun, describe areas nearly, proportional to the times, as is explained in Prop. LXV. If the sun was quiescent, and the other planets did not act mutually one upon another, their orbits would be elliptic, and the areas exactly proportional to the times (by Prop. XI, and Cor. 1, Prop. XIII). But the actions of the planets among themselves, compared with the actions of the sun on the planets, are of no moment, and produce no sensible errors. And those errors are less in revolutions about the sun agitated in the manner but now described than if those revolutions were made about the sun quiescent (by Prop. LXVI, and Cor. Prop. LXVIII), especially if the focus of every orbit is placed in the common centre of gravity of all the lower included planets; viz., the focus of the orbit of Mercury in the centre of the sun; the focus of the orbit of Venus in the common centre of gravity of Mercury and the sun; the focus of the orbit of the earth in the common centre of gravity of Venus, Mercury, and the sun; and so of the rest. And by this means the foci of the orbits of all the planets, except Saturn, will not be sensibly removed from the centre of the sun, nor will the focus of the orbit of Saturn recede sensibly from the common centre of gravity of Jupiter and the sun. And therefore astronomers are not far from the truth, when they reckon the sun's centre the common focus of all the planetary orbits. In Saturn itself the error thence arising does not exceed 1′ 45″. And if its orbit, by placing the focus thereof in the common centre of gravity of Jupiter and the sun, shall happen to agree better with the phænomena, from thence all that we have said will be farther confirmed.

If the sun was quiescent, and the planets did not act one on another, the aphelions and nodes of their orbits would likewise (by Prop. 1, XI, and Cor. Prop. XIII) be quiescent. And the longer axes of their elliptic orbits would (by Prop. XV) be as the cubic roots of the squares of their periodic times: and therefore from the given periodic times would be also given. But those times are to be measured not from the equinoctial points, which are moveable, but from the first star of Aries. Put the semi-axis of the earth's orbit 100000, and the semi-axes of the orbits of Saturn, Jupiter, Mars, Venus, and Mercury, from their periodic times, will come out 953806, 520116, 152399, 72333, 38710 respectively. But from the sun's motion every semi-axis is increased (by Prop. LX) by about one third of the distance of the sun's centre from the common centre of gravity of the sun and planet (p. 405, 406.) And from the actions of the exterior planets on the interior, the periodic times of the interior are something protracted, though scarcely by any sensible quantity; and their aphelions are transferred (by Cor. VI and VII, Prop. LXVI) by very slow motions in *consequentia*. And on the like account the periodic times of all, especially of the exterior planets, will be prolonged by the actions of the comets, if any such there are, without the orb of Saturn, and the aphelions of all will be thereby carried forwards in *consequentia*. But from the progress of the aphelions the regress of the nodes follows (by Cor. XI, XIII, Prop. LXVI). And if the plane of the ecliptic is quiescent, the regress of the nodes (by Cor. XVI, Prop. LXVI) will be to the progress of the aphelion in every orbit as the regress of the nodes of the moon's orbit to the progress of its apogeon nearly, that is, as about 10 to 21. But astronomical observations seem to confirm a very slow progress of the aphelions, and a regress of the nodes in respect of the fixed stars. And hence it is probable that there are comets in the regions beyond the planets, which, revolving in very eccentric orbs, quickly fly through their perihelion parts, and, by an exceedingly slow motion in their aphelions, spend almost their whole time in the regions beyond the planets; as we shall afterwards explain more at large.

The planets thus revolved about the sun (p. 413, 414, 415) may at the same time carry others revolving about themselves as satellites or moons, as appears by Prop. LXVI. But from the action of the sun our moon must move with greater velocity, and, by a radius drawn to the earth, describe an area greater for the time; it must have its orbit less curve, and therefore approach nearer to the earth in the syzygies than in the quadratures, except in so far as the motion of eccentricity hinders those effects.

For the eccentricity is greatest when the moon's apogeon is in the syzygies, and least when the same is in the quadratures; and hence it is that the perigeon moon is swifter and nearer to us, but the apogeon moon slower and farther from us, in the syzygies than in the quadratures. But farther; the apogeon has a progressive and the nodes a regressive motion, both unequable. For the apogeon is more swiftly progressive in its syzygies, more slowly regressive in its quadratures, and by the excess of its progress above its regress is yearly transferred in *consequentia*; but the nodes are quiescent in their syzygies, and most swiftly regressive in their quadratures. But farther, still, the greatest latitude of the moon is greater in its quadratures than in its syzygies; and the mean motion swifter in the aphelion of the earth than in its perihelion. More inequalities in the moon's motion have not hitherto been taken notice of by astronomers: but all these follow from our principles in Cor. II, III, IV, V, VI, VII, VIII, IX, X, XI, XII, XIII, Prop. LXVI, and are known really to exist in the heavens. And this may seen in that most ingenious, and if I mistake not, of all, the most accurate, hypothesis of Mr. *Horrox*, which Mr. *Flamsted* has fitted to the heavens; but the astronomical hypotheses are to be corrected in the motion of the nodes; for the nodes admit the greatest equation or prosthaphæresis in their octants, and this inequality is most conspicuous when the moon is in the nodes, and therefore also in the octants; and hence it was that *Tycho*, and others after him, referred this inequality to the octants of the moon, and made it menstrual; but the reasons by us adduced prove that it ought to be referred to the octants of the nodes, and to be made annual.

Beside those inequalities taken notice of by astronomers (p. 414, 445, 447,) there are yet some others, by which the moon's motions are so disturbed, that hitherto by no law could they be reduced to any certain regulation. For the velocities or horary motions of the apogee and nodes of the moon, and their equations, as well as the difference betwixt the greatest eccentricity in the syzygies and the least in the quadratures, and that inequality which we call the variation, in the progress of the year are augmented and diminished (by Cor. XIV, Prop. LXVI) in the triplicate ratio of the sun's apparent diameter. Beside that, the variation is mutable nearly in the duplicate ratio of the time between the quadratures (by Cor. I and II, Lem. X, and Cor. XVI, Prop. LXVI); and all those inequalities are something greater in that part of the orbit which respects the sun than in the opposite part, but by a difference that is scarcely or not at all perceptible.

By a computation (p. 422), which for brevity's sake I do not describe, I also find that the area which the moon by a radius drawn to the earth describes in the several equal moments of time is nearly as the sum of the number $237\frac{3}{10}$, and versed sine of the double distance of the moon from the nearest quadrature in a circle whose radius is unity; and therefore that the square of the moon's distance from the earth is as that sum divided by the horary motion of the moon. Thus it is when the variation in the octants is in its mean quantity; but if the variation is greater or less, that versed sine must be augmented or diminished in the same ratio. Let astronomers try how exactly the distances thus found will agree with the moon's apparent diameters.

From the motions of our moon we may derive the motions of the moons or satellites of Jupiter and Saturn (p. 413); for the mean motion of the nodes of the outmost satellite of Jupiter is to the mean motion of the nodes of our moon in a proportion compounded of the duplicate proportion of the periodic time of the earth about the sun to the periodic time of Jupiter about the sun, and the simple proportion of the periodic time of the satellite about Jupiter to the periodic time of our moon about the earth (by Cor. XVI, Prop. LXVI): and therefore those nodes, in the space of a hundred years, are carried 8° 24′ backwards, or in *antecedentia*. The mean motions of the nodes of the inner satellites are to the (mean) motion of (the nodes of) the outmost as their periodic times to the periodic time of this, by the same corollary, and are thence given. And the motion of the apsis of every satellite in *consequentia* is to the motion of its nodes in *antecedentia*, as the motion of the apogee of our moon to the motion of its nodes (by the same Corollary), and is thence given. The greatest equations of the nodes and line of the apses of each satellite are to the greatest equations of the nodes and the line of the apses of the moon respectively as the motion of the nodes and line of the apses of the satellites in the time of one revolution of the first equations to the motion of the nodes and apogeon of the moon in the time of one revolution of the last equations. The variation of a satellite seen from Jupiter is to the variation of our moon in the same proportion as the whole motions of their nodes respectively, during the times in which the satellite and our moon (after parting from) are revolved (again) to the sun, by the same Corollary; and therefore in the outmost satellite the variation does not exceed 5″ 12‴. From the small quantity of those inequalities, and the slowness of the motions, it happens that the motions of the satellites are found to be so regular, that the more modern astronomers either deny all motion to the nodes, or affirm them to be very slowly regressive.

(P. 404). While the planets are thus revolved in orbits about remote centres, in the mean time they make their several rotations about their proper axes; the sun in 26 days; Jupiter in $9^h.\ 56'$; Mars in $24\frac{2}{3}^h.$; Venus in $23^h.$; and that in planes not much inclined to the plane of the ecliptic, and according to the order of the signs, as astronomers determine from the spots or maculæ that by turns present themselves to our sight in their bodies; and there is a like revolution of our earth performed in $24^h.$;

and those motions are neither accelerated nor retarded by the actions of the centripetal forces, as appears by Cor. XXII, Prop. LXVI; and therefore of all others they are the most equable and most fit for the mensuration of time; but those revolutions are to be reckoned equable not from their return to the sun, but to some fixed star: for as the position of the planets to the sun is unequably varied, the revolutions of those planets from sun to sun are rendered unequable.

In like manner is the moon revolved about its axis by a motion most equable in respect of the fixed stars, viz., in 27^{d}. 7^{h}. 43′, that is, in the space of a sidereal month; so that this diurnal motion is equal to the mean motion of the moon in its orbit; upon which account the same face of the moon always respects the centre about which this mean motion is performed, that is, the exterior focus of the moon's orbit nearly; and hence arises a deflection of the moon's face from the earth, sometimes towards the east, and other times towards the west, according to the position of the focus which it respects; and this deflection is equal to the equation of the moon's orbit, or to the difference betwixt its mean and true motions; and this is the moon's libration in longitude: but it is likewise affected with a libration in latitude arising from the inclination of the moon's axis to the plane of the orbit in which the moon is revolved about the earth; for that axis retains the same position to the fixed stars nearly, and hence the poles present themselves to our view by turns, as we may understand from the example of the motion of the earth, whose poles, by reason of the inclination of its axis to the plane of the ecliptic, are by turns illuminated by the sun. To determine exactly the position of the moon's axis to the fixed stars, and the variation of this position, is a problem worthy of an astronomer.

By reason of the diurnal revolutions of the planets, the matter which they contain endeavours to recede from the axis of this motion; and hence the fluid parts rising higher towards the equator than about the poles (p. 405), would lay the solid parts about the equator under water, if those parts did not rise also (p. 405, 409): upon which account the planets are something thicker about the equator than about the poles; and their equinoctial points (p. 413) thence become regressive; and their axes, by a motion of nutation, twice in every revolution, librate towards their ecliptics, and twice return again to their former inclination, as is explained in Cor. XVIII, Prop. LXVI; and hence it is that Jupiter, viewed through very long telescopes, does not appear altogether round (p. 409), but having its diameter that lies parallel to the ecliptic something longer than that which is drawn from north to south.

And from the diurnal motion and the attractions (p. 415, 418) of the sun and moon our sea ought twice to rise and twice to fall every day, as well lunar as solar (by Cor. XIX, XX, Prop. LXVI), and the greatest height of the water to happen before the sixth hour of either day and after the twelfth hour preceding. By the slowness of the diurnal motion the flood is retracted to the twelfth hour; and by the force of the motion of reciprocation it is protracted and deferred till a time nearer to the sixth hour. But till that time is more certainly determined by the phænomena, choosing the middle between those extremes, why may we not conjecture the greatest height of the water to happen at the third hour? for thus the water will rise all that time in which the force of the luminaries to raise it is greater, and will fall all that time in which their force is less; viz., from the ninth to the third hour when that force is greater, and from the third to the ninth when it is less. The hours I reckon from the appulse of each luminary to the meridian of the place, as well under as above the horizon; and by the hours of the lunar day I understand the twenty-fourth parts of that time which the moon spends before it comes about again by its apparent diurnal motion to the meridian of the place which it left the day before.

But the two motions which the two luminaries raise will not appear distinguished, but will make a certain mixed motion. In the conjunction or opposition of the luminaries their forces will be conjoined, and bring on the greatest flood and ebb. In the quadratures the sun will raise the waters which the moon depresseth, and depress the waters which the moon raiseth; and from the difference of their forces the smallest of all tides will follow. And because (as experience tells us) the force of the moon is greater than that of the sun, the greatest height of the water will happen about the third lunar hour. Out of the syzygies and quadratures the greatest tide which by the single force of the moon ought to fall out at the third lunar hour, and by the single force of the sun at the third solar hour, by the compounded forces of both must fall out in an intermediate time that approaches nearer to the third hour of the moon than to that of the sun; and, therefore, while the moon is passing from the syzygies to the quadratures, during which time the third hour of the sun precedes the third of the moon, the greatest tide will precede the third lunar hour, and that by the greatest interval a little after the octants of the moon; and by like intervals the greatest tide will follow the third lunar hour, while the moon is passing from the quadratures to the syzygies.

But the effects of the luminaries depend upon their distances from the earth; for when they are less distant their effects are greater, and when more distant their effects are less, and that in the triplicate proportion of their apparent diameters. Therefore it is that the sun in the winter time, being then in its perigee, has a greater effect, and makes the tides in the syzygies something greater, and those in the quadratures something less, *cæteris paribus*, than in the summer season; and every month the moon, while in the perigee, raiseth greater tides than at the distance of 15 days before or after, when it is in its apogee. Whence it comes to pass that two highest tides do not follow one the other in two immediately succeeding syzygies.

The effect of either luminary doth likewise depend upon its declination or distance from the equator; for if the luminary was placed at the pole, it would constantly attract all the parts of the waters, without any intension or remission of its action, and could cause no reciprocation of motion; and, therefore, as the luminaries decline from the equator towards either pole, they will by degrees lose their force, and on this account will excite lesser tides in the solstitial than in the equinoctial syzygies. But in the solstitial quadratures they will raise greater tides than in the quadratures about the equinoxes; because the effect of the moon, then situated in the equator, most exceeds the effect of the sun; therefore the greatest tides fall out in those syzygies, and the least in those quadratures, which happen about the time of both equinoxes; and the greatest tide in the syzygies is always succeeded by the least tide in the quadratures, as we find by experience. But because the sun is less distant from the earth in winter than in summer, it comes to pass that the greatest and least tides more frequently appear before than after the vernal equinox, and more frequently after than before the autumnal.

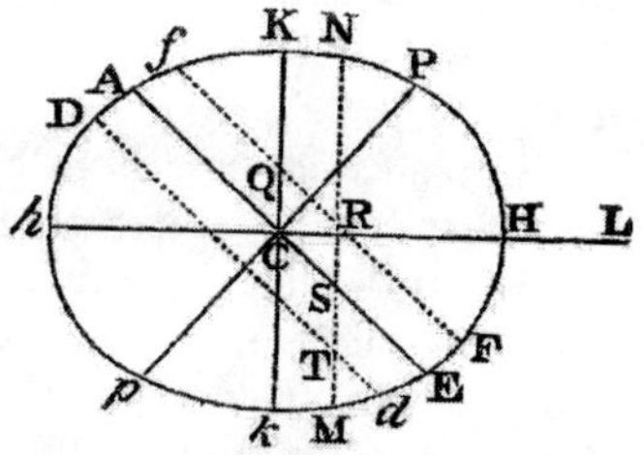

Moreover, the effects of the luminaries depend upon the latitudes of places. Let *Ap*EP represent the earth on all sides covered with deep waters; C its centre; P, *p*, its poles; AE the equator; F any place without the equator; F*f* the parallel of the place; D*d* the correspondent parallel on the other side of the equator; L the place which the moon possessed three hours before; H the place of the earth directly under it; *h* the opposite place; K, *k*, the places at 90 degrees distance; CH, C*h*, the greatest heights of the sea from the centre of the earth; and CK, C*k*, the least heights; and if with the axes H*h*, K*k*, an ellipsis is described, and by the revolution of that ellipsis about its longer axis H*h* a spheroid HPK*hpk* is formed, this spheroid will nearly represent the figure of the sea; and CF, C*f*, CD, C*d*, will represent the sea in the places F, *f*, D, *d*. But farther; if in the said revolution of the ellipsis any point N describes the circle NM, cutting the parallels F*f*, D*d*, in any places R, T, and the equator AE in S, CN will represent the height of the sea in all those places R, S, T, situated in this circle. Wherefore in the diurnal revolution of any place F the greatest flood will be in F, at the third hour after the appulse of the moon to the meridian above the horizon; and afterwards the greatest ebb in Q, at the third hour after the setting of the moon; and then the greatest flood in *f*, at the third hour after the appulse of the moon to the meridian under the horizon; and, lastly, the greatest ebb in Q, at the third hour after the rising of the moon; and the latter flood in *f* will be less than the preceding flood in F. For the whole sea is divided into two huge and hemispherical floods, one in the hemisphere KH*k*C on the north side, the other in the opposite hemisphere KH*k*C, which we may therefore call the northern and the southern floods: these floods, being always opposite the one to the other, come by turns to the meridians of all places after the interval of twelve lunar hours; and, seeing the northern countries partake more of the northern flood, and the southern countries more of the southern flood, thence arise tides alternately greater and less in all places without the equator in which the luminaries rise and set. But the greater tide will happen when the moon declines towards the vertex of the place, about the third hour after the appulse of the moon to the meridian above the horizon; and when the moon changes its declination, that which was the greater tide will be changed into a lesser; and the greatest difference of the floods will fall out about the times of the solstices, especially if the ascending node of the moon is about the first of Aries. So the morning tides in winter exceed those of the evening, and the evening tides exceed those of the morning in summer; at *Plymouth* by the height of one foot, but at *Bristol* by the height of 15 inches, according to the observations of *Colepress* and *Sturmy*.

But the motions which we have been describing suffer some alteration from that force of reciprocation which the waters [having once received] retain a little while by their *vis insita;* whence it comes to pass that the tides may continue for some time, though the actions of the luminaries should cease. This power of retaining the impressed motion lessens the difference of the alternate tides, and makes those tides which immediately succeed after the syzygies greater, and those which follow next after the quadratures less. And hence it is that the alternate tides at *Plymouth* and *Bristol* do not differ much more one from the other than by the height of a foot, or of 15 inches; and that the greatest tides of all at those ports are not the first but the third after the syzygies.

And, besides, all the motions are retarded in their passage through shallow channels, so that the greatest tides of all, in some strai's and mouths of rivers, are the fourth, or even the fifth, after the syzygies.

It may also happen that the greatest tide may be the fourth or fifth after the syzygies, or fall out yet later, because the motions of the sea are retarded in passing through shallow places towards the shores; for so the tide arrives at the western coast of *Ireland* at the third lunar hour, and an hour or two after at the ports in the southern coast of the same island; as also at the islands *Cassiterides*, commonly *Sorlings;* then successively at *Falmouth*, *Plymouth*, *Portland*, the isle of *Wight*, *Winchester*, *Dover*, the mouth of the *Thames*, and *London Bridge*, spending twelve hours in this passage. But farther; the propagation of the tides may be obstructed even by the channels of the ocean itself, when they are not of depth enough; for the flood happens at the third lunar hour in the *Canary* islands; and at all those western coasts that lie towards the Atlantic ocean, as of *Ireland*, *France*, *Spain*, and all *Africa*, to the *Cape* of *Good Hope*, except in some shallow places, where it is impeded, and falls out later; and in the straits of *Gibraltar*, where, by reason of a motion propagated from the Mediterranean sea, it flows sooner. But, passing from those coasts over the breadth of the ocean to the coasts of *America*, the flood arrives first at the most eastern shores of *Brazil*, about the fourth or fifth lunar hour; then at the mouth of the river of the *Amazons* at the sixth hour, but at the neighbouring islands at the fourth hour; afterwards at the islands of *Bermudas* at the seventh hour, and at port St. *Augustin* in *Florida* at seven and a half. And therefore the tide is propagated through the ocean with a slower motion than it should be according to the course of the moon; and this retardation is very necessary, that the sea at the same time may fall between *Brazil* and *New France*, and rise at the *Canary* islands, and on the coasts of *Europe* and *Africa*, and *vice versa:* for the sea cannot rise in one place but by falling in another. And it is probable that the *Pacific* sea is agitated by the same laws; for in the coasts of *Chili* and *Peru* the highest flood is said to happen at the third lunar hour. But with what velocity it is thence propagated to the eastern coasts of *Japan*, the *Philippine* and other islands adjacent to *China*, I have not yet learned.

Farther; it may happen (p. 418) that the tide may be propagated from the ocean through different channels towards the same port, and may pass quicker through some channels than through others, in which case the same tide, divided into two or more succeeding one another, may compound new motions of different kinds. Let us suppose one tide to be divided into two equal tides, the former whereof precedes the other by the space of six hours, and happens at the third or twenty-seventh hour from the appulse of the moon to the meridian of the port. If the moon at the time of this appulse to the meridian was in the equator, every six hours alternately there would arise equal floods, which, meeting with as many equal ebbs, would so balance one the other, that, for that day, the water would stagnate, and remain quiet. If the moon then declined from the equator, the tides in the ocean would be alternately greater and less, as was said; and from hence two greater and two lesser tides would be alternately propagated towards that port. But the two greater floods would make the greatest height of the waters to fall out in the middle time betwixt both, and the greater and lesser floods would make the waters to rise to a mean height in the middle time between them; and in the middle time between the two lesser floods the waters would rise to their least height. Thus in the space of twenty-four hours the waters would come, not twice, but once only to their greatest, and once only to their least height; and their greatest height, if the moon declined towards the elevated pole, would happen at the sixth or thirtieth hour after the appulse of the moon to the meridian; and when the moon changed its declination, this flood would be changed into an ebb.

Of all which we have an example in the port of *Batsham*, in the kingdom of *Tunquin*, in the latitude of 20° 50′ north. In that port, on the day which follows after the passage of the moon over the equator, the waters stagnate; when the moon declines to the north, they begin to flow and ebb, not twice, as in other ports, but once only every day; and the flood happens at the setting, and the greatest ebb at the rising of the moon. This tide increaseth with the declination of the moon till the seventh or eighth day; then for the seventh or eighth day following it decreaseth at the same rate as it had increased before, and ceaseth when the moon changeth its declination. After which the flood is immediately changed into an ebb; and thenceforth the ebb happens at the setting and the flood at the rising of the moon, till the moon again changes its declination. There are two inlets from the ocean to this port; one more direct and short between the island *Hainan* and the coast of *Quantung*, a province of *China;* the other round about between the same island and the coast of *Cochim;* and through the shorter passage the tide is sooner propagated to *Batsham*.

In the channels of rivers the influx and reflux depends upon the current of the rivers, which obstructs the ingress of the waters from the sea, and promotes their egress to the sea, making the ingress later and slower, and the egress sooner and faster; and hence it is that the reflux is of longer duration that the influx, especially far up the rivers, where the force of the sea is less. So *Sturmy* tells us, that in the river *Avon*, three miles below *Bristol*, the water flows only five hours, but ebbs seven; and without doubt the difference is yet greater above *Bristol*, as at *Caresham* or the *Bath*. This difference does likewise depend upon the quantity of the flux and reflux; for the more vehement motion of the sea near the syzygies of the luminaries more easily overcoming the resistance of the rivers, will make the ingress of the water to happen sooner and to continue longer, and will therefore diminish this difference. But while the moon is approaching to the syzygies, the rivers will be more plentifully filled, their currents being obstructed by the greatness of the tides, and therefore will something more retard the reflux of the sea a little after than a little before the syzygies. Upon which account the slowest tides of all will not happen in the syzygies, but precede them a little; and I observed above that the tides before the syzygies were also retarded by the force of the sun; and from both causes conjoined the retardation of the tides will be both greater and sooner before the syzygies. All which I find to be so, by the tide-tables which *Flamsted* has composed from a great many observations.

By the laws we have been describing, the times of the tides are governed; but the greatness of the tides depends upon the greatness of the seas. Let C represent the centre of the earth, EADB the oval figure of the seas, CA the longer semi-axis of this oval, CB the shorter insisting at right angles upon the former, D the middle point between A and B, and ECF or eCf the angle at the centre of the earth, subtended by the breadth of the sea that terminates in the shores E, F, or e, f. Now, supposing that the point

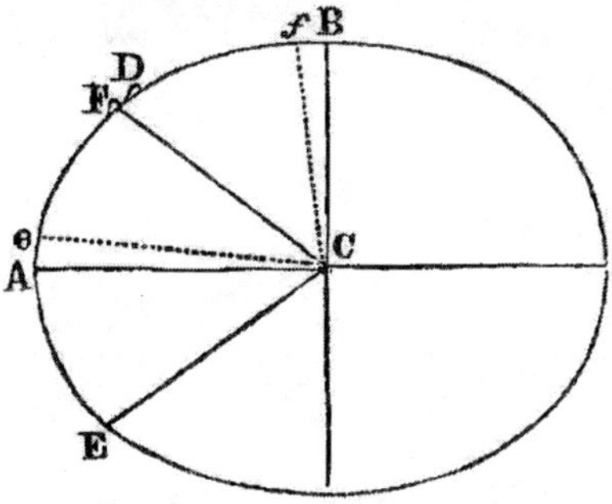

A is in the middle between the points E, F, and the point D in the middle between the points e, f, if the difference of the heights CA, CB, represent the quantity of the tide in a very deep sea surrounding the whole earth, the excess of the height CA above the height CE or CF will represent the quantity of the tide in the middle of the sea EF, terminated by the shores E, F; and the excess of the height Ce above the height Cf will nearly represent the quantity of the tide on the shores f of the same sea. Whence it appears that the tides are far less in the middle of the sea than at the shores; and that the tides at the shores are nearly as EF (p. 451, 452), the breadth of the sea not exceeding a quadrantal arc. And hence it is that near the equator, where the sea between *Africa* and *America* is narrow, the tides are far less than towards either side in the temperate zones, where the seas are extended wider; or on almost all the shores of the Pacific sea, as well towards *America* as towards *China*, and within as well as without the tropics; and that in islands in the middle of the sea they scarcely rise higher than two or three feet, but on the shores of great continents are three or four times greater, and above, especially if the motions propagated from the ocean are by degrees contracted into a narrow space, and the water, to fill and empty the bays alternately, is forced to flow and ebb with great violence through shallow places; as *Plymouth* and *Chepstow Bridge* in *England*, at the mount of *St. Michael* and town of *Avranches* in *Normandy*, and at *Cambaia* and *Pegu* in the *East Indies*. In which places, the sea, hurried in and out with great violence, sometimes lays the shores under water, sometimes leaves them dry, for many miles. Nor is the force of the influx and efflux to be broke till it has raised or depressed the water to forty or fifty feet and more. Thus also long and shallow straits that open to the sea with mouths wider and deeper than the rest of their channel (such as those about *Britain* and the *Magellanic Straits* at the eastern entry) will have a greater flood and ebb, or will more intend and remit their course, and therefore will rise higher and be depressed lower. On the coast of *South America* it is said that the *Pacific* sea in its reflux sometimes retreats two miles, and gets out of sight of those that stand on shore. Whence in these places the floods will be also higher; but in deeper waters the velocity of influx and efflux is always less, and therefore the ascent and descent is so too. Nor in such places is the ocean known to ascend to more than six, eight, or ten feet. The quantity of the ascent I compute in the following manner.

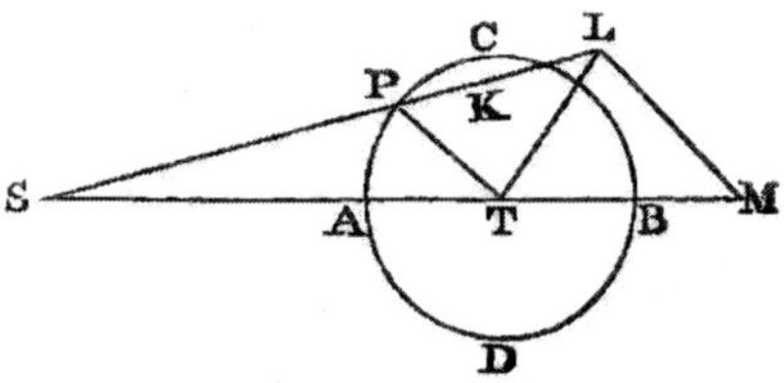

Let S represent the sun, T the earth (419, 420), P the moon, PAGB the moon's orbit. In SP take SK equal to ST and SL to SK in the duplicate ratio of SK to SP. Parallel to PT draw LM; and, supposing the mean quantity of the circum-solar force directed towards the earth to be represented by the distance ST or SK, SL will represent the quantity thereof directed towards the moon. But that force is compounded of the parts SM, LM; of which the force LM and that part of SM which is represented by TM, do disturb the motion of the moon (as appears from Prop. LXVI, and its Corollaries). In so far as the earth and moon are revolved about their common centre of gravity, the earth will be liable to the action of the like forces. But we may refer the sums as well of the forces as of the motions to the moon, and represent the sums of the forces by the lines TM and ML, which are proportional to them. The force LM, in its mean quantity, is to the force by which the moon may be revolved in an orbit, about the earth quiescent, at the distance PT in the duplicate ratio of the moon's periodic time about the earth to the earth's periodic time about the sun (by Cor. XVII, Prop. LXVI); that is, in the duplicate ratio of $27^d. 7^h. 43'$ to $365^d. 6^h. 9'$; or as 1000 to 178725, or 1 to $178\frac{29}{40}$. The force by which the moon may be revolved in its orb about the earth in rest, at the distance PT of $60\frac{1}{2}$ semi-diameters of the earth, is to the force by which it may revolve in the same time at the distance of 60 semi-diameters as $60\frac{1}{2}$ to 60; and this force is to the force of gravity with us as 1 to 60×60 nearly; and therefore the mean force ML is to the force of gravity at the surface of the earth as $1 \times 60\frac{1}{2}$ to $60 \times 60 \times 178\frac{29}{40}$, or 1 to

638092,6. Whence the force TM will be also given from the proportion of the lines TM, ML. And these are the forces of the sun, by which the moon's motions are disturbed.

If from the moon's orbit (p. 449) we descend to the earth's surface, those forces will be diminished in the ratio of the distances $60\frac{1}{2}$ and 1; and therefore the force LM will then become 38604600 times less than the force of gravity. But this force acting equally every where upon the earth, will scarcely effect any change on the motion of the sea, and therefore may be neglected in the explication of that motion. The other force TM, in places where the sun is vertical, or in their nadir, is triple the quantity of the force ML, and therefore but 12868200 times less than the force of gravity.

Suppose now ADBE to represent, the spherical surface of the earth, *a*D*b*E the surface of the water overspreading it, C the centre of both, A the place to which the sun is vertical, B the place opposite; D, E, places at 90 degrees distance from the former; ACE*mlk* a right angled cylindric canal passing through the earth's centre. The force TM in any place is as the distance of the place from the plane DE, on which a line from A to C insists at right angles, and therefore in the part of the canal which is represented by EC*lm* is of no quantity, but in the other part AC*lk* is as the gravity at the several heights; for in descending towards the centre of the earth, gravity is (by Prop. LXXIII) every where as the height; and therefore the force TM drawing the water upwards will diminish its gravity in the leg AC*lk* of the canal in a given ratio: upon which account the water will ascend in this leg, till its defect of gravity is supplied by its greater height; nor will it rest in an equilibrium till its total gravity becomes equal to the total gravity in EC*lm*, the other leg of the canal. Because the gravity of every particle is as its distance from the earth's centre, the weight of the whole water in either leg will increase in the duplicate ratio of the height; and therefore the height of the water in the leg AC*lk* will be to the height thereof in the leg C*lm*E in the subduplicate ratio of the number 12868201 to 12868200, or in the ratio of the number 25623053 to the number 25623052, and the height of the water in the leg EC*lm* to the difference of the heights, as 25623052 to 1. But the height in the leg EC*lm* is of 19615800 *Paris* feet, as has been lately

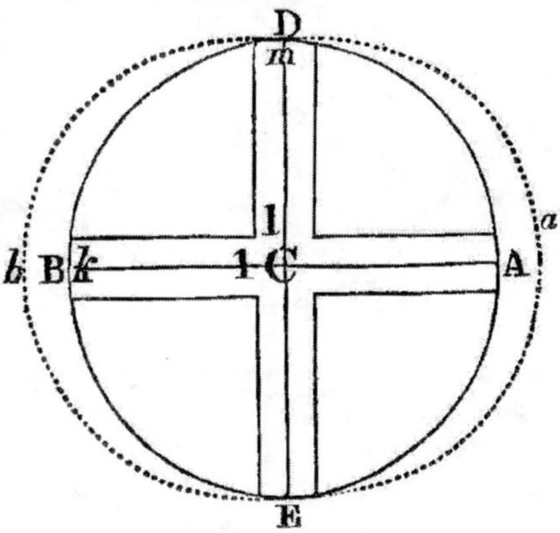

found by the mensuration of the *French*; and, therefore, by the preceding analogy, the difference of the heights comes out $9\frac{1}{5}$ inches of the *Paris* foot; and the sun's force will make the height of the sea at A to exceed the height of the same at E by 9 inches. And though the water of the canal ACE*mlk* be supposed to be frozen into a hard and solid consistence, yet the heights thereof at A and E, and all other intermediate places, would still remain the same.

Let A*a* (in the following figure) represent that excess of height of 9 inches at A, and *hf* the excess of height at any other place *h*; and upon DC let fall the perpendicular *f*G, meeting the globe of the earth in F; and because the distance of the sun is so great that all the right lines drawn thereto may be considered as parallel, the force TM in any place *f* will be to the same force in the place A as the sine FG to the radius AC. And, therefore, since those forces tend to the sun in the direction of parallel lines, they will generate the parallel heights F*f*, A*a*, in the same ratio; and therefore the figure of the water D*faeb* will be a spheroid made by the revolution of an ellipsis about its longer axis *ab*. And the perpendicular height *fh* will be to the oblique height F*f* as *f*G to *f*C, or as FG to AC: and therefore the height *fh* is to the height A*a* in the duplicate ratio of FG to AC, that is, in the ratio of the versed sine of double the angle DC*f* to double the radius, and is thence given. And hence to the several moments of the apparent revolution of the sun about the earth we may infer the proportion of the ascent and descent of the waters at any given place under the equator, as well as of the diminution of that ascent and descent, whether arising from the latitude of places or from the sun's declination; viz., that on account of the latitude of places, the ascent and descent of the sea is in all places diminished in the duplicate ratio of the co-sines of latitude; and on account of the sun's declination, the ascent and descent under the equator is diminished in the duplicate ratio of the co-sine of declination. And in places without the equator the half sum of the morning and evening ascents (that is, the mean ascent) is diminished nearly in the same ratio.

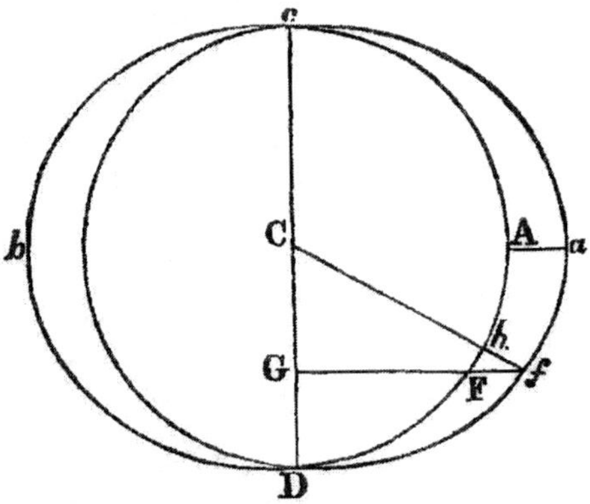

Let S and L respectively represent the forces of the sun and moon placed in the equator, and at their mean distances from the earth; R the radius; T and V the versed sines of double the complements of the sun and moon's declinations to any given time; D and E the mean apparent diameters of the sun and moon: and, supposing F and G to be their apparent diameters to that given time, their forces to raise the tides under the equator will be, in the syzygies $\frac{VG^3}{2RE^3}L + \frac{TF^3}{2RD^3}S$; in the quadratures, $\frac{VG^3}{2RE^3}L - \frac{TF^3}{2RD^3}S$. And if the same ratio is likewise observed under the parallels, from observations accurately made in our northern climates we may determine the proportion of the forces L and S; and then by means of this rule predict the quantities of the tides to every syzygy and quadrature.

At the mouth of the river *Avon*, three miles below *Bristol* (p. 450 to 453), in spring and autumn, the whole ascent of the water in the conjunction or opposition of the luminaries (by the observation of *Sturmy*) is about 45 feet, but in the quadratures only 25. Because the apparent diameters of the luminaries are not here determined, let us assume them in their mean quantities, as well as the moon's declination in the equinoctial quadratures in its mean quantity, that is, $23\frac{1}{2}°$; and the versed sine of double its complement will be 1682, supposing the radius to be 1000. But the declinations of the sun in the equinoxes and of the moon in the syzygies are of no quantity, and the versed sines of double the complements are each 2000. Whence those forces become L + S in the syzygies, and $\frac{1682}{2000}L - S$ in the quadratures, respectively proportional to the heights of the tides of 45 and 25 feet, or of 9 and 5 paces. And, therefore, multiplying the extremes and the means, we have $5L + 5S = \frac{15138}{2000}L - 9S$, or $L = \frac{28000}{5138}S = 5\frac{5}{11}S$.

But farther; I remember to have been told that in summer the ascent of the sea in the syzygies is to the ascent thereof in the quadratures as about 5 to 4. In the solstices themselves it is probable that the proportion may be something less, as about 6 to 5; whence it would follow that L is $= 5\frac{1}{6}S$ [for then the proportion is $\frac{1682}{2000}L + \frac{1682}{2000}S : L - \frac{1682}{2000}S :: 6 : 5$]. Till we can more certainly determine the proportion from observation, let us assume $L = 5\frac{1}{3}S$; and since the heights of the tides are as the forces which excite them, and the force of the sun is able to raise the tides to the height of nine inches, the moon's force will be sufficient to raise the same to the height of four feet. And if we allow that this height may be doubled, or perhaps tripled, by that force of reciprocation which we observe in the motion of the waters, and by which their motion once begun is kept up for some time, there will be force enough to generate all that quantity of tides which we really find in the ocean.

Thus we have seen that these forces are sufficient to move the sea. But, so far as I can observe, they will not be able to produce any other effect sensible on our earth; for since the weight of one grain in 4000 is not sensible in the nicest balance; and the sun's force to move the tides is 12868200 less than the force of gravity; and the sum of the forces of both moon and sun, exceeding the sun's force only in the ratio of $6\frac{1}{3}$ to 1, is still 2032890 times less than the force of gravity; it is evident that both forces together are 500 times less than what is required sensibly to increase or diminish the weight of any body in a balance. And, therefore, they will not sensibly move any suspended body; nor will they produce any sensible effect on pendulums, barometers, bodies swimming in stagnant water, or in the like statical experiments. In the atmosphere, indeed, they will excite such a flux and reflux as they do in the sea, but with so small a motion that no sensible wind will be thence produced.

If the effects of both moon and sun in raising the tides (p. 454), as well as their apparent diameters, were equal among themselves, their absolute forces would (by Cor. XIV, Prop. LXVI) be as their magnitudes. But the effect of the moon is to the effect of the sun as about $5\frac{1}{3}$ to 1; and the moon's diameter less than the sun's in the ratio of $31\frac{1}{2}$ to $32\frac{1}{5}$, or of 45 to 46. Now the force of the moon is to be increased in the ratio of the effect directly, and in the triplicate ratio of the diameter inversely. Whence the force of the moon compared with its magnitude will be to the force of the sun compared with its magnitude in the ratio compounded of $5\frac{1}{3}$ to 1, and the triplicate of 45 to 46 inversely, that is, in the ratio of about $5\frac{7}{10}$ to 1. And therefore the moon, in respect of the magnitude of its body, has an absolute centripetal force greater than the sun in respect of the magnitude of its body in the ratio to $5\frac{7}{10}$ to 1, and is therefore more dense in the same ratio.

In the time of $27^{d}.\ 7^{h}.\ 43'$, in which the moon makes its revolution about the earth, a planet may be revolved about the sun at the distance of 18,954 diameters of the sun from the sun's centre, supposing the mean apparent diameter of the sun to be $32\frac{1}{5}'$; and in the same time the moon may be revolved about the earth at rest, at the distance of 30 of the earth's diameters. If in both cases the number of diameters was the same, the absolute circum-terrestrial force would (by Cor. II, Prop. LXXII) be to the absolute circum-solar force as the magnitude of the earth to the magnitude of the sun. Because the number of the earth's diameters is greater in the ratio of 30 to 18,954, the body of the earth will be less in the triplicate of that ratio, that is, in the ratio of $3\frac{28}{29}$ to 1. Wherefore the earth's force, for the magnitude of its body, is to the sun's force, for the magnitude of its body, as $3\frac{28}{29}$ to 1; and consequently the earth's density to the sun's will be in the same ratio. Since, then, the moon's density is to the sun's density as $5\frac{7}{10}$ to 1, the moon's density will be to the earth's density as $5\frac{7}{10}$ to $3\frac{2}{2}\frac{2}{9}$, or as 23 to 16. Wherefore since the moon's magnitude is to the earth's magnitude as about 1 to $41\frac{1}{2}$, the moon's absolute centripetal force will be to the earth's absolute centripetal force as about 1 to 29, and the quantity of matter in the moon to the quantity of matter in the earth in the same ratio. And hence the common centre of gravity of the earth and moon is more exactly determined than hitherto has been done; from the knowledge of which we may now infer the moon's distance from the earth with greater accuracy. But I would rather wait till the proportion of the bodies of the moon and earth one to the other is more exactly defined from the phænomena of the tides, hoping that in the mean time the circumference of the earth may be measured from more distant stations than any body has yet employed for this purpose.

Thus I have given an account of the system of the planets. As to the fixed stars, the smallness of their annual parallax proves them to be removed to immense distances from the system of the planets: that this parallax is less than one minute is most certain; and from thence it follows that the distance of the fixed stars is above 360 times greater than the distance of Saturn from the sun. Such as reckon the earth one of the planets, and the sun one of the fixed stars, may remove the fixed stars to yet greater distances by the following arguments: from the annual motion of the earth there would happen an apparent transposition of the fixed stars, one in respect of another, almost equal to their double parallax; but the greater and nearer stars, in respect of the more remote, which are only seen by the telescope, have not hitherto been observed to have the least motion. If we should suppose that motion to be but less than 20″, the distance of the nearer fixed stars would exceed the mean distance of Saturn by above 2000 times. Again; the disk of Saturn, which is only 17″ or 18″ in diameter, receives but about $\frac{1}{2100000000}$ of the sun's light; for so much less is that disk than the whole spherical surface of the orb of Saturn. Now if we suppose Saturn to reflect about $\frac{1}{4}$ of this light, the whole light reflected from its illuminated hemisphere will be about $\frac{1}{4200000000}$ of the whole light emitted from the sun's hemisphere; and, therefore, since light is rarefied in the duplicate ratio of the distance from the luminous body, if the sun was $10000\sqrt{42}$ times more distant than Saturn, it would yet appear as lucid as Saturn now does without its ring, that is, something more lucid than a fixed star of the first magnitude. Let us, therefore, suppose that the distance from which the sun would shine as a fixed star exceeds that of Saturn by about 100,000 times, and its apparent diameter will be $7^{v}. 16^{vi}$. and its parallax arising from the annual motion of the earth 13″″: and so great will be the distance, the apparent diameter, and the parallax of the fixed stars of the first magnitude, in bulk and light equal to our sun.

Some may, perhaps, imagine that a great part of the light of the fixed stars is intercepted and lost in its passage through so vast spaces, and upon that account pretend to place the fixed stars at nearer distances; but at this rate the remoter stars could be scarcely seen. Suppose, for example, that $\frac{3}{4}$ of the light perish in its passage from the nearest fixed stars to us; then $\frac{3}{4}$ will twice perish in its passage through a double space, thrice through a triple, and so forth. And, therefore, the fixed stars that are at a double distance will be 16 times more obscure, viz., 4 times more obscure on account of the diminished apparent diameter; and, again, 4 times more on account of the lost light. And, by the same argument, the fixed stars at a triple distance will be $9 \times 4 \times 4$, or 144 times more obscure; and those at a quadruple distance will be $16 \times 4 \times 4 \times 4$, or 1024 times more obscure; but so great a diminution of light is no ways consistent with the phænomena and with that hypothesis which places the fixed stars at different distances.

The fixed stars being, therefore, at such vast distances from one another (p. 460, 461), can neither attract each other sensibly, nor be attracted by our sun. But the comets must unavoidably be acted on by the circumsolar force; for as the comets were placed by astronomers above the moon, because they were found to have no diurnal parallax, so their annual parallax is a convincing proof of their descending into the regions of the planets. For all the comets which move in a direct course, according to the order of the signs, about the end of their appearance become more than ordinarily slow, or retrograde, if the earth is between them and the sun; and more than ordinarily swift if the earth is approaching to a heliocentric opposition with them. Whereas, on the other hand, those which move against the order of the signs, towards the end of their appearance, appear swifter than they ought to be if the earth is between them and the sun; and slower, and perhaps retrograde, if the earth is in the other side of its orbit. This is occasioned by the motion of the earth in different situations. If the earth go the same way with the comet, with a swifter motion, the comet becomes retrograde; if with a slower motion, the comet becomes slower, however; and if the earth move the contrary way, it becomes swifter; and by collecting the differences between the slower and swifter motions, and the sums of the more swift and retrograde motions, and comparing them with the situation and motion of the earth from whence they arise, I found, by means of this parallax, that the distances of the comets at the time they cease to be visible to the naked eye are always less than the distance of Saturn, and generally even less than the distance of Jupiter.

The same thing may be collected from the curvature of the way of the comets (p. 462). These bodies go on nearly in great circles while their motion continues swift; but about the end of their course, when that part of their apparent motion which arises from the parallax bears a greater proportion to their whole apparent motion, they commonly deviate from those circles; and when the earth goes to one side, they deviate to the other; and this deflection, because of its corresponding with the motion of the earth, must arise chiefly from the parallax; and the quantity thereof is so considerable, as, by my computation, to place the disappearing comets a good deal lower than Jupiter. Whence it follows, that, when they approach nearer to us in their perigees and perihelions, they often descend below the orbits of Mars and the inferior planets.

Moreover, this nearness of the comets is confirmed by the annual parallax of the orbit, in so far as the same is pretty nearly collected by the supposition that the comets move uniformly in right lines. The method of collecting the distance of a comet according to this hypothesis from four observations (first attempted by *Kepler*, and perfected by Dr. *Wallis* and Sir *Christopher Wren*) is well known; and the comets reduced to this regularity generally pass through the middle of the planetary region. So the comets of the year 1607 and 1618, as their motions are defined by *Kepler*, passed between the sun and the earth; that of the year 1664 below the orbit of Mars; and that in 1680 below the orbit of Mercury, as its motion was defined by Sir *Christopher Wren* and others. By a like rectilinear hypothesis, *Hevelius* placed all the comets about which we have any observations below the orbit of Jupiter. It is a false notion, therefore, and contrary to astronomical calculation, which some have entertained, who, from the regular motion of the comets, either remove them into the regions of the fixed stars, or deny the motion of the earth; whereas their motions cannot be reduced to perfect regularity, unless we suppose them to pass through the regions near the earth in motion; and these are the arguments drawn from the parallax, so far as it can be determined without an exact knowledge of the orbits and motions of the comets.

The near approach of the comets is farther confirmed from the light of their heads (p. 463, 465); for the light of a celestial body, illuminated by the sun, and receding to remote parts, is diminished in the quadruplicate proportion of the distance; to wit, in one duplicate proportion on account of the increase of the distance from the sun; and in another duplicate proportion on account of the decrease of the apparent diameter. Hence it may be inferred, that Saturn being at a double distance, and having its apparent diameter nearly half of that of Jupiter, must appear about 16 times more obscure; and that, if its distance were 4 times greater, its light would be 256 times less; and therefore would be hardly perceivable to the naked eye. But now the comets often equal Saturn's light, without exceeding him in their apparent diameters. So the comet of the year 1668, according to Dr. *Hooke's* observations, equalled in brightness the light of a fixed star of the first magnitude; and its head, or the star in the middle of the coma, appeared, through a telescope of 15 feet, as lucid as Saturn near the horizon; but the diameter of the head was only 25″; that is, almost the same with the diameter of a circle equal to Saturn and his ring. The coma or hair surrounding the head was about ten times as broad; namely, $4\frac{1}{6}$ min. Again; the least diameter of the hair of the comet of the year 1682, observed by Mr. *Flamsted* with a tube of 16 feet, and measured with the micrometer, was 2′ 0″; but the nucleus, or star in the middle, scarcely possessed the tenth part of this breadth, and was therefore only 11″ or 12″ broad; but the light and clearness of its head exceeded that of the year 1680, and was equal to that of the stars of the first or second magnitude. Moreover, the comet of the year 1665, in *April*, as *Hevelius* informs us, exceeded almost all the fixed stars in splendor, and even Saturn itself, as being of a much more vivid colour; for this comet was more lucid than that which appeared at the end of the foregoing year, and was compared to the stars of the first magnitude. The diameter of the coma was about 6′; but the nucleus, compared with the planets by means of a telescope, was plainly less than Jupiter, and was sometimes thought less, sometimes equal to the body of Saturn within the ring. To this breadth add that of the ring, and the whole face of Saturn will be twice as great as that of the comet, with a light not at all more intense; and therefore the comet was nearer to the sun than Saturn. From the proportion of the nucleus to the whole head found by these observations, and from its breadth, which seldom exceeds 8′ or 12′, it appears that the stars of the comets are most commonly of the same apparent magnitude as the planets; but that their light may be compared oftentimes with that of Saturn, and sometimes exceeds it. And hence it is certain that in their perihelia their distances can scarcely be greater than that of Saturn. At twice that distance, the light would be four times less, which besides by its dim paleness would be as much inferior to the light of Saturn as the light of Saturn is to the splendor of Jupiter: but this difference would be easily observed. At a distance ten times greater, their bodies must be greater than that of the sun; but their light would be 100 times fainter than that of Saturn. And at distances still greater, their bodies would far exceed the sun; but, being in such dark regions, they must be no longer visible. So impossible is it to place the comets in the middle regions between the sun and fixed stars, accounting the sun as one of the fixed stars; for certainly they would receive no more light there from the sun than we do from the greatest of the fixed stars.

So far we have gone without considering that obscuration which comets suffer from that plenty of thick smoke which encompasseth their heads, and through which the heads always shew dull as through a cloud; for by how much the more a body is obscured by this smoke, by so much the more near it must be allowed to come to the sun, that it may vie with the planets in the quantity of light which it reflects; whence it is probable that the comets descend far below the orbit of Saturn, as we proved before from their parallax. But, above all, the thing is evinced from their tails, which must be owing either to the sun's light reflected from a smoke arising from them, and dispersing itself through the æther, or to the light of their own heads.

In the former case we must shorten the distance of the comets, lest we be obliged to allow that the smoke arising from their heads is propagated through such a vast extent of space, and with such a velocity of expansion, as will seem altogether incredible; in the latter case the whole light of both head and tail must be ascribed to the central nucleus. But, then, if we suppose all this light to be united and condensed within the disk of the nucleus, certainly the nucleus will by far exceed Jupiter itself in splendor, especially when it emits a very large and lucid tail. If, therefore, under a less apparent diameter, it reflects more light, it must be much more illuminated by the sun, and therefore much nearer to it. So the comet that appeared *Dec.* 12 and 15, O.S. *Anno* 1679, at the time it emitted a very shining tail, whose splendor was equal to that of many stars like Jupiter, if their light were dilated and spread through so great a space, was, as to the magnitude of its nucleus, less than Jupiter (as Mr. *Flamsted* observed), and therefore was much nearer to the sun: nay, it was even less than Mercury. For on the 17th of that month, when it was nearer to the earth, it appeared to *Cassini* through a telescope of 35 feet a little less than the globe of Saturn. On the 8th of this month, in the morning, Dr. *Halley* saw the tail, appearing broad and very short, and as if it rose from the body of the sun itself, at that time very near its rising. Its form was like that of an extraordinary bright cloud; nor did it disappear till the sun itself began to be seen above the horizon. Its splendor, therefore, exceeded the light of the clouds till the sun rose, and far surpassed that of all the stars together, as yielding only to the immediate brightness of the sun itself. Neither Mercury, nor Venus, nor the moon itself, are seen so near the rising sun. Imagine all this dilated light collected together, and to be crowded into the orbit of the comet's nucleus which was less than Mercury; by its splendor, thus increased, becoming so much more conspicuous, it will vastly exceed Mercury, and therefore must be nearer to the sun. On the 12th and 15th of the same month, this tail, extending itself over a much greater space, appeared more rare; but its light was still so vigorous as to become visible when the fixed stars were hardly to be seen, and soon after to appear like a fiery beam shining in a wonderful manner. From its length, which was 40 or 50 degrees, and its breadth of 2 degrees, we may compute what the light of the whole must be.

This near approach of the comets to the sun is confirmed from the situation they are seen in when their tails appear most resplendent; for when the head passes by the sun, and lies hid under the solar rays, very bright and shining tails, like fiery beams, are said to issue from the horizon; but afterwards, when the head begins to appear, and is got farther from the sun, that splendor always decreases, and turns by degrees into a paleness like to that of the milky way, but much more sensible at first; after that vanishing gradually. Such was that most resplendent comet described by *Aristotle*, Lib. 1, Meteor. 6. "The head thereof could not be seen, because it set before the sun, or at least was hid under the sun's rays; but the next day it was seen as well as might be; for, having left the sun but a very little way, it set immediately after it; and the scattered light of the head obscured by the too great splendour (of the tail) did not yet appear. But afterwards (says *Aristotle*), when the splendour of the tail was now diminished (the head of), the comet recovered its native brightness. And the splendour of its tail reached now to a third part of the heavens (that is, to 60°). It appeared in the winter season, and, rising to *Orion's* girdle, there vanished away." Two comets of the same kind are described by *Justin*, Lib. 37, which, according to his account, "shined so bright, that the whole heaven seemed to be on fire; and by their greatness filled up a fourth part of the heavens, and by their splendour exceeded that of the sun." By which last words a near position of these bright comets and the rising or setting sun is intimated (p. 494, 495). We may add to these the comet of the year 1101 or 1106, "the star of which was small and obscure (like that of 1680); but the splendour arising from it extremely bright, reaching like a fiery beam to the east and north," as *Hevelius* has it from *Simeon*, the monk of *Durham*. It appeared at the beginning of *February* about the evening in the south-west. From this and from the situation of the tail we may infer that the head was near the sun. *Matthew Paris* says, "it was about one cubit from the sun; from the third [or rather the sixth] to the ninth hour sending out a long stream of light." The comet of 1264, in *July*, or about the solstice, preceded the rising sun, sending out its beams with a great light towards the west as far as the middle of the heavens; and at the beginning it ascended a little above the horizon: but as the sun went forwards it retired every day farther from the horizon, till it passed by the very middle of the heavens. It is said to have been at the beginning large and bright, having a large coma, which decayed from day to day. It is described in *Append. Matth. Paris, Hist. Ang.* after this manner: "*An. Christi* 1265, there appeared a comet so wonderful, that none then living had ever seen the like; for, rising from the east with a great brightness, it extended itself with a great light as far as the middle of the hemisphere towards the west." The Latin original being somewhat barbarous and obscure, it is here subjoined. *Ab oriente enim cum magno fulgore surgens, usque ad medium hemisphærii versus occidentem, omnia perlucide pertrahebat.*

"In the year 1401 or 1402, the sun being got below the horizon, there appeared in the west a bright and shining comet, sending out a tail upwards, in splendor like a flame of fire, and in form like a spear, darting its rays from west to east. When the sun was sunk below the horizon, by the lustre of its own rays it enlightened all the borders of the earth, not permitting the other stars to shew their light, or the shades of night to darken the air, because its light exceeded that of the others, and extended itself to the upper part of the heavens, flaming," &c., *Hist. Byzant. Duc. Mich. Nepot.* From the situation of the tail of this comet, and the time of its first appearance, we may infer that the head was then near the sun, and went farther from him every day; for that comet continued three months. In the year 1527, *Aug.* 11, about four in the morning, there was seen almost throughout *Europe* a terrible comet in *Leo*, which continued flaming an hour and a quarter every day. It rose from the east, and ascended to the south and west to a prodigious length. It was most conspicuous to the north, and its cloud (that is, its tail) was very terrible; having, according to the fancies of the vulgar, the form of an arm a little bent holding a sword of a vast magnitude. In the year 1618, in the end of *November*, there began a rumour, that there appeared about sun-rising a bright beam, which was the tail of a comet whose head was yet concealed within the brightness of the solar rays. On *Nov.* 24, and from that time, the comet itself appeared with a bright light, its head and tail being extremely resplendent. The length of the tail, which was at first 20 or 30 deg., increased till *December* 9, when it arose to 75 deg., but with a light much more faint and dilute than at the beginning. In the year 1668, *March* 5, N. S., about 7 in the evening, *P. Valent. Estancius*, being in *Brazil*, saw a comet near the horizon in the south-west. Its head was small, and scarcely discernible, but its tail extremely bright and refulgent, so that the reflection of it from the sea was easily seen by those who stood upon the shore. This great splendor lasted but three days, decreasing very remarkably from that time. The tail at the beginning extended itself from west to south, and in a situation almost parallel to the horizon, appearing like a shining beam 23 deg. in length. Afterwards, the light decreasing, its magnitude increased till the comet ceased to be visible; so that *Cassini*, at *Bologna*, saw it (*Mar.* 10, 11, 12) rising from the horizon 32 deg. in length. In *Portugal* it is said to have taken up a fourth part of the heavens (that is, 45 deg.), extending itself from west to east with a notable brightness; though the whole of it was not seen, because the head in this part of the world always lay hid below the horizon. From the increase of the tail it is plain that the head receded from the sun, and was nearest to it at the beginning, when the tail appeared brightest.

To all these we may add the comet of 1680, whose wonderful splendor at the conjunction of the head with the sun was above described. But so great a splendor argues the comets of this kind to have really passed near the fountain of light, especially since the tails never shine so much in their opposition to the sun; nor do we read that fiery beams have ever appeared there.

Lastly, the same thing is inferred (p. 466, 467) from the light of the heads increasing in the recess of the comets from the earth towards the sun, and decreasing in their return from the sun towards the earth; for so the last comet of the year 1665 (by the observation of *Hevelius*), from the time that it was first seen, was always losing of its apparent motion, and therefore had already passed its perigee; yet the splendor of its head was daily increasing, till, being hid by the sun's rays, the comet ceased to appear. The comet of the year 1683 (by the observation of the same *Hevelius*), about the end of *July*, when it first appeared, moved at a very slow rate, advancing only about 40 or 45 minutes in its orbit in a day's time. But from that time its diurnal motion was continually upon the increase till *September* 4, when it arose to about 5 degrees; and therefore in all this interval of time the comet was approaching to the earth. Which is likewise proved from the diameter of its head measured with a micrometer; for, *August* the 6th, *Hevelius* found it only 6′ 5″, including the coma; which, *September* 2, he observed 9′ 7″. And therefore its head appeared far less about the beginning than towards the end of its motion, though about the beginning, because nearer to the sun, it appeared far more lucid than towards the end, as the same *Hevelius* declares. Wherefore in all this interval of time, on account of its recess from the sun, it decreased in splendor, notwithstanding its access towards the earth. The comet of the year 1618, about the middle of *December*, and that of the year 1680, about the end of the same month, did both move with their greatest velocity, and were therefore then in their perigees; but the greatest splendor of their heads was seen two weeks before, when they had just got clear of the sun's rays; and the greatest splendor of their tails a little more early, when yet nearer to the sun. The head of the former comet, according to the observations of *Cysatus, Dec.* 1, appeared greater than the stars of the first magnitude; and, *Dec.* 16 (being then in its perigee), of a small magnitude, and the splendor or clearness was much diminished. *Jan.* 7, *Kepler*, being uncertain about the head, left off observing. *Dec.* 12, the head of the last comet was seen and observed by *Flamsted* at the distance of 9 degrees from the sun, which a star of the third magnitude could hardly have been. *December* 15 and 17, the same appeared like a star of the third magnitude, its splendor being diminished by the bright clouds near the setting sun. *Dec.* 26, when it moved with the greatest swiftness, and was almost in its perigee, it was inferior to *Os Pegasi*, a star of the third magnitude. *Jan.* 3, it appeared like a star of the fourth; *Jan.* 9, like a star of the fifth. *Jan.* 13, it disappeared, by reason of the brightness of the moon, which was then in its increase. *Jan.* 25, it was scarcely equal to the stars of the seventh magnitude. If we take equal times on each hand of the perigee, the heads placed at remote distances would have shined equally before and after, because of their equal distances from the earth. That in one case they shined very bright, and in the other vanished, is to be ascribed to the nearness of the sun in the first case, and his distance in the other; and from the great difference of the light in these two cases we infer its great nearness in the first of them; for the light of the comets uses to be regular, and to appear greatest when their heads move the swiftest, and are therefore in their perigees; excepting in so far as it is increased by their nearness to the sun.

From these things I at last discovered why the comets frequent so much the region of the sun. If they were to be seen in the regions a great way beyond Saturn, they must appear oftener in those parts of the heavens that are opposite to the sun; for those which are in that situation would be nearer to the earth, and the interposition of the sun would obscure the others: but, looking over the history of comets, I find that four or five times more have been seen in the hemisphere towards the sun than in the opposite hemisphere; besides, without doubt, not a few which have been hid by the light of the sun; for comets descending into our parts neither emit tails, nor are so well illuminated by the sun, as to discover themselves to our naked eyes, till they are come nearer to us than Jupiter. But the far greater part of that spherical space, which is described about the sun with so small an interval, lies on that side of the earth which regards the sun, and the comets in that greater part are more strongly illuminated, as being for the most part nearer to the sun: besides, from the remarkable eccentricity of their orbits, it comes to pass that their lower apsides are much nearer to the sun than if their revolutions were performed in circles concentric to the sun.

Hence also we understand why the tails of the comets, while their heads are descending towards the sun, always appear short and rare, and are seldom said to have exceeded 15 or 20 deg. in length; but in the recess of the heads from the sun often shine like fiery beams, and soon after reach to 40, 50, 60, 70 deg. in length, or more. This great splendor and length of the tails arises from the heat which the sun communicates to the comet as it passes near it. And thence, I think, it may be concluded, that all the comets that have had such tails have passed very near the sun.

Hence also we may collect that the tails arise from the atmospheres of the heads (p. 487 to 488): but we have had three several opinions about the tails of comets; for some will have it that they are nothing else but the beams of the sun's light transmitted through the comets' heads, which they suppose to be transparent; others, that they proceed from the refraction which light suffers in passing from the comet's head to the earth;

and, lastly, others, that they are a sort of clouds or vapour constantly rising from the comets' heads, and tending towards the parts opposite to the sun. The first is the opinion of such as are yet unacquainted with optics; for the beams of the sun are not seen in a darkened room, but in consequence of the light that is reflected from them by the little particles of dust and smoke which are always flying about in the air; and hence it is that in air impregnated with thick smoke they appear with greater brightness, and are more faintly and more difficultly seen in a finer air; but in the heavens, where there is no matter to reflect the light, they are not to be seen at all. Light is not seen as it is in the beams, but as it is thence reflected to our eyes; for vision is not made but by rays falling upon the eyes, and therefore there must be some reflecting matter in those parts where the tails of comets are seen; and so the argument turns upon the third opinion; for that reflecting matter can be no where found but in the place of the tail, because otherwise, since all the celestial spaces are equally illuminated by the sun's light, no part of the heavens could appear with more splendor than another. The second opinion is liable to many difficulties. The tails of comets are never seen variegated with those colours which ever use to be inseparable from refraction; and the distinct transmission of the light of the fixed stars and planets to us is a demonstration that the æther or celestial medium is not endowed with any refractive power. For as to what is alledged that the fixed stars have been sometimes seen by the *Egyptians* environed with a coma or capillitium, because that has but rarely happened, it is rather to be ascribed to a casual refraction of clouds, as well as the radiation and scintillation of the fixed stars to the refractions both of the eyes and air; for upon applying a telescope to the eye, those radiations and scintillations immediately disappear. By the tremulous agitation of the air and ascending vapours, it happens that the rays of light are alternately turned aside from the narrow space of the pupil of the eye; but no such thing can have place in the much wider aperture of the object-glass of a telescope; and hence it is that a scintillation is occasioned in the former case which ceases in the latter; and this cessation in the latter case is a demonstration of the regular transmission of light through the heavens without any sensible refraction. But, to obviate an objection that may be made from the appearing of no tail in such comets as shine but with a faint light, as if the secondary rays were then too weak to affect the eyes, and for this reason it is that the tails of the fixed stars do not appear, we are to consider that by the means of telescopes the light of the fixed stars may be augmented above an hundred fold and yet no tails are seen; that the light of the planets is yet more copious without any tail, but that comets are seen sometimes with huge tails when the light of their heads is but faint and dull; for so it happened in the comet of the year 1680, when in the month of *De-*

cember it was scarcely equal in light to the stars of the second magnitude, and yet emitted a notable tail, extending to the length of 40°, 50°, 60°, or 70°, and upwards; and afterwards, on the 27th and 28th of *January*, the head appeared but as a star of the seventh magnitude; but the tail (as was said above), with a light that was sensible enough, though faint, was stretched out to 6 or 7 degrees in length, and with a languishing light that was more difficultly seen, even to 12° and upwards. But on the 9th and 10th of *February*, when to the naked eye the head appeared no more, I saw through a telescope the tail of 2° in length. But farther; if the tail was owing to the refraction of the celestial matter, and did deviate from the opposition of the sun, according as the figure of the heavens requires, that deviation, in the same places of the heavens, should be always directed towards the same parts: but the comet of the year 1680, *December* 28^d. $8\frac{1}{2}^h$. P. M. at *London*, was seen in *Pisces*, 8° 41′, with latitude north 28° 6′, while the sun was in *Capricorn* 18° 26′. And the comet of the year 1577, *December* 29, was in *Pisces* 8° 41′, with latitude north 28° 40′; and the sun, as before, in about *Capricorn* 18° 26′. In both cases the situation of the earth was the same, and the comet appeared in the same place of the heavens; yet in the former case the tail of the comet (as well by my observations as by the observations of others) deviated from the opposition of the sun towards the north by an angle of $4\frac{1}{2}$ degrees, whereas in the latter there was (according to the observation of *Tycho*) a deviation of 21 degrees towards the south. The refraction, therefore, of the heavens being thus disproved, it remains that the phænomena of the tails of comets must be derived from some reflecting matter. That vapours sufficient to fill such immense spaces may arise from the comet's atmospheres, may be easily understood from what follows.

It is well known that the air near the surface of our earth possesses a space about 1200 times greater than water of the same weight; and therefore a cylindric column of air 1200 feet high is of equal weight with a cylinder of water of the same breadth, and but one foot high. But a cylinder of air reaching to the top of the atmosphere is of equal weight with a cylinder of water about 33 feet high; and therefore if from the whole cylinder of air the lower part of 1200 feet high is taken away, the remaining upper part will be of equal weight with a cylinder of water 32 feet high. Wherefore at the height of 1200 feet, or two furlongs, the weight of the incumbent air is less, and consequently the rarity of the compressed air greater, than near the surface of the earth in the ratio of 33 to 32. And, having this ratio, we may compute the rarity of the air in all places whatsoever (by the help of Cor. Prop. XXII, Book II), supposing the expansion thereof to be reciprocally proportional to its compression; and this proportion has been proved by the experiments of *Hooke* and others. The result of the computation I have set down in the following table, in the first column of which you have the height of the air in miles, whereof 4000 make a semi-diameter of the earth; in the second the compression of the air, or the incumbent weight; in the third its rarity or expansion, supposing gravity to decrease in the duplicate ratio of the distances from the earth's centre. And the *Latin* numeral characters are here used for certain numbers of ciphers, as 0,xvii 1224 for 0,000000000000000001224, and 26956 xv for 26956000000000000000.

AIR's

Height.	Compression.	Expansion.
0	33	1
5	17,8515	1,8486
10	9,6717	3,4151
20	2,852	11,571
40	0,2525	136,83
400	0,xvii 1224	26956 xv
4000	0,cv. 4465	73907 cii
40000	0,cxcii 1628	20263 clxxxix
400000	0,ccx 7895	41798 ccvii
4000000	0,ccxii 9878	33414 ccix
Infinite.	0,ccxii 6041	54622 ccix

But from this table it appears that the air, in proceeding upwards, is rarefied in such manner, that a sphere of that air which is nearest to the earth, of but one inch in diameter, if dilated with that rarefaction which it would have at the height of one semi-diameter of the earth, would fill all the planetary regions as far as the sphere of Saturn, and a great way beyond; and at the height of ten semi-diameters of the earth would fill up more space than is contained in the whole heavens on this side the fixed stars, according to the preceding computation of their distance. And though, by reason of the far greater thickness of the atmospheres of comets, and the great quantity of the circum-solar centripetal force, it may happen that the air in the celestial spaces, and in the tails of comets, is not so vastly rarefied, yet from this computation it is plain that a very small quantity of air and vapour is abundantly sufficient to produce all the appearances of the tails of comets; for that they are indeed of a very notable rarity appears from the shining of the stars through them. The atmosphere of the earth, illuminated by the sun's light, though but of a few miles in thickness, obscures and extinguishes the light not only of all the stars, but even of the moon itself; whereas the smallest stars are seen to shine through the immense thickness of the tails of comets, likewise illuminated by the sun, without the least diminution of their splendor.

Kepler ascribes the ascent of the tails of comets to the atmospheres of their heads, and their direction towards the parts opposite to the sun to the action of the rays of light carrying along with them the matter of the comets' tails; and without any great incongruity we may suppose that, in so free spaces, so fine a matter as that of the æther may yield to the action of the rays of the sun's light, though those rays are not able sensibly to move the gross substances in our parts, which are clogged with so palpable a resistance. Another author thinks that there may be a sort of particles of matter endowed with a principle of levity as well as others are with a power of gravity; that the matter of the tails of comets may be of the former sort, and that its ascent from the sun may be owing to its levity; but, considering the gravity of terrestrial bodies is as the matter of the bodies, and therefore can be neither more nor less in the same quantity of matter, I am inclined to believe that this ascent may rather proceed from the rarefaction of the matter of the comets' tails. The ascent of smoke in a chimney is owing to the impulse of the air with which it is entangled. The air rarefied by heat ascends, because its specific gravity is diminished, and in its ascent carries along with it the smoke with which it is engaged. And why may not the tail of a comet rise from the sun after the same manner? for the sun's rays do not act any way upon the mediums which they pervade but by reflection and refraction; and those reflecting particles heated by this action, heat the matter of the æther which is involved with them. That matter is rarefied by the heat which it acquires, and because by this rarefaction the specific gravity, with which it tended towards the sun before, is diminished, it will ascend therefrom like a stream, and carry along with it the reflecting particles of which the tail of the comet is composed; the impulse of the sun's light, as we have said, promoting the ascent.

But that the tails of comets do arise from their heads (p. 488), and tend towards the parts opposite to the sun, is farther confirmed from the laws which the tails observe; for, lying in the planes of the comets' orbits which pass through the sun, they constantly deviate from the opposition of the sun towards the parts which the comets' heads in their progress along those orbits have left; and to a spectator placed in those planes they appear in the parts directly opposite to the sun; but as the spectator recedes from those planes, their deviation begins to appear, and daily becomes greater. And the deviation, *cæteris paribus,* appears less when the tail is more oblique to the orbit of the comet, as well as when the head of the comet approaches nearer to the sun; especially if the angle of deviation is estimated near the head of the comet. Farther; the tails which have no deviation appear straight, but the tails which deviate are likewise bended into a certain curvature; and this curvature is greater when the deviation is greater, and is more sensible when the tail, *cæteris paribus,* is longer; for in the shorter tails the curvature is hardly to be perceived. And the angle of deviation is less near the comet's head, but greater towards the other end of the tail, and that because the lower side of the tail regards the parts from which the deviation is made, and which lie in a right line drawn out infinitely from the sun through the comet's head. And the tails that are longer and broader, and shine with a stronger light, appear more resplendent and more exactly defined on the convex than on the concave side. Upon which accounts it is plain that the phænomena of the tails of comets depend upon the motions of their heads, and by no means upon the places of the heavens in which their heads are seen; and that, therefore, the tails of the comets do not proceed from the refraction of the heavens, but from their own heads, which furnish the matter that forms the tail; for as in our air the smoke of a heated body ascends either perpendicularly, if the body is at rest, or obliquely if the body is moved obliquely, so in the heavens, where all the bodies gravitate towards the sun, smoke and vapour must (as we have already said) ascend from the sun, and either rise perpendicularly, if the smoking body is at rest, or obliquely, if the body, in the progress of its motion, is always leaving those places from which the upper or higher parts of the vapours had risen before. And that obliquity will be less where the vapour ascends with more velocity, to wit, near the smoking body, when that is near the sun; for there the force of the sun by which the vapour ascends is stronger. But because the obliquity is varied, the column of vapour will be incurvated; and because the vapour in the preceding side is something more recent, that is, has ascended something more lately from the body, it will therefore be something more dense on that side, and must on that account reflect more light, as well as be better defined; the vapour on the other side languishing by degrees, and vanishing out of sight.

But it is none of our present business to explain the causes of the appearances of nature. Let those things which we have last said be true or false, we have at least made out, in the preceding discourse, that the rays of light are directly propagated from the tails of comets in right lines through the heavens, in which those tails appear to the spectators wherever placed; and consequently the tails must ascend from the heads of the comets towards the parts opposite to the sun. And from this principle we may determine anew the limits of their distances in manner following. Let S represent the sun, T the earth, STA the elongation of a comet from the sun, and ATB the apparent length of its tail; and because the light is propagated from the extremity of the tail in the direction of the right line TB, that extremity must lie somewhere in the line TB. Suppose it in D, and join DS cutting TA in C. Then, because the tail is always stretched out towards the parts nearly opposite to the sun, and therefore

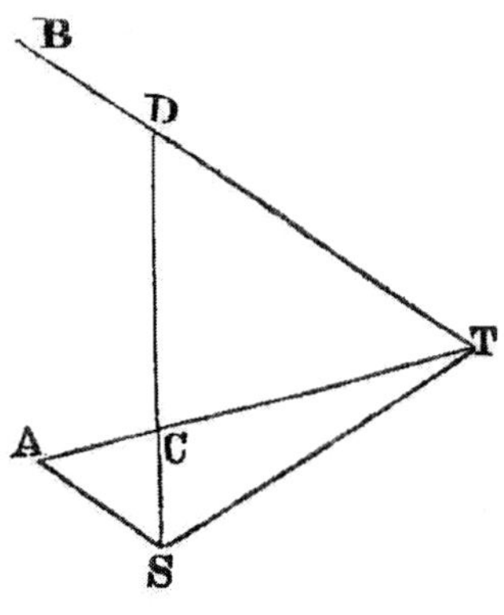

the sun, the head of the comet, and the extremity of the tail, lie in a right line, the comet's head will be found in C. Parallel to TB draw SA, meeting the line TA in A, and the comet's head C must necessarily be found between T and A, because the extremity of the tail lies somewhere in the infinite line TB; and all the lines SD which can possibly be drawn from the point S to the line TB must cut the line TA somewhere between T and A. Wherefore the distance of the comet from the earth cannot exceed the interval TA, nor its distance from the sun the interval SA beyond, or ST on this side the sun. For instance: the elongation of the comet of 1680 from the sun, *Dec.* 12, was 9°, and the length of its tail 35° at least. If, therefore, a triangle TSA is made, whose angle T is equal to the elongation 9°, and angle A equal to ATB, or to the length of the tail, viz., 35°, then SA will be to ST, that is, the limit of the greatest possible distance of the comet from the sun to the semi-diameter of the *orbis magnus*, as the sine of the angle T to the sine of the angle A, that is, as about 3 to 11. And therefore the comet at that time was less distant from the sun than by $\frac{3}{11}$ of the earth's distance from the sun, and consequently either was within the orb of Mercury, or between that orb and the earth. Again, *Dec.* 21, the elongation of the comet from the sun was $32\frac{2}{3}$°, and the length of its tail 70°. Wherefore as the sine of $32\frac{2}{3}$° to the sine of 70°, that is, as 4 to 7, so was the limit of the comet's distance from the sun to the distance of the earth from the sun, and consequently the comet had not then got without the orb of Venus. *Dec.* 28, the elongation of the comet from the sun was 55°, and the length of its tail 56°; and therefore the limit of the comet's distance from the sun was not yet equal to the distance of the earth from the same, and consequently the comet had not then got without the earth's orbit. But from its parallax we find that its egress from the orbit happened about *Jan.* 5, as well as that it had descended far within the orbit of Mercury. Let us suppose it to have been in its perihelion *Dec.* the 8th, when it was in conjunction with the sun; and it will follow that in the journey from its perihelion to its exit out of the earth's orbit it had spent 28 days; and consequently that in the 26 or 27 days following, in which it ceased to be farther seen by the naked eye, it had scarcely doubled its distance from the sun; and by limiting the distances of other comets by the like arguments, we come at last to this conclusion,—that all comets, during the time in which they are visible by us, are within the compass of a spherical space described about the sun as a centre, with a radius double, or at most triple, of the distance of the earth from the sun.

And hence it follows that the comets, during the whole time of their appearance unto us, being within the sphere of activity of the circumsolar force, and therefore agitated by the impulse of that force, will (by Cor. 1, Prop. XII, Book I, for the same reason as the planets) be made to move in conic sections that have one focus in the centre of the sun, and, by radii drawn to the sun, to describe areas proportional to the times; for that force is propagated to an immense distance, and will govern the motions of bodies far beyond the orbit of Saturn.

There are three hypotheses about comets (p. 466); for some will have it that they are generated and perish as often as they appear and vanish; others, that they come from the regions of the fixed stars, and are seen by us in their passage through the system of our planets; and, lastly, others, that they are bodies perpetually revolving about the sun in very eccentric orbits. In the first case, the comets, according to their different velocities, will move in conic sections of all sorts; in the second, they will describe hyperbolas, and in either of the two will frequent indifferently all quarters of the heavens, as well those about the poles as those towards the ecliptic; in the third, their motions will be performed in ellipses very eccentric, and very nearly approaching to parabolas. But (if the law of the planets is observed) their orbits will not much decline from the plane of the ecliptic; and, so far as I could hitherto observe, the third case obtains; for the comets do, indeed, chiefly frequent the zodiac, and scarcely ever attain to a heliocentric latitude of 40°. And that they move in orbits very nearly parabolical, I infer from their velocity; for the velocity with which a parabola is described is every where to the velocity with which a comet or planet may be revolved about the sun in a circle at the same distance in the subduplicate ratio of 2 to 1 (by Cor. VII, Prop. XVI); and, by my computation, the velocity of comets is found to be much about the same. I examined the thing by inferring nearly the velocities from the distances, and the distances both from the parallaxes and the phænomena of the tails, and never found the errors of excess or defect in the velocities greater than what might have arose from the errors in the distances collected after that manner. But I likewise made use of the reasoning that follows.

Supposing the radius of the *orbis magnus* to be divided into 1000 parts: let the numbers in the first column of the following table represent the distance of the vertex of the parabola from the sun's centre, expressed by those parts; and a comet in the times expressed in col. 2, will pass from its perihelion to the surface of the sphere which is described about the sun as a centre with the radius of the *orbis magnus;* and in the times expressed in col. 3, 4, and 5, it will double, triple, and quadruple, that its distance from the sun.

TABLE I.

The distance of a comet's perihelion from the Sun's centre.	The time of a comet's passage from its perihelion to a distance from the sun equal to											
	The radii of the *orbis magnus*.			To its double.			To its triple			To its Quadruple.		
	d.	h.	′	d.	h.	′	d.	h.	′	d.	h.	′
0	27	11	12	77	16	28	142	17	14	219	17	30
5	27	16	07	77	23	14						
10	27	21	00	78	06	24						
20	28	06	40	78	20	13	144	03	19	221	08	54
40	29	01	32	79	23	34						
80	30	13	25	82	04	56						
160	33	05	29	86	10	26	153	16	08	232	12	20
320	37	13	46	93	23	38						
640	37	09	49	105	01	28						
1280				106	06	35	200	06	43	297	03	46
2560							147	22	31	300	06	03

[This table, here corrected, is made on the supposition that the earth's diurnal motion is just 59′, and the measure of one minute loosely 0,2909, in respect of the radius 1000. If those measures are taken true, the true numbers of the table will all come out less. But the difference, even when greatest, and to the quadruple of the earth's distance from the sun, amounts only to 16ʰ. 55′.]

The time of a comet's ingress into the sphere of the *orbis magnus*, or of its egress from the same, may be inferred nearly from its parallax, but with more expedition by the following

TABLE II.

The apparent elongation of a comet from the sun.	Its apparent diurnal motion in its own orbit.		Its distance from the earth in parts whereof the radius of the *orbis magnus* contains 1000.
	Direct.	Retrog.	
60°	2° 18′	00° 20′	1000
65	2 33	00 35	845
70	2 55	00 57	684
72	3 07	01 09	618
74	3 23	01 25	551
76	3 43	01 45	484
78	4 10	02 12	416
80	4 57	02 49	347
82	5 45	03 47	278
84	7 18	05 20	209
86	10 27	08 19	140
88	18 37	16 39	70
90	Infi'ite	Infi'ite	00

The ingress of a comet into the sphere of the *orbis magnus*, or its egress from the same, happens at the time of its elongation from the sun, expressed in col. 1, against its diurnal motion. So in the comet of 1681, *Jan.* 4, O.S. the apparent diurnal motion in its orbit was about 3° 5′, and the corresponding elongation $71\frac{2}{3}$° ; and the comet had acquired this elongation from the sun *Jan.* 4, about six in the evening. Again, in the year 1680, *Nov.* 11, the diurnal motion of the comet that then appeared was about $4\frac{2}{3}$° ; and the corresponding elongation $79\frac{2}{3}$ happened *Nov.* 10, a little before midnight. Now at the times named these comets had arrived at an equal distance from the sun with the earth, and the earth was then almost in its perihelion. But the first table is fitted to the earth's mean distance from the sun assumed of 1000 parts; and this distance is greater by such an excess of space as the earth might describe by its annual motion in one day's time, or the comet by its motion in 16 hours. To reduce the comet to this mean distance of 1000 parts, we add those 16 hours to the former time, and subduct them from the latter; and thus the former becomes *Jan.* 4^d. 10^h. afternoon; the latter *Nov.* 10, about six in the morning. But from the tenor and progress of the diurnal motions it appears that both comets were in conjunction with the sun between *Dec.* 7 and *Dec.* 8; and from thence to *Jan.* 4^d. 10^h. afternoon on one side, and to *Nov.* 10^d. 6^h. of the morning on the other, there are about 28 days. And so many days (by Table 1) the motions in parabolic trajectories do require.

But though we have hitherto considered those comets as two, yet, from the coincidence of their perihelions and agreement of their velocities, it is probable that in effect they were but one and the same; and if so, the orbit of this comet must have either been a parabola, or at least a conic section very little differing from a parabola, and at its vertex almost in contact with the surface of the sun. For (by Tab. 2) the distance of the comet from the earth, *Nov.* 10, was about 360 parts, and *Jan.* 4, about 630. From which distances, together with its longitudes and latitudes, we infer the distance of the places in which the comet was at those times to have been about 280: the half of which, viz., 140, is an ordinate to the comet's orbit, cutting off a portion of its axis nearly equal to the radius of the *orbis magnus*, that is, to 1000 parts. And, therefore, dividing the square of the ordinate 140 by 1000, the segment of the axis, we find the *latus rectum* 19, 16, or in a round number 20; the fourth part whereof, 5, is the distance of the vertex of the orbit from the sun's centre. But the time corresponding to the distance of 5 parts in Tab. 1 is 27^d. 16^h. 7′. In which time, if the comet moved in a parabolic orbit, it would have been carried from its perihelion to the surface of the sphere of the *orbis magnus* described with the radius 1000, and would have spent the double of that time, viz., 55^d. $8\frac{1}{4}^h$. in the whole course of its motion within that sphere: and so in fact it did; for from *Nov.* 10^d. 6^h. of the morning, the time of the comet's ingress into the sphere of the *orbis magnus*, to *Jan.* 4^d. 10^h. afternoon, the time of its egress from the same, there are 55^d. 16^h. The small difference of 7$\frac{3}{4}^h$. in this rude way of computing is to be neglected, and perhaps may arise from the comet's motion being some small matter slower, as it must have been if the true orbit in which it was carried was an ellipsis. The middle time between its ingress and egress was *December* 8^d. 2^h. of the morning; and therefore at this time the comet ought to have been in its perihelion. And accordingly that very day, just before sunrising, Dr. *Halley* (as we said) saw the tail short and broad, but very bright, rising perpendicularly from the horizon. From the position of the tail it is certain that the comet had then crossed over the ecliptic, and got into north latitude, and therefore had passed by its perihelion, which lay on the other side of the ecliptic, though it had not yet come into conjunction with the sun; and the comet [see more of this famous comet, p. 475 to 486] being at this time between its perihelion and its conjunction with the sun, must have been in its perihelion a few hours before; for in so near a distance from the sun it must have been carried with great velocity, and have apparently described almost half a degree every hour.

By like computations I find that the comet of 1618 entered the sphere of the *orbis magnus December* 7, towards sun-setting; but its conjunction with the sun was *Nov.* 9, or 10, about 28 days intervening, as in the preceding comet; for from the size of the tail of this, in which it was equal to the preceding, it is probable that this comet likewise did come almost into a contact with the sun. Four comets were seen that year of which this was the last. The second, which made its first appearance *October* 31, in the neighbourhood of the rising sun, and was soon after hid under the sun's rays, I suspect to have been the same with the fourth, which emerged out of the sun's rays about *Nov.* 9. To these we may add the comet of 1607, which entered the sphere of the *orbis magnus Sept.* 14, O.S. and arrived at its perihelion distance from the sun about *October* 19, 35 days intervening. Its perihelion distance subtended an apparent angle at the earth of about 23 degrees, and was therefore of 390 parts. And to this number of parts about 34 days correspond in Tab. 1. Farther; the comet of 1665 entered the sphere of the *orbis magnus* about *March* 17, and came to its perihelion about *April* 16, 30 days intervening. Its perihelion distance subtended an angle at the earth of about seven degrees, and therefore was of 122 parts: and corresponding to this number of parts, in Tab. 1, we find 30 days. Again; the comet of 1682 entered the sphere of the *orbis magnus* about *Aug.* 11, and arrived at its perihelion about *Sep.* 16, being then distant from the sun by about 350 parts, to which, in Tab. 1, belong 33$\frac{1}{2}$ days. Lastly; that memorable comet of *Regiomontanus*, which in 1472 was carried through the circum-polar parts of our northern hemisphere with such rapidity as to describe 40 degrees in one day, entered the sphere of the *orbis magnus Jan.* 21, about the time that it was passing by the pole, and, hastening from thence towards the sun, was hid under the sun's rays about the end of ***Feb.;*** whence it is probable that 30 days, or a few more, were spent between its ingress into the sphere of the *orbis magnus* and its perihelion. Nor did this comet truly move with more velocity than other comets, but owed the greatness of its apparent velocity to its passing by the earth at a near distance.

It appears, then, that the velocity of comets (p. 471), so far as it can be determined by these rude ways of computing, is that very velocity with which parabolas, or ellipses near to parabolas, ought to be described; and therefore the distance between a comet and the sun being given, the velocity of the comet is nearly given. And hence arises this problem.

PROBLEM.

The relation betwixt the velocity of a comet and its distance from the sun's centre being given, the comet's trajectory is required.

If this problem was resolved, we should thence have a method of determining the trajectories of comets to the greatest accuracy; for if that relation be twice assumed, and from thence the trajectory be twice computed, and the error of each trajectory be found from observations, the assumption may be corrected by the Rule of False, and a third trajectory may thence be found that will exactly agree with the observations. And by determining the trajectories of comets after this method, we may come, at last, to a more exact knowledge of the parts through which those bodies travel, of the velocities with which they are carried, what sort of trajectories they describe, and what are the true magnitudes and forms of their tails according to the various distances of their heads from the sun; whether, after certain intervals of time, the same comets do return again, and in what periods they complete their several revolutions. But the problem may be resolved by determining, first, the hourly motion of a comet to a given time from three or more observations, and then deriving the trajectory from this motion. And thus the invention of the trajectory, depending on one observation, and its hourly motion at the time of this observation, will either confirm or disprove itself; for the conclusion that is drawn from the motion only of an hour or two and a false hypothesis, will never agree with the motions of the comets from beginning to end. The method of the whole computation is this.

LEMMA I.

To cut two right lines OR, TP, *given in position, by a third right line* RP, *so as* TRP *may be a right angle; and, if another right line* SP *is drawn to any given point* S, *the solid contained under this line* SP, *and the square of the right line* OR *terminated at a given point* O, *may be of a given magnitude.*

It is done by linear description thus. Let the given magnitude of the solid be $M^2 \times N$; from any point r of the right line OR erect the per-

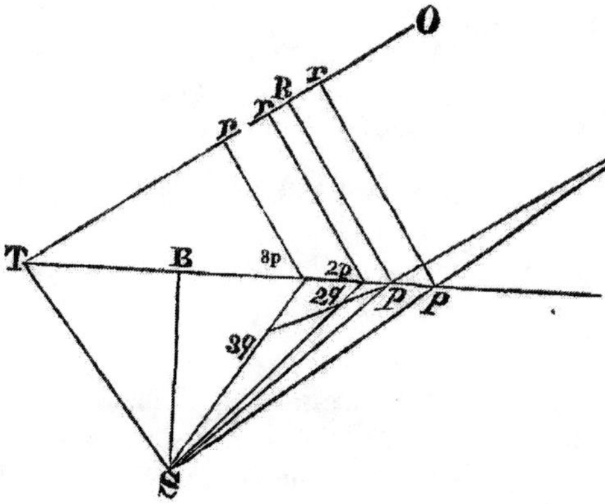

pendicular rp meeting TP in p. Then through the point Sp draw the line Sq equal to $\frac{M^2 \times N}{Or^2}$. In like manner draw three or more right lines S2q, S3q, &c.; and a regular line $q2q3q$, drawn through all the points $q2q3q$, &c., will cut the right line TP in the point P, from which the perpendicular PR is to be let fall. Q.E.F.

By trigonometry thus. Assuming the right line TP as found by the preceding method, the perpendiculars TR, SB, in the triangles TPR, TPS, will be thence given; and the side SP in the triangle SBP, as well as the error $\frac{M^2 \times N}{OR^2} - SP$. Let this error, suppose D, be to a new error, suppose E, as the error $2p2q \pm 3p3q$ to the error $2p3p$; or as the error $2p2q \pm D$ to the error $2p$P; and this new error added to or subducted from the length TP, will give the correct length TP $\pm$ E. The inspection of the figure will shew whether we are to add to or subtract; and if at any time there should be use for a farther correction, the operation may be repeated.

By arithmetic thus. Let us suppose the thing done, and let TP + e be the correct length of the right line TP as found out by delineation; and thence the correct lengths of the lines OR, BP, and SP, will be $OR - \frac{TR}{TP}e$, $BP + e$, and $\sqrt{SP^2 + 2BPe + ee} = \frac{M^2N}{OR^2 + \frac{2OR \times TR}{TP}e + \frac{TR^2}{TP^2}ee}$.

Whence, by the method of converging series, we have $SP + \frac{BP}{SP}e + \frac{SB^2}{2SP^3}ee$, &c., $= \frac{M^2N}{OR^2} + \frac{2TR}{TP} \times \frac{M^2N}{OR^3}e + \frac{3TR^2}{TP^2} \times \frac{M^2N}{OR^4}ee$, &c. For the given co-efficients $\frac{M^2N}{OR^2} - SP$, $\frac{2TR}{TP} \times \frac{M^2N}{OR^3} - \frac{BP}{SP}$, $\frac{3TR^2}{TP^2} \times \frac{M^2N}{OR^4} - \frac{SB^2}{2SP^3}$, putting F, $\frac{F}{G}$, $\frac{F}{GH}$, and carefully observing the signs, we find $F + \frac{F}{G}e + \frac{F}{GH}ee = 0$, and $e + \frac{ee}{H} = -G$. Whence, neglecting the very small term $\frac{e^2}{H}$, e comes out equal to $-G$. If the error $\frac{e^2}{H}$ is not despicable, take $-G - \frac{G^2}{H} = e$.

And it is to be observed that here a general method is hinted at for solving the more intricate sort of problems, as well by trigonometry as by arithmetic, without those perplexed computations and resolutions of affected equations which hitherto have been in use.

LEMMA II.

To cut three right lines given in position by a fourth right line that shall pass through a point assigned in any of the three, and so as its intercepted parts shall be in a given ratio one to the other.

Let AB, AC, BC, be the right lines given in position, and suppose D to be the given point in the line AC. Parallel to AB draw DG meeting BC

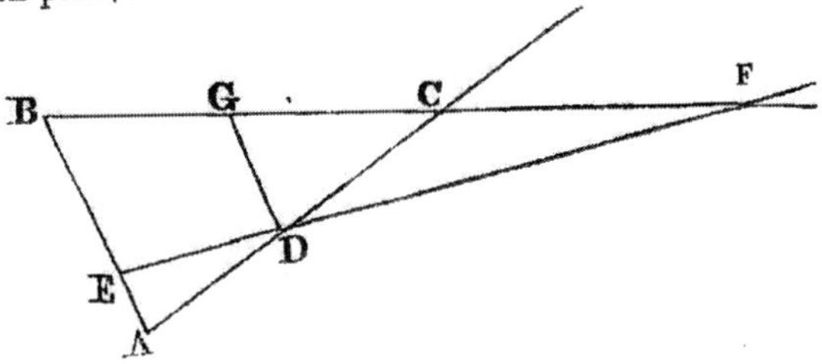

in G; and, taking GF to BG in the given ratio, draw FDE; and FD will be to DE as FG to BG. Q.E.F.

By trigonometry thus. In the triangle CGD all the angles and the side CD are given, and from thence its remaining sides are found; and from the given ratios the lines GF and BE are also given.

LEMMA III.

To find and represent by a linear description the hourly motion of a comet to any given time.

From observations of the best credit, let three longitudes of the comet be given, and, supposing ATR, RTB, to be their differences, let the hourly motion be required to the time of the middle observation TR. By *Lem.* II, draw the right line ARB, so as its intercepted parts AR, RB, may be

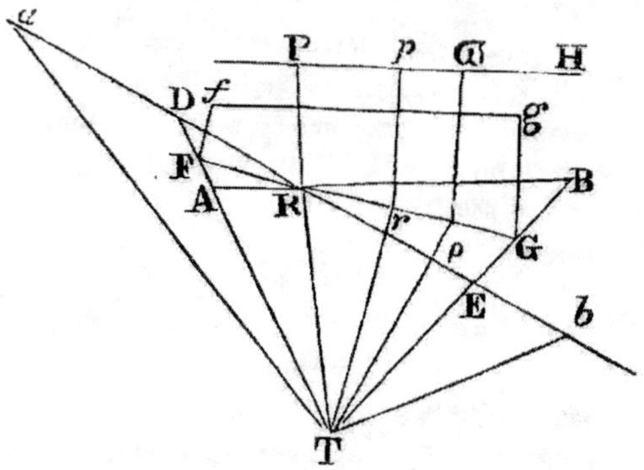

as the times between the observations; and if we suppose a body in the whole time to describe the whole line AB with an equal motion, and to be in the mean time viewed from the place T, the apparent motion of that body about the point R will be nearly the same with that of the comet at the time of the observation TR.

The same more accurately.

Let T*a*, T*b*, be two longitudes given at a greater distance on one side and on the other; and by *Lem.* II draw the right line *a*R*b* so as its intercepted parts *a*R, R*b* may be as the times between the observations *a*TR, RT*b*. Suppose this to cut the lines TA, TB, in D and E; and because the error of the inclination TR*a* increases nearly in the duplicate ratio of the time between the observations, draw FRG, so as either the angle DRF may be to the angle ARF, or the line DF to the line AF, in the duplicate ratio of the whole time between the observations *a*TB to the whole time between the observations ATB, and use the line thus found FG in place of the line AB found above.

It will be convenient that the angles ATR, RTB, *a*TA, BT*b*, be no less than of ten or fifteen degrees, the times corresponding no greater than of eight or twelve days, and the longitudes taken when the comet moves with the greatest velocity; for thus the errors of the observations will bear a less proportion to the differences of the longitudes.

LEMMA IV.

To find the longitudes of a comet to any given times.

It is done by taking in the line FG the distances Rr, Rρ, proportional to the times, and drawing the lines Tr, Tρ. The way of working by trigonometry is manifest.

LEMMA V.

To find the latitudes.

On TF, TR, TG, as radiuses, at right angles erect Ff, RP, Gg, tangents of the observed latitudes; and parallel to fg draw PH. The perpendiculars rp, $\rho\varpi$, meeting PH, will be the tangents of the sought latitudes to Tr and Tρ as radiuses.

PROBLEM I.

From the assumed ratio of the velocity to determine the trajectory of a comet.

Let S represent the sun; t, T, τ, three places of the earth in its orbit at equal distances; p, P, ϖ, as many corresponding places of the comet in

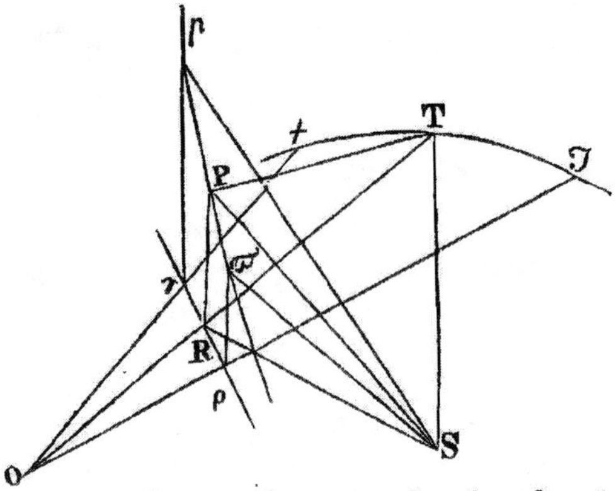

its trajectory, so as the distances interposed betwixt place and place may answer to the motion of one hour; pr, PR, $\varpi\rho$, perpendiculars let fall on the plane of the ecliptic, and rRρ the vestige of the trajectory in this plane. Join Sp, SP, Sϖ, SR, ST, tr, TR, $\tau\rho$, TP, and let tr, $\tau\rho$, meet in O, TR will nearly converge to the same point O, or the error will be inconsiderable. By the premised lemmas the angles rOR, ROρ, are given, as well as the ratios pr to tr, PR to TR, and $\varpi\rho$ to $\tau\rho$. The figure tTτO is likewise given both in magnitude and position, together with the distance ST, and the angles STR, PTR, STP. Let us assume the velocity of the comet in the place P to be to the velocity of a planet revolved about the sun in a circle, at the same distance SP, as V to 1; and we shall have a line $pP\varpi$ to be determined, of this condition, that the space $p\varpi$, described by the comet in two hours, may be to the space $V \times t\tau$ (that is, to the space which the earth describes in the same time multiplied by the number V) in the subduplicate ratio of ST, the distance of the earth from the sun, to SP, the distance of the comet from the sun; and that the space pP, described by the comet in the first hour, may be to the space $P\varpi$, described by the comet in the second hour, as the velocity in p to the velocity in P; that is, in the subduplicate ratio of the distance SP to the distance Sp, or in the ratio of 2Sp to SP + Sp; for in this whole work I neglect small fractions that can produce no sensible error.

In the first place, then, as mathematicians, in the resolution of affected equations, are wont, for the first essay, to assume the root by conjecture, so, in this analytical operation, I judge of the sought distance TR as I best can by conjecture. Then, by *Lem.* II. I draw $r\rho$, first supposing rR equal to Rρ, and again (after the ratio of SP to Sp is discovered) so as rR may be to Rρ as 2SP to SP + Sp, and I find the ratios of the lines $p\varpi$, $r\rho$, and OR, one to the other. Let M be to $V \times t\tau$ as OR to $p\varpi$; and because the square of $p\varpi$ is to the square of $V \times t\tau$ as ST to SP, we shall have, *ex æquo*, OR^2 to M^2 as ST to SP, and therefore the solid $OR^2 \times SP$ equal to the given solid $M^2 \times ST$; whence (supposing the triangles STP, PTR, to be now placed in the same plane) TR, TP, SP, PR, will be given, by *Lem.* I. All this I do, first by delineation in a rude and hasty way; then by a new delineation with greater care; and, lastly, by an arithmetical computation. Then I proceed to determine the position of the lines $r\rho$, $p\varpi$, with the greatest accuracy, together with the nodes and inclination of the plane S$p\varpi$ to the plane of the ecliptic; and in that plane S$p\varpi$ I describe the trajectory in which a body let go from the place P in the direction of the given right line $p\varpi$ would be carried with a velocity that is to the velocity of the earth as $p\varpi$ to $V \times t\tau$. Q.E.F.

PROBLEM II.

To correct the assumed ratio of the velocity and the trajectory thence found.

Take an observation of the comet about the end of its appearance, or any other observation at a very great distance from the observations used before, and find the intersection of a right line drawn to the comet, in that observation with the plane S$p\varpi$, as well as the comet's place in its trajectory to the time of the observation. If that intersection happens in this place, it is a proof that the trajectory was rightly determined; if otherwise, a new number V is to be assumed, and a new trajectory to be found; and then the place of the comet in this trajectory to the time of that probatory observation, and the intersection of a right line drawn to the comet with the plane of the trajectory, are to be determined as before; and by comparing the variation of the error with the variation of the other quantities, we may conclude, by the Rule of Three, how far those other quantities ought to be varied or corrected, so as the error may become as small as possible. And by means of these corrections we may have the trajectory exactly, providing the observations upon which the computation was founded were exact, and that we did not err much in the assumption of the quantity V; for if we did, the operation is to be repeated till the trajectory is exactly enough determined. Q.E.F.

END OF THE SYSTEM OF THE WORLD.

CONTENTS

OF

THE SYSTEM OF THE WORLD.

That the matter of the heavens is fluid, . . . 511
The principle of circular motion in free spaces, . . . 512
The effects of centripetal forces, . . . 512
The certainty of the argument, . . . 514
What follows from the supposed diurnal motion of the stars, . . . 514
The incongruous consequences of this supposition, . . . 514
That there is a centripetal force really directed to the centre of every planet, . . . 515
Centripetal forces decrease in duplicate proportion of distances from the centre of every planet, 516
That the superior planets are revolved about the sun, and by radii drawn to the sun describe areas proportional to the times, . . . 517
That the force which governs the superior planets is directed not to the earth, but to the sun, . 518
That the circum-solar force throughout all the regions of the planets decreaseth in the duplicate proportion of the distances from the sun, . . . 519
That the circum-terrestrial force decreases in the duplicate proportion of the distances from the earth proved in the hypothesis of the earth's being at rest, . . . 519
The same proved in the hypothesis of the earth's motion, . . . 520
The decrement of the forces in the duplicate proportion of the distances from the earth and planets, proved from the eccentricity of the planets, and the very slow motion of their apses, . 520
The quantity of the forces tending towards the several planets: the circum-solar very great, . 521
The circum-terrestrial force very small, . . . 521
The apparent diameters of the planets, . . . 521
The correction of the apparent diameters, . . . 522
Why the density is greater in some of the planets and less in others; but the forces in all are as their quantities of matter, . . . 524
Another analogy between the forces and bodies, proved in the celestial bodies, . . . 525
Proved in terrestrial bodies, . . . 525
The affinity of those analogies, . . . 526
And coincidence, . . . 526
That the forces of small bodies are insensible, . . . 527
Which, notwithstanding, there are forces tending towards all terrestrial bodies proportional to their quantities of matter, . . . 528
Proved that the same forces tend towards the celestial bodies, . . . 528
That from the surfaces of the planets, reckoning outward, their forces decrease in the duplicate; but, reckoning inward, in the simple proportion of the distances from their centres, . 529
The quantities of the forces and of the motions arising in the several cases, . . . 529
That all the planets revolve about the sun, . . . 529
That the common centre of gravity of all the planets is quiescent. That the sun is agitated with a very slow motion. This motion defined, . . . 531
That the planets, nevertheless, are revolved in ellipses having their foci in the sun; and by radii drawn to the sun describe areas proportional to the times, . . . 531
Of the dimensions of the orbits, and of the motions of their aphelions and nodes, . . . 532
All the motions of the moon that have hitherto been observed by astronomers derived from the foregoing principles, . . . 532
As also some other unequable motions that hitherto have not been observed, . . . 533
And the distance of the moon from the earth to any given time, . . . 533
The motions of the satellites of Jupiter and Saturn derived from the motions of our moon, . 534
That the planets, in respect of the fixed stars, are revolved by equable motions about their proper axes. And that (perhaps) those motions are the most fit for the equation of time, 534
The moon likewise is revolved by a diurnal motion about its axis, and its libration thence arises, 535
That the sea ought twice to flow, and twice to ebb, every day; that the highest water must fall out in the third hour after the appulse of the luminaries to the meridian of the place, . 535

The precession of the equinoxes, and the libratory motion of the axes of the earth and planets, 535
That the greatest tides happen in the syzygies of the luminaries, the least in their quadratures; and that at the third hour after the appulse of the moon to the meridian of the place. But that out of the syzygies and quadratures those greatest and least tides deviate a little from that third hour towards the third hour after the appulse of the sun to the meridian, . 536
That the tides are greatest when the luminaries are in their perigees, ... 536
That the tides are greatest about the equinoxes, ... 536
That out of the equator the tides are greater and less alternately, ... 537
That, by the conservation of the impressed motion, the difference of the tides is diminished; and that hence it may happen that the greatest menstrual tide will be the third after the syzygy, 538
That the motions of the sea may be retarded by impediments in its channels, ... 538
That from the impediments of channels and shores various phænomena do arise, as that the sea may flow but once every day, ... 539
That the times of the tides within the channels of rivers are more unequal than in the ocean, . 540
That the tides are greater in greater and deeper seas; greater on the shores of continents than of islands in the middle of the sea; and yet greater in shallow bays that open with wide inlets to the sea, ... 540
The force of the sun to disturb the motions of the moon, computed from the foregoing principles, 542
The force of the sun to move the sea computed, ... 543
The height of the tide under the equator arising from the force of the sun computed, . . 543
The height of the tides under the parallels arising from the sun's force computed, . . . 544
The proportion of the tides under the equator, in the syzygies and quadratures, arising from the joint forces of both sun and moon, ... 545
The force of the moon to excite tides, and the height of the water thence arising, computed, . 545
That those forces of the sun and moon are scarcely sensible by any other effect beside the tides which they raise in the sea, ... 546
That the body of the moon is about six times more dense than the body of the sun, . . . 547
That the moon is more dense than the earth in a ratio of about three to two, ... 547
Of the distance of the fixed stars, ... 547
That the comets, as often as they become visible to us, are nearer than Jupiter, proved from their parallax in longitude, ... 548
The same proved from their parallax in latitude, ... 549
The same proved otherwise by the parallax, ... 550
From the light of the comets' heads it is proved that they descend to the orbit of Saturn, . 550
And also below the orb of Jupiter, and sometimes below the orb of the earth, ... 551
The same proved from the extraordinary splendor of their tails when they are near the sun, . 551
The same proved from the light of their heads, as being greater, *cæteris paribus*, when they come near to the sun, ... 553
The same confirmed by the great number of comets seen in the region of the sun, . . . 555
This also confirmed by the greater magnitude and splendor of the tails after the conjunction of the heads with the sun than before, ... 555
That the tails arise from the atmospheres of the comets, ... 556
That the air and vapour in the celestial spaces is of an immense rarity; and that a small quantity of vapour may be sufficient to explain all the phænomena of the tails of comets, . . 558
After what manner the tails of comets may arise from the atmospheres of their heads, . . 559
That the tails do indeed arise from those atmospheres, proved from several of their phænomena, 559
That comets do sometimes descend below the orbit of Mercury, proved from their tails, . 560
That the comets move in conic sections, having one focus in the centre of the sun, and by radii drawn to that centre do describe areas proportional to the times, ... 561
That those conic sections are near to parabolas, proved from the velocity of the comets, . 561
In what space of time comets describing parabolic trajectories pass through the sphere of the *orbis magnus*, ... 562
At what time comets enter into and pass out of the sphere of the *orbis magnus*, . . . 563
With what velocity the comets of 1680 passed through the sphere of the *orbis magnus*, . . 564
That these were not two, but one and the same comet. In what orbit and with what velocity this comet was carried through the heavens described more exactly, ... 564
With what velocity comets are carried, shewed by more examples, ... 565
The investigation of the trajectory of comets proposed, ... 566
Lemmas premised to the solution of the problem, ... 567
The problem resolved, ... 570

INDEX TO THE PRINCIPIA.

ÆQUINOXES, their præcession—the cause of that motion shewn, ... 413
" the quantity of that motion computed from the causes, ... 458

AIR, its density at any height, collected by Prop. XXII, Book II, and its density at the height of one semi-diameter of the earth, shewn, ... 489
" its elastic force, what cause it may be attributed to, ... 302
" its gravity compared with that of water, ... 489
" its resistance, collected by experiments of pendulums, ... 315
" the same more accurately by experiments of falling bodies, and a theory, ... 355

ANGLES of contact not all of the same kind, but some infinitely less than others, ... 101

APSIDES, their motion shewn, ... 172, 173

AREAS which revolving bodies, by radii drawn to the centre of force describe, compared with the times of description, ... 103, 105, 106, 195, 200

As, the mathematical signification of this word defined, ... 100

ATTRACTION of all bodies demonstrated, ... 397
" the certainty of this demonstration shewn, ... 384
" the cause or manner thereof no where defined by the author, ... 507
" the common centre of gravity of the earth, sun, and all the planets, is at rest, confirmed by Cor. 2, Prop. XIV, Book III, ... 401
" the common centre of gravity of the earth and moon goes round the *orbis magnus*, 402
" its distance from the earth and from the moon, ... 452

CENTRE, the common centre of gravity of many bodies does not alter its state of motion or rest by the actions of the bodies among themselves, ... 87
" of the forces by which revolving bodies are retained in their orbits, how indicated by the description of areas, ... 107
" how found by the given velocities of the revolving bodies, ... 110

CIRCLE, by what law of centripetal force tending to any given point its circumference may be described, ... 108, 111, 114

COMETS, a sort of planets, not meteors, ... 465, 486
" higher than the moon, and in the planetary regions, ... 460
" their distance how collected very nearly by observations, ... 461
" more of them observed in the hemisphere towards the sun than in the opposite hemisphere; and how this comes to pass, ... 464
" shine by the sun's light reflected from them, ... 464
" surrounded with vast atmospheres, ... 463, 465
" those which come nearest to the sun probably the least, ... 495
" why they are not comprehended within a zodiac, like the planets, but move differently into all parts of the heavens, ... 502
" may sometimes fall into the sun, and afford a new supply of fire, ... 502
" the use of them hinted, ... 492
" move in conic sections, having their foci in the sun's centre, and by radii drawn to the sun describe areas proportional to the times. Move in ellipses if they come round again in their orbits, but these ellipses will be near to parabolas, ... 466

COMET'S parabolic trajectory found from three observations given, ... 472
" corrected when found, ... 495
" place in a parabola found to a given time, ... 466
" velocity compared with the velocity of the planets, ... 466

COMETS' TAILS directed from the sun, ... 489
" " brightest and largest immediately after their passage through the neighbourhood of the sun, ... 487
" " their wonderful rarity, ... 490
" " their origin and nature, ... 463
" " in what space of time they ascend from their heads, ... 490

Comet of the years 1664 and 1665—the observations of its motion compared with the theory, . 496
" of the years 1680 and 1681—observations of its motion, ... 474
" its motion computed in a parabolic orbit, ... 478
" in an elliptic orbit, ... 479
" its trajectory, and its tail in the several parts of its orbit, delineated, ... 484
" of the year 1682—its motion compared with the theory, ... 500
" seems to have appeared in the year 1607, and likely to return again after a period of 75 years, ... 501, 502
" of the year 1683—its motion compared with the theory, ... 499
" of the year 1723—its motion compared with the theory, ... 501
Conic Sections, by what law of centripetal force tending to any given point they may be described by revolving bodies, ... 125
" the geometrical description of them when the foci are given, ... 125
" when the foci are not given, ... 131
" when the centres or asymptotes are given, ... 147
Curvature of figures how estimated, ... 271, 423
Curves distinguished into geometrically rational and geometrically irrational, ... 157
Cycloid, or Epicycloid, its rectification, ... 184
" " its evoluta, ... 185
Cylinder, the attraction of a cylinder composed of attracting particles, whose forces are reciprocally as the square of the distances, ... 239
Descent of heavy bodies in vacuo, how much it is, ... 405
" and ascent of bodies in resisting mediums, ... 252, 265, 281, 283, 345
Descent or Ascent rectilinear, the spaces described, the times of decription, and the velocities acquired in such ascent or descent, compared, on the supposition of any kind of centripetal force, ... 160
Earth, its dimension by *Norwood*, by *Picart*, and by *Cassini*, ... 405
" its figure discovered, with the proportion of its diameters, and the measure of the degrees upon the meridian, ... 405, 409
" the excess of its height at the equator above its height at the poles, ... 407, 412
" its greatest and least semi-diameter, ... 407
" its mean semi-diameter, ... 407
" the globe of the earth more dense than if it was entirely water, ... 400
" the nutation of its axis, ... 413
" the annual motion thereof in the *orbis magnus* demonstrated, ... 498
" the eccentricity thereof how much, ... 452
" the motion of its aphelion how much, ... 404
Ellipses, by what law of centripetal force tending to the centre of the figure it is described by a revolving body, ... 114
" by what law of centripetal force tending to the focus of the figure it is described by a revolving body, ... 116
Fluid, the definition thereof, ... 108
Fluids, the laws of their density and compression shewn, ... 293
" their motion in running out at a hole in a vessel determined, ... 331
Forces, their composition and resolution, ... 84
" attractive forces of spherical bodies, composed of particles attracting according to any law, determined, ... 218
" attractive forces of bodies not spherical, composed of particles attracting according to any law, determined, ... 233
" the invention of the centripetal forces, when a body is revolved in a non-resisting space about an immoveable centre in any orbit, ... 103, 116
" the centripetal forces tending to any point by which any figure may be described by a revolving body being given, the centripetal forces tending to any other point by which the same figure may be described in the same periodic time are also given, ... 113
" the centripetal forces by which any figure is described by a revolving body being given, there are given the forces by which a new figure may be described, if the ordinates are augmented or diminished in any given ratio, or the angle of their inclination be any how changed, the periodic time remaining the same, ... 116
" centripetal forces decreasing in the duplicate proportion of the distances, what figures may be described by them, ... 120. 196

FORCE, centripetal force defined, . . . 74
" the absolute quantity of centripetal force defined, . . . 75
" the accelerative quantity of the same defined, . . . 76
" the motive quantity of the same defined, . . . 76
" the proportion thereof to any known force how collected, . . . 109
" a centripetal force that is reciprocally as the cube of the ordinate tending to a vastly remote centre of force will cause a body to move in any given conic section, . . 114
" a centripetal force that is as the cube of the ordinate tending to a vastly remote centre of force will cause a body to move in an hyperbola, . . . 243
" centrifugal force of bodies on the earth's equator, how great, . . . 405
GOD, his nature, . . . 506
GRAVITY mutual between the earth and its parts, . . . 94
" of a different nature from magnetical force, . . . 397
" the cause of it not assigned, . . . 507
" tends towards all the planets, . . . 393
" from the surfaces of the planets upwards decreases in the duplicate ratio of the distances from the centre, . . . 400
" from the same downwards decreases nearly in the simple ratio of the same, . . 400
" tends towards all bodies, and is proportional to the quantity of matter in each, . 397
" is the force by which the moon is retained in its orbit, . . . 391
" the same proved by an accurate calculus, . . . 453
" is the force by which the primary planets and the satellites of Jupiter and Saturn are retained in their orbits, . . . 393
HEAT, an iron rod increases in length by heat, . . . 412
" of the sun, how great at different distances from the sun, . . . 486
" how great in Mercury, . . . 400
" how great in the comet of 1680, when in its perihelion, . . . 486
HEAVENS are void of any sensible resistance, 401, 445, 492; and, therefore, of almost any corporeal fluid whatever, . . . 355, 356
" suffer light to pass through them without any refraction, . . . 485
HYDROSTATICS, the principles thereof delivered, . . . 293
HYPERBOLA, by what law of centrifugal force tending from the centre of the figure it is described by a revolving body, . . . 116
" by what law of centrifugal force tending from the focus of the figure it is described by a revolving body, . . . 117
" by what law of centripetal force tending to the focus of the figure it is described by a revolving body, . . . 118
HYPOTHESES of what kind soever rejected from this philosophy, . . . 508
JUPITER, its periodic time, . . . 388
" its distance from the sun, . . . 388
" its apparent diameter, . . . 386
" its true diameter, . . . 399
" its attractive force, how great, . . . 398
" the weights of bodies on its surface, . . . 399
" its density, . . . 399
" its quantity of matter, . . . 399
" its perturbation by Saturn, how much, . . . 403
" the proportion of its diameters exhibited by computation, . . . 409
" and compared with observations, . . . 409
" its rotation about its axis, in what time performed, . . . 409
" the cause of its belts hinted at, . . . 445
LIGHT, its propagation not instantaneous, . . . 246
" its velocity different in different mediums, . . . 245
" a certain reflection it sometimes suffers explained, . . . 245
" its refraction explained, . . . 243
" refraction is not made in the single point of incidence, . . . 247
" an incurvation of light about the extremities of bodies observed by experiments, . . 246
" not caused by the agitation of any ethereal medium, . . . 368
MAGNETIC force, . . . 94, 394, 397, 454

MARS, its periodic time, . . . 388
" its distance from the sun, . . . 389
" the motion of its aphelion, . . . 405
MATTER, its quantity of matter defined, . . . 73
" its *vis insita* defined, . . . 74
" its impressed force defined, . . . 74
" its extension, hardness, impenetrability, mobility, *vis inertiæ*, gravity, how discovered, 385
" subtle matter of *Descartes* inquired into, . . . 320
MECHANICAL POWERS explained and demonstrated, . . . 94
MERCURY, its periodic time, . . . 388
" its distance from the sun, . . . 389
" the motion of its aphelion, . . . 405
METHOD of first and last ratios, . . . 95
" of transforming figures into others of the same analytical order, . . . 141
" of fluxions, . . . 261
" differential, . . . 447
" of finding the quadratures of all curves very nearly true, . . . 448
" of converging series applied to the solution of difficult problems, . . . 271, 436
MOON, the inclination of its orbit to the ecliptic greatest in the syzygies of the node with the sun, and least in the quadratures, . . . 208
" the figure of its body collected by calculation, . . . 454
" its librations explained, . . . 405
" its mean apparent diameter, . . . 453
" its true diameter, . . . 453
" weight of bodies on its surface, . . . 453
" its density, . . . 453
" its quantity of matter, . . . 453
" its mean distance from the earth, how many greatest semi-diameters of the earth contained therein, . . . 453
" how many mean semi-diameters, . . . 454
" its force to move the sea how great, . . . 449
" not perceptible in experiments of pendulums, or any statical or hydrostatical observations, 452
" its periodic time, . . . 454
" the time of its synodical revolution, . . . 422
" its motions, and the inequalities of the same derived from their causes, . . . 413, 444
" revolves more slowly, in a dilated orbit, when the earth is in its perihelion; and more swiftly in the aphelion the same, its orbit being contracted, . . . 413, 444, 445
" revolves more slowly, in a dilated orbit, when the apogæon is in the syzygies with the sun; and more swiftly, in a contracted orbit, when the apogæon is in the quadratures, . 445
" revolves more slowly, in a dilated orbit, when the node is in the syzygies with the sun; and more swiftly, in a contracted orbit, when the node is in the quadratures, . . . 446
" moves slower in its quadratures with the sun, swifter in the syzygies; and by a radius drawn to the earth describes an area, in the first case less in proportion to the time, in the last case greater, . . . 413
" the inequality of those areas computed, . . . 420
" its orbit is more curve, and goes farther from the earth in the first case; in the last case its orbit is less curve, and comes nearer to the earth, . . . 415
" the figure of this orbit, and the proportion of its diameters collected by computation, . 423
" a method of finding the moon's distance from the earth by its horary motion, . . 423
" its apogæon moves more slowly when the earth is in its aphelion, more swiftly in the perihelion, . . . 414, 445
" its apogæon goes forward most swiftly when in the syzygies with the sun; and goes backward in the quadratures, . . . 414, 446
" its eccentricity greatest when the apogæon is in the syzygies with the sun; least when the same is in the quadratures, . . . 414, 446
" its nodes move more slowly when the earth is in its aphelion, and more swiftly in the perihelion, . . . 414, 445
" its nodes are at rest in their syzygies with the sun, and go back most swiftly in the quadratures, . . . 414

MOON, the motions of the nodes and the inequalities of its motions computed from the theory of gravity, . . . 427, 430, 434, 436
" the same from a different principle, . . . 437
" the variations of the inclination computed from the theory of gravity, . . . 441, 443
" the equations of the moon's motions for astronomical uses, . . . 445
" the annual equation of the moon's mean motion, . . . 445
" the first semi-annual equation of the same, . . . 446
" the second semi-annual equation of the same, . . . 447
" the first equation of the moon's centre, . . . 447
" the second equation of the moon's centre, . . . 448
MOON'S first variation, . . . 425
" the annual equation of the mean motion of its apogee, . . . 445
" the semi-annual equation of the same, . . . 447
" the semi-annual equation of its eccentricity, . . . 447
" the annual equation of the mean motion of its nodes, . . . 445
" the semi-annual equation of the same, . . . 437
" the semi-annual equation of the inclination of the orbit to the ecliptic, . . . 444
" the method of fixing the theory of the lunar motions from observations, . . . 464
MOTION, its quantity defined, . . . 73
" absolute and relative, . . . 78
" absolute and relative, the separation of one from the other possible, demonstrated by an example, . . . 82
" laws thereof, . . . 83
" of concurring bodies after their reflection, by what experiments collected, . . . 91
" of bodies in eccentric sections, . . . 116
" in moveable orbits, . . . 172
" in given superficies, and of the reciprocal motion of pendulums, . . . 183
" of bodies tending to each other with centripetal forces, . . . 194
" of very small bodies agitated by centripetal forces tending to each part of some very great body, . . . 233
" of bodies resisted in the ratio of the velocities, . . . 251
" in the duplicate ratio of the velocity, . . . 258
" partly in the simple and partly in the duplicate ratio of the same, . . . 280
" of bodies proceeding by their *vis insita* alone in resisting mediums, 251, 258, 259, 280, 281, 330
" of bodies ascending or descending in right lines in resisting mediums, and acted on by an uniform force of gravity, . . . 252, 265, 281, 283
" of bodies projected in resisting mediums, and acted on by an uniform force of gravity, 255, 268
" of bodies revolving in resisting mediums, . . . 287
" of funependulous bodies in resisting mediums, . . . 304
" and resistance of fluids, . . . 323
" propagated through fluids, . . . 356
" of fluids after the manner of a vortex, or circular, . . . 370
MOTIONS, composition and resolution of them, . . . 84
OVALS for optic uses, the method of finding them which Cartesius concealed, . . . 246
" a general solution of Cartesius's problem, . . . 247, 248
ORBITS, the invention of those which are described by bodies going off from a given place with a given velocity according to a given right line, when the centripetal force is reciprocally as the square of the distance, and the absolute quantity of that force is known, . 123
" of those which are described by bodies when the centripetal force is reciprocally as the cube of the distance, . . . 114, 171, 176
" of those which are described by bodies agitated by any centripetal forces whatever, 168
PARABOLA, by what law of centripetal force tending to the focus of the figure the same may be described, . . . 120
PENDULUMS, their properties explained, . . . 186, 190, 304
" the diverse lengths of isochronous pendulums in different latitudes compared among themselves, both by observations and by the theory of gravity, . . . 409 to 413
PLACE defined, and distinguished into absolute and relative, . . . 78
PLACES of bodies moving in conic sections found to any assigned time, . . . 153
PLANETS not carried about by corporeal vortices, . . . 378

PLANETS, their periodic times, ... 388
" their distances from the sun, ... 389
" the aphelia and nodes of their orbits do almost rest, ... 405
" their orbits determined, ... 406
" the way of finding their places in their orbits, ... 347 to 350
" their density suited to the heat they receive from the sun, ... 400
" their diurnal revolutions equable, ... 406
" their axes less than the diameters that stand upon them at right angles, ... 406
PLANETS, PRIMARY, surround the sun, ... 387
" " move in ellipses whose focus is in the sun's centre, ... 403
" " by radii drawn to the sun describe areas proportional to the times, . 388, 403
" " revolve in periodic times that are in the sesquiplicate proportion of the distances from the sun, ... 387
" " are retained in their orbits by a force of gravity which respects the sun, and is reciprocally as the square of the distance from the sun's centre, 389, 393
PLANETS, SECONDARY, move in ellipses having their focus in the centre of the primary, . 413
" " by radii drawn to their primary describe areas proportional to the times, ... 386, 387, 390
" " revolve in periodic times that are in the sesquiplicate proportion of their distances from the primary, ... 386, 387
PROBLEM KEPLERIAN, solved by the trochoid and by approximations, ... 157 to 160
" " of the ancients, of four lines, related by Pappus, and attempted by Cartesius, by an algebraic calculus solved by a geometrical composition, . 135
PROJECTILES move in parabolas when the resistance of the medium is taken away, 91, 115, 243, 273
" their motions in resisting mediums, ... 255, 268
PULSES of the air, by which sounds are propagated, their intervals or breadths determined, 368, 370
" these intervals in sounds made by open pipes probably equal to twice the length of the pipes, ... 370
QUADRATURES general of oval figures not to be obtained by finite terms, ... 153
QUALITIES of bodies how discovered, and when to be supposed universal, ... 384
RESISTANCE, the quantity thereof in mediums not continued, ... 329
" in continued mediums, ... 409
" in mediums of any kind whatever, ... 331
" of mediums is as their density, *cæteris paribus*, . . . 320, 321, 324, 329, 344, 355
" is in the duplicate proportion of the velocity of the bodies resisted, *cæteris paribus*, ... 258, 314, 324, 329, 344, 351
" is in the duplicate proportion of the diameters of spherical bodies resisted, *cæteris paribus*, ... 317, 318, 329, 344
" of fluids threefold, arises either from the inactivity of the fluid matter, or the tenacity of its parts, or friction, ... 286
" the resistance found in fluids, almost all of the first kind, ... 321, 354
" cannot be diminished by the subtilty of the parts of the fluid, if the density remain, 355
" of a globe, what proportion it bears to that of a cylinder, in mediums not continued, 327
" in compressed mediums, ... 343
" of a globe in mediums not continued, ... 329
" in compressed mediums, ... 344
" how found by experiments, ... 345 to 355
" to a frustum of a cone, how made the least possible, ... 328
" what kind of solid it is that meets with the least, ... 329
RESISTANCES, the theory thereof confirmed by experiments of pendulums, . . . 313 to 321
" by experiments of falling bodies, ... 345 to 356
REST, true and relative, ... 78
RULES of philosophy, ... 384
SATELLITES, the greatest heliocentric elongation of Jupiter's satellites, ... 387
" the greatest heliocentric elongation of the *Huygenian* satellite from Saturn's centre, 398
" the periodic times of Jupiter's satellites, and their distances from his centre, . 386, 387
" the periodic times of Saturn's satellites, and their distances from his centre, 387, 388
" the inequalities of the motions of the satellites of Jupiter and Saturn derived from the motions of the moon, ... 413
SESQUIPLICATE proportion defined, ... 101

SATURN, its periodic time, . . . 388
" its distance from the sun, . . . 388
" its apparent diameter, . . . 388
" its true diameter, . . . 399
" its attractive force, how great, . . . 398
" the weight of bodies on its surface, . . . 399
" its density, . . . 399
" its quantity of matter, . . . 399
" its perturbation by the approach of Jupiter how great, . . . 403
" the apparent diameter of its ring, . . . 388
SHADOW of the earth to be augmented in lunar eclipses, because of the refraction of the atmosphere, . . . 447
SOUNDS, their nature explained, . . . 360, 363, 365, 366, 367, 368, 369
" not propagated *in directum*, . . . 359
" caused by the agitation of the air, . . . 368
" their velocity computed, . . . 368, 369
" somewhat swifter by the theory in summer than in winter, . . . 370
" cease immediately, when the motion of the sonorous body ceases, . . . 365
" how augmented in speaking trumpets, . . . 370
SPACE, absolute and relative, . . . 78, 79
" not equally full, . . . 396
SPHEROID, the attraction of the same when the forces of its particles are reciprocally as the squares of the distances, . . . 239
SPIRAL cutting all its radii in a given angle, by what law of centripetal force tending to the centre thereof it may be described by a revolving body, . . . 107, 287, 291
SPIRIT pervading all bodies, and concealed within them, hinted at, as required to solve a great many phænomena of Nature, . . . 508
STARS, the fixed stars demonstrated to be at rest, . . . 404
" their twinkling what to be ascribed to, . . . 487
" new stars, whence they may arise, . . . 502
SUBSTANCES of all things unknown, . . . 507
SUN, moves round the common centre of gravity of all the planets, . . . 401
" the periodic time of its revolution about its axis, . . . 405
" its mean apparent diameter, . . . 453
" its true diameter, . . . 398
" its horizontal parallax, . . . 398
" has a menstrual parallax, . . . 403
" its attractive force how great, . . . 398
" the weight of bodies on its surface, . . . 399
" its density, . . . 399
" its quantity of matter, . . . 399
" its force to disturb the motions of the moon, . . . 391, 419
" its force to move the sea, . . . 448
TIDES of the sea derived from their cause, . . . 415, 448, 449
TIME, absolute and relative, . . . 78, 79
" the astronomical equation thereof proved by pendulum clocks, and the eclipses of Jupiter's satellites, . . . 79
A VACUUM proved, or that all spaces (if said to be full) are not equally full, . . . 396
VELOCITIES of bodies moving in conic sections, where the centripetal force tends to the focus, . 121
VELOCITY, the greatest that a globe falling in a resisting medium can acquire, . . . 344
VENUS, its periodic time, . . . 388
" its distance from the sun, . . . 388
" the motion of its aphelion, . . . 405
VORTICES, their nature and constitution examined, . . . 504
WAVES, the velocity with which they are propagated on the superficies of stagnant water, . 361
WEIGHTS of bodies towards the sun, the earth, or any planet, are, at equal distances from the centre, as the quantities of matter in the bodies, . . . 394
" they do not depend upon the forms and textures of bodies, . . . 395
" of bodies in different regions of the earth found out, and compared together, . . . 409